IEE POWER SERIES 24

Series Editors: Professor A. T. Johns
D. F. Warne

POWER SYSTEM COMMISSIONING and MAINTENANCE PRACTICE

Other volumes in this series:

POWER SYSTEM COMMISSIONING and MAINTENANCE PRACTICE

Keith Harker

The Institution of Electrical Engineers

Published by: The Institution of Electrical Engineers, London, United Kingdom

The Institution of Electrical Engineers,
Michael Faraday House,
Six Hills Way, Stevenage,
Herts. SG1 2AY, United Kingdom

British Library Cataloguing in Publication Data

A CIP catalogue record for this book
is available from the British Library

ISBN 0 85296 909 0

Typeset by Apek Typesetters, Nailsea
Printed in England by Bookcraft, Bath

Contents

Preface

Some 20 years ago, as a recent graduate in electrical engineering, I embarked on a career with the British electricity supply industry as an electrical power engineer. My early years were concerned with the commissioning of new equipment, maintenance and fault investigations. I thus commenced the slow, patient process of acquiring sufficient practical knowledge to enable me to possess a professional competency in the technology of my industry.

I obtained information from a variety of sources: equipment manuals; supply industry standards; site drawings; commissioning logs; specialist text books; and of course question and answer sessions with more experienced colleagues. From the outset, I consolidated the information acquired by recording explanations and drawings in note books. Over the years the books grew in number and became a constant source of reference. Relating to the practicalities of the work, I found commissioning to be the ultimate training ground for a power engineer, since it demands a complete knowledge of the new installation: the reason for the design; how it is constructed; how it is tested; and how it is put into service. To commission a new circuit or substation, armed with the knowledge of the 'whys and hows' of the whole commissioning process, is a measure of professional competence. It was during the commissioning process that the note books proved most useful.

The need to search out and record information, over many years, reflected the fact that I could not locate suitable publications which were concerned specifically with the commissioning and maintenance of power systems. This text, based on the notes produced over many years, seeks to redress that situation. It is aimed specifically at engineers who are at the outset of their careers and is therefore to be considered as an introduction to the subject. Qualifications, to first degree level or equivalent are assumed.

I have endeavoured to write in a simple, unambiguous and clear style, and as far as possible I have minimised the academic content and kept to the practicalities of the subject. The only exception is the inclusion of symmetrical components, and for this I make no apology. To my own mind, one can never be considered to be a fully accomplished power engineer unless the basic principles of symmetrical components are understood. The subject matter of necessity also covers aspects of design and application. I have throughout used site-based terminology, for example the term 'vector group test' is still used on site in preference to the more precise 'phasor group test'. Furthermore the text is mainly targeted on commissioning, since to a large measure maintenance is a repeat, or partial repeat, of the commissioning tests.

The material which I have drawn together reflects the technology and practice of the British power system, which may differ somewhat from that in other countries. It is derived from a huge range of sources, and I thank all those, too numerous to mention, who have assisted me, or pointed me in the right direction, over the years. Suffice it to say that any omissions or ambiguities in the text are merely a reflection of my own experience.

To those who read this text, I trust that you may have as much excitement and sense of achievement as I had in the early days of my career. There is certainly little to match standing in a control room, on a cold winter's night, with your commissioning team around you, when the final circuit breaker is closed and a new circuit is commissioned on to the system, while you take satisfaction from the knowledge that you have directed and understood the whole commissioning process. These are some of the ultimate memories in an electrical power engineer's career.

Symbols used in figures

The symbols used in the following chapters are in accordance with *BS 3939:* 1985 (which essentially corresponds to *IEC 617*) or predecessor documents. *BS 3939* is being superseded by *BS EM 60617.*

Drawing symbols associated with newer equipment will largely accord with *BS 3939:* 1985, but equipment of earlier vintage will have drawings which correspond to earlier standards, and many of these equipments and drawings are still in operational service. The author therefore makes no apology for the mixing of new and old symbols, since the practising engineer will have to contend with both.

The most commonly used symbols in this text are shown below. Note that the circuit breaker symbol described in *BS 3939:* 1985 is not employed, but rather the earlier symbols of either a square or a cross, the latter two being much more user friendly.

Note that, for simplicity, both three-phase and single-phase voltage transformers are shown as two-winding transformers (simplified).

	BS 3939: 1985	earlier symbols
circuit breaker		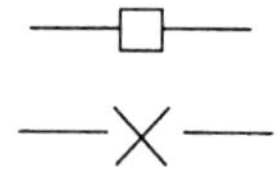
disconnector (isolator)	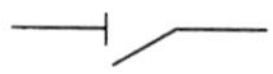	
two-winding transformer (simplified)	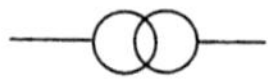	
two-winding transformer (general form)	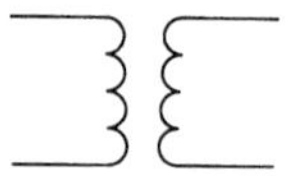	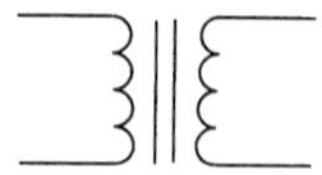
autotransformer		
AC generator		
current transformer	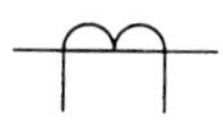	
fuse	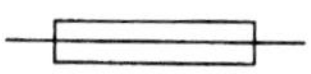	
link	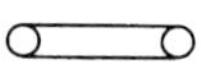	
relay	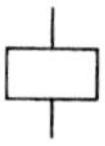	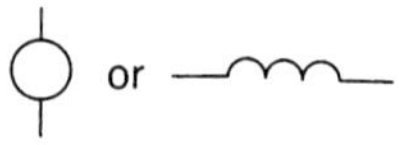
normally open contact (make)		
normally closed contact (break)		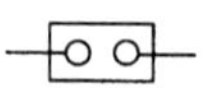

Chapter 1

Commissioning management

1.1 Introduction

The commissioning of electrical power systems may, in the first instance, be thought of as a purely engineering activity, associated with understanding and proving the technical integrity of the equipment; it is, however, also a management process, requiring such qualities as: setting of objectives; planning; organising; documentation control; and, most important, the creation of teamwork. This introductory chapter will therefore examine some fundamental concepts associated with commissioning management.

COMMISSIONING

1.2 Commissioning objectives

The objective of 'commissioning' is to prove that the individual items of equipment, which are interconnected to form systems, circuits and substations, are suitable for their intended purpose and acceptable for commercial service. This global objective can be subdivided into:

(i) verifying that the equipment has not been damaged in transit, that it has been correctly installed, and that it performs as specified;
(ii) obtaining test data as the basis for comparison with future maintenance and fault investigation results.

1.3 Definitions

There are a number of commonly used terms associated with commissioning, as follows:

Commissioning: activities which demonstrate that an item of equipment is acceptable for service. Such activities include inspections, off-load commissioning tests and on-load commissioning tests.
Off-load commissioning tests: tests which are carried out with the equipment disconnected from the power system.

The objective of off-load commissioning tests is to prove the integrity of the installed equipment, and hence its suitability for connection to the power

system, by observing its performance under test conditions which, as far as practicable, simulate in-service conditions and requirements. With some installations it is neither possible nor practicable to arrange tests which verify conclusively all aspects of the installation prior to energisation. In such instances, the tests may have to be limited to those which provide an acceptable level of confidence, with reliance being placed on on-load tests for more conclusive verification, or previous satisfactory experience of similar installations.
On-load commissioning tests: tests which are carried out with the equipment connected to the power system.
These tests utilise the power system itself to demonstrate that the installation is technically acceptable and hence suitable for commercial operation.
Switching programme: a step-by-step procedure which specifies how high-voltage equipment is to be energised and put into service, and at what stage the on-load commissioning tests are to be carried out.

Note that an alternative school of thought considers the term commissioning to cover only the on-load tests, and the off-load tests are termed 'pre-commissioning tests' or 'setting to work'.

1.4 Contractual responsibilities

Before commencing work, it is important to be clear about contractual responsibilities. Within this context the total project process from design to commissioning may be categorised as shown in Table 1.1. The commissioning process may be further subdivided into three basic types of responsibility:

In-house: the supply company does everything with no contractor involvement.
Witnessing: the contractor is responsible for carrying out the commissioning tests, which are witnessed and accepted by the supply company.
Turn key: the contractor is responsible for carrying out all tests with no involvement by the supply company. In practice, the supply authority is required to participate in the on-load tests.

1.5 Commissioning panel

The commissioning panel is responsible for managing the whole commissioning process. This consists of a number of key activities: planning the work, the equipment, the people, and the system requirements; providing the commissioning programme; defining the tests to be carried out, and the phasing of such tests; progressing the work; producing the switching programme; and collating the documentation.

The commissioning panel typically consists of the following representation:

(*a*) chairman, usually the project manager or supply company technical engineer;
(*b*) technical engineer from the supply company capable of defining the required tests and accepting the test results;
(*c*) contractor(s)' representatives;
(*d*) supply company power system operations department (mainly concerned with the energisation process);
(*e*) representatives of any companies whose equipment interfaces with that being commissioned.

Table 1.1 The project process

Activity	Responsibility
1 Prepare design specification	ESC
2 Issue tender document and award contract	ESC
3 Manufacture equipment	Contractor
4 Erect equipment on site	Contractor or ESC
5 Pre-commissioning inspection	Contractor and/or ESC
6 Off-load commissioning tests	Contractor or ESC
7 Pre-energisation inspection	Contractor and/or ESC
8 On-load commissioning test	Contractor and/or ESC
9 Post-commissioning inspection	Contractor and/or ESC

ESC = electricity supply company
Activities 5-9 comprise the commissioning programme.

1.6 Commissioning programme

The commissioning programme is the key activity to be progressed by the commissioning panel. It should ideally cover the following stages:

(i) pre-commissioning inspection (see Section 1.8)
(ii) off-load commissioning test programme (see Section 1.9)
(iii) pre-energisation inspection (see Section 1.10)
(iv) energisation and on-load commissioning tests (usually covered by the switching programme) (see Section 1.11)
(v) post-commissioning inspection (see Section 1.12)

1.7 Commissioning-panel agenda

It is usual to structure commissioning-panel meetings by use of a standard agenda. Typical format and subject matter for an agenda is as follows:

(*a*) project title;
(*b*) circuits and substations involved;
(*c*) commissioning-panel membership;
(*d*) terms of reference: a statement of what is to be achieved;
(*e*) scope: at this stage all members of the commissioning panel should be fully apprised of the extent of the work. To assist this process the activities may be categorised into:

  (i) civil works
  (ii) high-voltage equipment
  (iii) low-voltage equipment
  (iv) protection and control systems
  (v) substation control systems including interconnection with control centres
  (vi) metering
  (vii) DC and AC power supplies
  (viii) telecommunications equipment;

(*f*) commissioning-programme responsibilities: at this stage responsibilities for producing, executing and accepting the various stages of the commissioning programme should be agreed and minuted;
(*g*) outage requirements and changes to equipment nomenclature;
(*h*) commissioning certificates;
(*i*) method of control of drawings; i.e. who shall hold them and who has authority to modify them;
(*j*) commissioning-documentation responsibilities;
(*k*) decommissioning considerations;
(*l*) safety certificates and considerations;
(*m*) training requirements;
(*n*) temporary equipment (e.g. to facilitate on-load tests);
(*o*) switching programme.

1.8 Pre-commissioning inspection

This would consist of an inspection of the installation to determine whether work is complete and satisfactory to the point where off-load commissioning tests can commence.

1.9 Off-load commissioning tests

A programme of such tests would include

(*a*) Tests associated with AC circuit diagrams:

(i) tests on current and voltage transformers
(ii) primary injections;

(*b*) Tests associated with DC circuit diagrams:

(i) secondary injection of DC relays
(ii) circuit diagram logic tests;

(*c*) Equipment secondary injections;
(*d*) Power supplies:

(i) 110 V battery and distribution board
(ii) 48 V battery and distribution board
(iii) diesel generator
(iv) low-voltage AC supplies;

(*e*) High-voltage equipment tests:

(i) circuit breakers
(ii) switches
(iii) disconnectors
(iv) earth switches
(v) transformers
(vi) reactors
(vii) busbars
(viii) cables etc.;

(*f*) Common equipment:

(i) busbar protection
(ii) interlocking
(iii) synchronising
(iv) air systems
(v) substation control systems;

(*g*) Earthing systems:

(i) substation earth mat
(ii) interconnection of earthing system
(iii) tower earthing;

(*h*) High-voltage pressure tests;
(*i*) Auto-switching off-load simulation tests;
(*j*) End-to-end tests:

(i) power-line carrier
(ii) intertripping
(iii) protection systems
(iv) substation control systems;

(*k*) Insulation resistance tests associated with the above.

1.10 Pre-energisation inspection

The pre-energisation inspection must be carried out prior to the circuit being energised. Usually a proforma can be utilised; typical activities consist of:

(*a*) confirmation of circuit nomenclature;
(*b*) confirmation that equipment phase connections are in accordance with the phasing diagram;
(*c*) checking that there is no obvious infringement of safety clearances;
(*d*) application of locks;
(*e*) checking the correct position of all valves (e.g. transformer cooling equipment);
(*f*) ensuring satisfactory indication of all pressure gauges;
(*g*) checking that all equipment is satisfactorily earthed;
(*h*) ensuring the satisfactory condition of all battery supplies;
(*i*) checking that all test connections have been removed;
(*j*) confirmation of the tightness of all terminations;
(*k*) confirmation that all current transformer links are normal;
(*l*) confirmation that all relay settings have been applied correctly;
(*m*) confirmation that all commissioning documentation is complete.

1.11 Switching programme

This is a formal and precisely worded procedure which specifies the sequence of events required to energise the circuit, synchronise to the system (where required) and carry out the on-load commissioning tests. It needs to be circulated to all parties concerned to ensure unanimity of agreement and understanding. It should be provided and approved by the commissioning panel and it should usually culminate in on-load auto-switching tests.

1.12 Post-commissioning inspection

The purpose of this inspection is to double check that the commissioned equipment has been left in the in-service position. It would typically cover:
(i) ensuring that all temporary commissioning equipment has been removed;
(ii) ensuring that all current transformer links are normal;
(iii) checking the correct position of all IN/OUT switches and transformer tap positions;
(iv) ensuring that all relay settings are normal;
(v) general visual inspection;
(vi) ensuring that all commissioned equipment is clearly identified to prevent inadvertent interference.

1.13 Documentation

An essential requirement of the commissioning process is the formal documentation of both the management activities and the commissioning tests. Two distinct documentation systems, capable of standing up to a quality assurance audit, should be produced:

- commissioning report
- commissioning log.

The commissioning report is the formal record of the managerial process. It should consist of:

(*a*) a contents sheet, updated as the project progresses;
(*b*) a copy of the high level design specification (technical data sheet);
(*c*) formal minutes of commissioning meetings;
(*d*) the programme of work to be carried out, showing planned and actual completion dates;
(*e*) a schedule of the commissioning tests;
(*f*) a commissioning completion document, signed by the commissioning team leader, to certify that all work is complete.

The commissioning log is the record of the technical process, comprising the completed commissioning test schedules, test certificates, inspection reports, switching programme and other relevant technical documents. Documentation control and quality have assumed greater importance with the advent of safety legislation such as, in the UK, the 1994 Construction (Design and Management) Regulations.

1.14 Safety

Commissioning projects have great potential for hazardous occurrences, and the need to consider safety at each juncture cannot be stressed too highly. At an appropriate point in the commissioning programme, the high voltage equipment will need to be made subject to the electricity company's safety rules. If the substation is already energised, this may need to be done before erection commences. Placing new equipment under the safety rules needs to be carried out formally, and a 'Safety rules clearance certificate' or similar document is used by most electricity companies. Figure 1.1 provides an example of this type of document.

SAFETY-RULES CLEARANCE CERTIFICATE

Location.......................... Certificate No............

To: Contractor/
Department

....
PART 1: NOTICE

I HEREBY GIVE NOTICE that the PLANT/APPARATUS schedule below is being PUT INTO SERVICE/MADE LIVE or by reason of its proximity or relation to other plant/apparatus must be regarded as being in SERVICE/LIVE, and as from .... hours, on .... will come under the terms of the electricity company safety rules. No further work may be done on or near the PLANT/APPARATUS after this time and date unless the work is being undertaken in accordance with the safety rules and is authorised by a senior authorised person in the employ of the company. Please sign and return the master certificate and retain the duplicate and any attached drawings.

DESCRIPTION AND LOCATION OF PLANT/APPARATUS

(Signature)..................... Electricity Company Representative
(Date) ....

PART 2: CONTRACTOR'S ACKNOWLEDGMENT

To be returned by .... hours, on ....

I/we acknowledge receipt of the above notice and confirm that all persons in my/our charge have been duly warned of the above.

Signed	Contractor/Department	Time	Date
....	....	....	....
....	....	....	....

PART 3: CONFIRMATION OF ACKNOWLEDGMENT

I confirm that all contractors and/or electricity company departments in receipt of this SRCC have acknowledged its receipt.

Signature:......................Electricity Company Representative

Figure 1.1 Typical safety-rules clearance certificate

ACCEPTANCE CERTIFICATE

Location .... SERIAL NO............

PART 1

I certify that the following PLANT/APPARATUS, apart from any exceptions noted below, has been tested and inspected and is now ready for on-load commissioning tests.

Time........ Date............ Signed............... Electricity Company Representative

Time........ Date............ Signed............... Commissioning Engineer

DESCRIPTION OF PLANT/APPARATUS

PART 2

The PLANT/APPARATUS specified in Part 1 above has been subjected to on-load commissioning tests. These have been witnessed and the results approved. The PLANT/APPARATUS was commissioned at (time) .... on (date) .... with the following limitations:

LIMITATIONS

Time......... Date............. Signed................ Electricity Company Representative

Time........ Date............ Signed.............. Commissioning Engineer

Figure 1.2 Typical acceptance certificate

1.15 Commissioning certificates

(*a*) *Acceptance certificate*
This certificate declares that:

(i) the equipment is ready to be connected to the power system;
(ii) the equipment has been energised, and subject to on-load commissioning tests; it is therefore deemed to be commissioned and available for commercial operation.

Typical documentation which serves this purpose is shown in Figure 1.2.

(*b*) *Contract taking-over certificate*
When the work has been completed satisfactorily and the equipment commissioned, a contract taking-over certificate or similar document will require completion. This document notifies the contractor that the supply company has accepted and taken over the equipment, and that the (12 month) warranty period has commenced. Completion of this document usually signifies that financial payments can be concluded.

1.16 Commissioning practice

The following activities are items of good commissioning practice:

Philosophy: take nothing for granted. Periodically reflect on what is to be done, what might have been missed, what might alternatively be done. Employ lateral thinking. Assume everything is wrong until proven otherwise.
Extent of testing: proof of the satisfactory working of items of equipment and their interconnecting circuitry is required. Equipment tests should not replicate comprehensive factory tests but must be the minimum to verify satisfactory installation. Protection/measuring equipment should be tested at the proposed setting/range, and be within 5% accuracy.
Formality: commissioning requires formality to create a more disciplined, thoughtful and professional approach. It identifies accountability and creates a clear and auditable record of activities for future reference.
Advance prediction of results: a professional approach to testing requires that the results are predicted in advance of the test.
Lists: maintain a formal list of all items which require correcting (e.g. missing labels and links, faulty relays, broken terminals, leaking valves etc.), and of any temporary connections and arrangements.
Completion: very rarely is the task complete when the equipment is energised. Prepare a programme for resolving any items omitted (e.g. temporary labels to be replaced, drawings to be completed etc.). Ensure that the commissioning documentation is complete and properly filed away. Neatly file all drawings and produce an index.

1.17 Equipment commissioning

The importance of a highly professional approach to commissioning management cannot be overstated. Without it equipment may enter service with latent deficiencies with the potential to jeopardise the security of the power system. The following chapters will examine the commissioning of specific items of equipment which reflect the technology of power systems circa 1995. While it may refer to specific UK practices, the principles and much of the detail are relevant to all such power systems. Suffice it to say that, although the design of the equipment may change with time, the basic principles remain essentially constant. The commissioning techniques advocated are not necessarily the only methods of carrying out the work, but in the author's experience they represent typical main-stream industry practice.

Chapter 2

Maintenance and fault investigations

2.1 Introduction

This chapter will consider some of the salient considerations associated with the maintenance of electrical power systems. Suffice it to say that there is no single maintenance policy for power systems, but rather individual company policies based upon common areas of consideration, and experience gained through operating the equipment. Consideration will also be briefly given to investigations into the incorrect performance of equipment.

MAINTENANCE

2.2 Maintenance strategies

The prime objective of maintenance is to keep equipment in good working order, to maximise lifetime productivity. Maintenance strategies may be categorised as follows:

(*a*) *Preventative maintenance:* This is maintenance which is carried out at predetermined intervals to reduce the likelihood of an item of equipment failing in service.
(*b*) *Corrective maintenance:* This refers to maintenance which is carried out to restore an item of equipment which has failed.
(*c*) *Opportunity maintenance:* This is maintenance which is carried out when a higher priority item of equipment is out of service. For example, generator-circuit-breaker maintenance is carried out only when the generator is out of service.
(*d*) *Statutory maintenance:* This is maintenance which is carried out to facilitate statutory timescales and requirements usually associated with health and safety.

In general, the higher the voltage of the power system, the greater the tendency towards preventative maintenance.

2.3 Maintenance-policy considerations

The management task in maintenance is one of deciding the extent and periodicity against realistic and required standards of equipment performance. In financial terms this is the cost of maintenance compared with the cost benefit of carrying out the maintenance.

Traditional literature on maintenance theory shows that there is an optimum level, in terms of financial benefit, between preventative and corrective maintenance. However, this approach is more applicable to manufacturing industry. Similarly, maintenance performance indices can be produced for comparison purposes, such as

$$\frac{\text{maintenance cost}}{\text{asset value}} \quad \text{or} \quad \frac{\text{cost of preventative maintenance}}{\text{total cost of maintenance}}$$

Although these indices are helpful, they do not necessarily point to a maintenance policy for an electrical power system.

A starting point may be to recognise that the power system is not a collection of individual items of equipment or machines, but an interconnected system. Thus the loss of one item of equipment may have profound implications, not only locally, but at remote points in the system. Furthermore, experience has shown that many system incidents arise not so much from a lack of maintenance, as through equipment design limitations, poor commissioning practice, changing system conditions, unusual weather or unusual loading profiles. Finally, it needs to be recognised that the power system differs from manufacturing industry in that loss of supply has political, as well as economic, implications. A maintenance policy must take all the above into account. In particular, two basic questions relating to maintenance need to be addressed: why is it being done; what, specifically, must be done?

In general, most electricity companies tend towards a preventative maintenance policy based on equipment manufacturers' recommendations and operating experience. To accommodate periodic power system outage constraints which prevent access to the equipment to be maintained, or shortages of manpower resources to carry out maintenance (e.g. because resources are diverted to higher priority activities such as system fault rectification), maintenance periodicity needs to be flexible.

2.4 Maintenance activities

In many industries, maintenance is primarily associated with the replacement of parts, particularly moving parts, which have been subject to wear and tear. However in general the power system tends to be a static or quiescent system, whose parts are required to move infrequently. Thus replacement of worn parts due to mechanical movement is not a major maintenance activity.

In the past, the erosion of electrical contacts during the open/closing operation of equipment frequently required the contacts to be changed or 'dressed', but improved contact design and superior interruption mediums have reduced this requirement. The general state of the interruption medium or insulating medium, however, still requires frequent examination, particularly if the medium is oil: this is an important maintenance requirement.

The asset life of power-system equipment is long, typically 40–50 years for high-voltage equipment and 20–30 years for protection and control equipment. In addition, extensive parts of the power system are exposed to the elements (wind, rain, heat, frost, pollution). Within these parameters many items of equipment must operate instantaneously, after long periods of quiescence. A maintenance policy must take these considerations into account.

In general, power system maintenance consists of the following activities:

(i) periodic exercising of equipment
(ii) visual inspections
(iii) cleaning, dusting and painting
(iv) lubrication and periodic replacement of parts
(v) monitoring and examination of interrupting and insulating mediums
(vi) equipment performance tests
(vii) removal of foliage to maintain clearances.

2.5 Maintenance periodicity

Failure rates for equipment tend to follow what is termed a 'bath-tub curve', that is a high rate following installation (particularly of new designs), a levelling out of failures during the intermediate years of equipment life, and again a high rate as the equipment approaches the end of its life. It may thus be thought that a maintenance periodicity should match this profile; however, a significant number of power system equipment failures arise from design defects and would not necessarily be prevented by maintenance.

In practice, maintenance periodicity tends to follow one of the following patterns:

(*a*) regular intervals during the life of the equipment based on manufacturers' recommendations, and experience;
(*b*) in conjunction with (*a*), increased maintenance of older equipment where this can be shown to increase asset life, or to obviate health and safety concerns;
(*c*) statutory maintenance requirements;
(*d*) opportunity maintenance;
(*e*) targeted sampling of a population of equipment such that the target sample is considered representative of the general population;
(*f*) no maintenance. This would occur when the equipment is periodically operated (thereby proving its capability) and that it is known that operation does not result in wear and tear—or where the cost of maintenance cannot be justified. An example of the former may be some designs of solid state control equipment, and of the latter, rural distribution voltage transformers.

2.6 Post-fault maintenance

It has been the practice in former years, in some utilities, to carry out post-fault maintenance of circuit breakers which have interrupted or closed onto fault current. In more recent years such maintenance has only been carried out after taking into account the magnitude of fault current, and hence the estimated degree of wear and tear on the circuit breaker.

2.7 Maintenance specification

All maintenance should be carried out in accordance with a maintenance specification. This should consist of concise step by step instructions of the work to be carried out. The necessary instruments, tools and equipment required to carry out the maintenance should also be listed. Each stage of the maintenance instruction should be accompanied by a short summary of what that stage seeks to achieve. A formal tick list and results sheet should be provided and these should be neatly filed as a record of the work carried out.

2.8 Maintenance records

An essential feature of maintenance is the retention of records. Typical maintenance records will comprise:

(i) date of maintenance
(ii) extent of maintenance (e.g. as per standard specification)
(iii) what was found (if anything)
(iv) what extra was carried out (if anything)
(v) who carried out the maintenance
(vi) any specific requirements to be carried out in future
(vii) record of performance test results
(viii) planned date of next maintenance.

Maintenance records are usually held on a per-circuit basis, identifying the items of high-voltage plant and salient categories of protection/control equipment. Records of cables and lines are usually held separately. Database systems are eminently suitable for maintenance records, since they offer search facilities on selected data categories.

2.9 Stores

An essential consideration for both maintenance and fault breakdown management is the stores strategy. Requirements for both routine and strategic stores need to be identified: these in turn will depend on the volume of usage and the timescales for acquiring the items of stores.

2.10 Equipment maintenance

The following chapters will examine the maintenance of specific items of equipment. Where pertinent the text may recommend maintenance frequencies; however these should be taken as indicative since maintenance frequency is in a constant state of evolution. Generally speaking, maintenance tests tend to be a repeat or a partial repeat of the commissioning tests, thereby enabling a direct comparison of the results to be made.

INVESTIGATIONS

2.11 Investigation technique

The term 'investigation' refers to the analysis of a system incident, usually associated with a system fault. Examples of such activities range from a false operation of a protection relay, to a damaged transformer, or to a pole-slipping incident resulting in loss of generation.

No matter what the incident, the techniques for approaching it are essentially the same. There are effectively up to five stages to an investigation:

- data collection
- data analysis
- damage investigation
- diagnostic tests
- conclusion.

These will be discussed below.

2.12 Data collection

When collecting data the following should be contemplated:

(*a*) The first intimation of a system incident will often be by telephone. It is very easy to pick up the wrong information, either because it is poorly conveyed or because the recipient jumps to a wrong conclusion. Thus when receiving verbal information, request the provider to start at the beginning and reveal the facts slowly. Note down the key points and repeat them to the provider.

(*b*) Consider all possible sources of information, e.g. fault recorders, data loggers, recording instruments, manned control points, personal observations.

(*c*) Take a wide view of the problem. Consider the situation at the remote end of the circuit. Determine the load conditions. Find out whether there were any simultaneous incidents elsewhere on the system.
(*d*) Check the earthing arrangements.
(*e*) Check the ratio of, and connections to, the current transformers (CTs).
(*f*) Ensure that the relay settings are correct.

2.13 Data analysis

(*a*) Collate the information in chronological order and wherever possible in a diagrammatic form.
(*b*) Consider whether the incident has any precedents.
(*c*) Be clear about the circuitry involved and whether it deviates from standard circuitry and equipment.
(*d*) If the conclusion is not clear cut, list and contemplate possible sources of failure, e.g. equipment design limitations, faulty components, incorrect commissioning, wrong settings.
(*e*) Consider whether power supplies may have been lost temporarily.

2.14 Damage investigation

This will usually relate to damage to high-voltage equipment.

(*a*) Creep up on the problem; do not disturb anything unless you have fully thought out the situation.
(*b*) Take photographs, or produce drawings and notes, at each stage.
(*c*) Compare the situation with healthy equipment.

2.15 Diagnostic tests

(*a*) In the first instance repeat the commissioning tests and compare the results with the original commissioning results.
(*b*) Write down a brief schedule of any further investigational tests. Note the results at each stage. Note any temporary connections which may require removal when the tests are complete.
(*c*) Where it is considered appropriate, repeat the same test a number of times; some faults are transient and will not show up on the first test.
(*d*) Carry out tests which, as far as possible, repeat the incident conditions.

2.16 Conclusions

Produce a clear concise statement of:

- the incident
- the investigation carried out
- the conclusions and recommendations.

2.17 Post-investigation

Ensure that the recommendations are acted on. The momentum of implementing one set of recommendations often slows when a subsequent incident occurs!

Chapter 3

Power-system theory

3.1 Introduction

This chapter will examine fundamental aspects of electrical power system theory which have relevance to later chapters. In most instances, concepts and mathematical relationships will be stated without proof; if the latter is required, recourse can be made to more dedicated texts which are widely available. Suffice it to say that, for the purposes of commissioning, maintenance and fault investigations, knowledge of a theoretical conclusion, and the ability to apply that conclusion, is invariably more important than understanding its determination.

The subject matter covered will include symmetrical-components analysis, the synchronous generator, circuit-breaker fundamentals, theory of power transformers and the implications of electric and magnetic fields with reference to cables and overhead lines.

POWER-SYSTEM ANALYSIS

3.2 The vector operator *a*

In this text the vector operator a will be utilised where

$$a = 1\angle 120° = -\tfrac{1}{2} + j\sqrt{3}/2$$

$$a^2 = 1\angle 240° = -\tfrac{1}{2} - j\sqrt{3}/2$$

$$\text{and } 1 + a + a^2 = 0$$

3.3 Impedance

Impedance is given as $Z = R + jX$
(where Z=impedance, R=resistance, X=reactance).

With many items of power equipment the reactance is much greater than resistance, and to good approximation the impedance and reactance can be taken as one and the same. An exception to this general rule is feeders at 33 kV and below.

3.4 Per-cent impedance (%Z)

When utilising ohms in calculations, reference must be made to a common base voltage (i.e. one side or other of a transformer). When a number of voltages enter the consideration, this becomes tedious and open to calculation error. The use of per-cent impedance (or per-unit impedance) simplifies such calculations, since such values are additive irrespective of voltage level.

Per-cent impedance may be defined as that percentage of the applied voltage which is dropped across an item of equipment when a stated quantity of volt-amperes (VA) (and hence level of current), known as the base (VA), flows through the item of equipment. A commonly used and convenient base (VA) is 100 MVA. It should be noted that it is common practice for transformers to have the percentage impedance quoted at the nominal volt-ampere rating of the transformer.

The percentage impedance of an item of equipment is determined as follows:

$$(\%Z) = \frac{(\sqrt{3})IZ}{V_L} \times 100\% \tag{3.1}$$

where I = current corresponding to base (VA) flowing through the item of equipment
V_L = nominal system line voltage
Z = impedance of the equipment in ohms corresponding to the base voltage V_L

Eqn. 3.1 can be utilised to derive a more direct relationship between Z and $(\%Z)$, as follows:

$$(\%Z) = \frac{Z\ (\text{base VA})}{V_L^2} \times 100\% \tag{3.2}$$

Table 3.1 Percent impedance to ohms (100 MVA base)

At 400 kV	1% = 16 Ω
275 kV	1% = 7.5 Ω
132 kV	1% = 1.75 Ω
33 kV	1% = 0.11 Ω
11 kV	1% = 0.012 Ω

Table 3.1 gives actual values for per-cent impedance and ohms for the more common voltages. The relationship between three-phase fault level and per-cent impedance is expressed as follows:

$$\text{3-phase fault level (volt-amperes)} = \frac{\text{(base VA)}}{(\% Z)} \times 100 \tag{3.3}$$

where $(\% Z)$ = per-cent impedance of the item(s) of equipment between the voltage source and point of fault.

Table 3.2 summarises corresponding values of MVA and current for the more common voltages.

Table 3.2 MVA to amperes conversion table

At 400 kV	1 MVA corresponds to 1.44 A/phase
275 kV	1 MVA corresponds to 2.1 A/phase
132 kV	1 MVA corresponds to 4.38 A/phase
33 kV	1 MVA corresponds to 17.52 A/phase
11 kV	1 MVA corresponds to 52.54 A/phase

3.5 Three-phase equivalent fault level

Eqn. 3.3 determines the fault level for three-phase balanced conditions, based on a three-phase base (VA) quantity. However, for reasons of consistency and simplicity, the three-phase base (VA) is also utilised when calculating fault levels for faults other than three-phase, e.g. single-phase-to-earth or phase-to-phase. This introduces the concept of the three-phase equivalent fault level. Under such circumstances, the three-phase fault level determined in eqn. 3.3 becomes the three-phase equivalent fault level, and the fault current at the point of fault is simply calculated from the relationship

$$I = \frac{\text{three-phase equivalent fault level in (VA)}}{\sqrt{3} V_L} \tag{3.4}$$

3.6 Symmetrical components

The technique known as symmetrical components (alternatively termed phase-sequence components) is a mathematical tool for determining the voltages and currents which arise when power systems are subject to faults or unbalanced loading conditions. The technique is based on the premise that a set of unbalanced three-phase phasors can be represented by the summation of three sets of balanced, or symmetrical, phasors. The strength of this approach reflects the comparative simplicity of analysing a power system which is made up of balanced conditions, especially when it is considered that

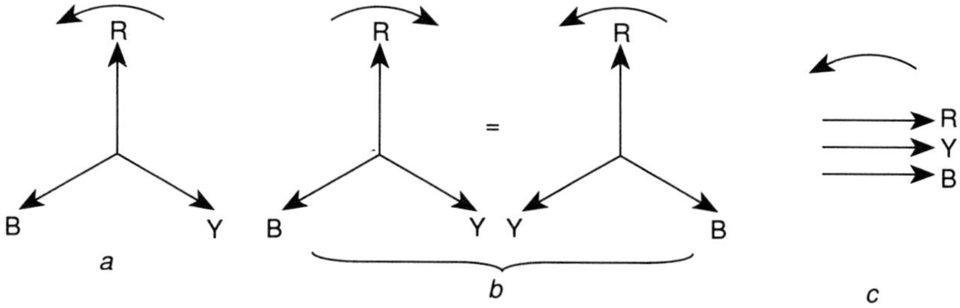

Figure 3.1 *Symmetrical-components phasor relationships*

a Positive-sequence
b Negative-sequence
c Zero-sequence

a three-phase balanced network can be represented by a single-phase equivalent circuit. The three sets of balanced vectors are termed positive, negative and zero-sequence and are illustrated in Figure 3.1.

3.7 Mathematical relationships

When utilising symmetrical components, the power system is considered to comprise three separate power systems which represent the positive, negative and zero-sequence components, the values of which are summated to determine the actual power system phase quantities. The power system quantities are derived from the symmetrical components as follows:

$$\begin{aligned} I_R &= I_{R1} + I_{R2} + I_{R0} \\ I_Y &= I_{Y1} + I_{Y2} + I_{Y0} = a^2 I_{R1} + a I_{R2} + I_{R0} \\ I_B &= I_{B1} + I_{B2} + I_{B0} = a I_{R1} + a^2 I_{R2} + I_{R0} \end{aligned} \tag{3.5}$$

where the subscripts *R*, *Y*, *B*, 1, 2 and 0 denote the phase colours and the positive, negative and zero-sequence components of current, respectively. Similarly,

$$\begin{aligned} V_R &= V_{R1} + V_{R2} + V_{R0} \\ V_Y &= V_{Y1} + V_{Y2} + V_{Y0} = a^2 V_{R1} + a V_{R2} + V_{R0} \\ V_B &= V_{B1} + V_{B2} + V_{B0} = a V_{R1} + a^2 V_{R2} + V_{R0} \end{aligned}$$

Similarly, from eqn. 3.5, complementary relationships may be deduced as follows:

$$\begin{aligned} \text{i.e. } I_{R1} &= \tfrac{1}{3}(I_R + a I_Y + a^2 I_B) \\ I_{R2} &= \tfrac{1}{3}(I_R + a^2 I_Y + a I_B) \\ I_{R0} &= \tfrac{1}{3}(I_R + I_Y + I_B) \end{aligned} \tag{3.6}$$

3.8 Practical determination of sequence impedances

The positive-sequence impedance of an item of equipment is obtained by applying a three-phase balanced voltage source, with anticlockwise (conventional) phase rotation, to one end of the item of equipment, applying a three-phase short circuit clear of earth to the other end of the equipment, and measuring the per-phase applied voltage and current. The positive-sequence impedance is then obtained by calculating as follows:

$$Z_1 = \frac{V_R}{I_R} \quad \text{(where the } R \text{ subscript denotes the red phase)}$$

The negative-sequence impedance is obtained similarly by applying a voltage with a clockwise phase rotation. As a general rule, the positive- and negative-sequence impedances are identical for all items of static plant.

The zero-sequence impedance is similarly obtained by applying a single-phase voltage source to all three terminals of the equipment and applying a three-phase short circuit to earth at the other end. The zero-sequence impedance is calculated from the measurements of the voltage applied to, and current into, the red phase (or any other phase)

Figure 3.2 illustrates the sequence impedance equivalent circuits for a range of the more common equipments. The format of the equivalent circuit must be such that it represents the impedance looking into any terminal when a short circuit is applied to any other terminal.

3.9 Transformer-core arrangement

With three-phase transformers, there is need to be aware of the implication of the structure of the core, i.e., whether it is a three-limbed or five-limbed transformer, as shown in Figure 3.3.

Positive- or negative-sequence currents are such that at any instant of time, the current in any phase is equal to the sum of the currents in the other two phases (conditions for balanced three-wire systems). Therefore the fluxes induced into transformer cores, from positive- or negative-sequence currents, are also equally balanced. Thus for the three-limbed transformer, the flux in any limb returns via the other two limbs; this similarly applies to the five-limbed transformer.

The situation is different for zero-sequence currents. If we consider zero-sequence currents creating zero-sequence fluxes in the five-limbed transformer, then, given that the three fluxes are equal in magnitude and phase, the return path for all three fluxes must be via the two outer limbs. The three-limbed transformer, however, provides no such return path and the flux is therefore caused to return via the oil and transformer tank, resulting in induced currents in the tank walls. The three-limbed transformer therefore

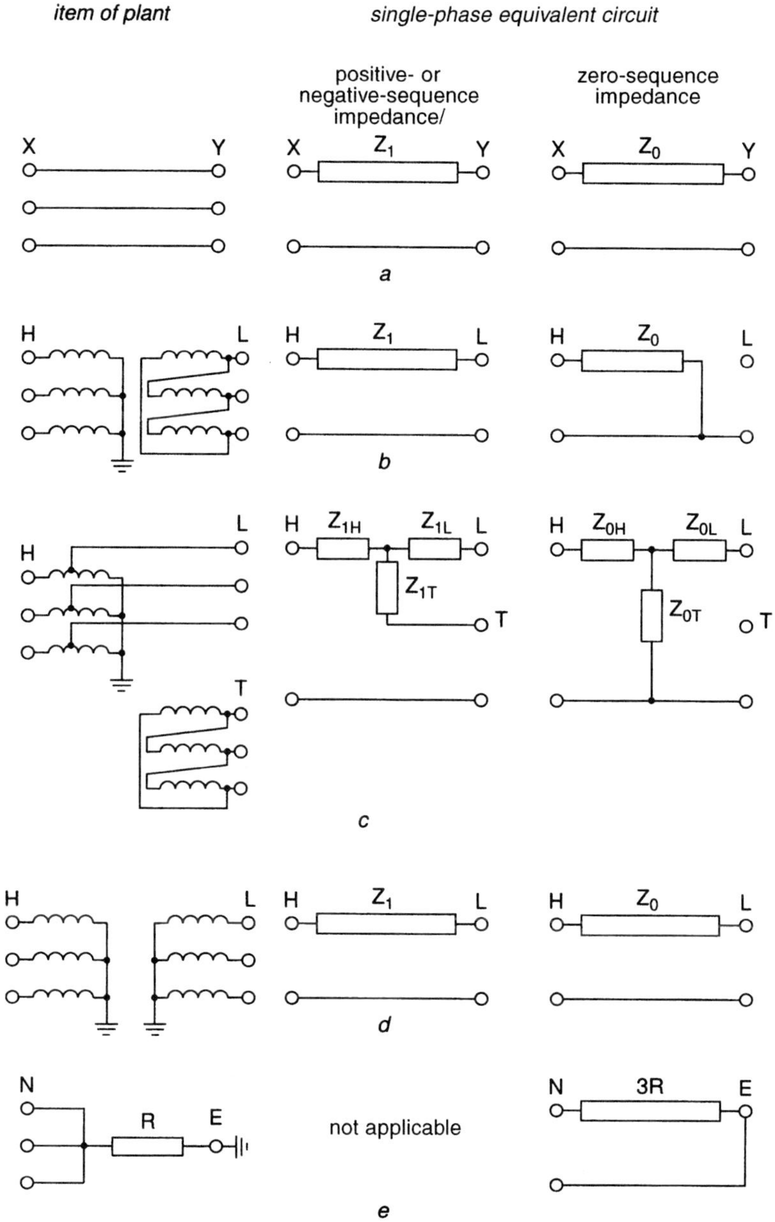

Figure 3.2 Sequence-impedance equivalent circuits

a Three-phase feeder
b Star–delta transformer
c Autotransformer with delta tertiary
d Star–star transformer
e Earthing resistor

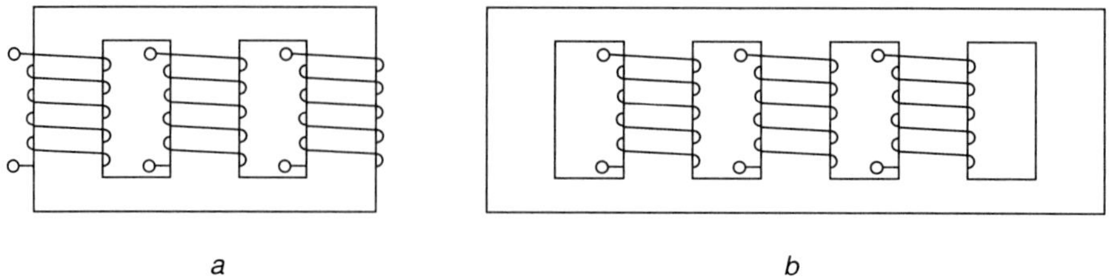

Figure 3.3 Transformer cores

a Three-limbed
b Five-limbed

provides a lower-reluctance path, and hence lower magnetising impedance and higher magnetising current, than the five-limbed transformer. However, magnetising current is usually small compared with load and fault currents, and for most practical purposes can usually be neglected. The situation is, however, different for voltage transformers, where accuracy is required. In such instances a five-limbed core must be employed.

Most power transformers are of the three-limbed variety. This is advantageous on economic grounds, and technically acceptable since load currents do not usually contain significant zero-sequence quantities. However very large transformers such as major generator transformers are often of the five-limbed variety. This arises because the five-limbed transformer has a lower height than its three-limbed counterpart, thereby making transport easier. More recently, generator transformers have comprised three single-phase units.

3.10 Reference phase

When deriving phase-sequence quantities it is usual to express them in terms of the red phase. The yellow- and blue-phase quantities are then obtained by applying the *a*-operator to the red-phase-sequence quantities, as indicated by eqn. 3.5. The red phase is termed the reference phase.

3.11 Sequence-network source voltage

The following sections on synchronous generators will show that the generator source (driving) voltages may differ considerably from the power system nominal voltages. Consideration of generator source voltages thus complicates system fault analysis.

A simpler analysis can be obtained by employing a Thevenin equivalent circuit. If the latter is applied to the point of fault, the equivalent-circuit driving voltage becomes the power system voltage prior to the fault occurring;

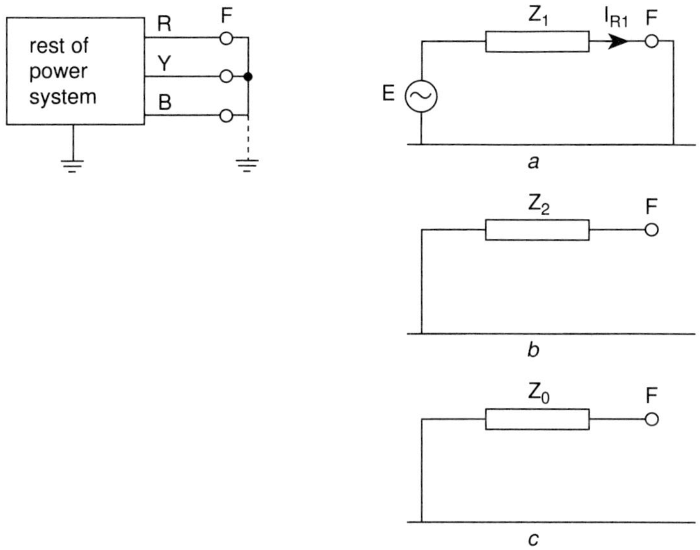

Figure 3.4 Sequence network of a three-phase fault

a Positive-sequence
b Negative-sequence
c Zero-sequence

for most purposes this can usually be considered as the system nominal voltage. This voltage then drives the fault current through a network consisting of the fault impedance (usually zero ohms) in series with the rest of the power system. Suffice it to say that, for most practical purposes, only small portions of the power system need to be considered.

3.12 Sequence networks for common fault conditions

This section will consider the sequence networks and current relationships for the more common fault conditions. It is advantageous to commit these relationships to memory.

(a) Three-phase fault
Figure 3.4 shows the sequence network for a three-phase fault which is either clear of earth or connected to earth. Salient current relationships at the point of fault are:

$$I_R = I_{R1},\ I_2 = 0,\ I_0 = 0$$

(b) Single-phase-to-earth fault
Figure 3.5 shows the sequence network for a single-phase-to-earth fault, with

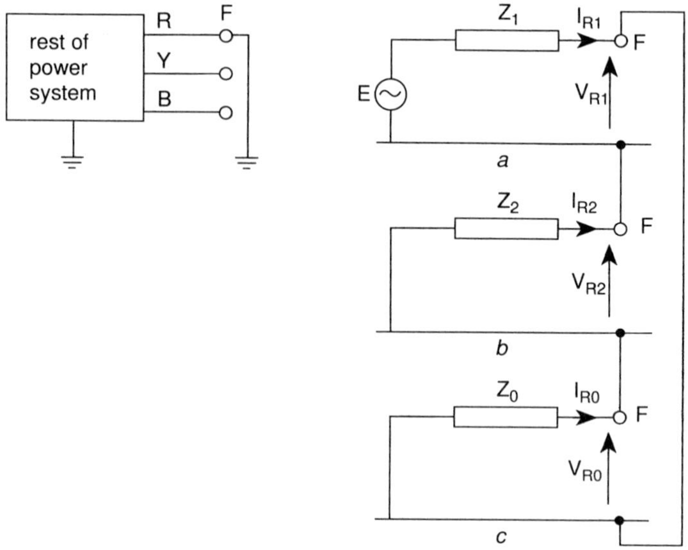

Figure 3.5 Sequence network of a single-phase-to-earth fault

a Positive-sequence
b Negative-sequence
c Zero-sequence

the fault applied to the red phase. Salient current relationships at the point of fault are:

$$I_R = I_{R1} + I_{R2} + I_{R0},\ I_{R1} = I_{R2} = I_{R0}$$

(*c*) *Phase-to-phase fault*
Figure 3.6 illustrates the sequence network for a phase-to-phase fault between the yellow and blue phases, with red as the reference phase. Salient current relationships at the point of fault are:

$$I_R = I_{R1} + I_{R2},\ I_{R1} = -\ I_{R2}$$

THE SYNCHRONOUS GENERATOR

3.13 Introduction

The detailed theoretical considerations which describe the behaviour of the synchronous machine are relatively complex and are well described elsewhere. Generally speaking a detailed theoretical knowledge is unnecessary for commissioning purposes. The following sections will therefore only provide a summary of those essential features of the synchronous machine, and its

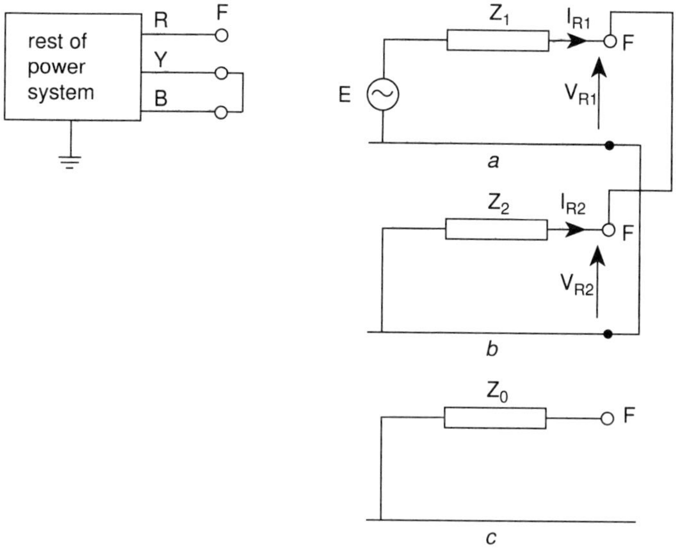

Figure 3.6 *Sequence network of a phase-to-phase fault*

a Positive-sequence
b Negative-sequence
c Zero-sequence

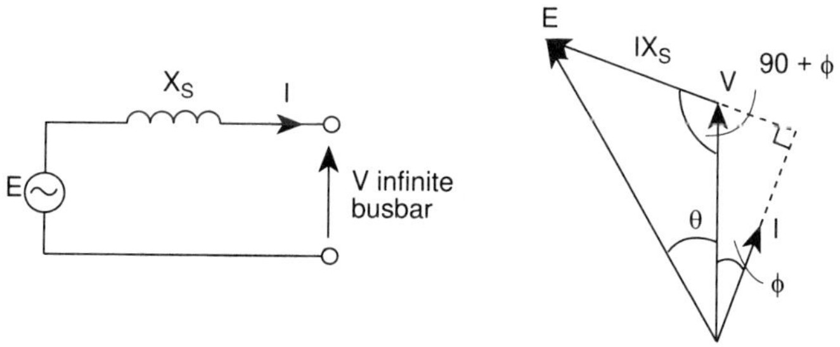

Figure 3.7 *Power generation into an infinite busbar*

associated generator transformer, with which an engineer concerned with commissioning, or power system fault analysis, should be familiar.

3.14 Power-transfer relationship

With reference to Figure 3.7 the power transferred from a generator into an infinite busbar is given, to good approximation, by

$$P = \frac{VE \sin \theta}{X_s} \tag{3.7}$$

where P = power transfer (w)
V = voltage of infinite busbar
E = generator (open-circuit) voltage
X_s = generator (synchronous) reactance
θ = angle between V and E (rotor angle).

Maximum power transfer is when $\theta = 90°$. If, when θ is at 90°, more mechanical driving power is input into the generator, the rotor angle will increase beyond 90°, and less electric power will be transferred to the infinite busbar. The excess mechanical driving power will therefore result in an ever increasing rotor angle, resulting in loss of synchronism (i.e. pole-slipping, see Chapter 14).

3.15 Generator on open circuit prior to fault

Consider a synchronous generator on open circuit, with generated voltage E, which is subject to a three-phase short circuit at its terminals. The current which flows into the short circuit is given by the expression:

$$I = \left(\frac{E}{X_{St}} - \frac{E}{X_t}\right) \exp\left(\frac{-t}{T_{St}}\right) + \left(\frac{E}{X_t} - \frac{E}{X_S}\right) \exp\left(\frac{-t}{T_t}\right) + \frac{E}{X_S} \tag{3.8}$$

where, with reference to Figures 3.8 and 3.9, X_{St}, the subtransient reactance, is equal to X_L in series with the parallel combination of X_a, X_F and X_{DW}; X_t, the transient reactance, is equal to X_L in series with the parallel combination of X_a and X_F; X_S, the synchronous reactance, is equal to X_L in series with X_a; and T_{St}, T_t are the subtransient and transient time constants, respectively.

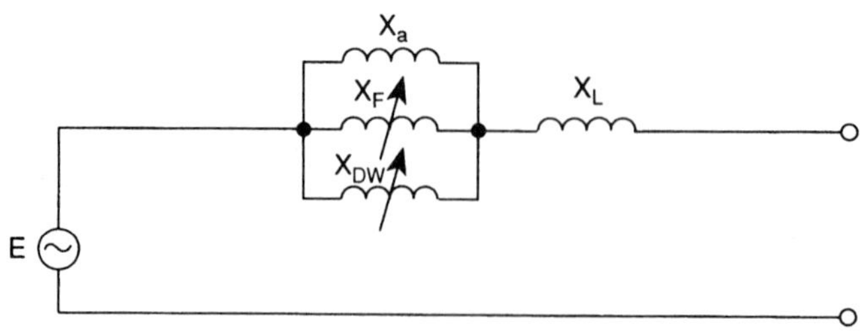

Figure 3.8 Generator equivalent circuit

X_{DW} = damping-winding leakage reactance
X_F = field winding leakage reactance
X_a = armature-reaction reactance
X_L = armature leakage reactance

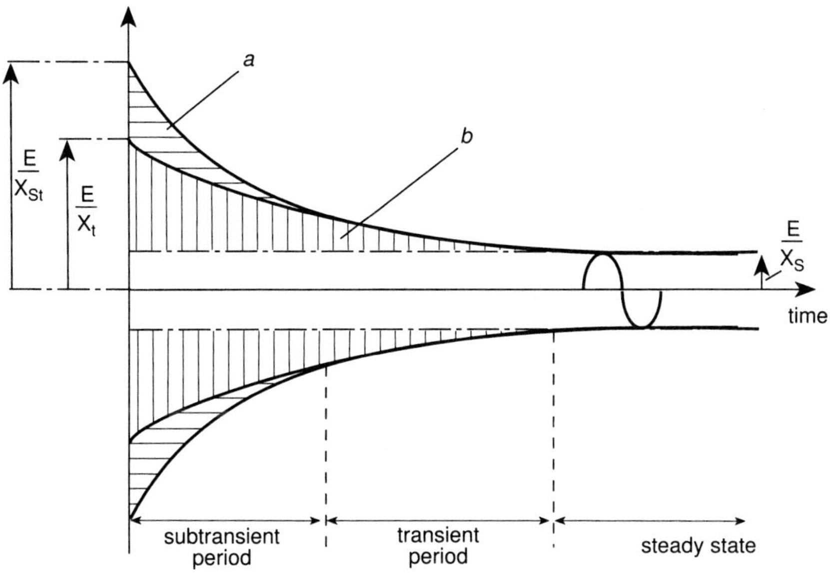

Figure 3.9 Generator fault-current envelope

$a\,\{(E/X_{St}) - (E/X_t)\}\,\exp(-t/T_{St})$
$b\,\{(E/X_t) - (E/X_S)\}\,\exp(-t/T_t)$

If we consider the situation a very significant time after fault inception, when the transient exponential terms have decayed, eqn. 3.8 reduces to:

$$I = \frac{E}{X_S}$$

At this time the three-phase stator currents produce a rotating magnetic field which rotates at synchronous speed, as such a voltage is induced into the stator but not the rotor; the induced stator voltage is interpreted in terms of the synchronous reactance X_S. The stator flux may be divided into two components:

(*a*) flux which crosses the air gap between stator and rotor to link both the field and damping windings. (The latter is fitted to damp rotor oscillations following system disturbances.)
(*b*) flux which does not cross the air gap and links only the stator conductors. This flux is responsible for the armature leakage reactance, X_L.

At the point of fault interception, the rotating magnetic field increases from zero, and in doing so induces voltages into both the field and damping windings, which result in induced currents and associated fluxes which oppose and virtually cancel the armature flux. The limitation to full flux cancellation arises from the existence of the leakage reactances X_F and X_{DW}

both of which are fully established and have minimum values at fault inception and increase towards infinity during the subtransient and transient periods, respectively. Thus, at fault inception, the fault current is limited by the subtransient reactance, X_{St}, which accords with eqn. 3.8 at time $t=0$.

Figure 3.9 illustrates a typical fault current envelope. Fault calculations are usually based on the impedance at the instant of fault, although a circuit breaker would typically be breaking the current that existed approximately 100 ms after fault inception. The time constants T_{St} and T_t are of the order of 20 ms and 1 s, respectively, with the time to steady state conditions being achieved in the order of 2–5 s.

3.16 Point of wave of fault

Figure 3.9 illustrates a symmetrical current envelope for a three-phase fault. However, it is well documented that, when an inductive circuit is switched, a DC transient occurs when the circuit is switched at some point other than at a natural current zero. The DC transient can result in a doubling of the fault current thereby causing an asymmetrical current envelope. Such a situation is relevant to a synchronous generator given that at the instant of fault the current is limited by the subtransient reactance, an inductive circuit.

3.17 Generator on-load prior to fault

If a generator is loaded prior to the instant of three-phase short circuit, eqn. 3.8 becomes modified as follows:

$$I=\left(\frac{E_{St}}{X_{St}}-\frac{E_t}{X_t}\right)\exp\left(\frac{-t}{T_{St}}\right)+\left(\frac{E_t}{X_t}-\frac{E}{X_S}\right)\exp\left(\frac{-t}{T_t}\right)+\frac{E}{X_S} \qquad (3.9)$$

where, with reference to Figure 3.10, E_{St}, E_t and E represent the voltages behind X_{St}, X_t and X_S, respectively, prior to the instant of fault. The changes from eqn. 3.8 to eqn. 3.9 reflect the fact that the prefault load current has already established stator fluxes.

Figure 3.10 shows that, for a previously loaded generator, the magnitude of E is much greater than V (for lagging-power-factor loads). This may theoretically result in fault-level currents up to 20% higher (or lower) than for the unloaded generator, although in practice they are likely to differ by more than 10%.

3.18 Positive-phase-sequence impedance

The preceding sections have examined the behaviour of the generator under three-phase-fault conditions, when only positive-phase-sequence impedances

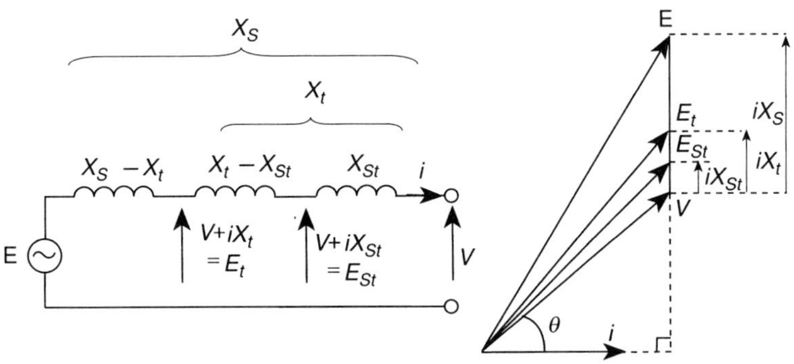

Figure 3.10 Voltage relationships for generator on-load prior to instant of fault

are experienced, and have shown that a synchronous generator has three positive-sequence impedances (which are usually considered as reactances, since the resistive component is comparatively small), namely:

subtransient reactance	X_{St}
transient reactance	X_t
synchronous reactance	X_S

3.19 Negative-phase-sequence impedance

The negative-phase-sequence impedance of a generator may be obtained by operating the generator with the rotor running at synchronous speed with the field winding unexcited, and applying negative-phase-sequence voltages to the generator terminals. Under such circumstances the rotating armature-reaction field produced by the stator rotates in the opposite direction to the rotor and therefore induces a voltage into the rotor. The induced voltage causes a rotor current to flow which produces a field which is in opposition to the armature-reaction field, virtually cancelling the latter (similar to transformer action). In a similar manner, the stator field induces a voltage into the damping windings which also produces a cancelling field. The flow of current from the NPS voltage source is thus limited by the subtransient component of the leakage flux; the negative-phase-sequence impedance is thus a similar value to the positive-sequence subtransient reactance.

At first sight it may seem somewhat puzzling that positive-phase-sequence impedances are determined by using the rotor field as the source of voltage, but the negative-phase-sequence impedance requires the use of an external generator. This, however, reflects the practical situation whereby negative-sequence voltages are not generated by the generator, but are imposed on the generator by the system, when the system three-phase impedances are unbalanced (i.e. fault conditions or unbalanced load).

3.20 Zero-phase-sequence impedance

The zero-phase-sequence impedance is determined in a similar manner to the negative, excepting that zero-sequence (in-phase) voltages are applied. In this instance the fluxes of the three-phases act along axes which are 120° displaced in space and thus cancel within the rotor body. The fluxes are thus limited to leakage paths around the stator conductors and this generally results in an impedance which is 20-50% less than the negative-sequence impedance. Most generator neutrals, in practice, are earthed via an impedance which limits the flow of zero-sequence current.

3.21 The automatic voltage regulator

The automatic voltage regulator (AVR) is an indispensable feature in the safe and successful operation of large generators; the main functions are as follows:

(*a*) to maintain the generator terminal voltage at a fixed value (within limits). This has a two fold benefit: first, it provides a stable voltage source to the unit transformer thus stabilising the generator auxiliary supply voltages; and secondly, it provides a reference and stabilising source for power system voltage;
(*b*) control of the field current to assist generator stability during changes in output power (MW);
(*c*) boosting the rotor field current under system fault conditions, to assist in maintaining synchronism with the rest of the system;
(*d*) reducing the field current at the time of load rejection to avoid over voltages.

Figure 3.11*a* shows an outline diagram of the AVR and its connections, where X_S represents the generator synchronous reactance and X_T the reactance of the generator transformer. The AVR compares the generator terminal voltage V_T with a required reference level, such that any departure of the terminal voltage from the reference results in an adjustment to the DC exciter output, and hence the rotor field and generated voltage, to restore the terminal voltage to the reference level.

Figure 3.11 also illustrates to a good approximation the performance of the AVR when a change in power output takes place. The initial conditions, shown in Figure 3.11*b*, consist of system voltage V, current I_1, generated voltage E_1 positioned on constant-power locus MW_1 and generator terminal voltage V_{T1}. Consider the impact of an increase in mechanical input power to the generator to increase the generated power output current by I_w. The locus of E must now lie on the constant power locus MW_2. However for E_1 to move to position E_2 it is necessary for I_1 to move to I_2, thus causing additional

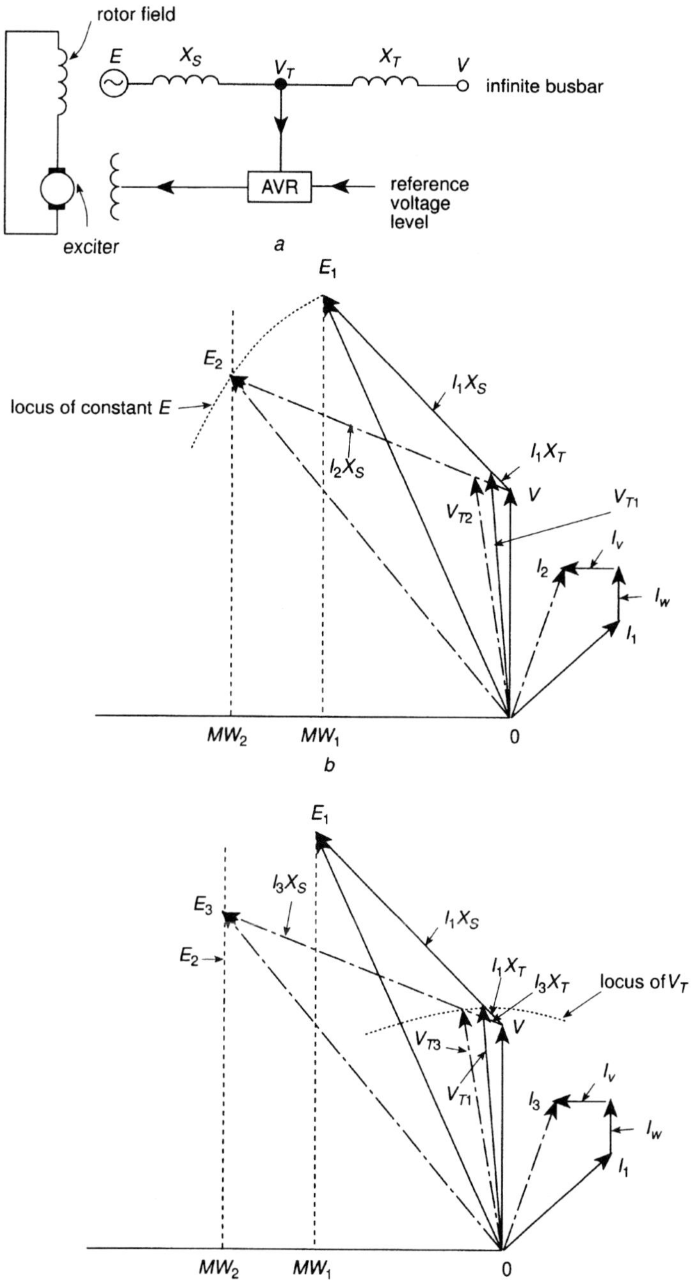

Figure 3.11 Action of automatic voltage regulator

a Outline of AVR connections
b ower increase without AVR action
c Power increase with AVR action

reactive current I_v. V_{T1} now moves to V_{T2}, such that the voltage drop ratio of IX_S to IX_T is maintained (and must remain constant for all machine-loading conditions). As can be seen, V_{T2} now has a smaller magnitude than V_{T1}.

The AVR now senses that V_{T2} is less than the reference level (i.e. less than V_{T1}) and corrects the situation by increasing excitation and hence E. The resulting situation is shown in Figure 3.11*c*. The new terminal voltage lies on a locus of radius V_{T1} with centre 0, and intercepts voltage I_3 (X_S+X_T) at a point such that the ratio of IX_S to IX_T is maintained. As can be seen, the rotor angle between E_3 and V is less than between E_2 and V and thus the AVR has assisted in both restoring generator rotor angle, and hence stability, and maintaining the terminal voltage. Note also that the effect of an increase in generated power is also to change the generated reactive power.

3.22 AVR speed of response

Figure 3.12 shows a typical response curve for a modern AVR. The greater the deviation in terminal voltage from the reference, the faster is the response time. A consequence of a fast response is that a change in rotor field flux causes a transient induced voltage in the stator due to flux rate of change. In addition, there is a tendency towards an overshoot of terminal voltage, resulting in an oscillatory settling down period. These transient effects are undesirable. However, when a significant reduction in terminal voltage occurs, restoration of voltage takes priority. The factor limiting speed of response is the finite time it takes for the flux to change in the excitation system. As observed from Figure 3.12, the speed of response for short-circuit conditions could be as fast as 100 ms.

3.23 AVR control and range

Control of the rotor-field-excitation system may be via AVR or manual control. A typical range of excitation (rotor currents) would be from zero to about 250% nominal for manual control and, transiently, up to 400% nominal for AVR control. Nominal excitation is that required to produce the nominal machine terminal voltage with the generator on open-circuit.

3.24 The generator transformer

Given that the AVR is used to maintain a fixed generator-terminal voltage, it cannot be used directly to control reactive-power generation. This task is left to the generator-transformer tap changer, as will now be illustrated.

With reference to Figure 3.13, consider the generator transformer reactance X_T as being referred to the generator side of the transformer. X_S will

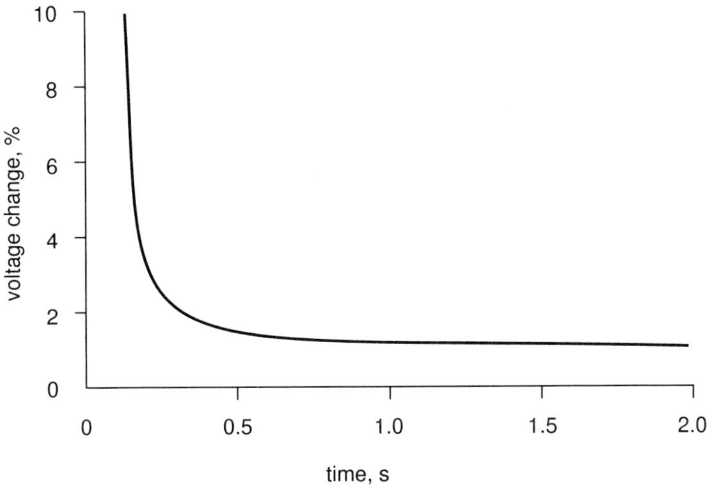

Figure 3.12 AVR response curve

typically be about 10 times greater in magnitude than X_T. Initial starting conditions are generated voltage E_1, terminal voltage V_{T1}, transformer voltage V_1 and generated current I_1. If, for illustration purposes, the transformer tap-changer is adjusted to obtain a 10% increase in voltage (an unusually large step change), V_1 becomes replaced by V_2. Given that the generated power has not changed, the product of V and the component of I in phase with V must remain constant. Thus if V_1 increases to V_2, I_1 decreases by an amount I_W. The voltage drop resulting from $I_W(X_S+X_T)$ now causes the locus of E to lie on the MW_2 constant-power locus. In addition the new generator terminal voltage V_{T2} must lie upon a locus of radius V_{T1} (given that V_{t1} is maintained constant via the AVR) drawn from a centre of 0, and at a point such that the interception of the locus with the voltage $I_2(X_S+X_T)$ maintains the ratio of I_2 X_S to I_2 X_T identical to that of I_1 X_S to I_1 X_T (i.e. the percentage voltage drops across X_S and X_T, respectively, are the same for all working conditions). E_2 and I_2 as shown on Figure 3.13 satisfy these conditions. As can be seen, the effect of the tap change is to increase the generation of leading VARs by an amount I_V. Tapping in the other direction would conversely result in lagging VARs. The tap changer is usually located on the HV side of the transformer. Note that there are some approximations in the above analysis.

3.25 Transformer overfluxing

A condition can arise during operation and testing of transformers, in which the flux induced in the core is such that the latter is driven significantly in to saturation. This results in the following undesirable conditions:

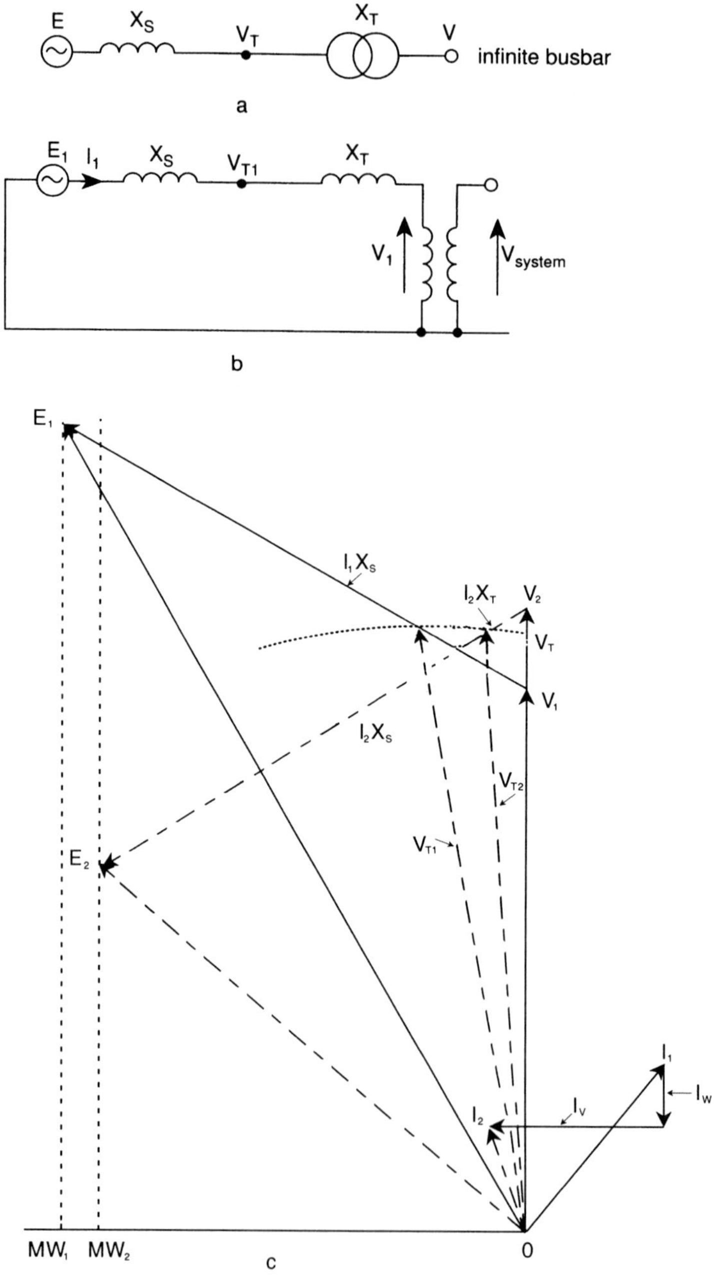

Figure 3.13 Effect of generator-transformer tap-changer

a System under consideration
b Equivalent circuit (showing conditions prior to tap change)
c Effect of tap-changer producing 10% voltage change

(i) a large increase in magnetising current;
(ii) an extension of the flux path into the transformer tank, causing eddy currents, losses and heating;
(iii) an increase in transformer noise, vibration and temperature.

An explanation of overfluxing is given with reference to the voltage and flux conditions within the transformer, as follows:

$$E = KfB$$

where E = induced voltage
B = flux density
f = frequency
K = a constant

Rearranging gives

$$B = K\frac{E}{f}$$

Thus to limit the magnitude of flux density, the ratio of voltage to frequency must not exceed a certain ratio.

The chances of experiencing overfluxing are perhaps greatest in generator transformers when the generator is being run up or run down. A modern AVR should have additional features in-built to avoid this problem, but machines with older AVRs, or manual control, may be subject to overfluxing, unless caution is exercised.

Overfluxing may also be experienced in system transformers if they are energised from the HV side whilst on a low tap with the LV circuit breaker open.

CIRCUIT BREAKERS

3.26 Current interruption

The salient purpose of the circuit breaker is to interrupt, or close onto, all currents, including fault current. However, the commissioning and maintenance processes are not required to verify this major capability: this is because it is impracticable to carry out such tests on site, and experience has shown it to be unnecessary. Recourse, for such verification, must therefore be made to the factory test data.

With reference to on-load commissioning tests, and fault analysis, there is however a need to have a rudimentary understanding of circuit breaker capability and the interruption process. The following sections will summarise the key theoretical considerations.

3.27 Circuit-breaker arc

When the current-carrying contacts of a circuit breaker commence to open there is an increase in resistance, and thus temperature rise, which is sufficient to cause thermionic emission of electrons from the contact surface, thereby maintaining current flow via an arc.

Twice per cycle the current passes through zero and the arc is momentarily extinguished; successful arc interruption requires the rate of rise of dielectric strength in the contact gap, during and after current zero, to exceed the ability of the voltage across the gap to breakdown the dielectric, to re-establish the arc.

3.28 Post-fault voltages

Figure 3.14 shows the circuitry to be considered when evaluating circuit breaker performance resulting from system faults. The inductances, resistances and capacitances represent the lumped impedances on the source and feeder side, with the capacitances also including that of the circuit breaker. Fault-arc extinctions will take place at a current zero, when the source voltage E will be at a peak (since the impedance to fault is essentially inductive).

Consider first a fault at the circuit breaker terminals (i.e. short-circuit across points T). At the instance of current zero, and following opening of the circuit breaker, the situation may be likened to that of the source voltage E being applied suddenly to an *LRC* circuit. This will result in the voltage across the circuit breaker (i.e. the voltage across C_s) consisting of the addition of a steady-state and a decaying transient component. The frequency of the transient voltage is a function of both L_s and C_s, and the amplitude usually undergoes significant exponential decay within the first 10 ms after arc interruption. However, the oscillatory transient when added to the steady state voltage may result in the voltage across C_s, and hence across the circuit breaker, rising to a theoretical maximum magnitude of $2E_{max}$. The voltage which appears across the circuit breaker immediately after arc interruption is known as the 'restriking voltage' (i.e. up to $2E_{max}$), since it is this voltage which may cause restriking of the arc. If the arc does not restrike, the steady-state voltage which subsequently appears across the circuit breaker contacts is known as the 'recovery voltage'.

3.29 Short-line faults

A potentially onerous situation, in terms of arc restrike, occurs when the fault is a short distance along the feeder (overhead line), i.e. across points L in Figure 3.14. In this instance two oscillatory transients are created: one across C_s as described in Section 3.28, and the other across C_F arising from the decaying stored energy in the feeder.

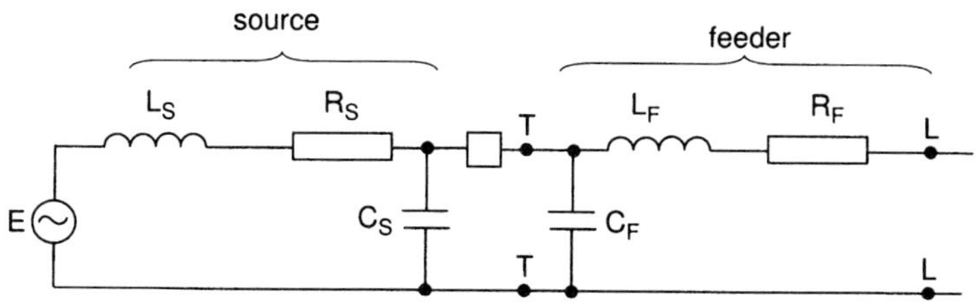

Figure 3.14 Circuitry for considering fault-current interruption

The combined effect of the transients is to cause a rapid rate of rise of voltage across the circuit breaker contacts. At 'supergrid' voltages, the most onerous fault position is at a distance of about 1 km from the circuit breaker, termed the 'kilometric fault'. Air-blast circuit breakers in particular have been prone to restrike as a result of this type of fault.

3.30 Current chopping

Current chopping occurs when a circuit breaker is breaking inductive current considerably below its rating (load current), so that the arc is extinguished before the natural current zero. This leaves stored energy in inductance L_s (see Figure 3.14) which produces a high-frequency, high-voltage transient across the circuit-breaker contacts, which may rise up to three times nominal voltage. A correctly designed circuit breaker will undergo repeated restrike until normal current zero is reached, thereby preventing damage to insulation.

3.31 Resistors and resistance switching

Resistance switching is utilised to limit overvoltages which follow circuit-breaker arc extinction. With reference to Figure 3.15, the resistance is inserted across the circuit-breaker main interrupter contacts prior to the

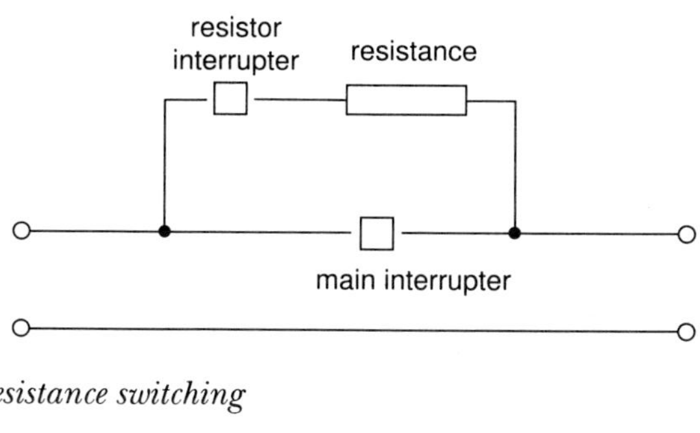

Figure 3.15 Resistance switching

latter opening. The effect is to dampen and diminish the transient voltages. A short time delay after the main interrupters open, the resistance is switched out of circuit. Insertion and removal of the resistance may be via resistor interruptors. Main applications of resistance switching are:

(*a*) reduction of the rate of rise of restrike voltage, particularly due to the short-line fault. This would require low-value resistors in the range 300–2000 Ω.
(*b*) to limit overvoltages due to current chopping. This would require resistors typically in the range 2000–10 000 Ω.

Resistors may also be utilised to ensure even sharing of the recovery voltage across circuit breakers with a number of interruptors in series. Such resistors would have a typical value of 100 kΩ per phase, and may be of a nonlinear type.

3.32 Capacitance switching

Switching a capacitive load may be a particularly onerous duty. On arc extinction (current zero), the capacitance is left fully charged at E_{max}. As the source voltage subsequently varies, the voltage difference across the interruptors rises to a peak of 2 E_{max}. Should the arc restrike across the contact gap a high frequency oscillation takes place resulting in the capacitive load being charged to $2E_{max}$. This process may be repeated with, theoretically, ever increasing voltages on the capacitance. In practice, circuit resistance and leakage paths limit the voltage rise.

3.33 Asynchronous switching

This refers to the opening of a circuit breaker which is connecting two systems which have fallen out of synchronism. Under the most onerous conditions, the voltage across the circuit-breaker contacts, upon opening, would be twice the phase-to-earth voltage. Generally speaking, only circuit breakers operating on the transmission system (but not all) are rated for this purpose.

3.34 Switching asymmetrical currents

Eqn. 10.1 (Section 10.10) identifies fault currents as having two components: an RMS steady-state component, and an exponentially decaying component which is a function of the reactance-to-resistance ratio (X/R) of the impedance between source voltage and fault position. The exponential is a DC quantity which influences the points at which current zeros, and hence fault current interruption, take place. Modern technical standards usually require circuit breakers to accommodate a DC quantity caused by a system X/R ratio of 18. The most onerous duty imposed on the circuit breaker is a function of the product of maximum system X/R ratio and maximum system short-circuit current, although it is unusual in practice to find both of these maxima occurring simultaneously. However, to clear successfully a fault current containing a DC component, a circuit breaker may be required which has a higher-rated RMS short-circuit breaking current than would normally be the case. The manufacturer would need to verify the capability of the circuit breaker for this purpose.

When a circuit breaker positioned between a generator and its generator transformer is subject to a fault on the transformer side of the circuit breaker, the fault currents supplied from the generator will have an exceptionally high X/R ratio (see Table 10.2), as such a current zero may not occur in the first 300 ms following fault inception. In such instances the tripping of the circuit breaker may have to be delayed by use of a time delay relay.

TRANSFORMERS

3.35 Equivalent circuit

The practising power engineer should have a good appreciation of the transformer equivalent circuit. Figure 3.16*a* shows the complete equivalent circuit; however, since R_H and R_L are relatively small, and R_0 is relatively large, a simpler equivalent circuit, as shown in Figure 3.17*b*, is usually employed. This latter figure requires the reactances (impedances) to be referred to one side of the transformer as follows:

$$\text{low voltage referred to high voltage} = \frac{(N_H)^2}{(N_L)^2} X_L \qquad (3.10)$$

or

$$\text{high voltage referred to low voltage} = \frac{(N_L)^2}{(N_H)^2} X_H \qquad (3.11)$$

When fault levels are considered, the current through X_0 is relatively small and may be neglected, resulting in an even simpler equivalent as shown in Figure 3.17*c* where $X = X_H + X_L$.

Note that the current flowing into the magnetising impedance X_0 (i.e. the majority of I_0) typically ranges from 3% of transformer full-load current for small distribution transformers (e.g. 11 kV/415 V), to 0.5% for 400/132 kV and similar supergrid transformers.

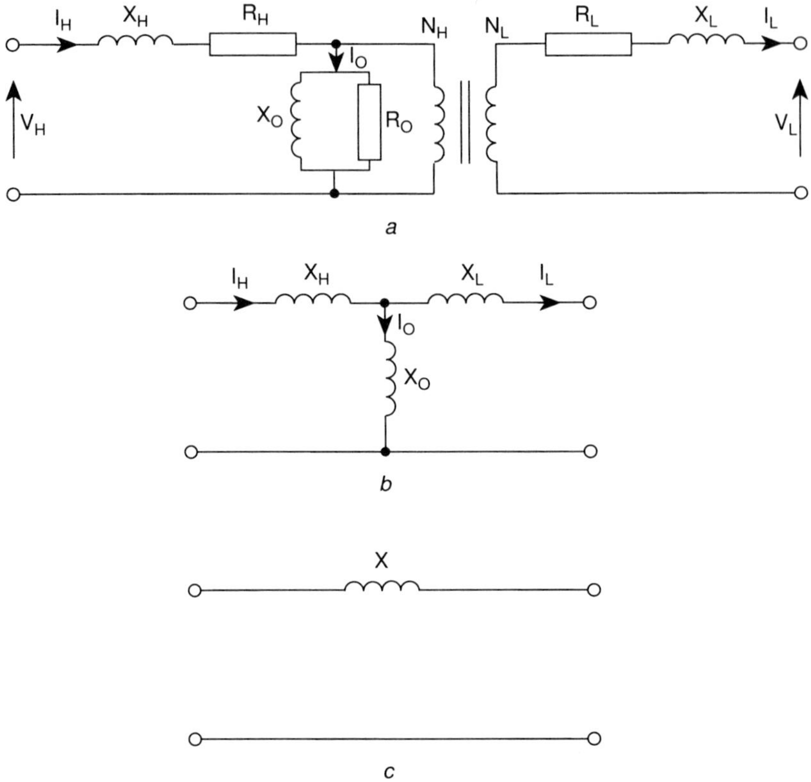

Figure 3.16 Transformer equivalent circuits

R_H, X_H, R_L, X_L=resistances and reactances of HV and LV windings, respectively
X_0=magnetising impedance
R_0=iron losses (eddy currents and hysteresis)
N_H, N_L=turns on HV and LV windings, respectively

3.36 Magnetising inrush current

When the primary side of a transformer is energised from the supply system, with the secondary on open-circuit, it acts as a simple inductive reactor. Ignoring iron losses the transformer input current is thus the magnetising current I_0, as shown in Figure 3.16, which is limited by the magnetising impedance X_0.

At any instant, the voltage induced in the magnetising impedance must be equal and opposite to the applied voltage, according to the relationship:

$$e = \frac{-Nd\Phi}{dt} \tag{3.12}$$

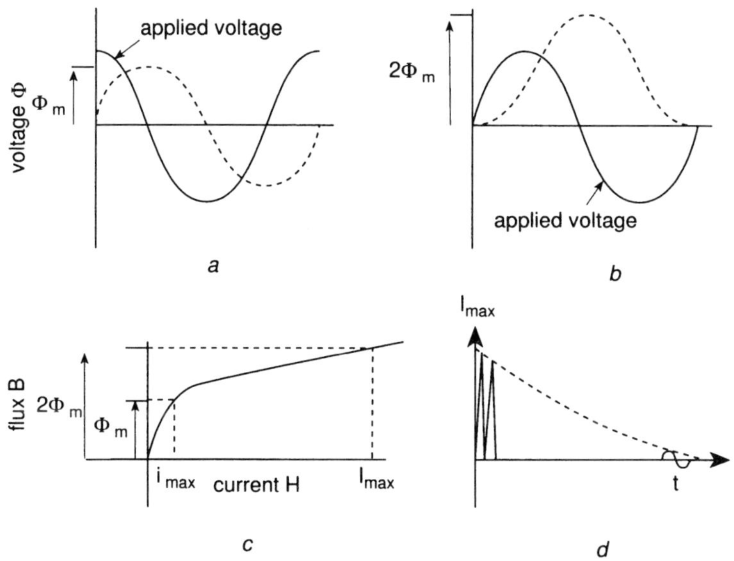

Figure 3.17 Magnetising inrush current

a Voltage-to-flux relationship at peak voltage
b Voltage-to-flux relationship at minimum voltage
c Current-to-flux relationships
d Envelope of the inrush current

where e=induced voltage
N=number of turns on primary
Φ=flux

Now if the transformer is energised at the instant of peak voltage, the voltage-to-flux relationship is as shown in Figure 3.17*a* and the resulting current i_m into the magnetising impedance is shown in Figure 3.17*c*. As can be seen, the transformer is operating on the linear portion of the *B/H* curve, and the current is therefore directly proportional to the flux (with peak value Φ_m), and in phase with it.

If, alternatively, the transformer is energised at the point of minimum voltage, the induced voltage-to-flux relationship is as shown in Figure 3.17*b*, and the flux now rises to a peak value of $2\Phi_m$. This is because the flux must adopt a shape which satisfies eqn. 3.12. In so doing the flux enters the saturated portion of the *B/H* curve, so resulting in a large, but transient, increase in current known as the magnetising inrush current. The envelope of the inrush current is shown in Figure 3.17*d* and decays to a normal steady-state level after several seconds, the rate of decay being a function of the resistance to inductance (R/L) ratio of the circuit.

Peak magnetising inrush currents range typically from one to seven times RMS full-load current, depending on transformer voltage and size. The

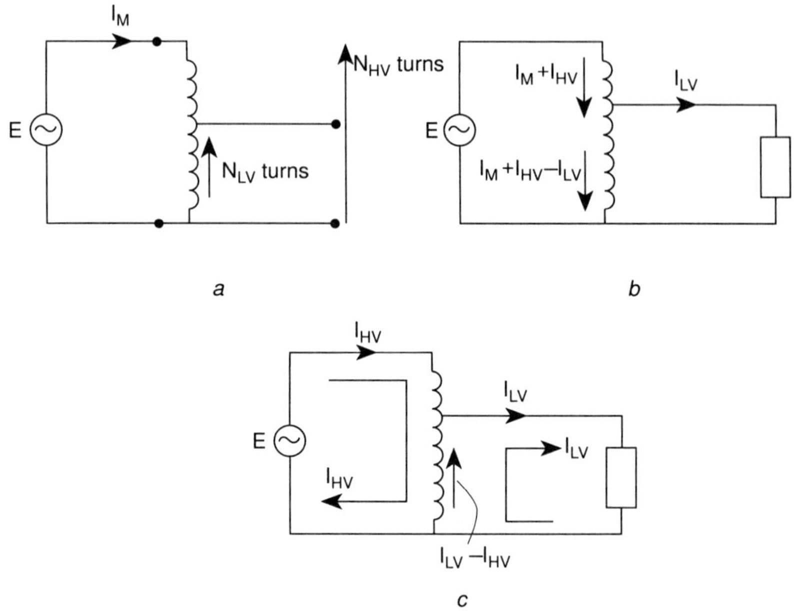

Figure 3.18 Analysis of autotransformer

a Single-phase autotransformer
b Load connected to primary
c Circulating currents

inrush current is expressed in peak values because the large harmonic content makes an RMS value somewhat meaningless. As a consequence of circuit breakers switching each phase at a different point on the waveform, the level of inrush current for each phase will differ. For star-connected three-phase transformers this imbalance will cause spill current to flow down the transformer neutral to earth. An important point to note is that magnetising inrush current contains a significant proportion of harmonics, the most predominant of which is the second harmonic.

3.37 The autotransformer

Autotransformers are economically advantageous where the voltage ratio is of the order of 3:1 or less; they are extensively installed on the UK transmission system, with typical voltage ratios of 400/132 kV, 275/132 kV and 400/275 kV. Practising engineers need to be familiar with autotransformer theory and a simple explanation is therefore included here.

Figure 3.18*a* shows a single-phase autotransformer with a winding of N_{HV} turns and a low-voltage tapping of N_{LV} turns. Consider, in the first instance, the secondary being open-circuit. Under such circumstances a magnetising current I_M flows such that the applied voltage E is equal to the voltage drop across the magnetising impedance, as given by

$$E = KN_{HV}I_M$$

where KN_{HV} represents the inductive reactance of the winding.

If a load is now connected to the secondary, additional currents I_{HV} and I_{LV} are assumed to flow, as shown in Figure 3.18*b*. Now given that the ampere-turns produced by the magnetising current are proportional to the applied voltage E, the additional ampere-turns produced by I_{HV} and I_{LV} must sum to zero, i.e.

$$I_{HV}(N_{HV} - N_{LV}) + (I_{HV} - I_{LV})\,N_{LV} = 0$$

or

$$I_{HV}N_{HV} - I_{LV}N_{LV} = 0$$

or

$$I_{HV}N_{HV} = I_{LV}N_{LV}$$

which is consistent with two-winding transformer theory.

If the magnetising current is considered to be negligible compared with load currents, the current distribution within the autotransformer can be considered as being due to two circulating currents as shown in Figure 3.18*c*. A point worthy of note is that because the magnitude of I_{LV} will exceed I_{HV}, the relative direction of current in the LV component of the winding is towards the tapping point.

ELECTRIC AND MAGNETIC FIELDS

3.38 Electric and magnetic fields

Electric and magnetic fields arising from the power system have implications for both safety and testing, during commissioning and maintenance. The following sections summarise some of the salient considerations.

3.39 The electric field

An electric field emanates from positive charges and flows to corresponding negative charges; it may be described in terms of volts per metre or as an equivalent capacitance.

Figure 3.19*a* shows the effect of the electric field from a charged overhead line (single-phase for simplicity) on an unearthed conductor (e.g. an adjacent overhead line which is out of service). The latter assumes an intermediate voltage, the magnitude of which depends on the relative capacitances shown.

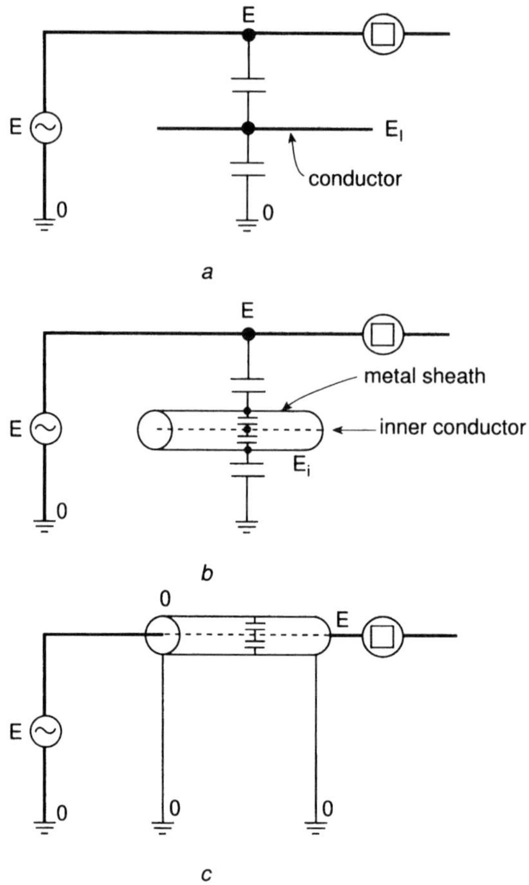

Figure 3.19 The electric field

a Conductor in electric field
b Screening effect of cable sheath from external electric field
c Screening effect of cable sheath from internal electric field

Figure 3.19*b* shows the conductor described in Figure 3.19*a* now contained within a metallic sheath. The metallic sheath now assumes the intermediate voltage; no electric field exists within the sheath so that the sheath effectively screens the inner conductor. Earthing of the metallic sheath would bring its voltage to that of the ground and the shielding of the inner conductor would be retained.

Figure 3.19*c* illustrates a cable with an insulated oversheath (i.e. a metal sheath surrounded by insulated sheath) whose metallic sheath is earthed. The electric field is contained entirely within the cable, with the earthed metallic sheath acting as a screen to the outside world.

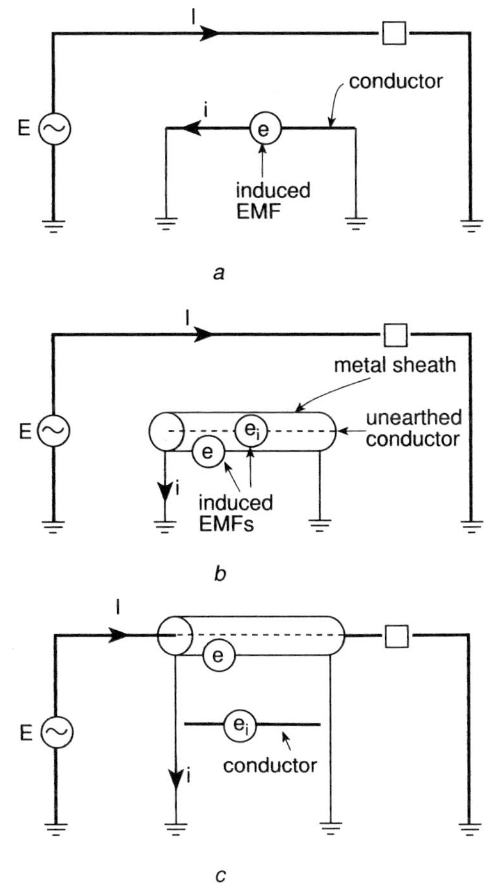

Figure 3.20 The magnetic field

a Conductor in magnetic field
b Partial screening effect of cable sheath from external source of magnetic field
c Partial screening effect of cable sheath from internal source of magnetic field

3.40 The magnetic field

A magnetic field is produced when a current flows in a conductor. In accordance with Faraday's law, an induced voltage exists around any closed path around a varying magnetic field.

Figure 3.20*a* shows an overhead power line (single-phase for reasons of simplicity) subject to a short circuit at its remote end. The latter removes the electric field from the line but results in a current *I* and associated magnetic field. *I* induces a voltage *e* into an adjacent conductor with circulating current *i*. The magnetic fields produced by *I* and *i* are in opposition.

Figure 3.20*b* shows the implications of a conductor being enclosed by a metallic sheath. The induced current which circulates around the sheath and earth loop causes a flux which opposes that arising from current I, thus partly screening the internal conductor. The smaller the impedance of the sheath loop, the greater the value of i, and in turn the greater the degree of shielding and the lower the value of induced voltage e_i.

Figure 3.20*c* illustrates the screening effect of a cable sheath on an external conductor. As with Figure 3.20*b* the degree of screening of the conductor from induced voltages is dependent on the impedance of the sheath loop.

Chapter 4
Circuit breakers

4.1 Introduction

Circuit breakers fall within the category of equipment known as switchgear, a term which also embraces switches, switch disconnectors and disconnectors. Switchgear is essentially concerned with the making and breaking of electrical circuits. Table 4.1 identifies the functions of the different items of switchgear.

This chapter will examine those aspects of circuit breaker design which are relevant to commissioning and maintenance, in addition to describing the essential requirements of the latter.

DESIGN

4.2 Circuit-breaker specification

Aspects of the design specification worthy of familiarisation are:

(*a*) *The rated RMS voltage*; this refers to the maximum continuous voltage (e.g. 420 kV, 300 kV, 145 kV, 36 kV, 12 kV).
(*b*) *The rated RMS short-circuit breaking current*, both symmetrical three-phase and single-phase. The necessity for the single-phase quantity reflects the fact that in some instances the single-phase to earth fault level of the system may exceed the three-phase fault level by up to 20%. This is more likely to arise on transmission systems with multiple earthed neutrals (thus reducing the zero sequence impedance). An asymmetrical figure is also quoted.

The circuit breaker's short circuit breaking current capability must equal or exceed the system maximum fault level (usually expressed in MVA). Typical system maximum fault levels are identified in Table 4.2, although some parts of the power system may have fault levels in excess of those stated so requiring a circuit breaker with a larger short-circuit breaking-current capability.
(*c*) *The peak making current*
(*d*) *Rated RMS continuous current and the rated short term currents* (e.g. 20 min, 10 min, 5 min etc.)

Table 4.1 Switchgear capability

	Circuit breaker	Switch	Switch disconnector	Disconnector
• Make, break and carry busbar-charging current	Yes	Yes	Yes	Yes*
• Carry load current	Yes	Yes	Yes	Yes
• Make and break load current	Yes	Yes	Yes	No
• Make fault current	Yes	Yes	Yes	No
• Break fault current	Yes	No	No	No
• Carry short-term fault current	Yes	Yes	Yes	Yes

* Some disconnectors are not rated for busbar on-load change-over

Table 4.2 System maximum fault levels

Nominal voltage (kV)	Fault level (MVA)	Fault level* (A)
400	35 000	50 000
275	15 000	31 500
132	3500 †	15 300
33	1000	17 500
11	250	13 000

* Figures are rounded † Rapidly becoming 5000 MVA, 21800 A

(*e*) *Rated duration of short circuit with the circuit breaker closed*; this refers to the period of time for which maximum current (usually equal to the short-circuit breaking current) can be carried. For example, this is usually 1 s for 400 or 275 kV circuit breakers and 3 s for 132 kV circuit breakers. The factors which limit this time are both mechanical (forces) and thermal.

(*f*) *Maximum tripping time* (i.e. trip-coil initiation to arc extinction), under rated RMS short-circuit breaking-current conditions. Tripping time is particularly important for transmission circuit breakers, for the retention of system stability.

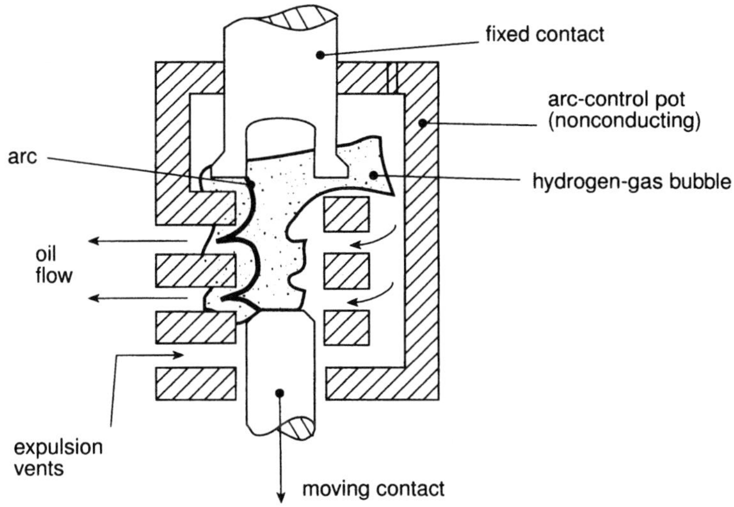

Figure 4.1 Principle of arc interruption: oil circuit breaker

(*g*) *Close–open time*; this is the maximum time to open if the circuit breaker is closed onto an already operated trip relay.
(*h*) *Rated line and cable-charging breaking current* (i.e. the ability to interrupt capacitance current)
(*i*) *Rated small inductive breaking current*

Note that the requirements of some power systems, including that in the UK, are such that circuit breakers are specified as having simultaneous operation of all three phases. Some other countries have single-phase tripping and reclosure arrangements.

4.3 Methods of arc interruption

The common circuit-breaker arc-interrupting media, from which the circuit breaker type designation is usually taken, are:

- oil
- air blast
- sulphur hexafluoride (SF_6)
- vacuum
- air break

These will be briefly examined.

(*a*) *Oil:* Figure 4.1 illustrates one of the most common arrangements where the contacts are housed in an arc-control pot (or chamber), alternatively known as an explosion pot, which is in turn immersed in a tank of oil. When the moving and fixed contacts separate, an arc is created, the heat of which decomposes the adjacent oil into a number of gases, some 80% of which is hydrogen. The hydrogen gas efficiently conducts the heat of the arc into the cool surrounding oil.

The limited space within the arc-control pot causes a build up of pressure in the expanding gas. This drives the oil within the control pot through the labyrinth of plates to direct cool oil across the arc and out of the vents. This process takes place each time the moving contact (poker) moves down and uncovers a vent. The labyrinth plate design is very intricate, to optimise this process, and great care must be taken when assembling the plates, to ensure that the oil blasts across the arc take place at the desired points in the stroke. As the downward movement of the poker continues, the arc length is increased and extended into the vents. Ultimately, at a current zero, when the pressure within the arc control pot is sufficiently high, and the arc is sufficiently extended and cooled, the arc products are expelled and replaced by cool oil, which is of sufficient dielectric strength to prevent restrike of the arc.

At low currents not much gas pressure is built up to drive the oil across the arc, and thus longer arcing times are experienced than at higher currents. At higher currents the labyrinth of plates must allow any excess gas out of the vents at appropriate stages in the poker stroke, or the arc-control pot will suffer overpressurisation and rupture or explode.

(*b*) *Air blast:* In an air-blast circuit breaker the separation of the contacts allows a blast of high pressure air to flow through the nozzle provided by the contact gap, and then to atmosphere. The principle is illustrated in Figure 4.2. As the air passes through the nozzle it expands, to an extent that its velocity reaches sonic level. At the same time the arc is extended and transferred to an arcing probe. The air conducts heat very rapidly and efficiently away from the arc to achieve extinction at current zero. The arc-extinguishing power of air-blast circuit breakers is independent of the current to be interrupted, and they are thus prone to causing current chopping.

(*c*) *SF_6:* SF_6 is an odourless, colourless, nontoxic, noninflammable gas. It is five times heavier than air with approximately twice the electric strength. It is termed an electronegative gas, which means that its molecules rapidly absorb the free electrons in the arc; there is thus a rapid build up of dielectric strength after current zero. The mechanism of arc interruption is similar to that of the air-blast circuit breaker, as shown in Figure 4.3, except that the gas is not exhausted to atmosphere, because of its cost, and because some of the byproducts resulting from the arcing process are toxic. On contact separation, a high pressure flow of SF_6 gas is forced through the interrupting nozzles. The pressurisation may be from a reservoir of high pressure SF_6, which is maintained by a compressor system, or via a piston mechanism which compresses the gas as the moving contacts move to the fully open position (the 'puffer' mechanism). SF_6 circuit breakers, like air-blast circuit breakers, are prone to causing current chopping, but at much lower currents. The

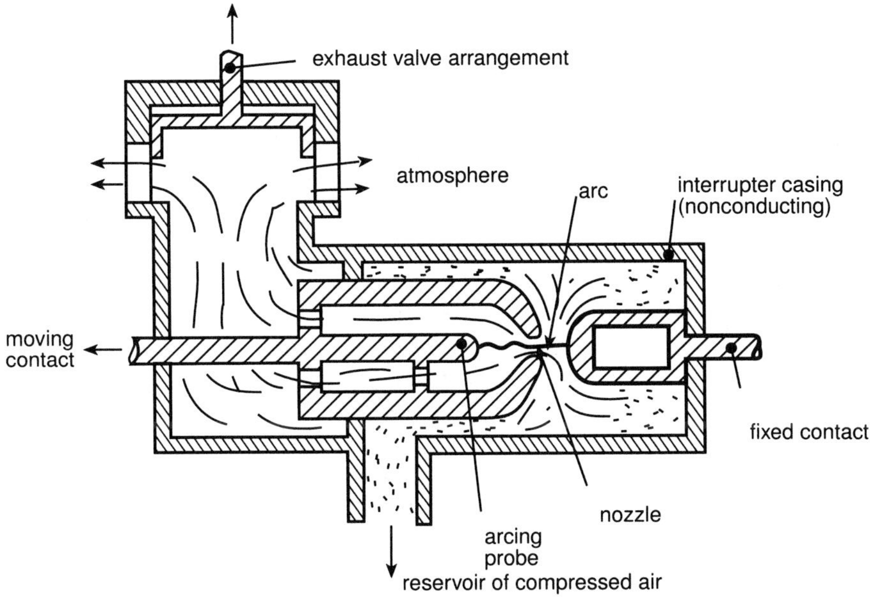

Figure 4.2 Principle of arc interruption: air blast circuit breaker

resulting oscillatory transients are thus less severe than those caused by an air-blast circuit breaker.

(*d*) *Vacuum:* The vacuum interrupter consists of a fixed and moving contact assembly contained within a sealed chamber under high vacuum. Opening of the contacts causes an arc arising out of vaporised metal from the contacts. The vapour rapidly diffuses away from the gap at current zero, to be absorbed by 'sputter' shields specially designed for this purpose, as illustrated in Figure 4.4. Vacuum circuit breakers are finding favour at distribution voltages, mainly 11 kV, but also at 33 and 132 kV, and particularly in power stations.

(*e*) *Air break*: Figure 4.5 illustrates the essential features of an air-break circuit breaker. When the contacts open, the arc is forced into the splitter plates, both by convection and electromagnetic interaction between the field arising from the current in the arc runners and the current in the arc. Interruption is caused by elongation of the arc and natural cooling in the splitter plates. Air-break circuit breakers find their greatest application where fire risk (e.g. arising from the oil in an oil circuit breaker) must be kept to a minimum. They have been widely used in power stations, but are now being displaced by vacuum circuit breakers.

4.4 Circuit-breaker types

There is no absolute system of categorisation of circuit-breaker types, but they may be roughly summarised as in Table 4.3.

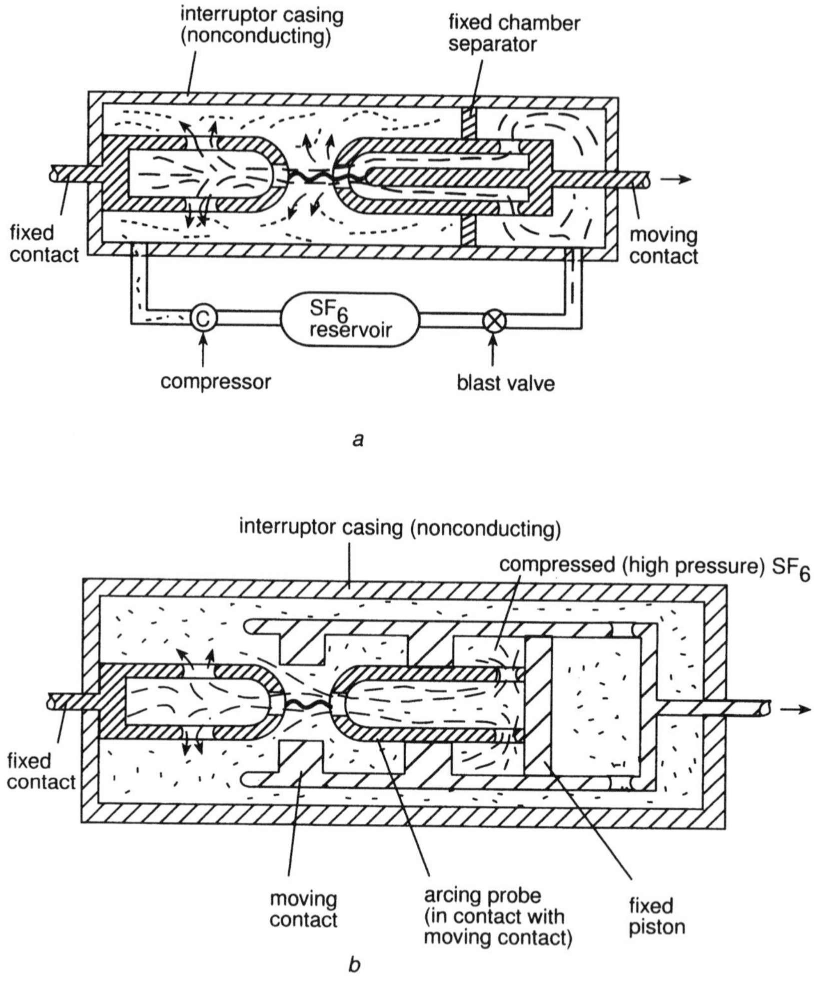

Figure 4.3 Principle of arc interruption: SF_6 circuit breaker

a Compressor-and-reservoir type
b Puffer type

In Table 4.3 the term 'metalclad' relates to a construction whereby all live parts of the circuit breaker and its incoming terminations are enclosed by earthed metal, as shown in Figure 4.6*a*. They are normally constructed as an integrated suite of circuit breakers and are usually (but not exclusively) only suitable for erection within a weatherproof structure. The SF_6-gas type of metalclad circuit breaker is often termed 'gas insulated switchgear' (GIS).

The term 'dead tank' refers to a type of construction where the interruptor is enclosed in an earthed metal tank, but with the terminations exposed.

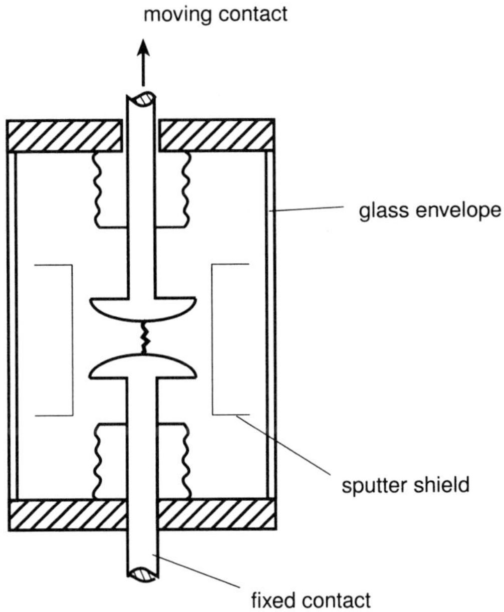

Figure 4.4 Principle of arc interruption: vacuum circuit breaker

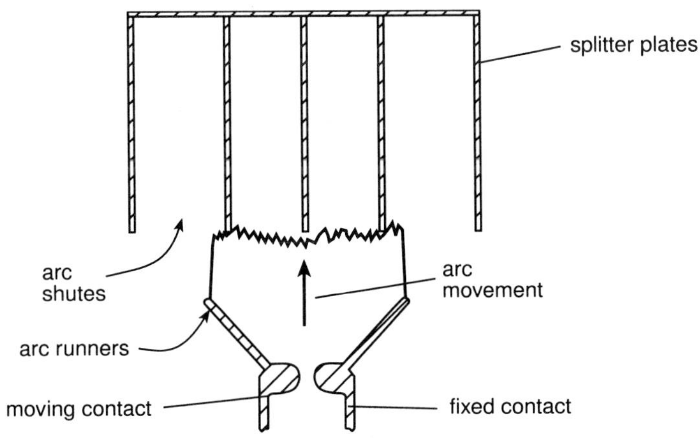

Figure 4.5 Principle of arc interruption: air-break circuit breaker

These circuit breakers are individual (i.e. not part of a suite) and are suitable for erection outdoors (see Figure 4.6*b*).

The term 'live tank' refers to the type of construction shown in Figure 4.6*c*, where the interruptor tank operates at system voltage and is insulated from

Table 4.3 Circuit-breaker types

Interrupting medium	Tank and terminal arrangement	Typical voltages (kV)
1. Bulk oil	Metalclad	Up to 33
2. Bulk oil	Dead tank, open terminal	Up to 275
3. Air blast	Live tank, nonpressurised head, open terminal	Up to 275
4. Air blast	Live tank, pressurised head, open terminal	400, 275, 132
5. SF_6	Metalclad	400, 275, 132
6. SF_6	Dead tank, open terminal	400, 275, 132
7. SF_6	Live tank, open terminal	400, 275, 132
8. Vacuum	Metalclad	132, 33, 11
9. Air break	Metalclad	Up to 11

earth. Again, these are a stand-alone type of circuit breaker with open terminals, suitable for erection outdoors.

Note that in most circuit breakers, the interrupting medium is also the insulating medium.

4.5 The bulk-oil circuit breaker (OCB)

The oil circuit breaker is one of the oldest designs of circuit breaker. The term bulk oil refers to the fact that the interruption process takes place in a large tank of oil. The oil acts as an insulating medium in addition to being an interrupting medium. Figure 4.6*a* and *b* illustrate types of bulk-oil circuit breakers. Note that another type of oil circuit breaker exists known as a 'small oil volume' circuit breaker, though this type of circuit breaker has not found favour in all countries.

Most bulk-oil circuit breakers consist of two breaks (interruptors) per phase. At 132 kV and 275 kV, switching resistors are frequently employed, as illustrated in Figure 4.7. The resistors have a damping effect on transient overvoltages during both opening and closing. At 275 kV it has been usual to employ multiple (usually four) breaks per phase.

Switchgear oil, like transformer oil, is required to conform to *BS 148.* Chapter 5 identifies oil requirements and deterioration problems. Oil circuit breakers impose a more onerous duty on oil than transformers, since the arc products are left to diffuse through the mass of the oil.

At voltages of 132 kV and above, oil filled bushings are used for entry into the circuit-breaker tank. These are very similar to the bushings used for

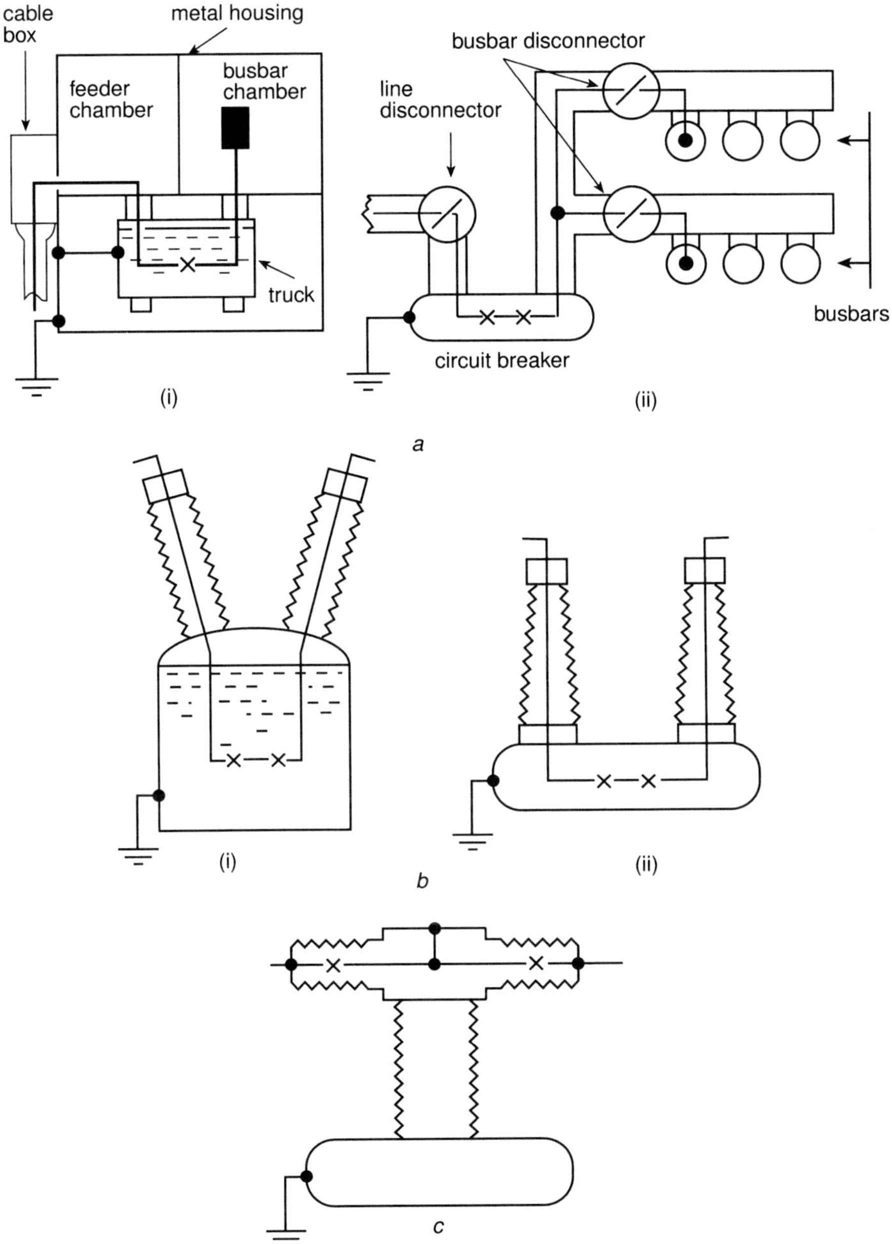

Figure 4.6 *Circuit-breaker types*

a Metalclad (i) Bulk oil (ii) SF_6
b Dead-tank open-terminal (i) Bulk oil (ii) SF_6
c Live-tank open-terminal (air-blast or SF_6)

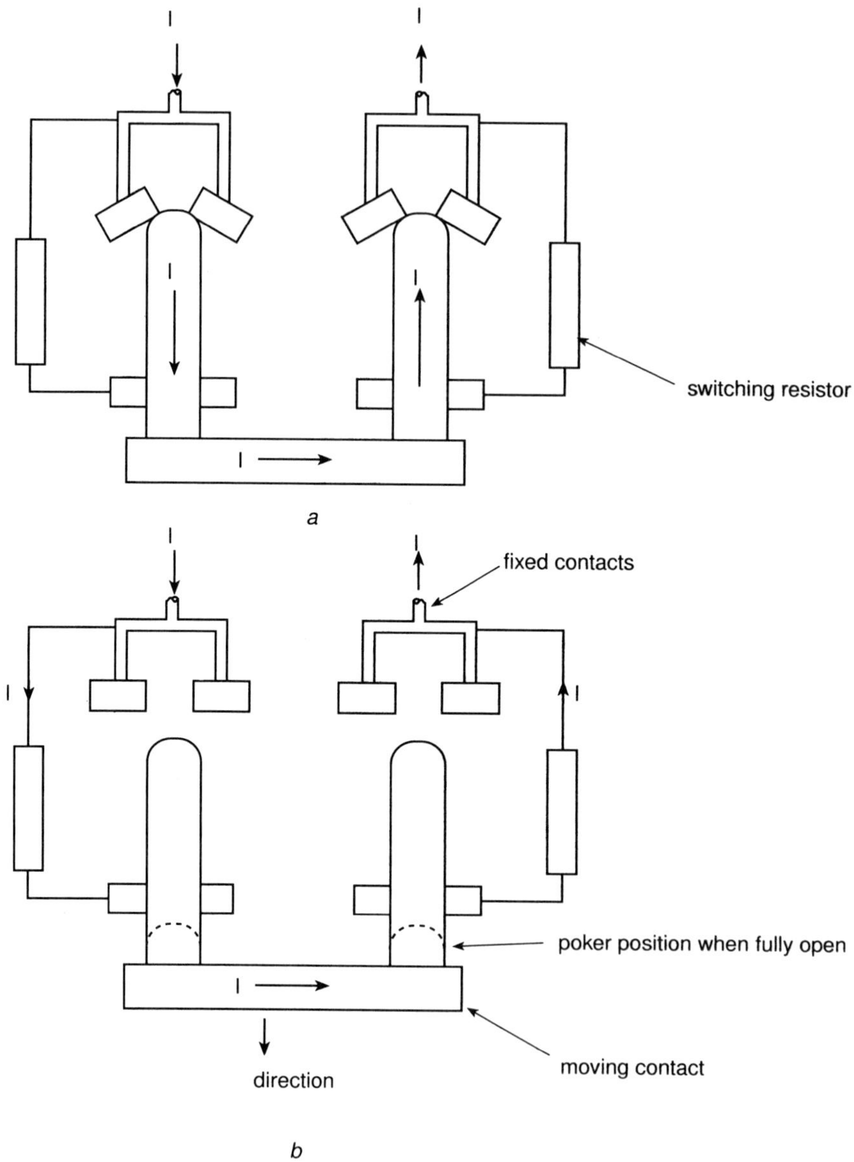

Figure 4.7 Switching-resistor arrangement for oil circuit breaker

a Closed
b Opening

transformers except that they are more robust in design to accommodate the high mechanical shock loadings during short-circuit faults.

At the lower voltages, the operating mechanism tends to be spring operated, but at higher voltages the tendency is towards a solenoid or compressed air.

At the time of writing oil circuit breakers are being replaced at 132 kV and 275 kV by SF_6 circuit breakers, but still remain in use in significant numbers.

4.6 The air-blast circuit breaker

The air-blast circuit breaker (ABCB) was once the most common type of circuit breaker, by far, on 275 kV and 400 kV transmission systems, but they are gradually being replaced by SF_6. This design of circuit breaker may be categorised into two types:

(*a*) *Nonpressurised head:* This is an earlier design of circuit breaker, where the interruptors are subject to a blast of high pressure air during the opening sequence. All designs include an integral series-connected sequence disconnector ('isomaker') whereby, during an open operation, the main interrupters open, immediately followed by the sequence disconnector, immediately followed by closing of the main interrupters. The closing operation only involves closing the sequence disconnector.

(*b*) *Fully pressurised head:* This is illustrated in Figure 4.2 where the interrupter is permanently pressurised from the air receiver. The compressed air in this instance also acts as an insulating medium. This type of circuit breaker usually has a series-connected circuit sequential disconnector (i.e. not an integral part of the circuit breaker) which opens some seconds after the main interrupter, to provide electrical isolation, specifically to protect against a loss of air, and additionally to prevent a lightning strike to the line impinging on an open interruptor. Some circuit breakers use the compressed air supply to hold the main interruptors open; thus loss of air would result in the interruptors closing. Earlier pressurised head designs functioned with an air pressure of about 400 lb/in^2, whilst later designs increased the pressure up to 1000 lb/in^2.

Many circuit breakers on 400 and 275 kV systems are pressurised-head types, and these epitomise the ultimate in air-blast circuit-breaker design. Figure 4.8 illustrates the essential features of such a circuit breaker, typically consisting of an air receiver and 12 series-connected interruptors per phase. Each interrupter has in parallel a capacitor, and a resistor with resistor interrupter. The function of the capacitors is to ensure an equal voltage division across each interrupter when the circuit breaker is open. The resistor is utilised to damp potential excessive overvoltages, caused by short-line faults or by current chopping. During a tripping sequence, the resistor interrupters close before the main interrupters commence to open, and open after the main interrupters are fully open. In some designs of pressurised-head circuit breaker, the resistor and resistor interrupter and/or the capacitor are located in the same housing as the main interrupter.

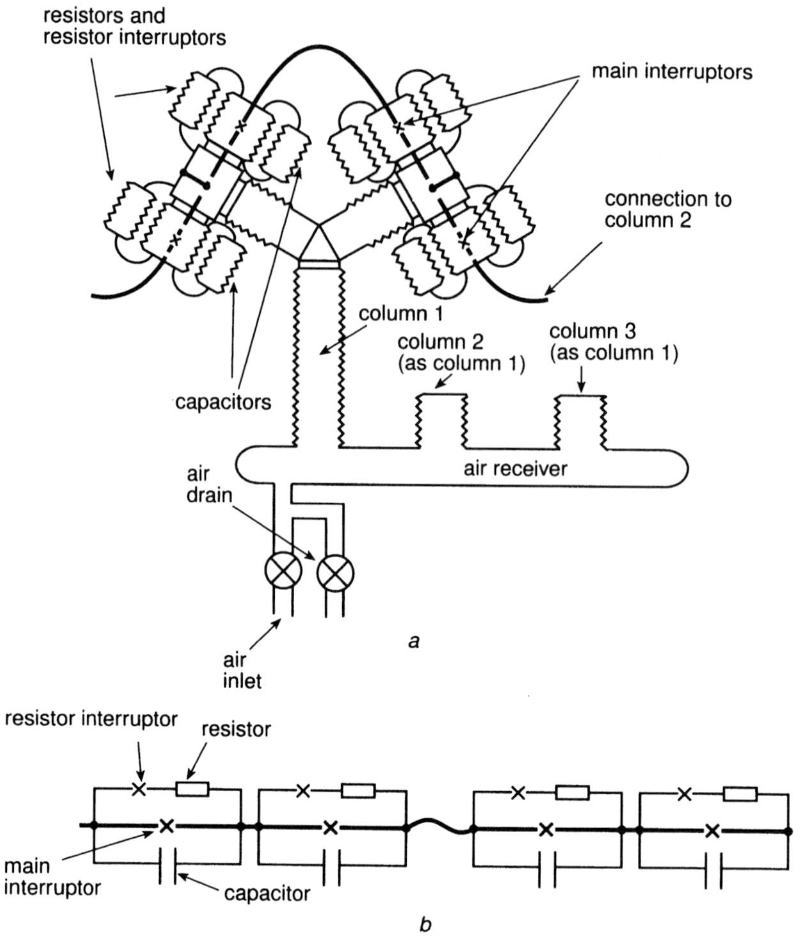

Figure 4.8 400 kV pressurised-head air-blast circuit breaker

a Physical arrangement of one phase
b Electrical representation of arrangement on one column

Once the trip or close coil, of the circuit breaker, has been energised, latches are released which result in the circuit-breaker mechanism being powered from the compressed air within the receiver. The receivers normally contain enough air to facilitate a trip-close-trip sequence in quick succession (i.e. a trip, followed by an autoreclose onto a permanent fault, and hence another trip). The air pressure is monitored and alarmed. Two main alarms exist:

(*a*) low air pressure—which is solely an alarm
(*b*) trip and close air-pressure lockout—which also inhibits tripping or closing on the basis that the air pressure has dropped to an unacceptable level.

Within an ABCB there are many spaces, between the pressurised chamber and the porcelains, which must be kept dry to avoid insulation failure. This is accomplished by bleeding air from the inlet valve and passing a continuous flow over all surfaces concerned. This low-pressure air flow is termed conditioning air.

ABCBs are generally known, with reference to tripping time, as a three-cycle breaker, comprising 50 ms from trip initiation to contact separation and 10 ms arcing time. The arcing time can however be as long as 18 ms. This longer time can arise if the contacts commence to open as the voltage is approaching a maximum, i.e. the contact gap is too small to prevent the arc restriking. Arc extinction then takes place after the next half cycle. This latter point also applies to other circuit breaker types.

At the time of writing, many ABCBs are approaching obsolescence, and are being replaced by SF_6 circuit breakers. However, significant numbers will remain in use for considerable years to come.

4.7 Compressed-air system

An essential requirement for air-blast circuit breakers is a supply of compressed air. Figure 4.9 illustrates a typical system for the manufacture and storage of compressed air. Air is compressed to a pressure of 3000 lb/in^2 and passed through a filter and dryer to a storage receiver. It then passes through a pressure-reducing valve, which reduces pressure to about 400 lb/in^2, and then into an air busbar system. It finally passes through nonreturn valves into the circuit-breaker air-storage receivers. The whole process of compression and storage is automatic and instigated via compressor-control gauges. The system is fully alarmed to detect states of sustained under or over pressure.

4.8 The SF_6 circuit breaker

The SF_6 circuit breaker is often referred to as a gas circuit breaker (GCB). Virtually all new and replacement circuit breakers at 132 kV and above are SF_6. Its advantages are:

(i) relatively low capital cost;
(ii) interrupters are virtually maintenance-free due to the effectiveness of the current-interruption process;
(iii) it does not suffer from the noise problems when operating, or require significant ancillary equipment such as compressed-air systems, which are features of ABCBs;
(iv) the operating medium is not inflammable, as with the oil in an oil circuit breaker;
(v) a limitation of the ABCB is that its fastest tripping time is about 60 ms (50 ms to contact separation + 10 ms arcing). SF_6 circuit breakers have

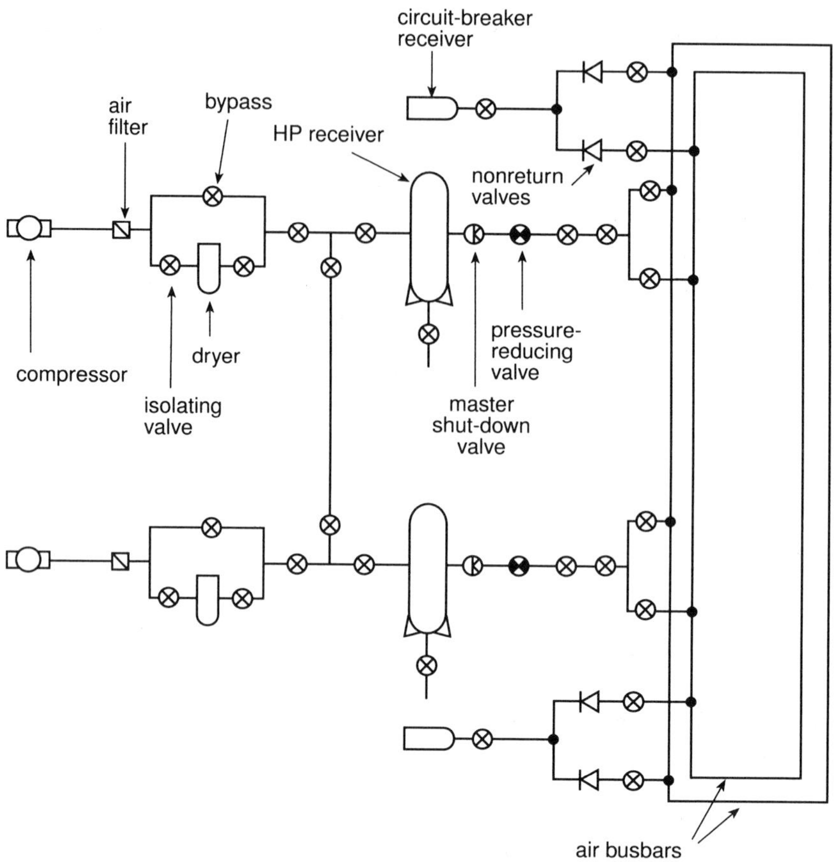

Figure 4.9 Typical features of a compressed-air system

⊗=manually operated valve

been designed to operate down to 40 ms, a present requirement for maintaining system stability for some power systems.

Figure 4.6 illustrates the three forms of SF_6 switchgear, namely metalclad (GIS), dead-tank and live-tank. The number of interrupters per phase is less than for an ABCB, being typically four at 400 kV (or two if asynchronous switching capability is not required). SF_6 circuit breakers are not as severe in terms of current chopping as are ABCBs, and are more capable of withstanding the effects of the short-line fault. It is therefore not usual to fit switching resistors to facilitate the opening sequence. It is, however, usual to fit voltage-grading capacitors in parallel with each interrupter.

The sulphur hexafluoride acts as both an interrupting and an insulating medium. It is, however, most important that the SF_6 does not become subject to impurities or contamination, since they may constitute a hazard to the reliable operation of the equipment. Impurities/contaminants of particular concern are:

- oxygen
- acidity
- dew point

Such contamination may arise from imperfect filling techniques or from impurities left in the SF_6 receiver prior to filling, the effect of which may be in the production of corrosive products which affect mechanical operation, or cause a reduction in the insulating and interrupting properties of the SF_6.

Pure sulphur hexafluoride is nontoxic, but the sulphur-fluoride powders and gases produced by electrical breakdown of the gas are toxic if inhaled in sufficient quantities. Experience, however, has shown that exposure of personnel to the breakdown products causes symptoms such as irritation of the skin, eyes, nose and mouth within seconds, and well prior to any toxic effects taking place. In such circumstances, the personnel concerned should quickly get into fresh air and obtain medical assistance.

Pungent and unpleasant odours are indications of impurities. Note also that pure sulphur hexafluoride, like nitrogen, will not support life and if breathed in sufficient quantities will lead to asphyxiation.

The operating mechanism has tended to be powered either by compressed air or via a pressurised oil/nitrogen accumulator arrangement.

4.9 Earthing facilities

Open-terminal circuit breakers are usually earthed via portable earths, or mechanism-operated fixed-position earths which are clamped to the busbars connected to the circuit-breaker terminations.

The earthing of metalclad circuit breakers is usually an integral feature of the circuit breaker itself, and is termed 'integral earthing'. In its simplest form, integral earthing consists of closing an earth switch which is within the circuit breaker housing. This can usually only be accomplished when certain interlocks have been satisfied, such as the circuit breaker being open and isolated.

A form of integral earthing termed 'transfer earthing', consists of physically repositioning or transferring the circuit breaker, within the metalclad housing, to an earth position (feeder earth or busbar earth) and closing the circuit breaker to effect the earth. Figure 4.10 illustrates the arrangement.

On some older designs of distribution-voltage metalclad switchgear, the circuit breaker is withdrawn from the service position, and a set of three earthing bushings is fitted to the circuit-breaker feeder, or busbar, plug contacts. A crossbar (conductor) is fitted to the opposite set of circuit-breaker plug contacts and a connection is made from the crossbar to the main circuit breaker earthbar. The circuit breaker is then racked into position such that

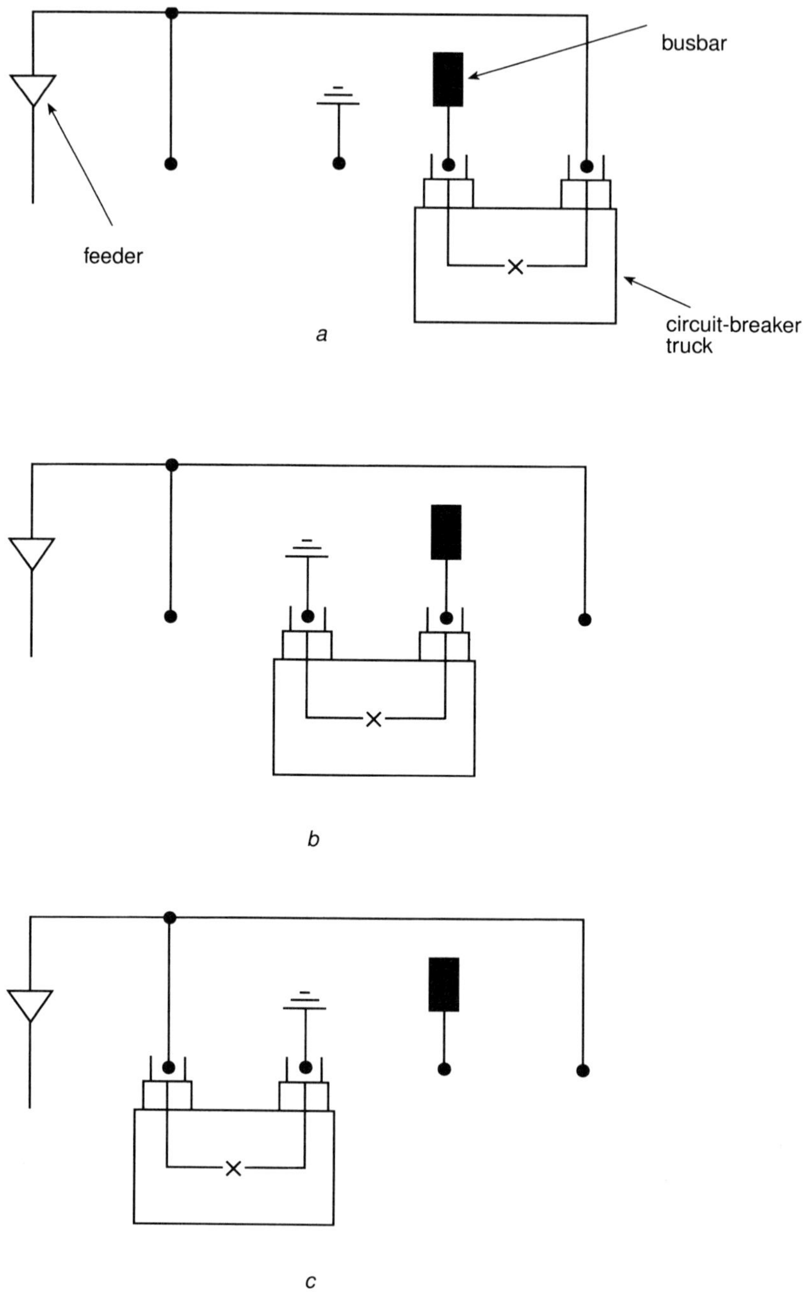

Figure 4.10 Transfer earthing

a Service position
b Busbar-to-earth position
c Feeder-to-earth position

the earthing bushing makes with the feeder or busbar contacts to be earthed. The circuit breaker is then closed to create the connection to earth.

4.10 Horizontal and vertical isolation

The terms horizontal and vertical isolation describe the way in which a metalclad switchgear circuit-breaker truck (see Figure 4.6*a*) is racked from the service position to the isolated position. In some designs of switchgear the truck is withdrawn vertically, and in other designs horizontally.

4.11 Red, right, rear

It is standard practice, with distribution-voltage metalclad switchgear, to terminate incoming feeders in the same way. When standing at the rear of the circuit breaker, and looking at the circuit breaker, the right-hand phase is always red, the middle yellow and the left-hand blue, the arrangement being known as RED, RIGHT, REAR.

4.12 Circuit-breaker auxiliary switches

Circuit-breaker auxiliary switches mimic the position of the circuit breaker interrupter contacts, and are utilised in protection and control circuitry. They are generally operated via a simple linkage from the circuit-breaker operating mechanism. In its simplest form, the arrangement consists of a butt contact, but more recent designs of circuit breakers tend to utilise rotary contact arrangements. Figure 4.11*a* illustrates a typical auxiliary contact arrangement.

Distribution-voltage metalclad circuit breakers of the truck type have the auxiliary switches mounted within the truck (but usually not immersed within the oil of OCBs). The wiring connected to the contacts is usually run into the circuit-breaker housing via a plug and socket arrangement, often utilising flexible trunking as illustrated in Figure 4.11*b*.

The trunking can usually be disconnected manually from the truck, to enable it to be withdrawn from the housing. A trunking extension piece can then be inserted and the trunking reconnected to the truck to enable the circuit breaker to be power operated (for testing purposes) in the withdrawn position.

As stated above, auxiliary switches are intended to mimic the position of the interrupters. However in moving from one position to another (i.e. open to closed) there is inevitably a time discrepancy between the two. This usually does not exceed 5 ms, but in some older designs it has been known to be as long as 10 ms. Engineers should be aware of this discrepancy when evaluating fault recorder charts.

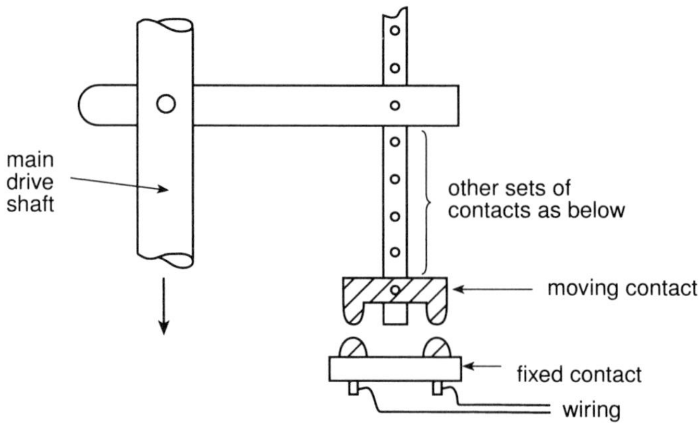

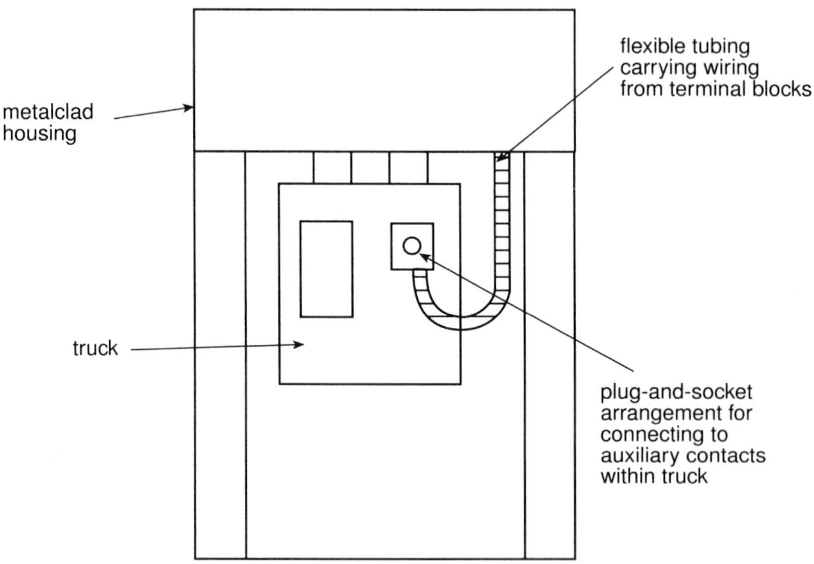

Figure 4.11 Auxiliary switches

a Simple representation of auxiliary switch
b Typical arrangement for metalclad OCB

4.13 The slow-closing jack

Circuit breakers are usually designed to accommodate the fitting of a slow-closing jack (handle). This is used to operate the circuit breaker mechanism manually. Cranking the jack results in a slow movement of the mechanism which facilitates checks to be made on linkage positions, and the alignment of auxiliary switches, under controlled conditions. In some instances it is used to aid assembly of the mechanism. With designs of circuit breaker which function solely via mechanical linkages, the jack can be used to close the interrupter contacts (e.g. oil circuit breakers) but with other designs, particularly those which operate partially via hydraulic or gaseous stored energy (e.g. ABCBs), it may be necessary for safety or technical reasons to drain the source of stored energy prior to application of the jack; thus closure of the contacts via the jack cannot be achieved.

4.14 Trip-free mechanism

If a circuit breaker is closed onto an operated trip relay, the mechanism is termed 'trip free' if it is collapsed into the open state, the instant the trip coil, via its auxiliary contacts, becomes re-energised. This feature has a bearing on determining circuit breaker fail protection times. It may also influence the speed of fault clearance if a circuit breaker is closed onto a fault. Most designs of circuit breakers possess this feature, but not all. Alternatively, some circuit breakers cannot collapse the mechanism to the open position until it reaches the fully (or near) closed position.

4.15 Circuit-breaker erection

Distribution-voltage circuit breakers (e.g. 11 kV) would normally be shipped to site with their interrupters, mechanism and housings assembled and complete. Metalclad switchgear bays would, of course, be interconnected on site.

At transmission-voltages, circuit breakers are physically much larger than at lower voltages and therefore a greater degree of site assembly takes place. Interrupter modules, support columns, base frames and mechanism boxes would thus typically be dispatched to site for assembly and erection.

It is usual to dispatch equipment to site devoid of the operating/insulating medium, e.g. oil, air, SF_6 (except for sealed units). However, to avoid moisture ingress into units which are required to be kept dry, it is usual before dispatch to fill and seal these units with a gas, usually nitrogen, under a slight pressure. Care should be taken when this gas is removed on site, since even a slight pressurisation can be a potential source of danger.

COMMISSIONING

4.16 Commissioning requirements

Commissioning requirements for circuit breakers are generally as follows:

- data collection
- general checks and inspections
- contact resistance
- auxiliary switch checks
- trip and close timing tests
- OCB specific tests
- OCB oil tests
- ABCB specific tests
- ABCB air dryness
- ABCB air leakage
- ABCB pressure switches
- ABCB pressure drop with operations
- GCB (SF_6) specific tests
- GCB (SF_6) leakage detection
- GCB (SF_6) dew point
- GCB (SF_6) oxygen content
- GCB (SF_6) acidity
- operational tests

Sections 4.17 to 4.25 examine these requirements in further detail.

4.17 Data collection

This would typically consist of:

- type reference and serial number
- data specified in Section 4.2.

4.18 Checks/inspections

This is largely a visual check and would typically consist of the following:

- circuit identification and phase labels
- earth strap connections
- tightness of fastenings
- arcing horns and rings
- control and operation of heaters
- insulators clean and undamaged
- inspection of breathers
- oil levels—for OCBs
- correct position of any valves
- acceptable readings on any gauges

4.19 Contact resistance

This is a measure of resistance across each set of contacts with the circuit breaker closed. A digital micrometer as described in Section 8.33 should be used. Manufacturers will quote maximum values of resistance, which will typically be in the range of a few tens of micro-ohms. All results should be recorded as a source for comparison during maintenance.

4.20 Auxiliary switches

Auxiliary switches are typically of the following types:

- make coincident with the main contacts closing (normally open);
- break coincident with the main contacts closing (normally closed);
- make (or break) in advance of the main contacts closing;
- make (or break) after the main contacts close;
- momentarily make during the open or closing sequence (fleeting contact).

The auxiliary switches are usually proved by closing the circuit breaker manually with the slow-close jack, while observing an open- or short-circuit across the auxiliary contacts with a meter. The point at which the circuit breaker main contacts close can also be obtained by measuring the resistance across the latter with a meter. If the auxiliary switches are unable to be directly compared with the main interrupter contacts (i.e. when there is no direct mechanical linkage between close jack and main interrupter contacts, such as with an ABCB), then the comparison is made with a known point in the mechanism linkage movement. There is usually some means of adjusting auxiliary switches to the required type. Verification that auxiliary switches and main contacts are in reasonable unison should be carried out during the timing test; see Section 4.21.

4.21 Timing tests

The timing test is an important commissioning test since it proves the acceptability of the whole electromechanical process from trip initiation to contact separation. Figure 4.12*a* illustrates the essential features of a timing test. The timing equipment usually consists of a switch which can be selected to energise either the trip or close coil, and inputs which monitor the state of the main circuit-breaker contacts.

When the switch is selected to trip, or close, a graphical record (usually on paper) is produced by the timing equipment, showing the changing state of the circuit breaker main contacts; see Figure 4.12*b*. A timing mark on the record is synchronised with the instant that the trip, or close pulse, is output, such that the exact time from trip initiation to contact separation can be determined, to check whether the times are within specification. The range of tests carried out would normally embrace the following:

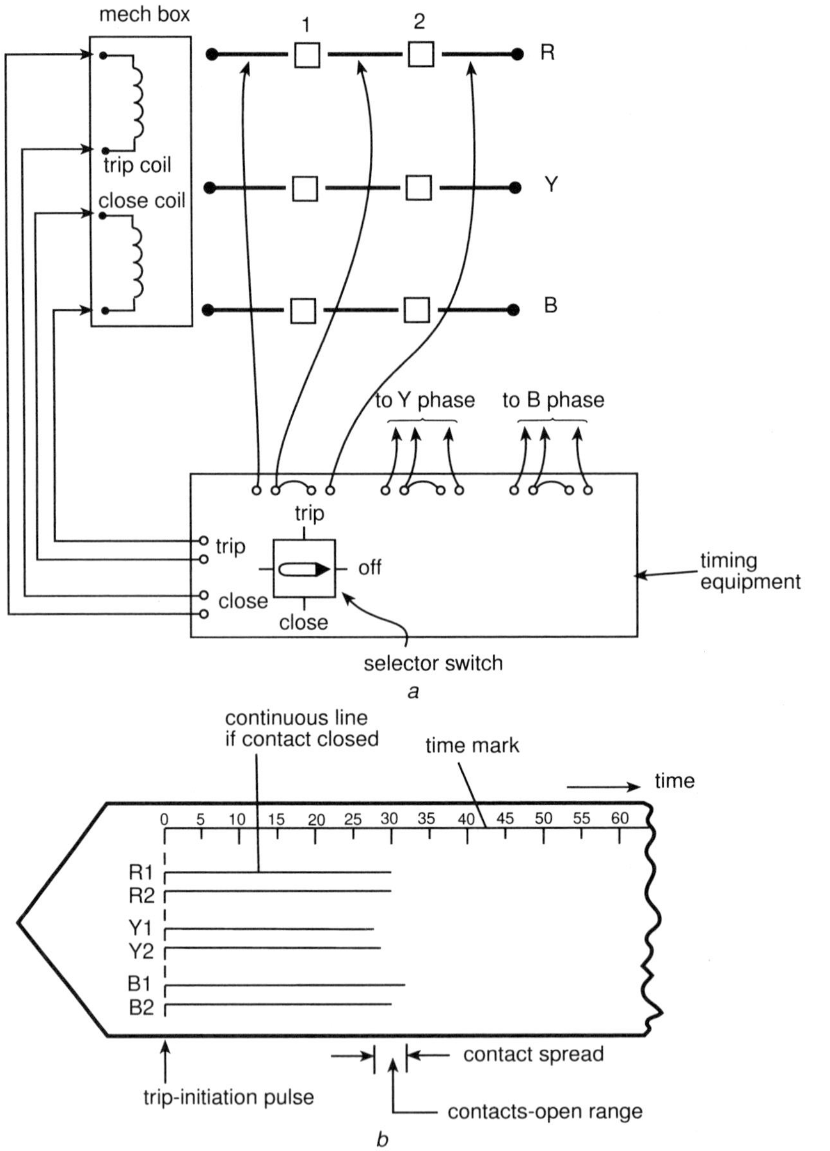

Figure 4.12 Timing test

a Timing equipment and connections
b Timing record for a circuit-breaker trip

- trip test—initiation to contacts part
- close test—initiation to contacts touch
- trip on close test—close onto an already operated trip relay
- auxiliary switch test—comparison of auxiliary switch with main contacts.

In addition, the contact spread time should also be checked against the manufacturer's specification.

For a modern SF_6 transmission-voltage circuit breaker, typical times would be:

Trip test	=30 ms±3 ms with a contact spread of 5 ms
Close test	=100 ms±5 ms with a contact spread of 5 ms
Trip on close test	=70 ms±5 ms with a contact spread of 6 ms

The design of some circuit breakers is such that it is difficult, if not impossible, to gain access to each individual interrupter contact. This is particularly true of SF_6 metal-clad circuit breakers. In these instances monitoring of contact position is achieved by reference to the drive shaft position—the position of which may be determined by light from light-emitting diodes, either passing through slots, or being reflected via reflecting patches, to photo-detecting transistors.

Timing tests are less important for distribution-voltage circuit breakers, where fault-clearance times are less critical. In such instances, and particularly if the circuit breaker and its mechanisms are factory assembled, the timing test may be dispensed with. At transmission voltages the test is imperative. More recent types of timing equipment provide a direct numeric read out of results.

4.22 OCB tests

Commissioning tests specific to OCBs generally relate to verifying the satisfactory condition of the oil. The tests required are specified in *BS 148* and are detailed in Section 5.33.

4.23 ABCB tests

Given that the manufacture of ABCBs has almost ceased, it is unlikely that an ABCB will require commissioning in future. However, recommissioning may be required in the context of major refurbishment work. The required tests are as follows:

(*a*) *Air dryness:* Relatively dry air is obviously important for the satisfactory performance of a pressurised head ABCB. In particular, it is imperative that dew-point conditions do not arise within the ABCB receiver. Air under pressure has a more concentrated moisture content than at atmospheric pressure, and therefore the dewpoint temperature increases with air pressure. Thus unless the receiver air is reasonably dry, a winter cold spell could cause moisture in the air within the breaker to condense onto insulation surfaces resulting in insulation breakdown. At a receiver operating pressure of, say, 450 lb/in^2, the required dewpoint temperature should be $-20°C$ or lower; this would correspond to a dewpoint of $-50°C$ at atmospheric pressure.

Dewpoint can be obtained by passing the compressed air through a dewpoint meter.

(*b*) *Air leakage:* This is a check to verify that the seals and seats are reasonably airtight. This can be done by closing the ABCB inlet valve, and observing the pressure gauge drop over a number of hours.

(*c*) *Pressure switches:* In general three pressure switches are fitted:

- low-air-pressure alarm
- close lockout
- trip lockout

The operation and resetting of these switches should be checked against a calibrated pressure gauge. This is usually accomplished by isolating a section of pipework to which the switch is connected, and slowly draining (and then refilling) the section.

(*d*) *Pressure drop with operations:* The air receiver usually contains sufficient air for a trip–close–trip operation in close succession. This is checked by closing the air-inlet valve and manually tripping, closing and tripping the circuit breaker. The receiver air pressure should be taken some two to three minutes later and recorded.

4.24 GCB (SF_6) tests

The tests specific to SF_6 circuit breakers are concerned with the SF_6 gas. These tests involve a chemical/physical analysis of the gas, and a comprehensive understanding of the procedures and results requires a reasonable knowledge of physical chemistry. The tests will thus be examined at a superficial level.

(*a*) *SF_6 leakage detection:* This should commence with the SF_6 compartments filled to full pressure and an observation made of whether the pressure is then stable or falls over a period of time. Leaks can be detected by use of an SF_6-leakage detector. This generally consists of a hand-held probe which should be slowly moved over the surface of the SF_6 containment vessel.

(*b*) *SF_6 dewpoint:* The dewpoint can be obtained by use of a dewpoint cell. This consists of a closed cell in which there is an exposed metallic mirror, through which SF_6 is passed at a controlled rate while the mirror temperature is slowly lowered. Temperature control is achieved by exposing the reverse side of the mirror to a mix of acetone and solid carbon dioxide, the latter being added slowly to reduce temperature. As the mirror temperature falls, a point is reached where water condenses on the surface forming a dew or a frost. The temperature T_1 at which this takes place is recorded. The temperature is then allowed to rise and point T_2 at which the dew clears is recorded. The dewpoint is then taken as $1/2\ (T_1+T_2)$.

Given that SF_6 is often held under pressure in the circuit breaker, the dewpoint temperature at working pressure must be obtained; this can be derived from formulae and tables using the above data; *BS 5209* describes the process in detail. The dewpoint temperature at working pressure must be lower than the lowest expected ambient temperature. More recently, electronic meters for obtaining dewpoint have appeared on the market.
(*c*) *SF_6 oxygen content:* This can be determined by use of an oxygen-measuring instrument, which may function on the principle of magnetic susceptibility (oxygen is paramagnetic, which is unusual for a gas) or gas chromatography. SF_6 is passed through the measuring device at a controlled rate. The maximum permitted oxygen concentration is 0.01% by mass.
(*d*) *SF_6 acidity:* The acidity is measured by passing SF_6 through a dilute alkali of known strength and volume until the alkali is neutralised by the acid. The alkali chosen is usually sodium hydroxide, to which an indicator (such as bromocresol purple) is added. Neutralisation is indicated by a change in the colour of the solution. The acidity content is obtained by calculation, which is well described in *BS 5207*. The acidity is expressed in terms of hydrofluoric acid (HF) content, in mg of acid/kg of SF_6. The maximum permitted concentration = 0.3 mg/kg.

SF_6 circuit breakers differ from other circuit-breaker types discussed in this chapter in the number of alarms generated. Whereas OCBs and ABCBs would typically have fewer than five alarms, modern SF_6 circuit breakers may well have as many as 20. These arise because of the relative complexity of the mechanism which may include oil and nitrogen under pressure, and a separate mechanism for each phase.

4.25 Operational tests

The following operational tests should be carried out during commissioning:

(*a*) manual closure of the circuit breaker via the close handle/switch while a trip relay is operated. The manual closure should be executed for about 10 s. The circuit should commence to close, trip and remain open. Once the circuit breaker is in the open position, the trip signal should be automatically removed from the trip coil, via the breaker auxiliary contacts (see Chapter 8) and thus it may seem, at first sight, that the circuit breaker should commence to close again. However, the mechanism should be designed in such a way that a second closure cannot be initiated until the first close signal is removed and then reapplied. This is usually termed the 'antihunting' feature;
(*b*) functioning of the circuit breaker from the local, standby and supervisory (main) positions;
(*c*) opening and closing under reduced battery voltage - see Chapter 9;
(*d*) proving of interlocks. Most circuit breakers have interlocks; some are equipped with 'castel' type keys such that they cannot be operated if the key is removed; see Chapter 11;
(*e*) check on/off-position indicator.

MAINTENANCE

4.26 Circuit-breaker maintenance

Maintenance can only commence after the circuit breaker is switched out of service, isolated and earthed—although metal clad circuit breakers of the 'truck' type do not require earthing once they are withdrawn from service and discharged.

To gain access to some circuit breakers (such as bulk-oil OCBs) or to facilitate safe working (such as on ABCBs), the interrupting, insulating or operating medium may have to be removed. Access may then require the erection of scaffolding. Maintenance can then commence and would generally consist of the following:

- inspection, cleaning and lubricating
- periodic replacement of parts
- mechanism checks
- contact resistance measurement
- assessment of the interrupting and insulating mediums
- tests specific to circuit breaker types
- timing tests
- operational tests.

Sections 4.27 to 4.34 examine these requirements in further detail.

4.27 Maintenance inspection/cleaning/lubricating

This should be done against a well documented schedule of requirements. The lubricating requirements will be specified by the manufacturer. The inspection should address the state of the insulators, inspection windows, labels, gauges, heaters, terminal blocks, paintwork, support structures, earthing, connections, general state of the mechanism and periodic inspection of the interrupters.

4.28 Mechanism checks

Manufacturers' maintenance instructions usually recommend checks on various gaps, or extent of travel, of the mechanism. These are usually carried out using 'go–no go' gauges, feeler gauges or a metal rule. The slow-close jack is usually required to move the mechanism to the required position for obtaining the various measurements.

4.29 Contact resistance

Contact resistance measurement should be a repeat of the test specified in Section 4.19. The results should be compared with the commissioning results. This is particularly important for a vacuum interrupter, which is a sealed unit and cannot be inspected visually.

4.30 OCB maintenance

Specific tests pertaining to OCBs relate to checking the condition of the oil. A sample of oil should be taken from each tank and tested as specified in Section 5.33. OCBs with oil-filled bushings should have oil samples taken from each bushing helmet. If a sample fails the tests, further samples should be taken from the base of the bushing to determine whether the contaminated oil is confined to the helmet or is found throughout the whole bushing.

If the circuit-breaker-tank oil is found to be contaminated, or if it is necessary to examine the interrupter contacts, the tank will need to be drained. With smaller circuit breakers the tank may be drained into drums using a portable pump and pipes. Larger circuit breakers have fixed pipes, valves and pumps, draining into a fixed dirty-oil storage tank. The dirty-oil tank is usually connected via an oil filter to an associated clean oil tank, which in turn is connected via clean oil pipework to the circuit breaker tank.

Inspection of the interrupters and contacts is usually straightforward. With smaller OCBs the contact assembly and mechanism can usually be lifted out of the tank by use of portable lifting gear. With larger OCBs it is possible to climb into the tank.

Larger OCBs usually have a contact-linkage buffer to provide a buffer for the mechanism at the end of its travel. This device is usually oil filled—the oil level should be checked during maintenance.

4.31 ABCB maintenance

Maintenance of ABCBs invariably requires draining the air from the circuit breaker for reasons of safety. Draining and refilling the circuit breaker may, with some types of circuit breakers, take some hours to achieve.

Access to the interrupters is usually more involved than with an OCB and is only carried out when there is a need, as opposed to a matter of routine; selected interrupters would therefore be withdrawn for inspection at approximately six-yearly periods.

Once the circuit breaker has been refilled with air, towards the end of the maintenance procedure the various pressure switches, identified in Section 4.23, should be recalibrated.

A check on the dryness of the compressed air in the high-pressure receiver should be carried out about twice per year. If the air does not possess the required degree of dryness it is usually an indication of unsatisfactory performance of the air-dryer system—see Figure 4.9. Both the compressors and the dryer system have had a history of being particularly troublesome over the years, and they are a significant weakness of this type of circuit breaker.

The value of the voltage-grading capacitors should be measured using a capacitance-measuring bridge. The value of any resistors should also be determined.

4.32 GCB (SF_6) maintenance

Routine maintenance would normally require a check of SF_6 dewpoint, acidity and oxygen as described in Section 4.24. All alarms should additionally be functioned.

If access is required to interrupters or chambers containing SF_6, specific safety precautions should be taken with reference to SF_6 gas not supporting life, and the toxicity of the sulphur-fluoride powders. Prior to commencing access, personnel should be clad in protective clothing—i.e. boilersuits having elastic ankle and wrist grips which overlap both boots and gloves, chemical-type box goggles and a respirator with dust filter and activated charcoal. In many instances, the SF_6 is drained from the chamber by discharging through a filter to remove toxic impurities. When access is gained to the chamber, all breakdown products should be removed using a vacuum cleaner, with disposal bags especially for the purpose. All disposals should totally adhere to the manufacturer's instructions, including that of breakdown products, filters and protective clothing.

4.33 Maintenance timing tests

This is a repeat of the commissioning test described in Section 4.21. This test would not usually be carried out as a routine on distribution-voltage circuit breakers, where speed of circuit-breaker operation is not as imperative.

A dilemma associated with this test is whether it should be carried out at the beginning of the maintenance, thereby assessing performance before maintenance, or whether it should be carried out after maintenance, or both. One school of thought argues that the 'before' maintenance test is imperative to facilitate diagnosis of possible in-service circuit-breaker problems. Another school of thought argues that it is too labour intensive to carry out a 'before' and 'after' test and the 'before' test need only be carried out on a small representative sample which would act as a general indication of performance deterioration. The 'after' maintenance test is always necessary to ensure that a healthy circuit breaker is returned to service.

It is usually necessary to remove maintenance safety earths from one side of the circuit breaker to prevent a short circuit corrupting the test results. This may result in significant induced voltages, from adjacent circuits into parts of the circuit breaker, and hence the test leads, when the circuit breaker is open. The test equipment must be capable of withstanding these voltages.

4.34 Operational tests following maintenance

This would usually include the following:
(*a*) open and close from reduced battery voltage (see Chapter 9). The case in favour of this as a maintenance test is increasingly weak;
(*b*) tripping from all trip relays;
(*c*) opening and closing from local, standby and main (remote) positions.

4.35 Frequency of maintenance

The frequency of maintenance will depend on the operating conditions of the circuits controlled and the track record of the type of circuit breaker. In general, three periods of maintenance can be categorised, as follows:

(*a*) *Periodic exercising* (open and close) of the circuit breakers at about six-monthly intervals, The objective is to avoid stiffness in the mechanism, arising from over-long periods of inactivity;
(*b*) *Routine maintenance* of the circuit breaker which would include most of the above activities, but not including any major strip down, or evaluation of the condition of the interrupters. This would probably be carried out every two to three years;
(*c*) *Major maintenance*, which would also include a strip down of selected features and examination of some, if not all, interrupters. This would usually be carried out approximately every 6–12 years. A factor determining the frequency of interrupter inspections is the number of times the circuit breaker is required to break fault current, coupled with the magnitude of the fault current.

The above is an idealised maintenance strategy and would not always be feasible in practice. Generator circuit breakers, for example, must have their maintenance period aligned to generator outages, and it may be necessary for distribution circuit breakers controlling circuits without alternative supplies to have longer maintenance periods.

4.36 Operating problems

The main operating problem (although in practice infrequent) with circuit breakers is loss of interrupting, insulating or operating medium while in service. This may relate to:

- loss of oil from a circuit breaker tank, i.e. beneath the level of the sight glass;
- loss of oil from a bushing;
- loss of air or SF_6 below the trip lock-out level;
- a break in the trip coil circuitry: this would cause operation of the trip circuit supervision alarm;
- loss of oil, nitrogen or air from an SF_6 circuit-breaker operating mechanism.

If the condition cannot be rectified, it is usually necessary to evacuate personnel from the site, switch out autoreclose and make the circuit breaker dead as soon as possible. This requires isolation of the circuit breaker on a busbar, and de-energisation of the busbar via a bus coupler or section circuit breaker (i.e. the suspect breaker should not be operated).

It may be necessary to retain switchgear in service if the fault level exceeds its rated breaking capacity. In such instances the tripping should be removed.

4.37 Fault investigations

The circuit breaker exists to disconnect faults on other parts of the system (e.g. overhead lines). Occasionally, however, the fault is associated with the circuit breaker itself. These may be categorised as:

(i) faults due to a failure of the interrupting or insulating medium, e.g. flashover through the oil of an OCB;
(ii) failure of the mechanism.

A failure of the interrupting or insulating medium is usually associated with depletion or contamination of the medium. Mechanism problems can occur for a whole variety of reasons.

Failures which result in catastrophic or significant damage to the circuit breaker can be difficult to diagnose. Before such a failure is investigated, the workings and physical arrangements of a healthy circuit breaker should be fully understood. Photographs should be taken at each stage of the investigation for future reference.

Chapter 5
The power transformer

5.1 Introduction

The power transformer is a key item of equipment in electrical power systems and is relatively complex to both commission and maintain. This chapter begins by considering the practical aspects of winding arrangements, transformer oil, terminations and overvoltages, as a prelude to defining the commissioning and maintenance requirements.

WINDING ARRANGEMENTS

5.2 Terminal identity

British practice is to bring the transformer terminals out in rows, the HV on one side and the LV on the other. They are lettered from left to right facing the HV side. The HV terminals have capital letters (A,B,C) and the LV terminals small letters (a,b,c). Tertiary windings, where provided, are lettered with capitals enclosed in circles. Neutral terminals precede line terminals. Each winding has two ends designated by subscript numbers 1 and 2 or, if there are intermediary tappings, these are numbered in order of separation from end 1.

If the induced EMF in the HV phase A_1A_2 is in the direction from A_1 to A_2 at any given instant, then the induced EMF in the corresponding LV phase at the same instant will be from a_1 to a_2, as shown in Figure 5.1.

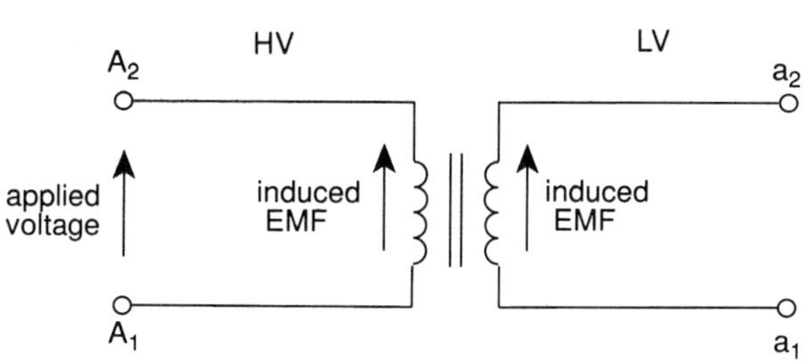

Figure 5.1 Criterion for induced EMF

5.3 Vector-group symbol

Three-phase transformers are allotted symbols giving the type of phase connection and the phase-angle difference between the HV and LV terminals. The angle is described by a clockface hour figure. The HV vector (phase) is taken as 12 o'clock, the reference, and the corresponding LV vector is represented by the hour hand. For example a Dy11 represents:

D = HV delta winding
y = LV star winding
11 = clockface reference indicating that the LV vector is at 11 o'clock (30° lead) with reference to the HV vector, at 12 o'clock

Windings fall into three main types: star, delta and zig-zag. From these combinations, phase shifts of 0°, 180°, +30°, −30° can be obtained. They are categorised into four groups as follows:

Group 1:	0°	phase shift:	Yy0, Dd0, Dz0, Zd0
Group 2:	180°	phase shift:	Yy6, Dd6, Zd6, Dz6
Group 3:	−30°	phase shift:	Dy1, Yd1, Yz1, Zy1
Group 4:	+30°	phase shift:	Dy11, Yd11, Yz11, Zy11

Most power transformers have links termed vector-group links which allow the windings to be configured either as Group 1 or Group 2, or as Group 3 or Group 4. For example a star–delta transformer can be configured as a Yd1 or a Yd11.

The phase shift across a transformer can be verified by use of a vector diagram. When the vector diagram is constructed, it is assumed that the EMFs induced in the three phases are equal, balanced and mutually displaced by 120° and have a definite sequence of rotation. Figures 5.2, 5.3 and 5.4 show the vector diagrams for a Yd11, Dy1 and Zy1 transformer, respectively.

5.4 Rotation of phases

In the preceding section the transformer HV terminals A, B, C were fed with supply voltages R, Y, B, and the LV terminals a, b, c were connected to supply voltages r, y, b. As a consequence, the phase shifts as categorised by Groups 1–4 were obtained. However, by rotating the phases on the LV side of the transformer, with respect to the HV side, it is possible to obtain additional phase shifts of multiples of 120° (i.e. clockface multiples of 4). Figure 5.5 shows how this can be achieved for a Yd1 transformer.

5.5 Negative phase-sequence rotation

The previous section examined the case of the HV terminals A, B, C and the LV terminals a, b, c being connected to supply voltages with a positive-phase-

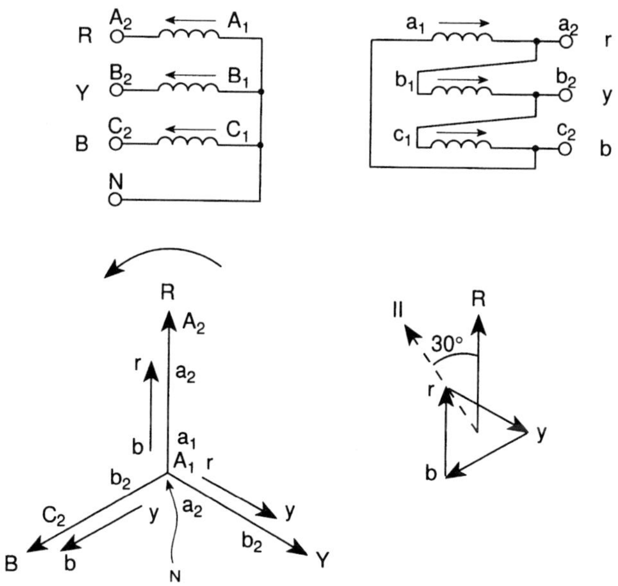

Figure 5.2 *Vector relationship for Yd11 transformer*

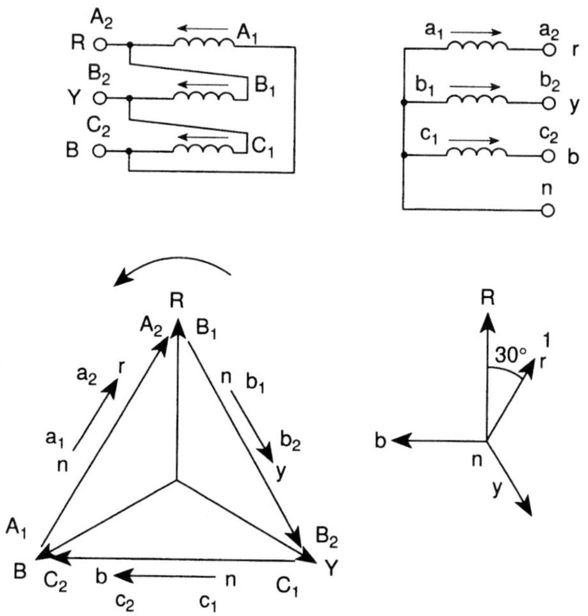

Figure 5.3 *Vector relationship for Dy1 transformer*

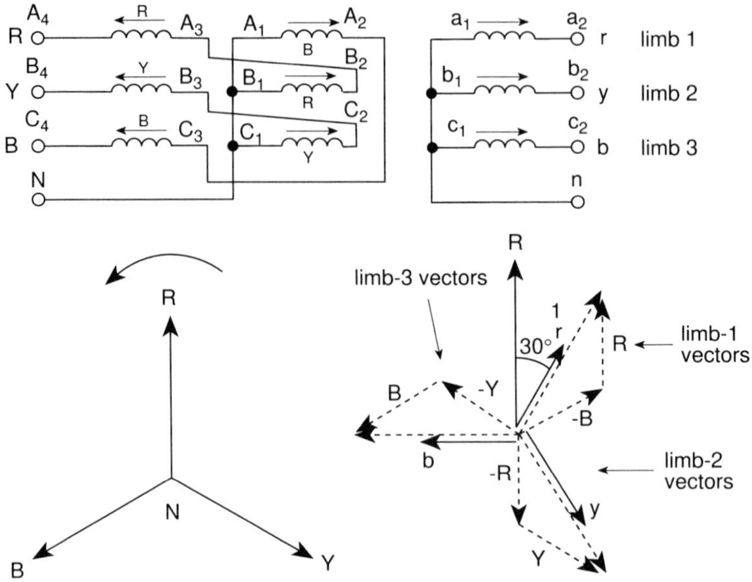

Figure 5.4 Vector relationship for Zy1 transformer

sequence rotation. The effect of connecting supply voltages, which have of themselves a positive-phase sequence of rotation, in a negative-phase-sequence manner to the transformer, is to multiply the angle of phase shift by -1. For example a Yd11 transformer with a $+30°$ phase shift between HV and LV becomes a Yd1 with -30° of phase shift.

This is illustrated in Figure 5.6. Note that both the HV and LV supply voltages are positive-sequence rotation and the effect of a negative-sequence connection to the transformer is only to change the HV-to-LV phase shift. The negative-phase-sequence connection also causes a similar phase shift in the HV-to-LV current.

5.6 Vector group: practical considerations

Transformer data plates, usually fixed to the transformer tank, normally identify the vector group in terms of positive-phase-sequence inputs with the assumption that the HV and LV phase colours are connected to the same limb (e.g. R to A2, r to a2 etc.)

However, as seen in previous sections, rotation of phases and negative-phase-sequence connections to the transformer, can result in other (apparent) vector groups being obtained. Such manipulation of transformer

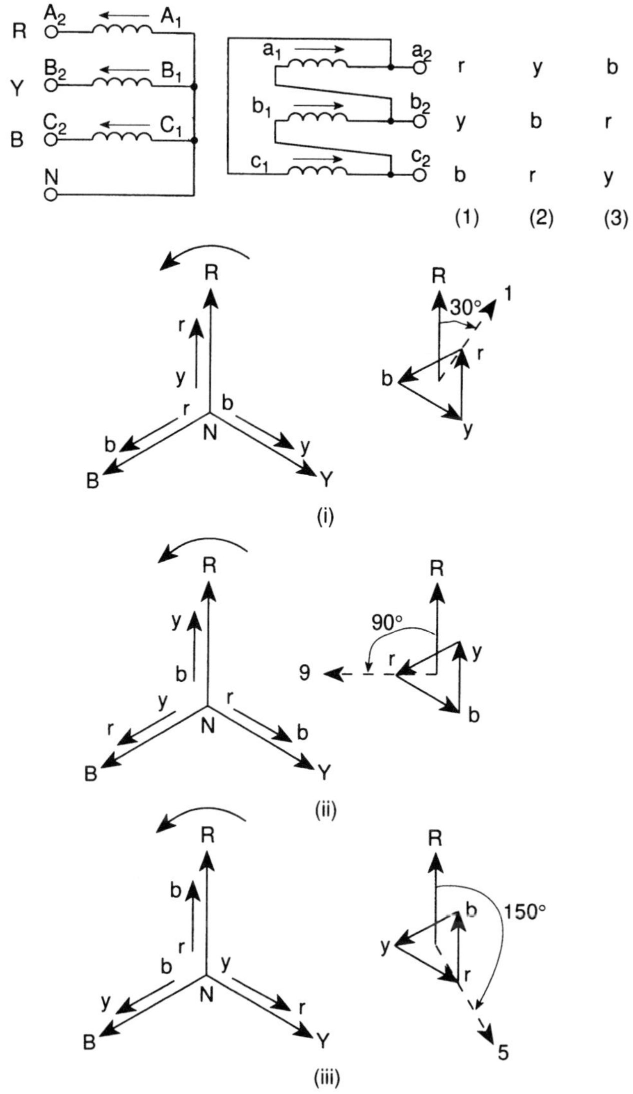

Figure 5.5 Positive-phase-sequence vector rotation

(i) Vector arrangement for (1): Yd1
(ii) Vector arrangement for (2): Yd9
(iii) Vector arrangement for (3): Yd5

inputs and outputs enables transformers with different internal connections to be operated in parallel. Often recourse has to be made to tables of connections to determine which transformers can be operated in parallel.

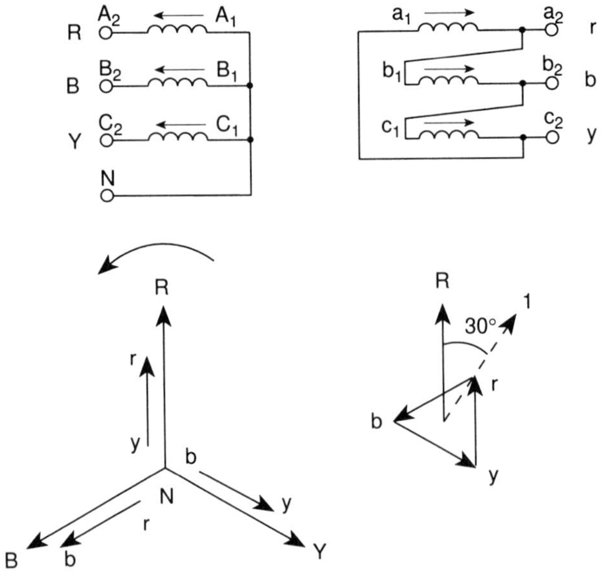

Figure 5.6 *Vector relationship showing how a negative-phase-sequence connection converts a Yd11 transformer into a Yd1 transformer*

However, the purpose of this text is to enable the reader to deduce his/her own connections.

Figure 5.7 demonstrates the solution to a practical problem. Transformer T_1 is an existing transformer and a second transformer T_2 is required to be connected in parallel. The connections from the HV busbar to T_2 are given. The required internal connections, and LV connections to transformer 2, are to be determined: the solution is shown dotted.

As can be seen, transformer T_1 is nominally a Yd11 transformer, but it has a negative phase-sequence input, therefore making it appear as a Yd1. The LV connections are then subject to a 120° rotation anticlockwise; thus the transformer is made to appear as a Yd9, i.e. HV vector at 12 o'clock, LV vector at 9 o'clock. Now transformer T_2 HV connections are given as a conventional positive-phase-sequence input; thus to obtain a Yd9 connection a nominal Yd1 transformer is required with the LV vectors subject to a 120° rotation anti-clockwise.

With reference to Figure 5.7, it may, at first sight, seem somewhat puzzling that the connections to the HV side of each transformer are different. However, this may be dictated by the physical disposition of the busbars and the transformers.

British power system practice, which may differ from that of other countries, is to standardise the relative vector positions of each voltage level:

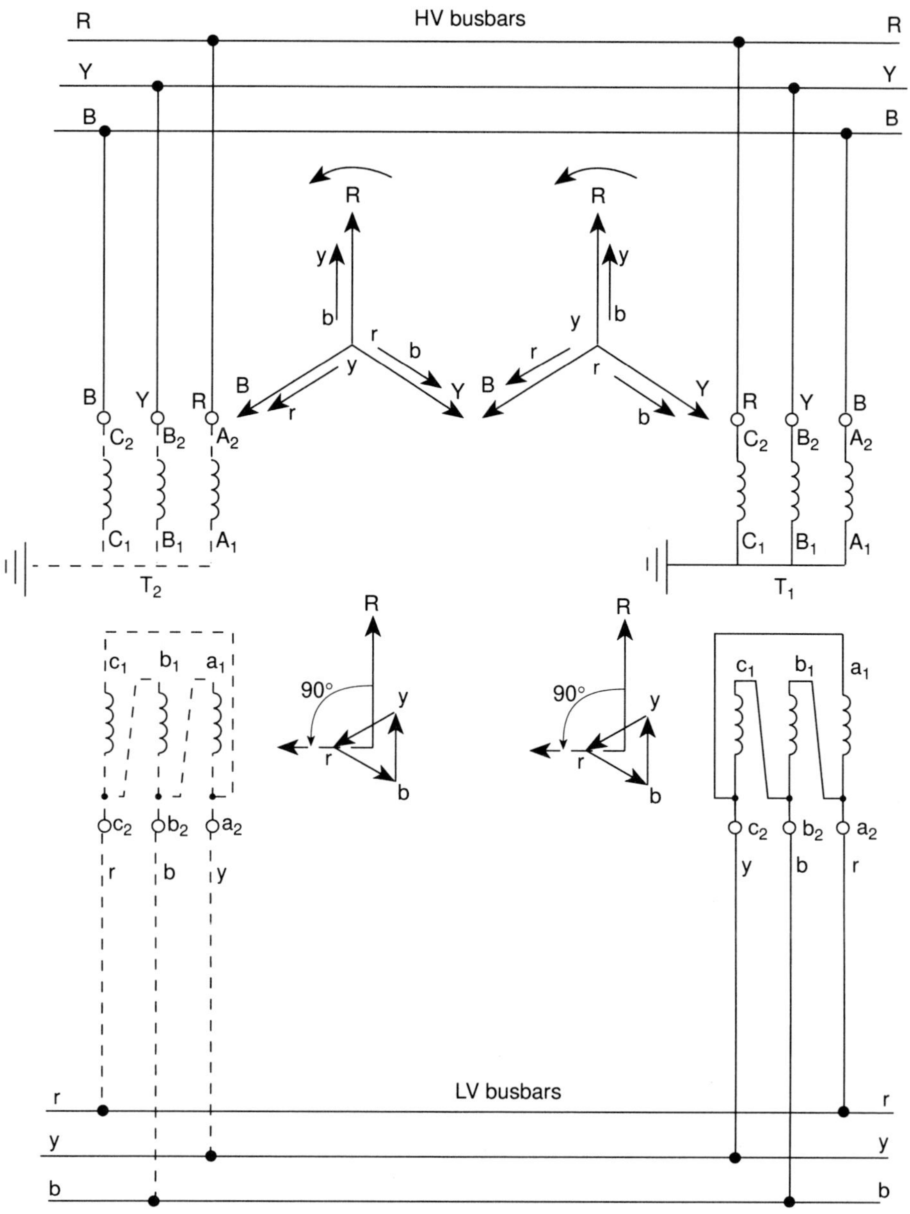

Figure 5.7 Example of parallel operation of transformers with different connections

i.e. 400 kV, 275 kV, 132 kV and 11 kV red phase are each at 12 o'clock, 33 kV is at one o'clock and 415 V at 11 o'clock. There are, however, numerous exceptions to this standard currently in service, largely arising from historical factors, e.g. some 33 kV systems are positioned at 9 o'clock.

5.7 Vector group: autotransformer

Figure 5.8 shows the terminal nomenclature for a three-phase autotransformer with tappings, with a delta tertiary, and the corresponding vector relationship. The autotransformer is designated as YyO.

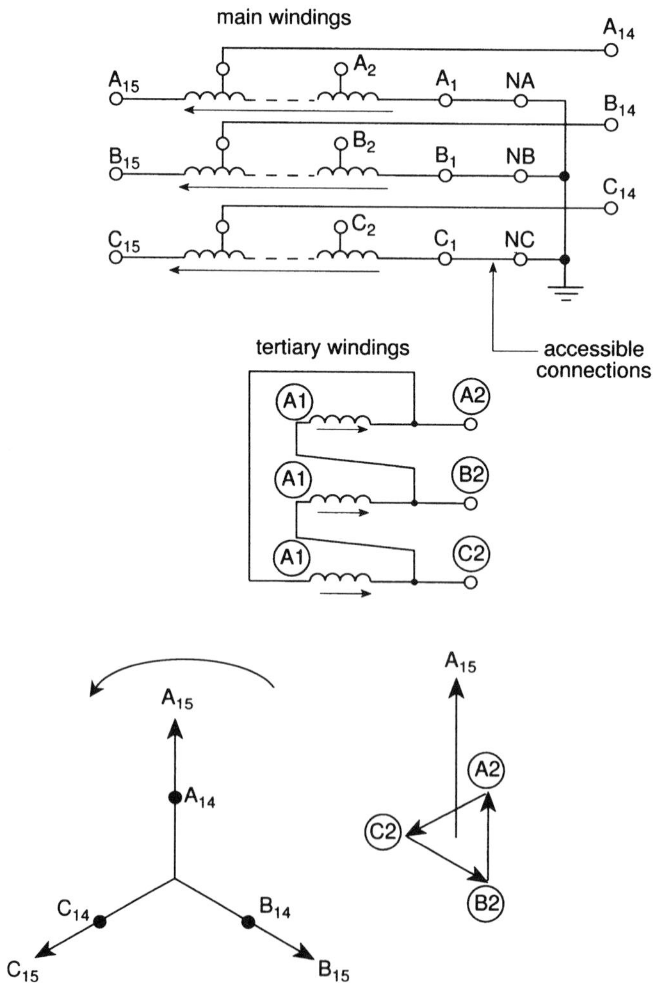

Figure 5.8 Autotransformer-terminal identity and vector group

OIL CONSIDERATIONS

5.8 Transformer oil

The oil within the transformer tank serves two prime functions:

- as a source of insulation
- to provide a cooling medium

Oil deterioration occurs largely as a result of two processes: oxidisation and thermal decomposition; these will be discussed briefly.

(*a*) *Oxidisation of oil*
There is a free flow of air and hence oxygen through the transformer breather system. The oil reacts with oxygen slowly over a very long period to form a hydrocarbon sludge. This becomes deposited on the winding tank walls. Overheating of the transformer exacerbates the situation. The effect of sludge is to insulate the transformer and make it hotter, thus aggravating the problem. The sludge may even block the cooling ducts. Oxidisation also results in the formation of hydrocarbon acids and water. This causes corrosion of the various metals within the transformer, particularly the iron. The effects of oxidisation contribute towards the ageing process of the transformer. Experience has, however, shown that the timescales for oxidation to become problematic, within modern transformers, are very long. Transformers with some 30+ years of service life are currently showing little sign of deterioration due to oxidation.

(*b*) *Thermal decomposition of oil*
Thermal decomposition, as a result of heating, causes breakdown of the oil resulting in carbon formation, sludge and insulation deterioration. Prolonged operation at temperatures above 98°C reduces transformer life.

(*c*) *Contamination*
A third area of concern is that of contamination, particularly from moisture. Careful attention to the breather system is significant in reducing moisture contamination.

5.9 Transformer cooling

Transformer cooling systems can be categorised as either natural or forced. Natural cooling occurs as a result of convection. Forced cooling implies a combination of fans and pumps. The main categories are as follows:

Symbol	*Meaning*	*Interpretation*
ONAN	oil natural, air natural	oil-immersed transformer with natural air cooling
ONAF	oil natural, air forced	oil-immersed transformer with fanned air cooling
OFAN	oil forced, air natural	oil-immersed transformer with oil pumped around the cooling system, with natural air cooling
OFAF	oil forced, air forced	oil-immersed transformer with oil pumped around the cooling system, with fanned air cooling
OFWF	oil forced, water forced	oil-immersed transformer with oil pumped around the cooling system, which is further cooled by a heat-exchange pumped-water system.

As can be seen, the first letter indicates the cooling medium which is in contact with the windings; the second letter the type of circulation; the third letter indicates the external cooling medium; and the fourth letter the type of circulation of the external cooling medium.

The above abbreviations can be linked together to form mixed cooling systems, e.g.

ONAN/OFAF = an oil-natural, air-natural system which can be switched to an oil-forced, air-forced system.

Forced cooling enables the transformer to have a higher rating, which can be of the order of twice the rating for natural cooling. When quoting transformer ratings it is usual to quote the higher rating.

Figure 5.9 illustrates some typical cooling systems. Figure 5.9*a* shows an ONAN system which employs cooling tubes attached to the transformer tank. This type of cooling is found on many of the 11 kV/415 V distribution transformers. Figure 5.9*b* shows an ONAN system which utilises a radiator system, as typically installed on a 33/11 kV distribution transformer, although most of these transformers are now assisted (i.e. equipped with forced systems).

Figure 5.9*c* shows an ONAN/OFAF type of transformer such as might be found on transmission systems. At low loads, the transformer will be ONAN, but as the temperature rises the winding temperature monitoring device will switch in the fans and pump to become an OFAF system.

5.10 Oil systems for large transformers

With reference to Figure 5.9, oil systems for large transformers embrace the following design considerations:

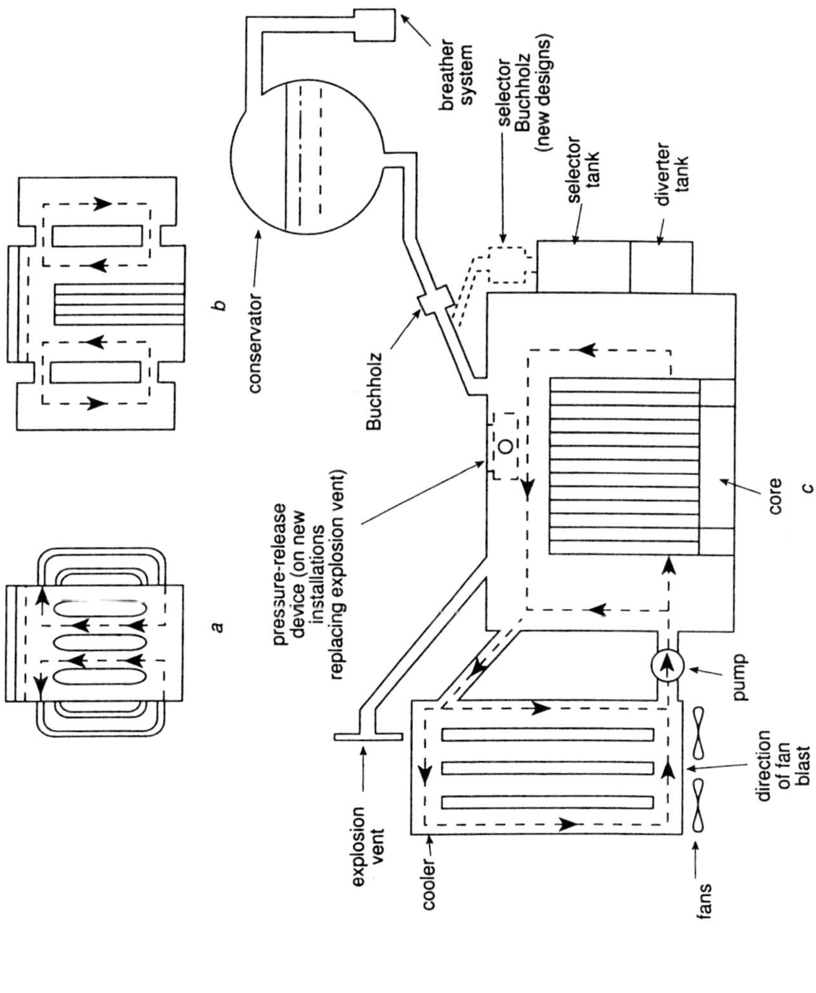

Figure 5.9 *Transformer cooling systems*

a ONAN system using cooling tubes; *b* ONAN system using radiator; *c* ONAN/OFAF transformer

(*a*) *The conservator*: This is essentially an oil reservoir and expansion vessel. The conservator exhausts to atmosphere via a breathing device which is usually of the silica gel desiccant type or the 'drycol' refrigeration-dryer type. By this means the intake of moisture during the cooling period of the transformer load cycle is limited.
(*b*) *The explosion vent*: This usually consists of a thin and relatively weak diaphragm which, under normal circumstances, seals the transformer from the atmosphere. Under transformer internal fault conditions, the excess pressure in the tank bursts the explosion vent to release the pressure, to safeguard the tank from damage. More recent transformers have a spring-operated oil-pressure-release device which sits near the top of the transformer, in place of the explosion vent.
(*c*) *Selector tank*: This houses the tap-change-selector mechanism. There is usually a pipe, via an isolating valve, connecting the top of the selector tank to the main tank; this permits filling of the selector tank following maintenance, but the valve must be closed when the transformer is in service. Earlier designs suffered from a weak barrier between selector and main tank, through which the tap-changer tappings passed. If, when oil was pumped out of the selector tank, during maintenance, a partial vacuum resulted, the pressure differential could rupture this barrier. More recently, consideration has been given to a selector Buchholz (see Section 5.11) to afford better protection.
(*d*) *The diverter tank*: This is usually a self contained tank bolted to the main tank; it contains the tap-change diverter mechanism. It is necessary to segregate the diverter tank, since the diverter both makes and breaks current, thus creating arc products which, over a period of time, contaminate the oil. The diverter-tank oil must therefore be regularly monitored and changed.

5.11 The Buchholz

The Buchholz is a protective device whose operating principle depends on the fact that transformer electrical faults cause violent generation of gases, which flow to the highest point, namely the conservator. In doing so they force an oil surge up the pipework to the conservator, so causing operation of the Buchholz trip facility. Gas can also be liberated more slowly as a result of minor discharges due to insulation weakness. Again, such gases make their way to the conservator and in doing so become trapped in the Buchholz chamber, resulting in the generation of an alarm.

Figure 5.10 illustrates a typical Buchholz system using floats and mercury switches. With the chamber full of oil, the positions of the floats are such that the mercury switches are open. As the oil level falls the floats change position and the switches close. The upper float is the alarm and the lower float the trip. The lower float often has a baffle attached to it to cause more rapid operation of the mercury switch under oil surge conditions.

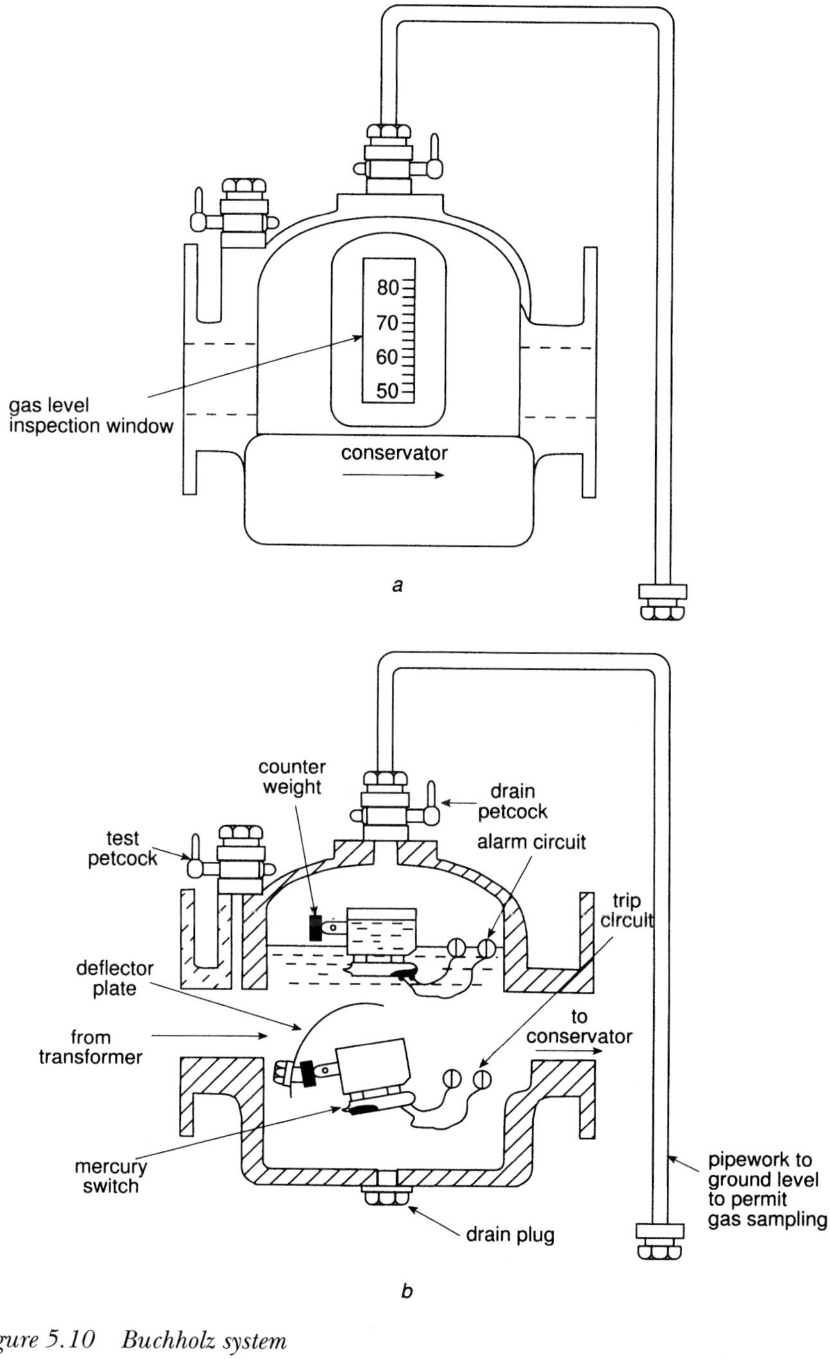

Figure 5.10 Buchholz system

a External view
b Internal view

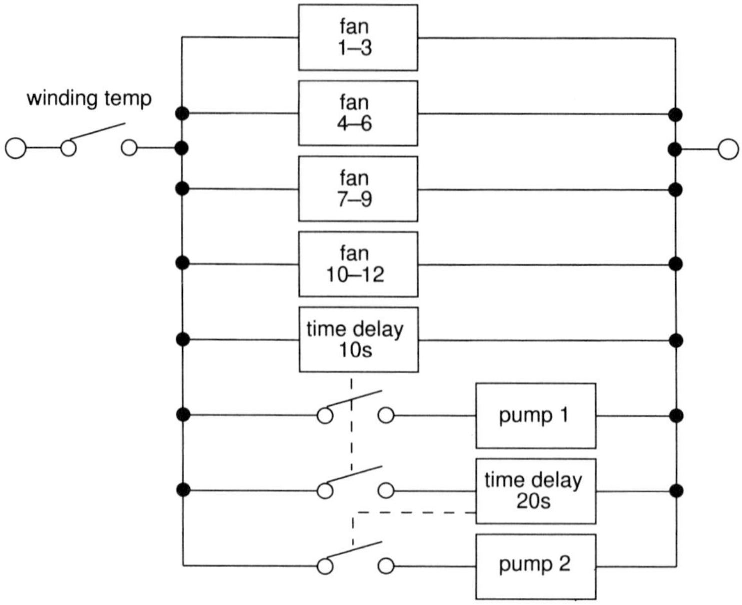

Figure 5.11 Typical cooling control system for a transmission transformer

The top of the chamber usually has a test petcock fitted to permit testing. A drain petcock is also usually fitted and connected via pipework to ground level to facilitate filling of the Buchholz chamber and gas analysis testing.

5.12 Cooling-system circuitry

Figure 5.11 illustrates a typical cooling control system for a transformer with fan and pump cooling. As can be seen there are time delays in switching certain facilities. The time delays are particularly important in relation to the pumps, in:

(*a*) preventing undue turbulence in the cooling oil when the pumps start which could cause operation of the Buchholz or bursting of the diaphragm of the explosion vent;
(*b*) spreading the electrical loading of the control system caused by motor-starting currents.

5.13 Winding temperature

Transformers with forced cooling usually have a winding-temperature monitoring device fitted which has two prime purposes:

(*a*) to start and stop the fans and pumps on rising and falling transformer temperature;
(*b*) to provide an alarm and trip signal should the transformer become subject to excessive overloading, or suffer failure of the cooling equipment, thereby causing excessive heating.

A typical winding-temperature-monitoring device is illustrated in Figure 5.12. As can be seen, an isolated pocket of oil, located at the top of the transformer tank, reflects the top-of-tank oil temperature. In addition, the pocket of oil is also heated by a heater which is fed from a current transformer such that the additional heat is proportional to load. This combined heating effect is arranged to bring the pocket of oil to a temperature which reflects the transformer-winding hot spot. The hot spot is the location of the point of highest temperature within the transformer. Some modern instruments incorporate the heater directly into the instrument itself.

The pocket of oil heats a thermometer bulb which, via a capillary tube and bellows mechanism, operates a winding-temperature indicator (calibrated in degrees Celsius). Rotation of the indicator results in the closing of mercury-switch contacts when certain preset temperature settings are reached.

Winding-temperature settings for starting and stopping the fans and pumps are usually of the order of 75°C and 55°C, respectively. This avoids prolonged overheating of the oil which would cause loss of transformer life. The winding-temperature trip is usually in the range 120–140°C. This is to avoid undue aging of the transformer insulation given that the aging of insulation doubles for every 6°C increase beyond 98°C.

TERMINATIONS AND OVERVOLTAGES

5.14 Terminations

The methods employed to bring the ends of the transformer windings out through the transformer tank, for connection to the external power system, involve the use of either bushings or cable boxes.

Where cable boxes are used, the connection from the winding is usually brought through the tank wall by means of a plain porcelain bushing into a termination chamber which contains the cable ends, and which is usually compound filled. Figure 5.13*a* illustrates a typical arrangement.

Cable boxes are found chiefly on distribution- and low-voltage-system transformers. At higher voltages it is usual to utilise bushings which connect to short lengths of busbar. For voltages up to 33 kV a simple bushing is employed consisting of a glazed porcelain insulator with a conductor through the centre. Bushings for higher voltages are usually of the oil filled condenser pattern as illustrated in Figure 5.13*b*: in this design, oil impregnated paper is interleaved with thin layers of metal foil, to form a series of capacitors to provide voltage grading throughout the radial depth of the insulation. Oil

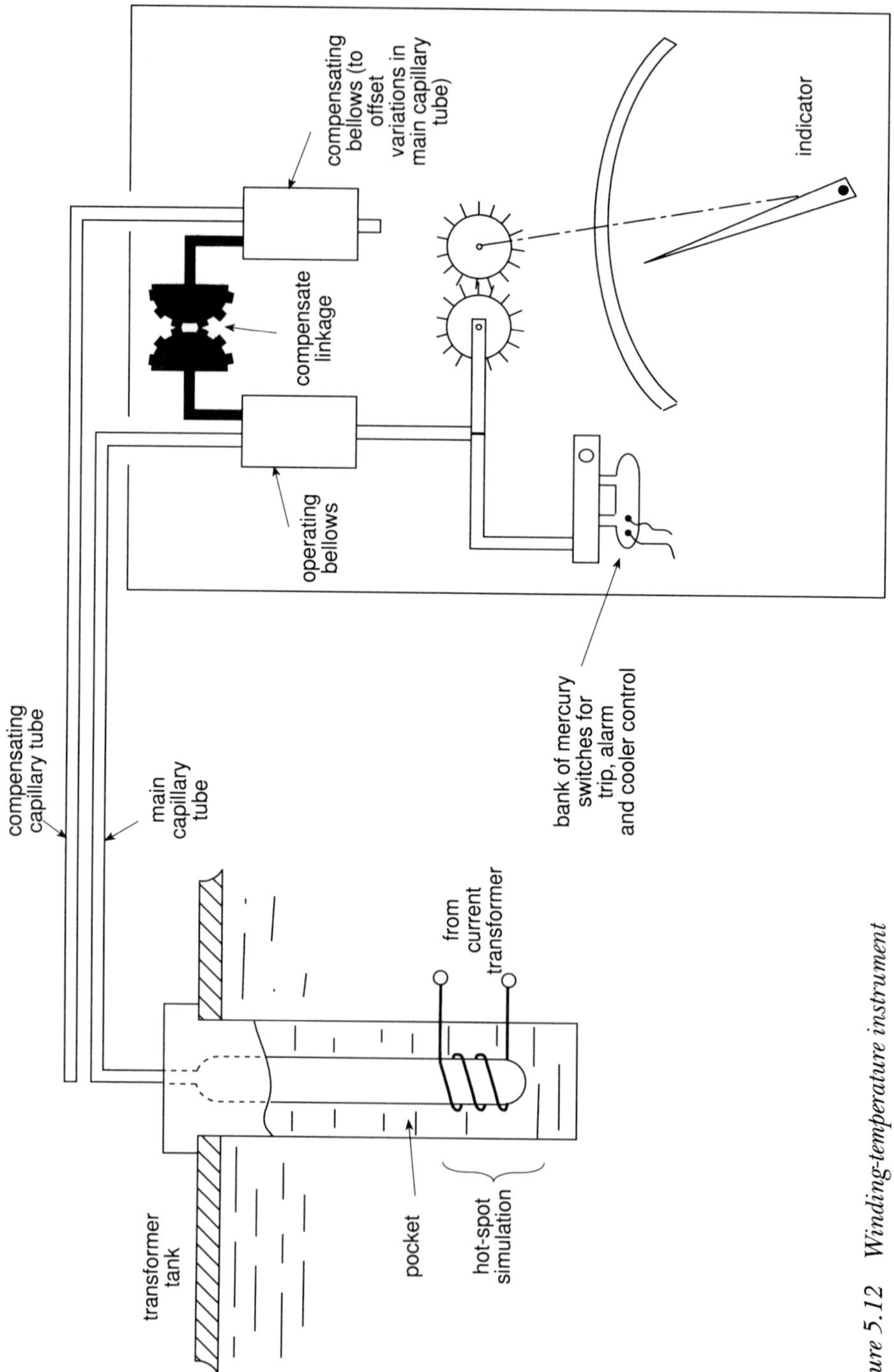

Figure 5.12 *Winding-temperature instrument*

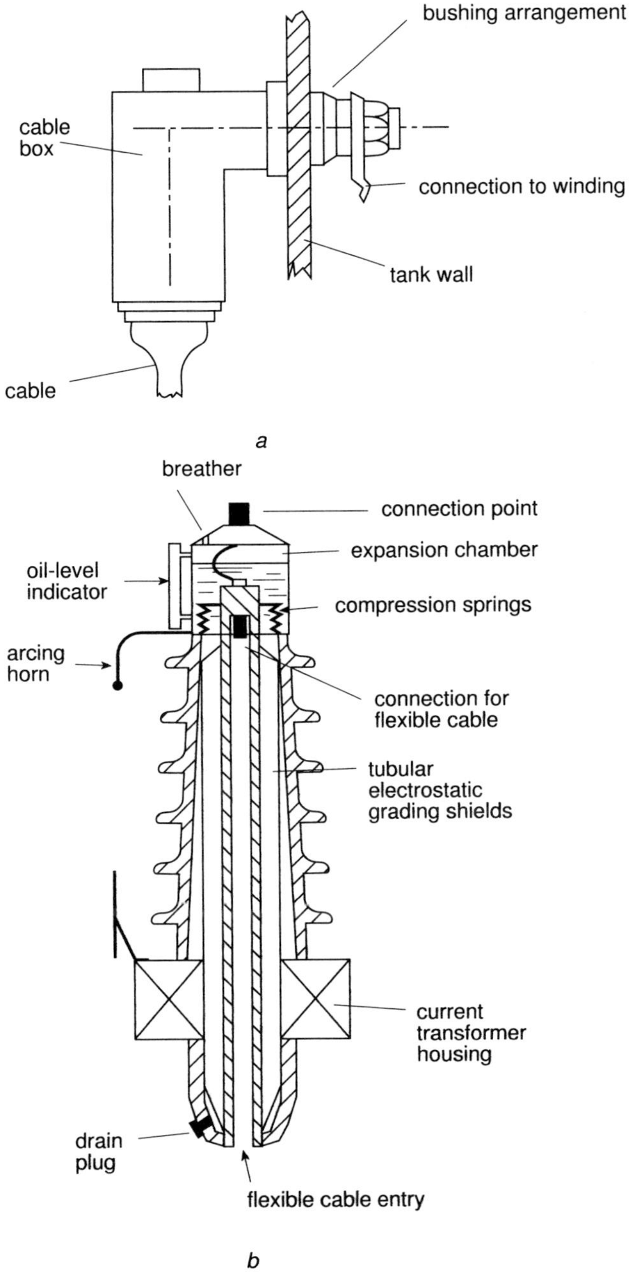

Figure 5.13 *Transformer terminations*

a Cable-box arrangement
b Condenser-type oil-filled bushing

within the bushing is sealed from the transformer-tank oil, and an expansion chamber at the top of the bushing seals the oil-impregnated paper from the atmosphere and possible ingress of moisture. The connection to the transformer windings, known as the 'pigtail', passes through the centre of the bushing and is terminated in the bushing helmet (expansion chamber). There is also a modern tendency to use epoxy-resin-impregnated paper (ERIP) capacitor bushings, which are smaller in diameter than oil-filled bushings.

5.15 Co-ordinating gaps

Co-ordinating gaps of the straight-rod type, as illustrated in Figure 5.14, are usually fitted to higher-voltage transformers. Their objective is to prevent any voltage surge which appears at the transformer terminals from damaging the transformer insulation. This is accomplished by the surge flashing over the gap. The size of the co-ordinating gap must therefore be graded carefully with the insulation it is protecting.

In general, co-ordinating gaps are very successful in affording protection, but they are not perfect. The imperfection arises as a result of the length of time it takes the gap to break down, during which the surge is impinging upon the insulation. The great advantage of the co-ordinating gap is its simplicity and cheapness. In practice it is usual to paint the tips of the co-ordinating gap rods, so that when flashover occurs the arcing burns the paint, which is subsequent proof of the flashover. Table 5.1 identifies some typical co-ordinating gap settings.

In recent years there has been a move from straight-rod gaps to screened-rod gaps, as illustrated in Figure 5.14*b*. These have the advantage of responding equally to both positive- and negative-going surges, whereas the straight-rod gap is more sensitive to positive-going surges.

5.16 Insulator greasing

Figure 5.14 also illustrates an arcing horn connected to the 132 kV bushing. Its main function is to protect the live-side and earth-side metal fittings, and the porcelain insulation, by diverting arcs, arising from overvoltages or as a result of bushing pollution, through the arcing-horn rods.

Surface pollution of the porcelain bushing arises largely from industrial sources or salt deposits. At times of high humidity or heavy rain, significant pollution can cause flashover at the normal system voltage. The quality of insulation provided by a bushing can be improved by coating the porcelain with either hydrocarbon or silicone grease. Greases have a two-fold effect. First they tend to enclose pollution particles to prevent the formation of a continuous conducting path, and secondly they are water repellent, causing

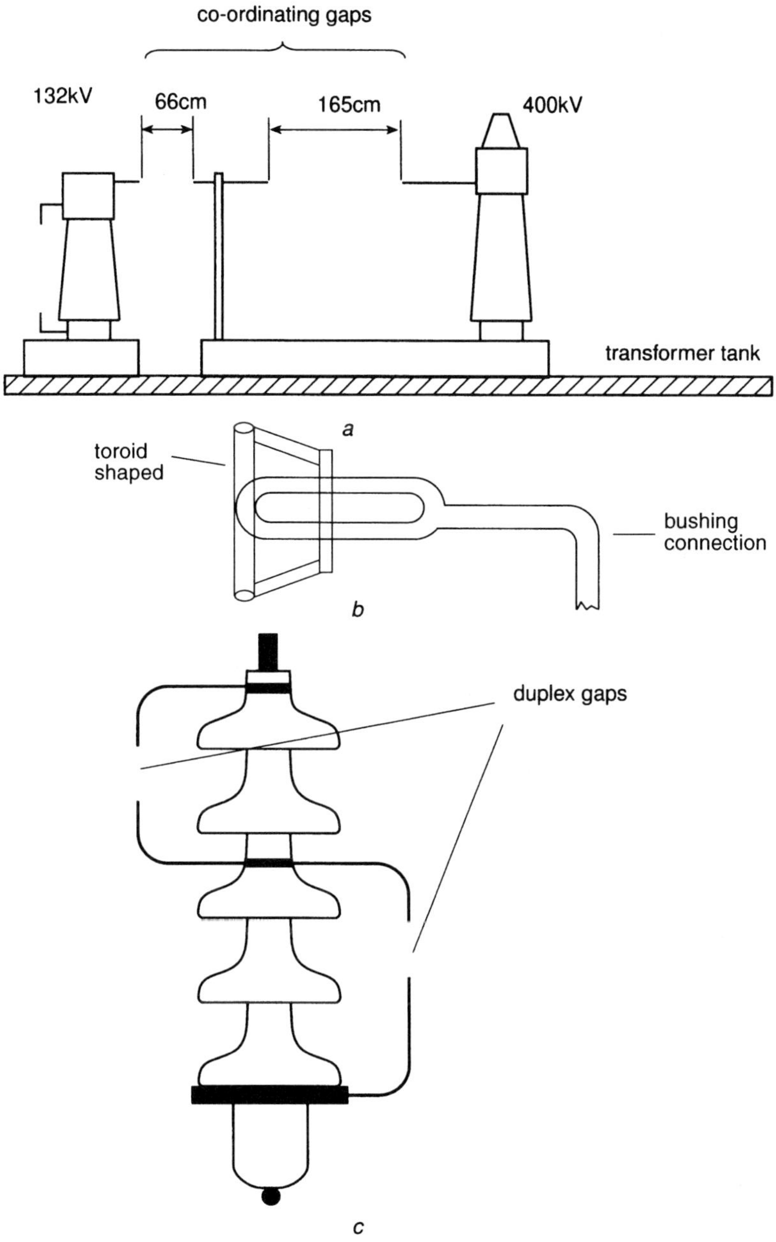

Figure 5.14 Co-ordinating and duplex gaps

a Co-ordinating gaps - straight-rod type
b Co-ordinating gap - screened-rod type
c Duplex gaps

Table 5.1 Typical co-ordinating-gap settings—straight-rod type

System voltage	Co-ordinating-gap setting
(kV)	(mm)
66	380
132	660
275	1240
400	1650

water to separate into individual droplets, thus preventing the formation of a continuous film of water. Greasing can be effective for up to three years, although in practice it often ranges from six months to two years, depending upon the degree of pollution. The application and removal of grease is a dirty, labour intensive job and tends to be limited to equipment connected at only the hightest voltages, in areas of significant pollution.

5.17 Duplex gaps

At voltages of 33 kV and below the single-co-ordinating gap, used at higher voltages, is of such a relatively small setting that it could be bridged by birds and animals. This problem is overcome by the use of duplex gaps, where the gap consists of two gaps in series, one either side of the bushing insulator, as shown in Figure 5.14*c*. Table 5.2 identifies some typical duplex-gap settings.

Table 5.2 Typical duplex-gap settings

System voltage	Duplex-gap setting
(kV)	(mm)
6.6	2×25
11	2×32
33	2×95

5.18 Ferroresonance

Consider the plant arrangement shown in Figure 5.15, where a double-circuit overhead line consists of a feeder, which is energised, and a transformer feeder, which is to be de-energised. At the instant of de-energisation the

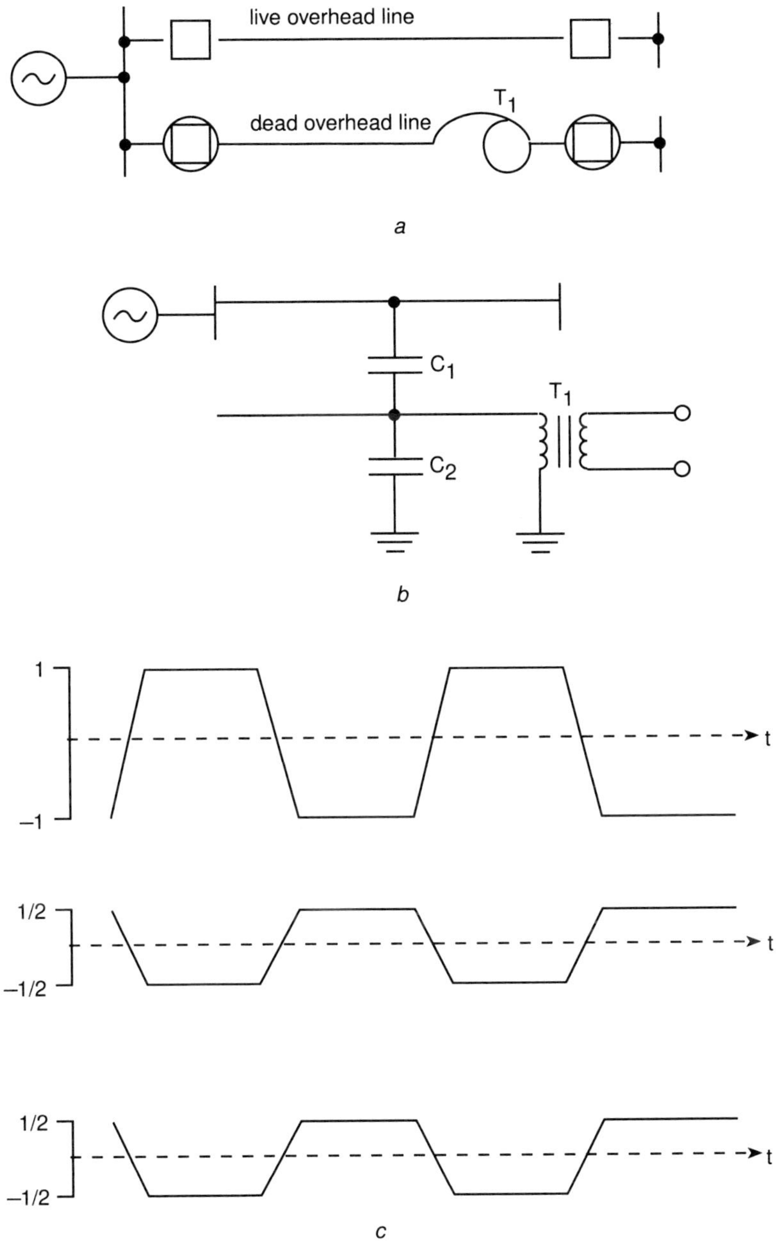

Figure 5.15 Ferroresonance conditions

a System under consideration
b Equivalent circuit
c Three-phase voltage waveform

stored energy in capacitance C_2 discharges into the transformer winding, resulting in a damped oscillation between C_2 and the winding. However, depending on the system parameters and the point on wave of switching, it is possible for the oscillation to be sustained as a result of energy transfer through C_1, from the energised line. This phenomenon is termed ferroresonance. The resulting voltage waves are shown in Figure 5.15*c*. As can be seen, they are flat topped; this arises as a result of the transformer core being pushed into saturation.

When 50 Hz ferroresonance occurs, the three voltages tend to exhibit amplitudes of 1 per-unit, $-1/2$ per-unit, $-1/2$ per-unit, respectively. However, many instances of ferroresonance have also occurred with a frequency of $16\frac{2}{3}$ Hz, i.e. one third of fundamental, with voltage amplitudes of $1/3$ per-unit, $-1/6$ per-unit, $-1/6$ per-unit, respectively.

Linear resonance can also occur without the transformer core being pushed into saturation, so resulting in near perfect sinusoidal voltages. There is no simple mathematical model for determining whether ferroresonance will exist, and if so in what shape. However, if a circuit is subject to ferroresonance, the same type of phenomenon will be experienced in each instance. In general, ferroresonance is more likely to be experienced in equipment connected at transmission voltages than in that connected at distribution voltages.

When a transformer undergoes ferroresonance, the peak flux density within the core will usually exceed that which occurs during normal operation. This results in overheating of the core, local heating of the tank due to an increase in flux path outside the core (a consequence of saturation), and excessive mechanical vibration. Sustained ferroresonance can damage the transformer. In some instances, it has caused operation of Buchholz relays. A period of 10 min is considered to be a maximum time a transformer should be subject to ferroresonance. It is quenched simply by opening the transformer disconnector, and without any significant damage being suffered by the disconnector, although some newer disconnectors are of a less robust design and not rated for this purpose, and as a result high-speed earth switches are being investigated as an alternative.

5.19 Energisation overvoltages

Significant, and troublesome, overvoltages can be experienced when energising transformer feeders, especially when the other side of the transformer has a length of cable, and hence capacitance, connected to it.

Consider the system in Figure 5.16*a* and the single-phase equivalent circuit as shown in Figure 5.16*b*. When the circuit breaker at substation A is closed to energise the circuit, a voltage surge travels down the overhead line. On reaching the transformer it encounters a surge impedance (the transformer winding) which is significantly higher than the line surge impedance and, in

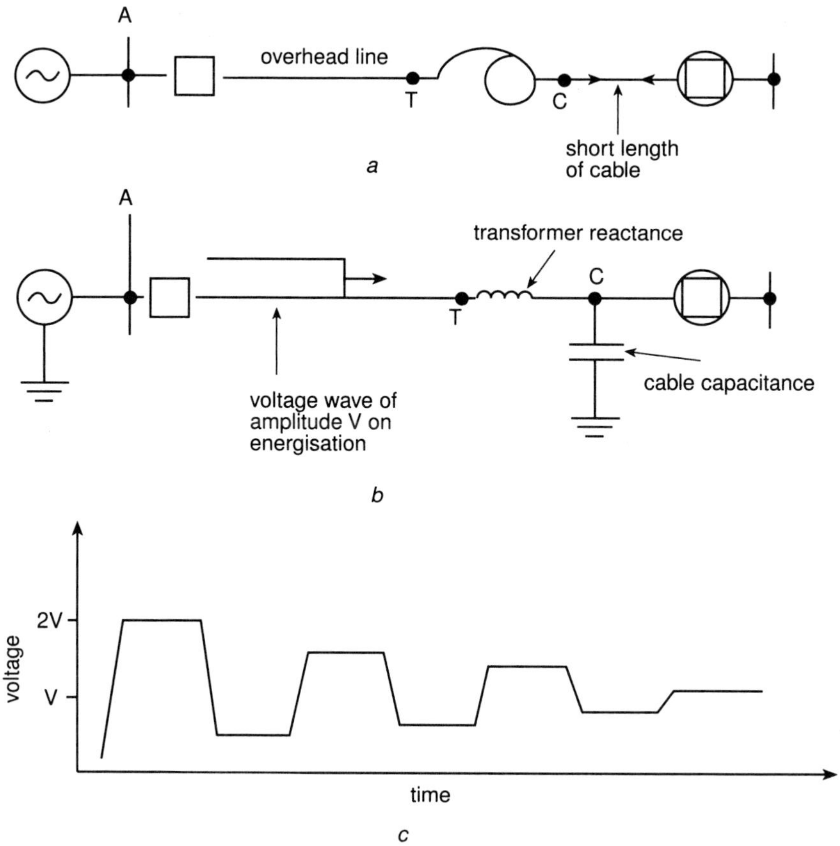

Figure 5.16 Overvoltages

a System under consideration
b Equivalent circuit
c Typical plot of voltage at point T

accordance with classical surge theory, the voltage wave doubles in magnitude. The voltage wave is then reflected back to the source, following which it undergoes multiple reflections between source and transformer.

The net effect of these reflections, when the effect of damping is taken into account, is to produce a damped stepped wave, as shown in Figure 5.16*c*. When the surge first meets the transformer and doubles in magnitude, it is applied to the *LC* circuit comprising the transformer leakage reactance and the cable capacitance. This circuit is thus caused to oscillate. If the frequency of the wave shown in Figure 5.16*c* matches, or is close to, the resonant frequency of the *LC* circuit, the repeated shock excitation causes the *LC* circuit oscillation to increase in voltage amplitude up to a theoretical maximum of 4 per-unit, resulting in significant overvoltage at point C in

Figure 5.16*b*, and eventual flashover of the co-ordinating gap. This phenomenon has proved to be a significant problem on transmission systems. The probability of experiencing these overvoltages is dependent on the circuit parameters and point of wave switching.

Remedies to this problem are twofold:

(*a*) energise the circuit from the side remote from the line, which is usually the low-voltage side;
(*b*) apply surge diverters to point C of Figure 5.16*a*. This option obviously has cost implications. It is worthy of note that modern surge diverters form a low-impedance path to overvoltages, but once the overvoltage is dissipated they reseal to form a high impedance to system voltages. Thus they do not impose a permanent fault on the system.

Significant overvoltages can also be experienced if a transformer feeder is energised while in a state of ferroresonance, especially if the system voltage and ferroresonance voltage become additive. If this is likely, delayed autoreclose on such a circuit may have to be inhibited until the ferroresonance is quenched.

Before a circuit is commissioned these problems should have been well explored and resolved.

TRANSPORT AND ASSEMBLY

5.20 Transport to site

Small distribution transformers are usually shipped to site by lorry, complete and full of oil. Larger transformers, with their increased weight, are shipped via flat bottomed low loaders either partially full or completely drained of oil. The radiators and auxiliary equipment are transported separately and assembled on site. The assembled transformer and cooling equipment are then filled with oil, on site, which is delivered either in bulk tankers or in drums. A point to note when transporting transformers is the weight and height capability of any bridge which has to be passed over or under.

When a transformer is shipped to site devoid of oil, it is usual to pressurise the tank with dry air, or nitrogen, at a pressure of a few pounds per square inch above atmospheric, to ensure that no moisture can enter the tank during transport. Pressurisation can be a potential source of danger, and during the installation process the tank must be depressurised before any blanking plates can be totally removed. Movement of the transformer from the low loader to its plinth is usually accomplished by use of winches and steel bar rollers, or more recently by power lifting equipment. Advance thought should be given to this process, since the transformer's final position on its plinth must be reasonably exact.

5.21 Assembly on site

Prior to the assembly of coolers, bushings, conservators etc., the transformer should be inspected for damage.

Many transformers, particularly at the higher voltages, have current transformers within the power transformer bushing, as shown in Figure 5.13*b*. The current transformers are usually shipped to site separately from the bushings and installed when the bushing is positioned in the transformer. It is usual to carry out a flick test to prove the polarity of each current transformer before it is installed in the bushing. This is because once the current transformer is installed, any flick-test current would have to pass through the high impedance of the power-transformer winding, which is of such a relatively high impedance as to render the test impracticable. The flick test is usually carried out with the current transformers standing on the ground adjacent to the point at which they are to be located in the power transformer. Particular care should be taken at this time to ensure that they are stacked in the order that they are to be installed in the bushing (given that a number may be installed in one bushing) and with the correct physical orientation in relation to the bushing, i.e. to be absolutely certain of position and polarity. Immediately after the flick test, the current transformers should be lifted into position. Modern transformers often contain a primary test winding through the current transformers which facilitates both flick tests and primary injections.

After the transformer and its fittings have been assembled, they should be filled with oil. Larger transformers must be filled under vacuum and then the oil circulated around a continuous oil-vacuum-treatment system, which can be provided as a mobile facility by the major oil manufacturers. The transformer and cooler bleed points should then be utilised to remove air from the oil system. When the oil pumps have been commissioned, they should be run for a number of hours per day, for a few days, to expel any air via the conservator. In passing to the conservator, air will become trapped in the Buchholz, which must then be drained. This process may cease when air is no longer collecting in the Buchholz.

COMMISSIONING

5.22 Commissioning tests

The transformer off- and on-load commissioning tests should be carried out in accordance with a test schedule. Sections 5.23–5.34 outline the detailed tests to be carried out. In summary they consist of:

- data collection
- checks and inspections
- insulation tests

- winding-resistance tests (preferred)
- vector-group test
- ratio, magnetising-current and tap-changer full-range-continuity test
- phasing diagram
- Buchholz and winding-temperature indicator
- oil tests
- tap changer and automatic voltage control (see Chapter 6)
- fans and pumps
- neutral-earthing resistance
- breather tests (larger transformers)

5.23 Data collection

The collection of data is required for both on-site and central records systems. This is often an area which can be neglected, but it is important not only for future reference purposes, but to make the commissioning engineer fully aware of the parameters of the transformer to be commissioned. Typical information would include:

- manufacturer and date of manufacture
- serial number
- rating—normal and assisted
- nominal voltages
- full-load current, usually at nominal tap
- percentage impedance
- vector group
- tap-changer type, range and number of tap positions

5.24 General checks and inspection

Typical items would include:

- co-ordinating gaps (alignment and gap measurement)
- labelling
- earth connections
- explosion-vent/pressure-relief device
- breathers
- valves in operational positions
- tightness of fastenings
- insulators undamaged and clean
- no oil leaks
- correct oil levels in conservator and bushings

It is also wise at this stage, if it has not already been done, to check the routing and proposed connections of each phase to busbars/cable-sealing ends/

overhead-line down droppers. Within this context the commissioning engineer should also check that all connections will give adequate safety clearance to permit the issue of safety documents and to allow safe passage of personnel and vehicles.

5.25 Insulation tests

The core and clamping structure of the transformer is usually connected to earth at one point only to prevent induced circulating currents. In older transformers the connection is directly to the inside of the tank, but modern transmission transformers usually have a connection, from the core and the clamping structure, brought through bushings in the tank wall, and then connected to earth via test links, which are mounted in a box on the tank. At the commissioning stage it is necessary to verify the single-point connection by measuring the insulation resistance of the core to earth, by breaking the connection and connecting an insulation resistance measuring device (e.g. a 'Megger') across the break. Obviously, for older transformers this must be done within the tank itself and therefore before it is filled with oil. It is usual to carry out this test before the transformer is moved from the low loader to the plinth, to ensure that no changes have taken place during transport, and again after transfer to the plinth, using a test voltage of at least 1000 V DC. For smaller distribution transformers the factory test results may suffice as evidence of this test.

It is also necessary to check the insulation resistance between the HV and LV windings (for two-winding transformers) and between all windings and earth, using a preferred test voltage of 1000 V DC.

Typical results for the above tests should be several hundreds of megohms. The temperature at which the tests are carried out should be recorded, since the results can vary slightly with temperature.

5.26 Winding resistance

This test is not always carried out, but the results are helpful during fault investigations. The resistance of each winding should be obtained and compared with the manufacturer's data. The test may be carried out with a low-voltage source such as a 6 V battery (plus voltmeters and ammeters for measuring and calculating resistance). The current should be injected via a switch capable of breaking the potentially dangerous sources of stored energy in the transformer winding. Alternatively, a measuring device such as a Kelvin bridge may be used.

The resistance should be measured with the transformer on the highest, nominal and lowest tap positions. Star-connected windings should be measured between phase and neutral, and delta-connected windings between phase and phase. Autotransformers should be measured between HV and LV, HV and neutral, and LV and neutral. Although only two of these measurements would suffice, obtaining the third measurement verifies the results. Some actual results for a 750 MVA 400/275 kV autotransformer are as follows (note that this transformer does not usually have a tap changer):

HV to LV	$=0.119\ \Omega$	average of the three phases
HV to N	$=0.535\ \Omega$	
LV to N	$=0.412\ \Omega$	
Tertiary (line-to-line)	$=0.0076\ \Omega$	

5.27 Vector-group test

The vector-group test is carried out to determine the phase shift between the HV and LV terminals to obtain the vector-group symbol. The test method is illustrated in Figure 5.17. Connections are taken from a 415 V, three-phase supply to one side of a test block, the other side of which is connected to the transformer HV terminals. The HV terminal A_2 is then connected to the LV terminal a_2, and connections are brought back from the LV terminals to the test block. Note that the LV neutral must be disconnected from earth to prevent circulating currents arising. It is advisable to label the test block to ensure that each wire is easily identified during the test. When the fuses/links are inserted, the current which flows is magnetising current. For example, if the transformer is, say, 10 MVA rating,

$$\text{the HV rated current at full voltage} = \frac{10\ \text{MVA}}{(\sqrt{3}) \times 33\ \text{kV}} = 175\ \text{A}$$

$$\text{Magnetising current is typically 1\% of rated current} \quad = \frac{1}{100} \times 175$$

$$= 1.75\ \text{A}$$

but with an applied voltage of 415 V the magnetising current reduces to

$$\frac{0.415}{33} \times 1.75 = 22\ \text{mA}$$

The test is carried out by measuring the voltages between all terminals to produce a table of results. A vector diagram can then be drawn, to scale, to show relative vector positions, as shown in Figure 5.17*b*. The angular difference between R–N and r–n can then be obtained, so proving, in this instance, that the transformer is a Dy11. Note that although the three-phase supply is not perfectly balanced the vector diagram still conclusively proves that this particular transformer is a Dy11. This test should be carried out with the transformer on nominal tap. Table 5.3 shows an actual set of results.

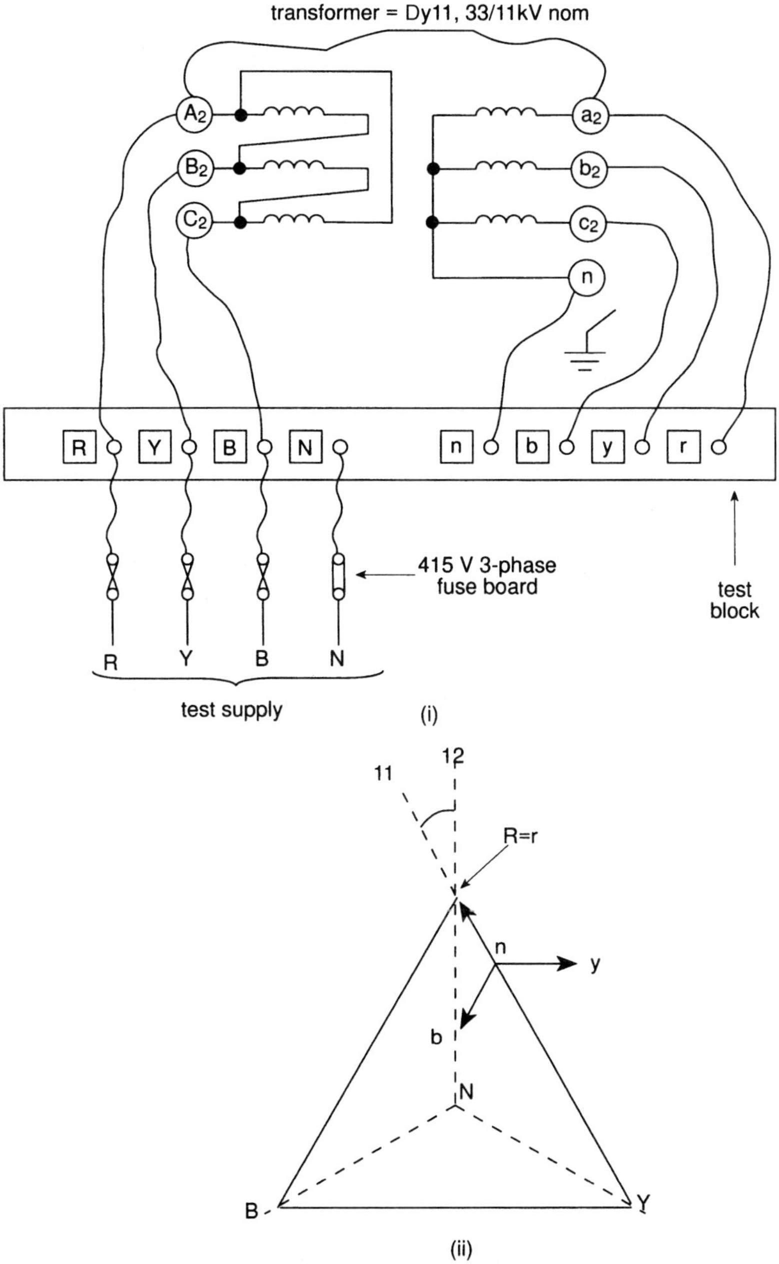

Figure 5.17 Vector-group test

(i) Test connections
(ii) Graphical representation

Table 5.3 Vector group test voltage results

	R	Y	B	N	r	y	b
R	X	X	X	X	X	X	X
Y	425	X	X	X	X	X	X
B	425	420	X	X	X	X	X
N	250	250	245	X	X	X	X
r	0	425	425	250	X	X	X
y	139	317	435	213	139	X	X
b	141	319	309	107	141	137	X
n	82	343	365	165	82	81	83

X = not required

Figure 5.18 illustrates a vector-group test which is commonly carried out on an autotransformer, with the HV terminal connected to the LV terminal. This, however, is a more complex situation than might at first sight seem the case. Given that the autotransformer has a point of commonality at the neutral (i.e. N_T in Figure 5.18), the HV-to-LV short-circuit results in a short-circuited turns situation which produces a circulating current I_c. The current I_R consists of two components: a component to balance I_c and a component which represents the magnetising current for the A limb. In addition I_R is limited in magnitude since it must equal, at any instant, the magnetising currents I_Y and I_B. This complex relationship results in the phasor diagram shown in Figure 5.18, which is drawn for an actual test on a 400/132 kV autotransformer, the results of which are given in Table 5.4. As can be seen, the voltage of N_T is no longer symmetrical with respect to the three applied voltages.

An alternative, simpler and more readily understandable vector-group test on an autotransformer is illustrated in Figure 5.19. As can be seen, the transformer star point is taken as the point of common connection and this results in a symmetrical phasor diagram. A test which embraces the delta tertiary requires a connection between HV and tertiary terminals as shown.

5.28 Ratio, magnetising current and tap-changer-continuity test

This test effectively utilises the test equipment shown in Figures 5.17 and 5.18 except that the test connection, connecting the two windings, is not required. The objective of this test is to measure the transformer ratio on all taps and to observe that there is continuity during the full range of the tap changer (i.e. that the selectors and diverters are not causing either an open- or short-circuit) by observing the HV magnetising current and LV voltage during the tap-change process. For each tap position, the HV and LV voltage and the

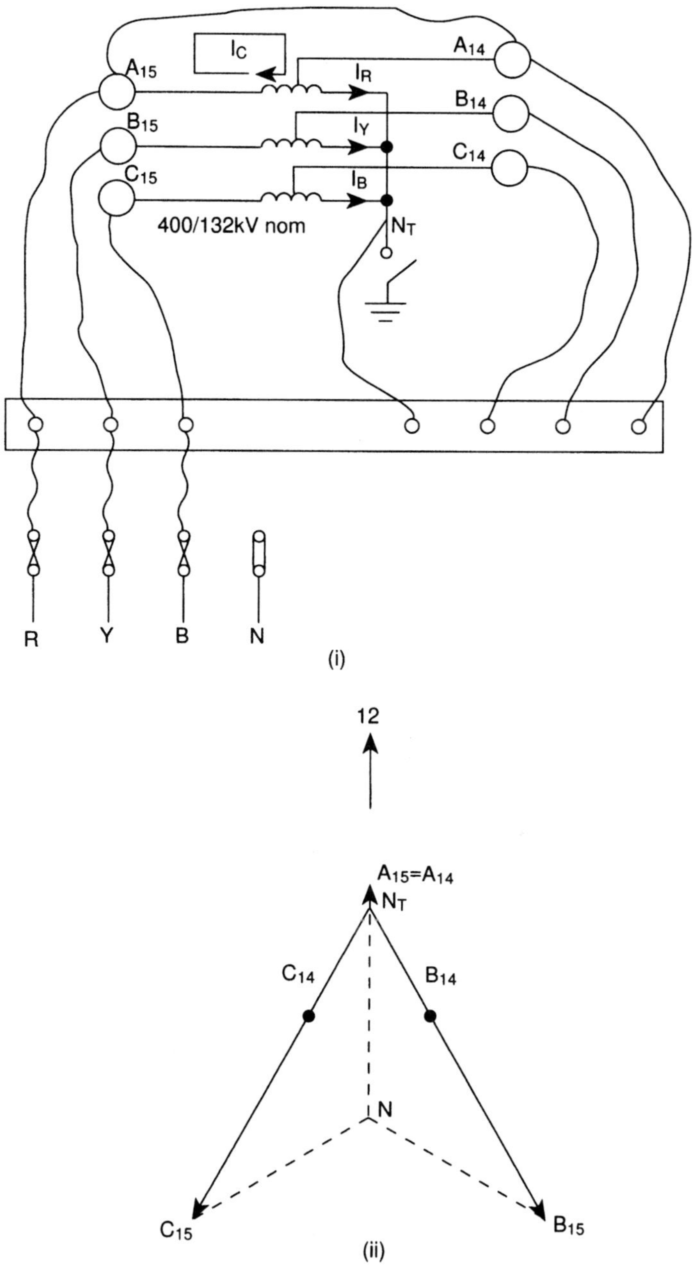

Figure 5.18 *Vector-group test for an autotransformer by connecting HV and LV terminals*

(i) Test connections
(ii) Phasor diagram

Table 5.4 Vector-group test-voltage result for an autotransformer as shown in Figure 5.18

	A15	B15	C15	A14	B14	C14
A15	X	X	X	X	X	X
B15	420	X	X	X	X	X
C15	418	420	X	X	X	X
A14	0	420	418	X	X	X
B14	163	260	362	162	X	X
C14	160	361	261	163	150	X
NT	28	394	394	28	135	135

X = not required

magnetising current should be recorded. With three-phase, three-limbed transformers, the currents in the phases connected to the two outer limbs will usually be higher than that in the phase connected to the central limb. This arises because the magnetic reluctance, and hence reactance, presented to each phase differs between the outer and central limbs. This information is useful data for any future fault investigations. Figure 5.20 illustrates typical test connections and instrumentation for a two-winding transformer.

5.29 Phase diagram

Building on the points made in Section 5.24, it is imperative that there is no ambiguity about the per-phase connections to each side of the transformer. A phasing diagram, as shown in Figure 5.21(i), should thus be produced. This shows the physical connections, as seen from above, between the HV and LV busbars. In some instances, however, the connections may be via cables into metal-clad switchgear. In these circumstances it is necessary to show the switchgear cable-box connections as shown in Figure 5.21(ii).

It is also necessary to show the phase-conductor arrangement within the busbar chamber. Alternatively, in those instances where a transformer may be connected directly to an overhead line, such as with a transformer feeder, the overhead line down-dropper connections should be shown as illustrated in Figure 5.21(iii).

5.30 Buchholz

Figures 5.10 and 5.9 illustrate a typical Buchholz relay and its position on the transformer. When the Buchholz is installed, it is important that the slope of

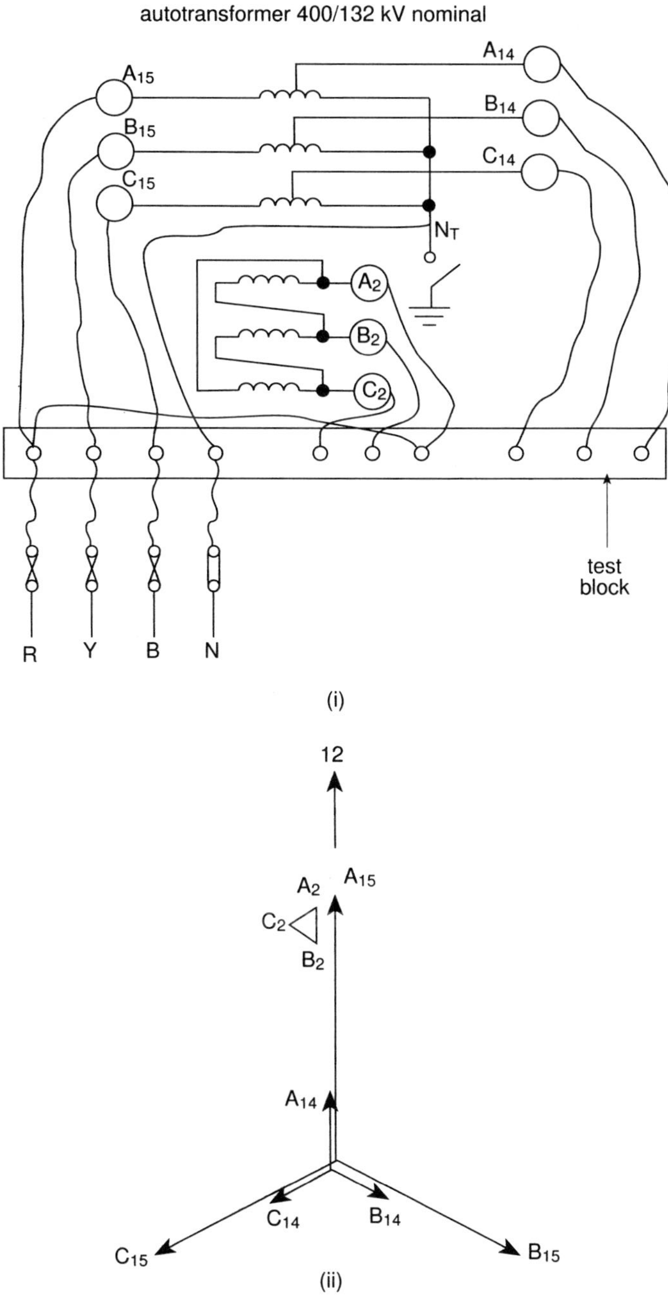

Figure 5.19 Alternative vector-group test for an autotransformer (including delta tertiary)

(i) Test connections
(ii) Graphical representation

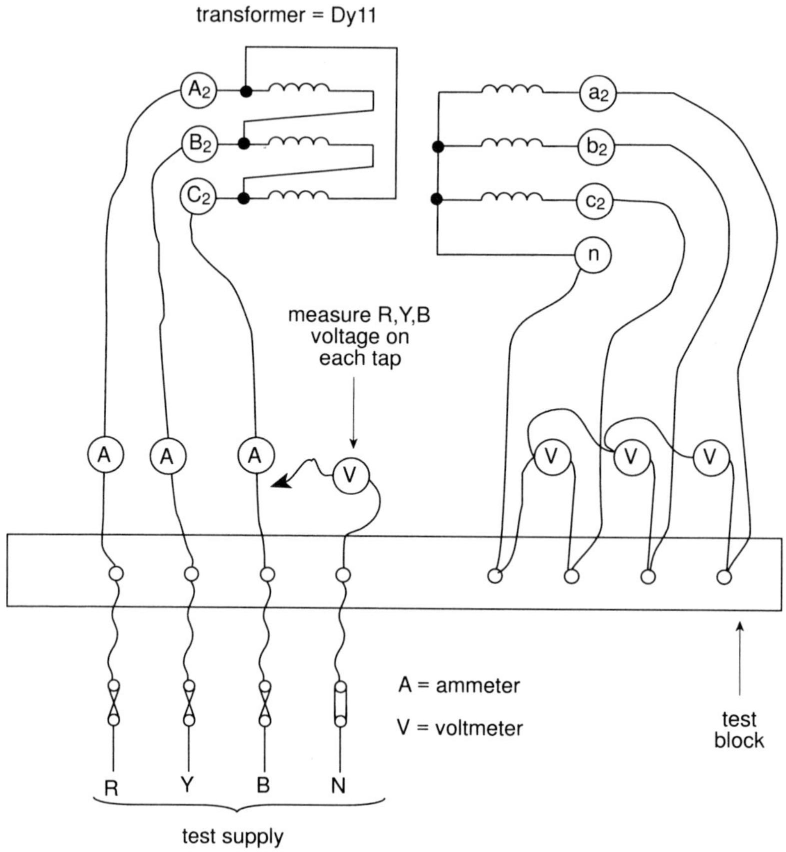

Figure 5.20 Ratio and magnetising-current test

the pipework to the conservator is between 3° and 10° to the horizontal (as will be described in the manufacturer's instructions). This requirement is to ensure optimum collection of gas in the Buchholz. It is additionally important to ensure that the Buchholz relay is fitted the right way about. To assist with fitting, the relay usually has an arrow pointing in the direction of the conservator.

Testing of the relay is usually carried out with the assistance of a pressure vessel, which is filled with air from a foot pump. The pressure vessel is connected via flexible pipework to the test petcock. The output valve of the pressure vessel is opened quickly to allow a surge of air into the Buchholz chamber. Operation of the Buchholz trip is observed by monitoring the trip contacts with an ohmmeter. If the relay does not operate, the test petcock is closed and the drain petcock is opened to remove the air from the Buchholz chamber. The pressure vessel is then refilled with air to a pressure of 18–27 kPa, say, above the previous test pressure, and the process is repeated

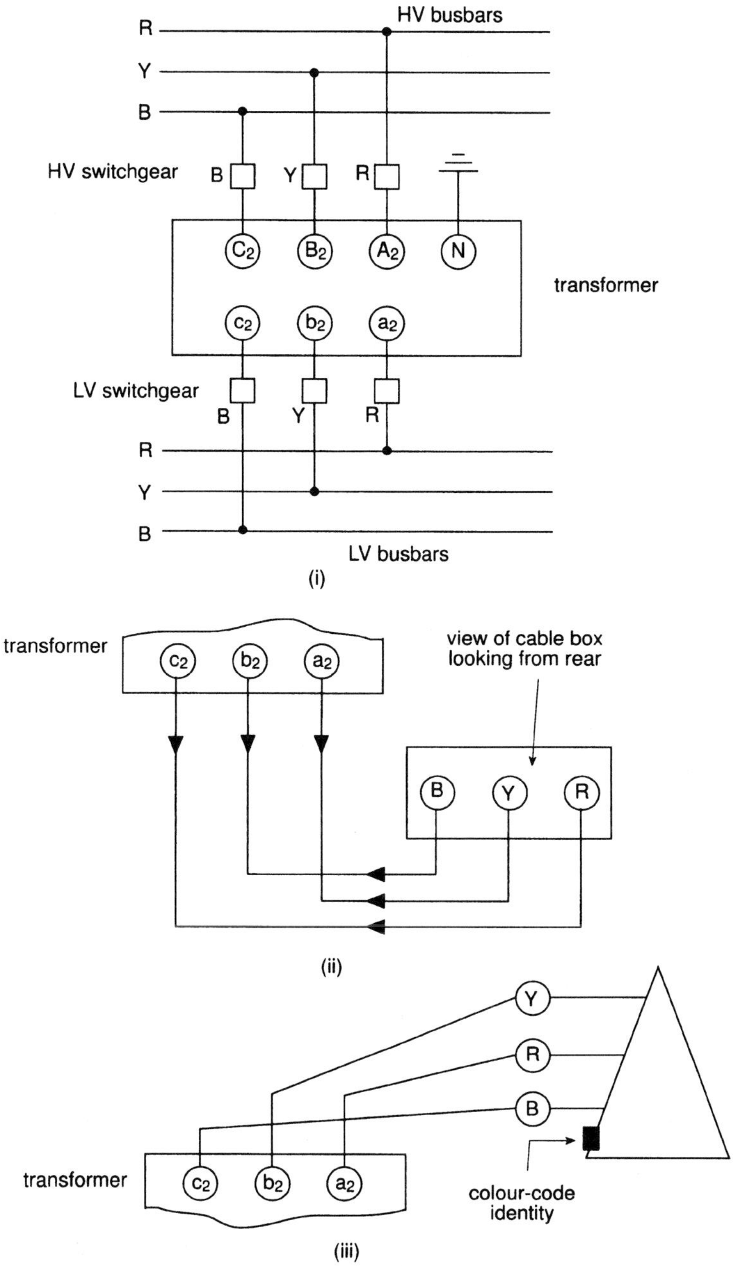

Figure 5.21 Phasing diagram

(i) Outdoor-type substation
(ii) Connections to metalclad switchgear
(iii) Connections to overhead line

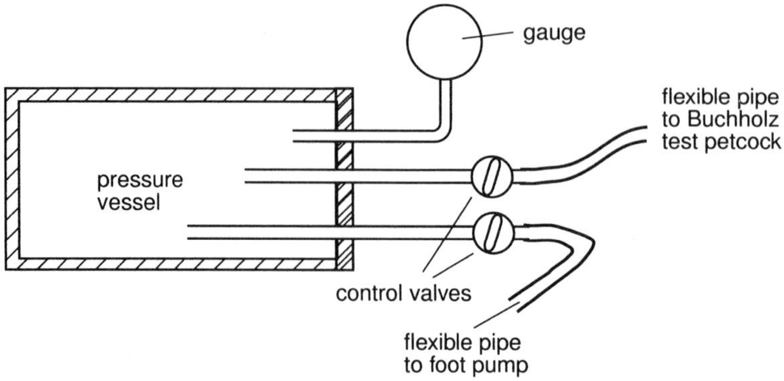

Figure 5.22 Typical Buchholz test vessel

until eventually the Buchholz trip operates. This pressure should be recorded.

The Buchholz gas alarm is evaluated in a similar manner to the trip, except that in this instance the alarm contacts are monitored and the gas is let in slowly from the pressure vessel, until the alarm relay operates. The level of gas, as shown in the inspection window, is then recorded. The air within the relay is then slowly drained via the drain petcock until the alarm relay resets (as indicated on the ohmmeter). The level of gas as shown in the inspection window is then, again, recorded. The recorded values are used as a basis of comparison for future maintenance tests. Figure 5.22 shows a typical pressure vessel arrangement.

5.31 Winding-temperature indicator

A typical winding-temperature indicator (WTI) is illustrated in Figure 5.12. To check the calibration of the device, the bulb should be withdrawn from its pocket and immersed in a small oil reservoir (usually purpose built for this test). The reservoir should then be heated and its temperature taken at, say, 10°C intervals with a thermometer.

A comparison should then be made between the winding-temperature indicator and the thermometer temperature. The discrepancy should, ideally, not exceed 1°C. During both the heating and cooling processes, the temperature at which the mercury switches operate (or reset) should be noted and compared with the setting.

A check of the WTI heater circuit should also be carried out. This would normally be carried out with the bulb placed in its pocket. Current equal to the full-load current of the transformer should then be injected into the heater. The temperature should be recorded at the beginning of the injection and again some 25 min or so later when the temperature of the bulb has stabilised. The temperature increment between start and end of the test should then be compared with the manufacturer's heat-run tests.

5.32 Fans and pumps

An essential feature to check on the fans and pumps is the direction of rotation. Most fans and pumps also have an overload device which disconnects the power supply and it is therefore necessary to check its correct functioning. The majority of fans and pumps are driven from a three-phase supply and simple checks of the overload can be carried out as follows:

(i) remove one fuse with the fan or pump running;
(ii) start the fan or pump with one fuse removed.

The time to operation of the overload should be recorded and where possible checked against the manufacturer's data. A check of the sequential starting of the fans and pumps should also be recorded.

5.33 Oil tests

BS 148 specifies the required characteristics for new insulating oils for transformers and switchgear. This encompasses limiting values for properties such as sludge, acid, flash point, viscosity, electric strength, water content, loss angle and resistivity. New oil for transformers should already be certified as acceptable by the manufacturer, with reference to these characteristics, before the transformer is filled. Transmission and larger distribution transformers, once filled with oil on site, should have samples of oil taken from the bottom of the transformer tank, and from every selector and diverter compartment, for detailed analysis by a laboratory. In addition, on-site oil tests should also be carried out.

A sample of oil should ideally be obtained by filling a glass (not plastics) sampling bottle in accordance with Figure 5.23. This method ensures that the sample is free from contamination.

The most usual on-site test is a check on the electrical strength of the oil. This should be carried out in accordance with the requirements of *BS 5874* (now superseded by *BS EM 60156*). The test requires the use of a standard test cell with electrodes consisting of spheres 2.5 mm apart. The sample is required to withstand 30 kV AC for 1 min. The test voltage should then be increased to find the breakdown voltage of the sample.

A resistivity test should also be carried out. This again requires the use of standard test cells. Typical cells are described in *BS 5737*.

When a laboratory analysis is not carried out, which is often the case for distribution-voltage transformers, it is recommended that an acidity test is carried out on-site. In this test a measured sample of oil is placed in a

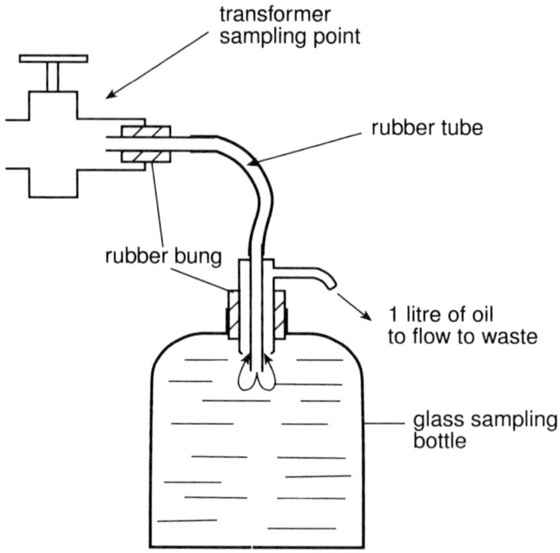

Figure 5.23 Obtaining an oil sample

graduated tube, and a few drops of the indicator alkali blue are added. KOH (potassium hydroxide) solution is then added from a burette until the acid content of the oil is neutralised by the alkaline KOH. This is indicated by a colour change to a pinkish colour.

The oil and KOH solution in this test are required to be measured in terms of weight. New oil should have a maximum acidity level of 0.03 mg of KOH per gramme of oil. However, for transformers in service, it is acceptable for the level to rise to 0.5 mg KOH/g. Oil with measured levels above 1.0 mg KOH/g should be discarded.

To complete the complement of on-site tests, mention must be made of the crackle test, which is a test for moisture. This test is more likely to be carried out during fault investigations in the absence of the more sophisticated test methods mentioned above. It consists simply of boiling a partially full test tube of oil in a flame. If any crackling is heard then moisture is present. Note that this test shows up only undissolved water in the oil down to a level of about 100 parts in 10^6, whereas *BS 148* specifies a maximum of 50 parts in 10^6. In general, if this test is positive the oil is in an extremely poor state.

5.34 Neutral earthing resistance

In some countries, including the UK, the 11 kV and 33 kV or equivalent systems (and some generators) are usually resistance-earthed systems. This is usually accomplished through a neutral-earthing resistance (NER). Most NERs consist of an electrode immersed in a large metallic tank of distilled, or deionised, water whose resistance is reduced by the addition of sodium carbonate (washing soda).

The NER is usually calibrated by filling the tank to the required level with the water, to which is added a small amount of sodium carbonate solution from a bowl, using a cup (or some other small container). After each addition of sodium carbonate, a current of about 5 A should be injected into the NER and the voltage, and hence the resistance, measured. This process should continue until the required resistance is obtained. Some correction is usually required for temperature. The use of a cup, as mentioned above, is required to ensure that no accidents take place whereby too much sodium carbonate is added. If this occurs and the resistance becomes too low, the tank will require draining and the whole process repeating. Figure 5.24 shows the earthing arrangement for the 11 kV and 33 kV systems. The latter, being delta connected, requires the use of an interconnected-star earthing transformer to create an artificial neutral. Figure 5.24 also shows the method of calibration.

MAINTENANCE

5.35 Maintenance

The degree of maintenance carried out on a transformer will depend on its size. Obviously the greater the capital cost of the transformer, and the greater its significance to the power system, the greater the case for maintenance.

Maintenance thus ranges from virtually zero on pole-mounted distribution transformers (usually only a periodic external visual inspection) to a repeat of many of the original commissioning tests for a transmission, generator or major distribution transformer. This section will concentrate on the maintenance of the larger transformers.

Experience over the years has shown that the following periodicity is probably an optimum for the UK system:

Annually
- oil tests
- maintenance of diverter switches which are called upon
- to operate regularly, e.g. on generator transformers;

2-yearly
- full maintenance excluding selector switches;

6-yearly
- maintenance of selector switches.

The annual on-site oil tests should be a repeat of those carried out during commissioning, as described in Section 5.32. Tests carried out on oil in a laboratory have, over the last two decades, become very sophisticated and if the dissolved gases in the oil are analysed, trends in deterioration, incipient problems and causes of fault can be diagnosed. The major gases investigated are hydrogen (H_2), methane (CH_4), ethylene (C_2H_4), acetylene (C_2H_2), ethane (C_2H_6), carbon dioxide (CO_2) and carbon monoxide (CO). Quantities of these gases can be liberated and absorbed into the oil without

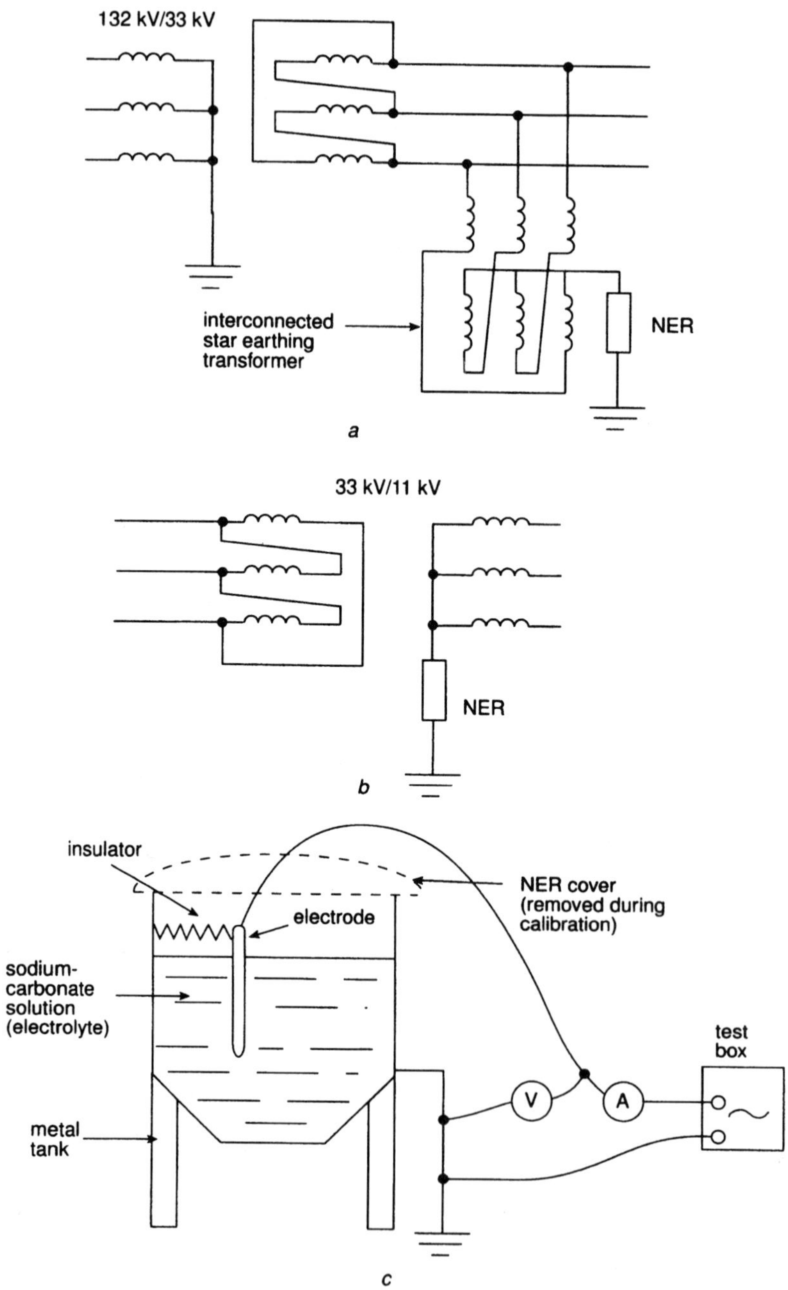

Figure 5.24 Neutral-earthing resistor

a 33 kV system NER
b 11 kV system NER
c Calibration of NER

Table 5.5 Problems which can be determined from dissolved gas in oil analysis (OGA)

Normal deterioration
Partial discharge
Slight overheating
Significant overheating
Winding circulating currents
Core and tank circulating currents
Flashover without power follow through
Flashover with power follow through
Continuous sparking
Partial discharge with tracking

operation of the Buchholz. Analysis of the gas in oil can thus lead to the identification of developing faults, thereby enabling prompt remedial action to be taken. Table 5.5 identifies the range of problems that can be determined from gas in oil analysis.

Note that a good appreciation of the general state of oil can be obtained by visual inspection. Healthy oil should be clear and light brown/yellow in colour. Oil which is cloudy, a very dark colour, exhibits sludge or has a grassy or acid smell is probably in an unsatisfactory condition.

The maintenance of diverter switches normally consists of periodically changing the oil in the diverter tanks and either cleaning up (if recommended by the manufacturer) or replacing the diverter-switch contacts.

A two-yearly maintenance programme would normally consist of the following elements:

(*a*) an external inspection of any bushings, and where required the topping up of oil levels;
(*b*) an external examination of the conservator for oil leaks. A sample of oil should be taken for testing and the oil level topped up if necessary. The amount of added oil should be recorded;
(*c*) a visual inspection of the pressure relief device;
(*d*) a check on the correct gap size of any coordinating gaps;
(*e*) a visual inspection of all cubicles, terminations and valves;
(*f*) an insulation-resistance test of any tap change, pump and fan motors;
(*g*) a check on the integrity of the earthing;
(*h*) lubrication as per manufacturers' recommendations;
(*i*) a test of the Buchholz as described in Section 5.30;
(*j*) a test of the winding-temperature device as described in Section 5.31;
(*k*) a check that the fans and pumps are running smoothly and start sequentially;
(*l*) a check on the sensitivity of any AVC equipment;
(*m*) calibration of any neutral-earthing resistor.

Given that selector switches, unlike diverter switches, do not make or break load current, and therefore do not suffer the same degree of wear, they do not need to be examined as frequently. A point worth noting is that the valve connecting the selector tank to the main tank must be closed when the selector tank is drained of oil. Generally speaking a visual examination of the selector switches is all that is required; these will usually prove to be in satisfactory condition.

FAULTS

5.36 Investigation of faulty transformer

Faults on small distribution transformers will normally become apparent as a result of a loss of supply. Faults on larger transformers will usually be notified as a result of a transformer alarm being generated. In general, the fault investigation procedure is similar, no matter what the size of the transformer, although there are obviously more features to check on large transformers. The following summary relates to recommended actions to be taken on large transformers:

(*a*) Before investigation commences, ensure that the transformer is isolated from sources of supply.
(*b*) Check any associated fault recorders for evidence of an electrical fault.
(*c*) Check and record the operation of protective devices, alarm systems and winding-temperature actual and maximum settings.
(*d*) Make an external examination, paying particular attention to oil expulsion, oil-level gauges, bulging of the tank, the state of the explosion vent or oil-pressure-relief device, any smell from the oil indicating gas or burning, and the position of any on-load tap changer.
(*e*) Oil samples should be taken from the transformer main tank and tap-changer diverter and selector tanks for analysis. In the 1960s and 1970s, it was fashionable to carry out a test of any gas collected in the Buchholz by connecting a series of gas analysis tubes (draeger tubes) to the sampling pipe (see Figure 5.24) and releasing the gas into the sampling tubes via the drain petcock. The test proved time consuming, cumbersome and somewhat unsuccessful due to its complexity, and has been largely abandoned in favour of laboratory analysis of oil. Given that it will take some time to obtain laboratory results, on-site oil tests as described in Section 5.32 should also be carried out.
(*f*) Check the insulation resistance between: each winding and earth; between windings (for a two-winding transformer); and between core and earth.
(*g*) Check the winding resistances and compare with the works-test (or commissioning) results (see Section 5.26).
(*h*) Carry out a magnetising-current, ratio and tap-changer continuity test and compare with the commissioning results (see Section 5.28).

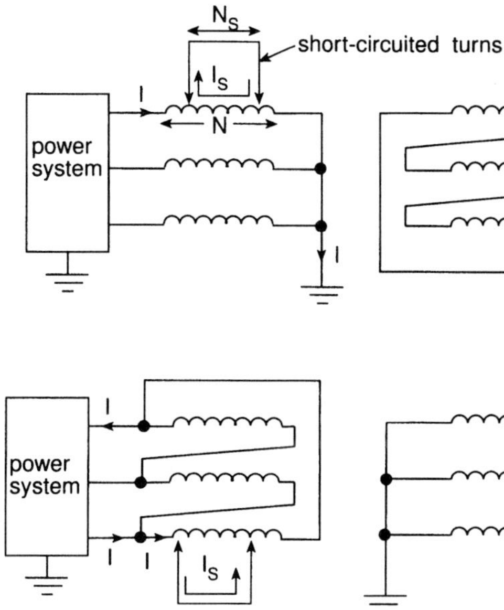

Figure 5.25 Effect of an interturn fault

N_s = short-circuited turns
N = turns on full winding
$I_sN_s = IN$ (autotransformer action)
or $I_s = NI/N_s$
If $N/N_s = 50$ (say)
$I_s = 50I$

The foregoing tests should enable a decision to be made as to whether there is a genuine transformer problem. The transformer should not be returned to service unless all the results are satisfactory. Should the test reveal a fault within the tank, the oil will require removal to enable an internal investigation to be carried out.

5.37 Interturn fault

The tripping of a transformer from the Buchholz without any accompanying operation of electrical protection, or without any indication of fault current on a fault recorder, may well be indicative of an interturn fault. Figure 5.25 shows a typical interturn fault situation. It can be seen that the faulted turns result in an autotransformer action, such that a large current flowing in the short-circuit is balanced by a small current flowing in the rest of the winding. The large current can result in local burning, overheating of the oil and liberation of gas to operate the Buchholz relay.

5.38 Removal of earthing

Incidents have taken place where transformer protection has operated, coincident with a fault elsewhere on the system, as a result of the copper connection between transformer neutral and earth bar being removed by copper thieves. In such an instance the transformer neutral rises in voltage, as a result of unbalanced conditions arising from a through fault, to a level which results in breakdown of insulation.

5.39 Transformer replacement

The exercise of replacing a faulted transformer essentially consists of ensuring a like-for-like replacement. The important points to consider are:

(*a*) the voltage ratio;
(*b*) the rating of the transformer (MVA) and the adequacy of any existing cooling equipment. Heat dissipation and pump flow rates may need to be checked;
(*c*) physical considerations relating to: reconnecting of busbars, cables and connections; clearance distances to other items of plant; physical size and load-bearing capability of the plinth; connecting of the cooling equipment; access availability; and the position of current transformers etc.;
(*d*) the percentage impedance of the transformer—for determining the fault level and the sharing of load with any parallel-connected transformers;
(*e*) the number and position of current transformers within bushings with reference to polarity, ratio, magnetising-impedance characteristics, overlap of zones of protection, matching to existing protection relays and systems, and continuous and short-time ratings. In many instances CTs have to be recovered from the faulted transformer and re-used;
(*f*) the suitability of the tap changer in terms of range, voltage steps, matching to the existing tap-change control scheme and external tap-position indication;
(*g*) the suitability of any existing winding-temperature instrument for recording the winding-temperature hot spot, particularly with respect to the winding-temperature heater's associated current-transformer ratio.

Chapter 6

Tap change and automatic voltage control

6.1 Introduction

Transformers are utilised to interconnect the different voltage levels which make up the power system. Within this context transformer tap changers alter the winding (and hence voltage) ratio to maintain system nominal voltage within prescribed limits; they are therefore an essential feature in the control of the power system. This chapter examines the mechanical process of tap changing, the effect of tap changing on the power system, and automatic-voltage-control systems (AVC), alternatively known as automatic tap-change-control systems (ATCC). Commissioning and maintenance requirements are discussed later in the chapter.

TAP CHANGERS

6.2 Tap changers

Most power-system transformers are fitted with tap changers for the purpose of adjusting the secondary voltage so that it may be maintained within prescribed limits. An exception is 400/275 kV transmission transformers with ratings generally between 500 and 1000 MVA: in the UK these transformers are often referred to as 'tapless wonders'.

With two-winding transformers it is usual to find the tap changer located on the high-voltage side of the transformer; this is usually for both economic and technical reasons. 11 kV/415 V distribution transformers are usually equipped with off-load tap changers. To change the taps, the transformer must therefore be removed from service and the tap connection repositioned manually. At higher voltages, on-load tap changing is carried out.

Figure 6.1 illustrates a typical sequence of events for a high-speed on-load resistor tap changer, moving from tap 1 to tap 2. The three essential components of the tap changer are the selector switches S_1 and S_2 which connect to the tapping point, and are non-load-breaking or making; the diverter switches D_1, D_2, D_3 and D_4 which make and break the intertap circulating current; and the resistors R_1 and R_2 which limit the circulating current. The salient steps in Figure 6.1 to move from tap position 1 to tap position 2 are as follows:

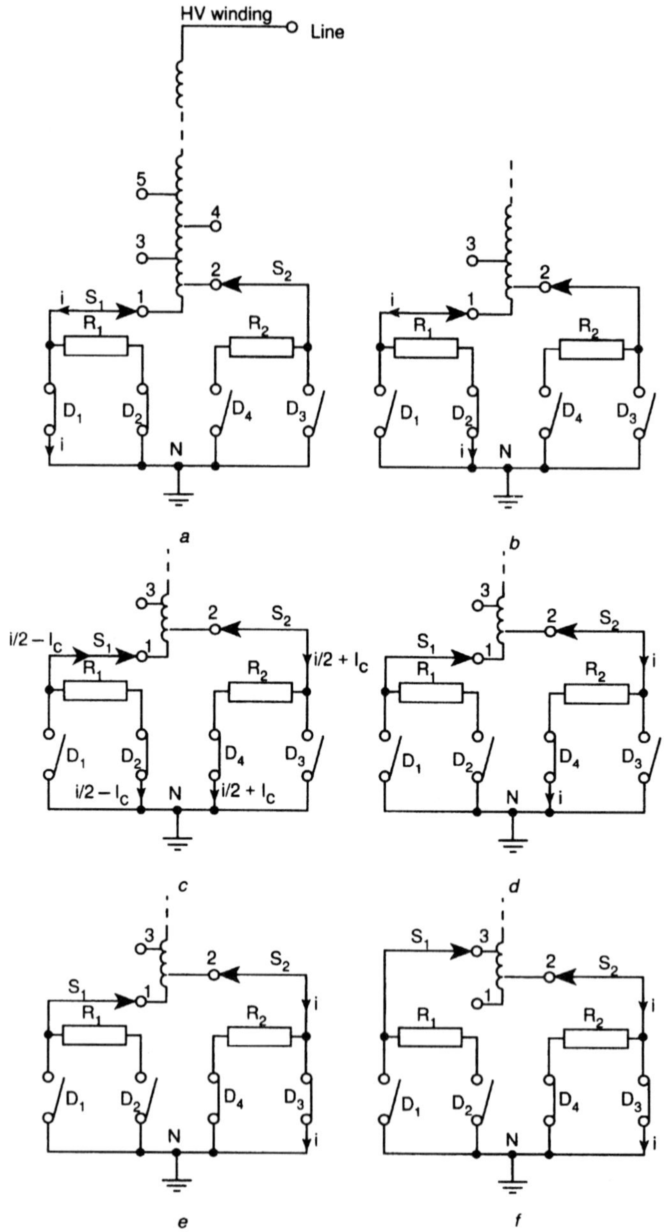

Figure 6.1 Tap-change sequence of events

a Selected to tap 1
b Step 1: D_1 opens
c Step 2: D_4 closes
d Step 3: D_2 opens
e Step 4: D_3 closes
f Step 5: S_1 repositions

Step 1: D_1 opens, resulting in load-current flow from S_1 (tap position 1) through R_1 to N.

Step 2: D_4 closes, resulting in load current dividing through resistors R_1 and R_2, and an intertap circulating current flowing around, and limited by, R_1 and R_2.

Step 3: D_2 opens, thus transferring all load onto tap 2.

Step 4: D_3 closes, short-circuiting R_2. The transition from tap 1 to tap 2 is now complete.

At the start of the next tap change, S_1 moves to tap position 3 (Step 5) and the cycle recommences with D_3 opening. The above sequence of events is repeated for each tap change up the winding; note, however, that for the first operation in the opposite direction the selectors do not move, since they are already in the correct position. This is mechanically accomplished by use of the 'lost-motion' part of the mechanism which is a feature of most tap changers.

6.3 Tap-change control

Figure 6.2 illustrates some features of a typical tap-change control scheme. The sequence of events for, say, tap-change raise would be as follows:

(*a*) Raise push button operated to energise raise contactor.

(*b*) Raise contactor energises both the tap-change motor, to commence the tap change, and the brake coil, which releases the brake.

(*c*) Early in the cycle the tap-change mechanism closes both the maintain contact and the sequence contact (to energise the step relay).

(*d*) Towards the end of the tap-change cycle, the maintain contact opens, and the tap-change motor, and the brake coil to apply the brake, are de-energised. The sequence contact also opens towards the end of the tap change to de-energise the step relay.

Two typical alarms are generated:

(i) 'Tap change in progress' alarm which is initiated by the step relay;

(ii) 'Tap change faulty alarm' which is initiated by a timer which in turn is energised by the step relay. This alarm caters for the instance of the tap changer sticking part way through a tap change, or the raise/lower push buttons sticking in the operated position.

Note that most tap-change-control schemes are driven from an AC supply.

6.4 Tap-change conventions

The terms 'tap up' and 'tap down' are often thought of as referring to an increase and decrease in secondary voltage, respectively. However, a closer

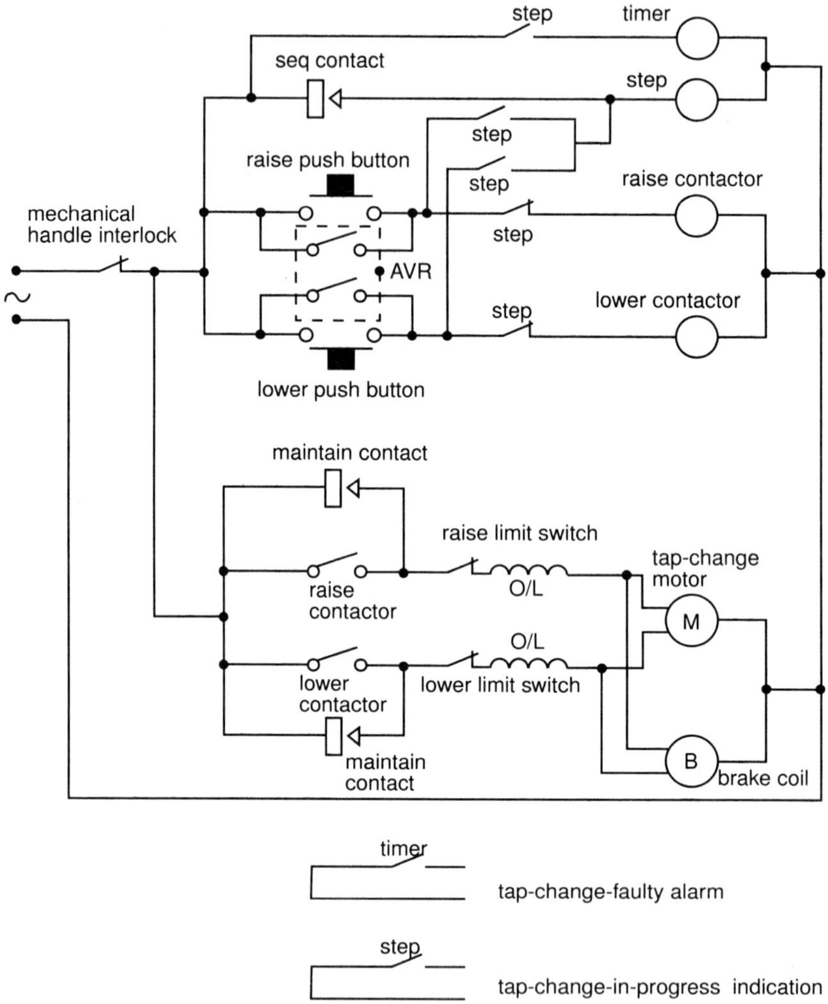

Figure 6.2 Tap-change control circuitry

examination of the tapping arrangements reveals that care must be exercised with this terminology.

Figure 6.3*a* shows the typical tapping arrangements for a 132/33 kV two-winding transformer, although the situation covers the general case for two-winding transformers. The general and classical relationship between HV and LV voltage is

$$V_{LV} = \frac{N_{LV}}{N_{HV}} V_{HV}$$

From this it can be seen that for a fixed value of V_{HV}, an increase in V_{LV} can

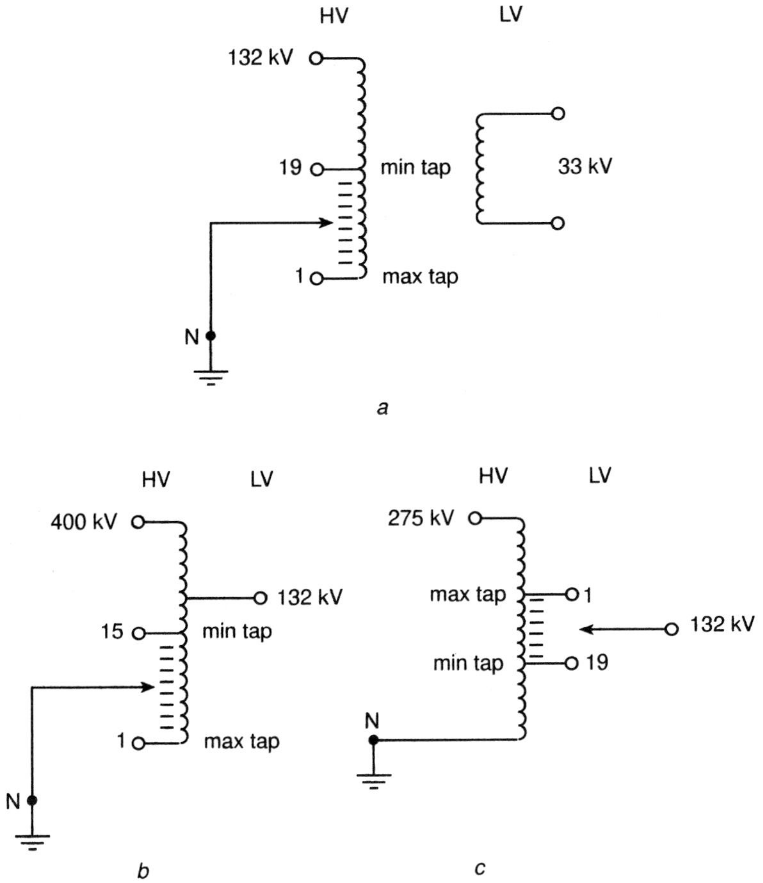

Figure 6.3 Tapping arrangements

a Typical two-winding transformer, 132 kV +10/-20%, nominal tap 7
b 400/132 kV autotransformer, 132 kV +15/-5%, nominal tap 7
c 275/132 kV autotransformer, 132 kV ±15%, nominal tap 10

only be achieved by a decrease in N_{HV} (the turns on the HV winding). Thus the minimum tap position is given the highest tap number and thus 'tapping-up' (in number) increases the LV voltage.

For the autotransformer, illustrated in Figure 6.3*b*, the situation is more or less reversed. In this instance the HV and LV windings are directly connected, such that an increase in tap position reduces both the HV and LV turns. From Figure 3.18, if we refer to the portion of the winding which solely pertains to the HV winding as N_F, then $N_{HV} - N_{LV} = N_F$. Thus we can say that

$$V_{LV} = \frac{N_{LV}}{N_{LV} + N_F} V_{HV}$$

If we divide both the top and bottom of this expression by N_{LV}, we obtain

$$V_{LV} = \frac{1}{1+\frac{N_F}{N_{LV}}} V_{HV}$$

From this it can clearly be seen that any reduction in N_{LV}, for a fixed V_{HV}, would lead to a reduction in V_{LV}. Thus an increase in tap number results in a decrease in V_{LV}, which is the opposite of the two-winding situation. Thus some care must be taken with the meaning of the term 'tapping up'.

Figures 6.1 and Figure 6.3*a* and *b* show the tap-changer tappings as being at the neutral end of the winding. This is advantageous in terms of cost, since the insulation requirement is minimised. However, for the autotransformer, any removal of turns from the common winding affects both the HV and LV turns, as stated above. It can therefore be shown that the tapped winding has to be larger, and hence more expensive, than in a two-winding transformer for the same voltage buck/boost. Further, it can be shown that, at voltage ratios of 2:1 and below, it is more economic to place the tapped winding towards the line end. This is usually the arrangement for 275/132 kV transformers, as shown in Figure 6.3*c*.

6.5 Effect of taps on magnetising current

When a voltage is applied to a transformer, the relationship between the applied voltage V and the flux Φ in the core of the transformer is given by

$$V = KN\Phi$$

where N is the number of turns on the winding and K is a constant.
Now

$$\Phi = \frac{I_m N}{S}$$

where I_m = magnetising current and S = reluctance of the magnetic circuit.

Thus

$$V = \frac{KI_m N^2}{S}$$

or

$$I_m = \frac{SV}{KN^2}$$

For V and S fixed, we can say that:

$$I_m = \frac{C}{N^2} \text{ where } C = \frac{SV}{K}$$

Thus the magnetising current varies with the square of the inverse of the number of turns. This is a useful point to be aware of when carrying out tap-changer tests.

6.6 Transformer circulating current

When transformers are connected in parallel and have different voltage ratios, due to different tap positions, a circulating current is produced which effectively circulates around the transformers. With reference to Figure 6.4*a* and *b*, transformer T_1 is on a higher tap than T_2 and the following is assumed:

V = HV-side voltage of both transformers
V_1 = LV-side voltage of transformer 1
V_2 = LV-side voltage of transformer 2
Z_1 = impedance (leakage reactance) in ohms of transformer 1, referred to the low-voltage side
Z_2 = impedance (leakage reactance) in ohms of transformer 2, referred to the low-voltage side
Z = load impedance in ohms

$$\text{Now } V_1 = I_1 Z_1 + (I_1 + I_2) Z$$
$$\text{and } V_2 = I_2 Z_2 + (I_1 + I_2) Z$$

Solving for I_1 and I_2 we obtain:

$$I_1 = \frac{V_1 Z_2 + (V_1 - V_2) Z}{Z_1 Z_2 + (Z_1 + Z_2) Z} \qquad I_2 = \frac{V_2 Z_1 - (V_1 - V_2) Z}{Z_1 Z_2 + (Z_1 + Z_2) Z}$$

For the usual case of $Z \gg Z_1$ or Z_2,

$$I_1 = \frac{V_1 Z_2 + (V_1 - V_2) Z}{(Z_1 + Z_2) Z} \qquad I_2 = \frac{V_2 Z_1 - (V_1 - V_2) Z}{(Z_1 + Z_2) Z}$$

Now both I_1 and I_2 have a common component which we shall call I_C, where

$I_C = \dfrac{(V_1 - V_2)}{(Z_1 + Z_2)}$ and is termed circulating current.

In addition, the remaining component of I_1

$$\text{i.e. } \frac{V_1 Z_2}{(Z_1 + Z_2) Z}$$

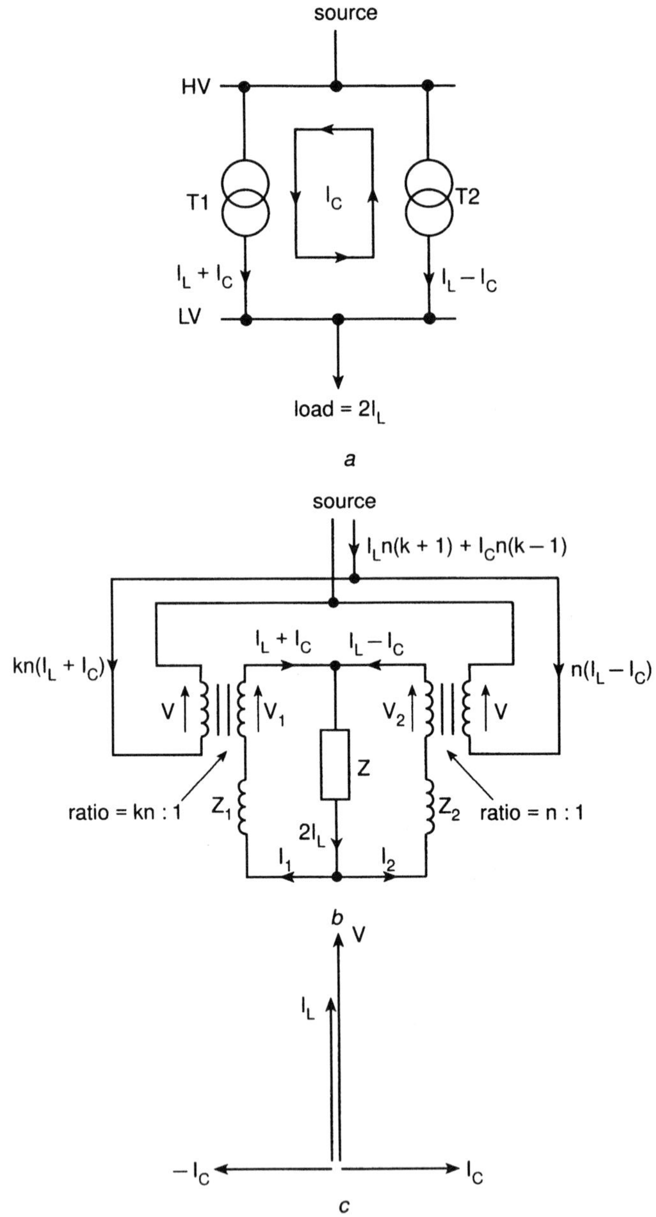

Figure 6.4 Transformer circulating current

a T_1 on higher tap than T_2
b Equivalent circuit
c Phasor diagram

may be termed I_{L1} and the remaining component of I_2

$$\text{i.e. } \frac{V_2 Z_1}{(Z_1 + Z_2)\ Z}$$

may be termed I_{L2}. Thus $I_1 = I_{L1} + I_C$ and $I_2 = I_{L2} - I_C$.

It can further be shown to good approximation that, within the limit of the tap-changer range,

$I_{L1} \approx L_{L2}$, which we shall now term I_L. Thus $I_1 = I_L + I_C$ and $I_2\ I_L - I_C$.

As seen in Figure 6.4*a*, I_C appears as a current which circulates around the two transformers. I_C is determined by Z_1 and Z_2 and thus lags the voltage by approximately 90°. This is in contrast to I_L which, being a load current, usually tends towards 0° (unity power factor). See Figure 6.4*c*. For the case of no load current, the circulating current I_C still continues to flow.

Transformer circulating current is usually considered as flowing around the transformer HV and LV connections as shown in Figure 6.3*a* and this is a reasonable approximation. However, although the circulating current has the same magnitude in the LV winding, it differs in the HV winding by virtue of the different transformation ratios, and, as can be seen in Figure 6.4*c*, a component of circulating current flows into the source. Thus deliberately staggering the taps can be a source of VAR generation. This technique is occasionally utilised as a source of voltage control on the power system; however, it suffers from both the limitation and the disadvantage of requiring a significant tap stagger to generate an appreciable flow of VARs.

6.7 Magnitude of circulating current

Consider, with reference to Figure 6.4, two transformers connected in parallel but operating on different taps. Then from the above, the circulating current is given by

$$I_C = \frac{V_1 - V_2}{Z_1 + Z_2}$$

If

$$Z_1 = Z_2 = (Z), \text{ then } I_C = \frac{V_1 - V_2}{2(Z)}$$

Let

$$V_1 - V_2 = V, \text{ then } I_C = \frac{V}{2(Z)} \qquad (6.1)$$

and let

$$\frac{\sqrt{3}V}{V_{NOM}} \times 100 = (\%V)$$

where V_{NOM}=system nominal line voltage.

Then

$$V = \frac{(\%V)\,V_{NOM}}{100\sqrt{3}} \qquad (6.2)$$

Now from Section 3.4, $(Z) = \dfrac{(\%Z)\;V_{NOM}^2}{100(\text{base VA})}$ where (Z) is in ohms.

If we assume that $(\%Z)$ is quoted at transformer full load I_{NOM}, then

$$(Z) = \frac{\%Z\;V_{NOM}^2}{100\;(\sqrt{3}V_{NOM}\,I_{NOM})} \qquad (6.3)$$

Substituting eqns. 6.2 and 6.3 into eqn. 6.1, we obtain

$$I_c = \frac{(\%V)\;I_{NOM}}{2\;(\%Z)}$$

or I_C as a percentage of full load is given by

$$\%\text{ full load} = \frac{I_C}{I_{NOM}} \times 100 = \frac{50(\%V)}{(\%Z)} \qquad (6.4)$$

Applying eqn. 6.4 to a 400/132 kV 240 MVA transformer, with $\%Z$=20, and a $\%V$ of 1.5% per tap, the circulating current per tap would be $\dfrac{50 \times 1.5\%}{20}$ = 3.75% of full load.

In general, the circulating current pertaining to most transformers connected to European power systems is likely to fall within the range of 1–10% of full load per tap difference, for a two-transformer system.

AUTOMATIC VOLTAGE CONTROL

6.8 Automatic voltage control

The automatic adjustment of the transformer-tap position is required when a close control of voltage is required as the load varies. Automatic voltage control (AVC) is achieved by use of an automatic voltage-regulating relay (AVR). In practice, the terms AVC and AVR are often both used as simple interchangeable abbreviations for both the process and the relay. In the past AVC equipment was generally applied only to transformers connected to LV busbars in the range 11 to 33 kV, but in recent years there has been an increasing application to higher-voltage transformers.

AVC equipment has now developed to a state of sophistication such that new equipment is almost entirely solid state or computer based. However, the high-tech solutions still embody the basic principles of their electromagnetic predecessors. For the sake of simplicity, and for illustrative purposes, the AVC principles described in the following sections are based on electromagnetic-type technology. Indeed, many electromagnetic systems are still in service.

6.9 Simple application of AVC

Figure 6.5 shows a simple application of an AVC which is required to maintain LV-busbar voltage. A voltage transformer supplies the relay with a voltage which is, in effect, the LV-busbar voltage. The AVR will have a nominal voltage setting corresponding to the required LV-busbar voltage. It will also have a deadband setting of, typically, ±2% of nominal voltage. Any excursion of the voltage outside the deadband will initiate a timer with a setting typically in the range 30–60 s. Should the timer-delay elapse, a tap change will be initiated which will either raise or lower the busbar voltage. The purpose of the timer is to prevent unnecessary tap changing as a result of transient voltage variations. The nominal voltage setting, dead band and timer are all part of the AVR.

Figure 6.5(*a*) illustrates a plot of voltage variation, with periods of time outside the dead band which exceed the timer setting, resulting in a tap change. Typical instances of a need to tap to increase busbar voltage are shown at points *a* and *b* in Figure 6.5(*b*), whilst a need to tap to decrease busbar voltage is shown at point *c*.

6.10 The master-follower system

When two (or more) transformers are connected in parallel, as shown in Figure 6.6, some means is required of keeping the two as near as possible on the same tap. If both transformers were allowed to operate independently

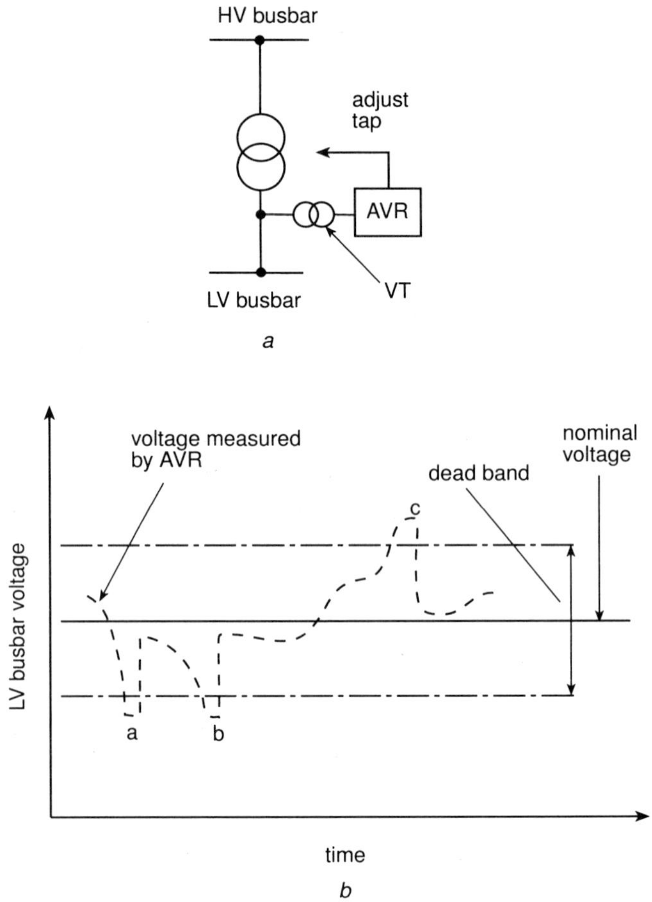

Figure 6.5 Simple application of AVC

a Application of AVC to a single transformer
b Example of voltage control

using the simple AVC of Figure 6.5, the more sensitive of the two AVRs would always tap change, resulting in significant tap differences, which in turn would result in significant, unwanted, circulating current.

The master–follower system has facilities which enable either AVC to be selected as 'master' (i.e. operational) with the other as 'follower' (i.e. nonoperational). If the master AVC concludes that a tap change is required, it instructs its associated transformer to carry out a tap change, after which, it immediately instructs the other transformer to carry out a tap change. Thus the transformers remain in step.

The master–follower system has been applied extensively to distribution systems. Generally speaking, its application is limited to radial systems, i.e.

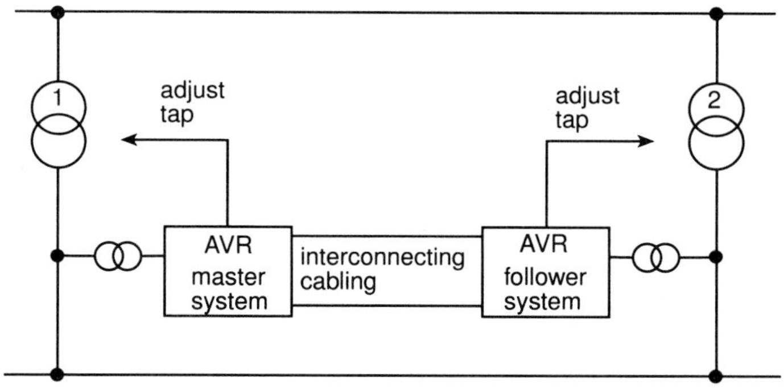

Figure 6.6 Master–follower system

systems which have no connection to other sources of supply on the LV side, but this is not universally so. Obviously both transformers need to be on the same site, as shown in Figure 6.6, since interconnecting cabling between the two AVCs is required. In recent years, master–follower types of systems have also been applied to supergrid transformers as an integral part of a substation control system.

The master–follower system has the advantage of being simple to understand and to operate. Its disadvantage is that the AVC equipment and interconnecting cabling are more complex than might appear at first sight. In particular, it is not flexible in the context of adding/removing transformers, and it is not directly applicable to interconnected power systems.

6.11 Compensation systems

The term 'compensation', alternatively known as 'compounding', is utilised to raise the busbar voltage, as load current increases, to maintain constant voltage at a point remote from the busbars. It is alternatively utilised as a method of tap-change control. Compensation methods therefore influence the voltage applied to the AVR through application of additional voltages which are derived from, and proportional to, the current through the transformer.

Figure 6.8*b* illustrates some of the general features associated with a compensation system. As can be seen, the applied voltage is $V_{R\text{-}B}$ and the compensation current is I_Y. This arrangement is a puzzle to many engineers since it would seem simpler to use V_Y and I_Y. However, some voltage transformers do not have their neutral brought out, and are earthed via the yellow phase; thus Figure 6.8 reflects a consistency of approach. $V_{R\text{-}B}$ and I_Y are 90° displaced and the two components designated X and R are therefore, in fact, the converse, namely a resistance and reactance respectively. This is necessary to obtain the correct phase shifts, as a consequence of the 90°

displacement between $V_{R\text{–}B}$, and I_Y. Another puzzle to many engineers is why $V_{R\text{-}B}$ is used in preference to the more conventional $V_{B\text{–}R}$. The answer is that $V_{B\text{–}R}$ could equally well be used, and the same results obtained, but the use of $V_{R\text{–}B}$ simplifies the analysis, as will shortly be seen.

From Figure 6.8*b*, the main features of a typical scheme which employs compensation are:

(*a*) *Zero adjustment:* The zero adjustment enables the AVR to be set to the required busbar voltages with the transformer on no load and $V_{R\text{–}B}$ at nominal voltage. The zero, which usually has a scale of 0–10, is set to midpoint of 5. The setting resistor is then adjusted until the voltage-control relay (AVR) becomes centred within the deadband (for an electromagnetic scheme this would be via visual observation). Subsequent adjustment of the zero control enables the busbar voltage to be adjusted by up to +10% (zero position 10) or −10% (zero position 0), should some other nominal value of busbar voltage be required.
(*b*) *Reactive compensation:* This is an adjustable potentiometer labelled *X*, scaled 0–10 and giving 0–15% voltage compensation when full-load current flows in the compensation-CT circuitry. Reactive compensation has maximum effect on the AVR at zero power-factor currents.
(*c*) *Active compensation:* This is an adjustable inductance labelled *R*, scaled 0–10 and giving 0–15% voltage compensation when full-load current flows in the compensation-CT circuitry. Active compensation has maximum effect on the AVR at unity-power-factor currents.
(*d*) *Voltage reduction:* Voltage reduction is brought into operation under emergency situations when there are load restrictions. Operation of voltage reduction causes resistors to be short-circuited, resulting in a higher voltage being applied to the AVR, thus causing it to tap change to reduce LV-busbar voltage.

In more modern equipments, many of the above features are hidden in a 'black box'; however, the basic facilities as described above are still provided. Sections 6.12–6.15 will discuss specific applications.

6.12 Line-drop compensation

Referring to Figure 6.7, the purpose of line-drop compensation is to enable a required voltage to be maintained at a remote busbar. This is accomplished by setting the values of the compensation quantities *X* and *R* to be effectively the same magnitude as the feeder quantities X_L and R_L. By passing the feeder current *I* through *X* and *R*, via a current transformer, the voltage imposed on the AVR is arranged to be of the same magnitude as the remote busbar voltage V_{RBB}, as shown in Figure 6.7*c*. The AVR then instructs the tap changer to compensate for voltage variations at the remote busbar.

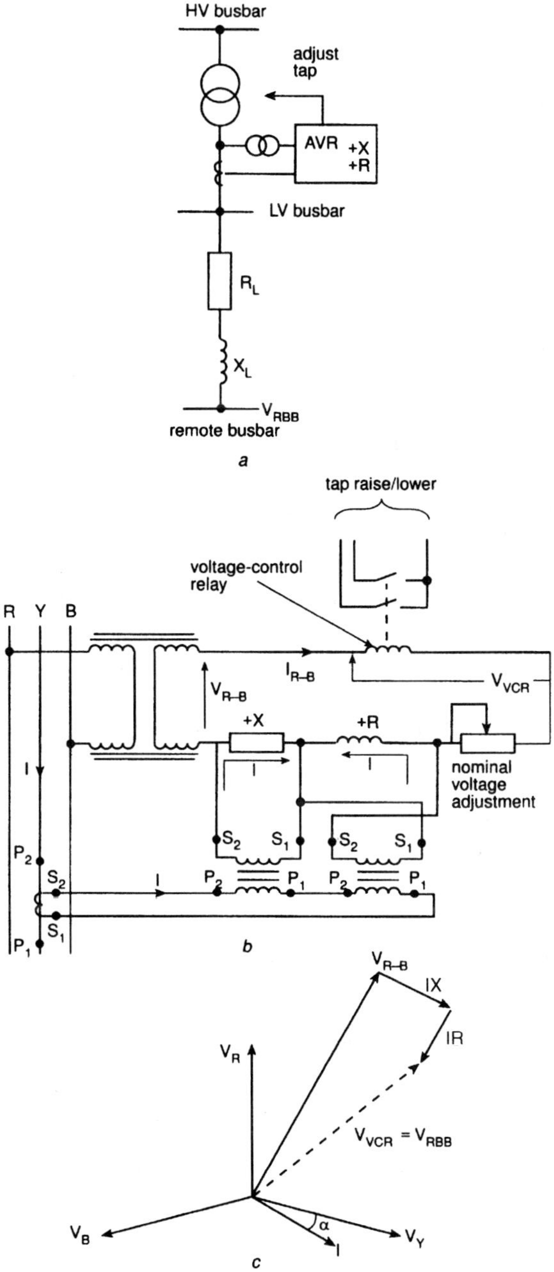

Figure 6.7 Application of line-drop compensation

a System under consideration
b Automatic voltage control circuitry
c Phasor diagram

6.13 Application of reversed reactive compensation to transformers in a radial system

The compensation equipment shown in Figure 6.8*b* can be applied to transformers in a radial system, see Figure 6.8*a*, to achieve tap consistency between the transformers and hence limit circulating current, in addition to maintaining the required busbar voltage. To achieve this, it is necessary to reverse the effect of the reactive compensation X by reversing the flow of current through it. This is known as 'reversed reactive compensation'.

If, in Figure 6.8*a*, transformer 1 is on a higher tap than transformer 2, circulating currents flow as shown. If, in Figure 6.8*b*, we set the R component of compensation to zero, then only the X component will be available to influence the AVR.

Figures 6.8*c* and *d* show the relative phasor position of the voltages and currents for transformers 1 and 2, respectively. In Figure 6.8*b*, it can be seen that the load current I_L flows around the wiring interconnecting the two transformers and therefore does not flow into the compensation equipment. Thus the circulating current $\pm I_C$ flows into the X compensation only, and the effect on the AVR is shown in Figures 6.8*e* and *f*. As can be seen the AVR for transformer 1 is caused to see a larger voltage than $V_{R\text{-}B}$, while the AVR for transformer 2 is caused to see a lower voltage. In consequence, as I_C increases, transformer 1 is caused to tap to reduce voltage and transformer 2 to tap to increase voltage, so causing the two tap changers to keep in step. Note that

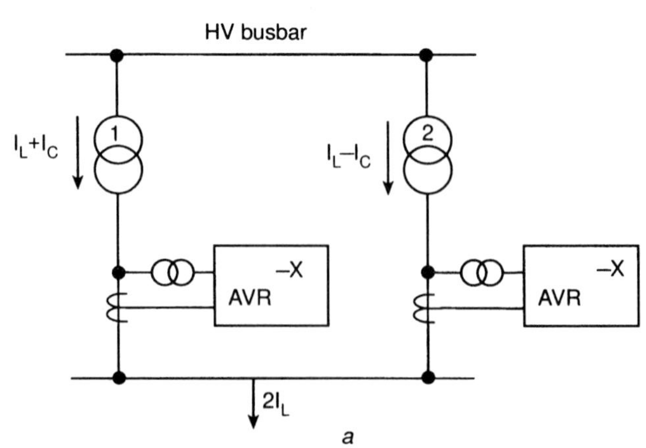

Figure 6.8 Application of compensation equipment to a radial power system

a Transformers in a radial system

Notes
X behaves as $-X$ by virtue of change of relative current compared with Figure 6.7
R is set to zero for a radial system

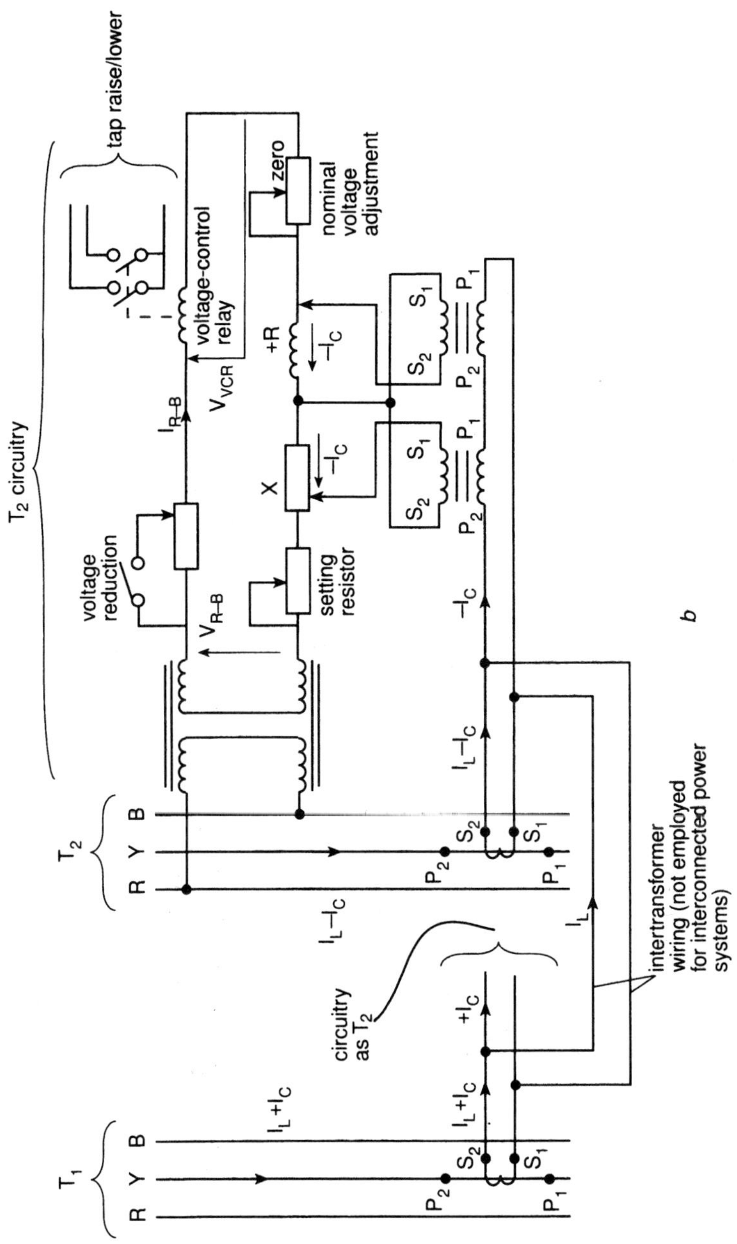

Figure 6.8 continued

b Compensation equipment

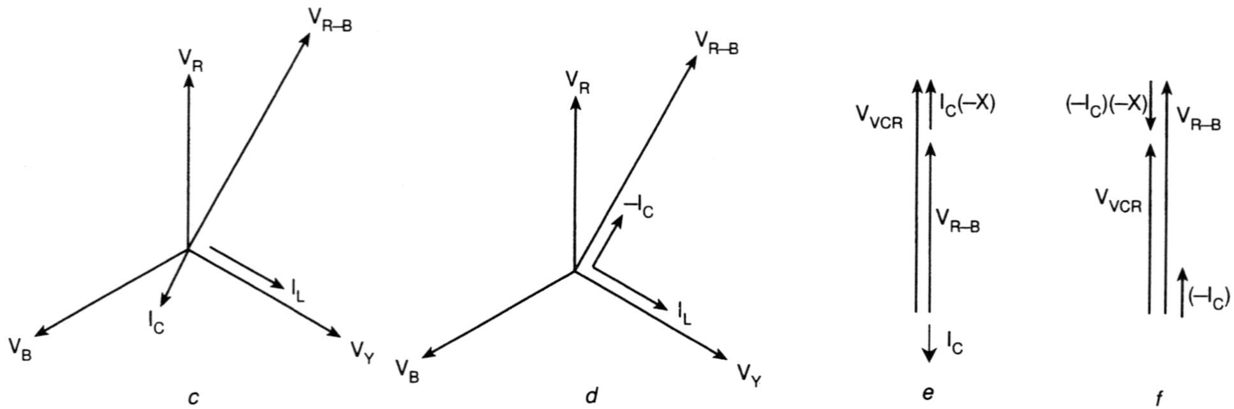

Figure 6.8 *continued*

c Relative phasor position of the voltage and current for transformer 1
d Relative phasor position of the voltage and current for transformer 2
e Effect on AVR of transformer 1
f Effect on AVR of transformer 2

both AVRs remain individually responsive to changes in busbar voltage, and tap change to maintain busbar voltage accordingly.

The question which next arises is the value to which X should be set. If we assume a tap-change difference of 1.5% between taps, a reasonable strategy may be to develop a voltage across the compensation equipment equal to about half of the tap difference, i.e. about 0.8%, when one tap apart. If we assume a circulating current equal to about 7% of full load per tap, which is a typical value for 132/33 kV transformers, the X setting would require to be about 12%, (i.e. $12\% \times 7\% = 0.8\%$), which is position 8 on the 0–10 setting range of X.

6.14 Application of reversed reactive compensation to transformers in an interconnected power system

An interconnected system is one where two or more supply points are geographically separate, their LV busbars being connected via interconnectors. The solution shown in Figure 6.8, whereby intertransformer wiring eliminates load current from the compensation equipment, is therefore not practicable.

Figure 6.9*a* shows an interconnected system with both transformers on the same tap and carrying the same load current I_L. The busbar voltages V are identical and therefore there is no circulating current. For satisfactory tap-change control of the AVR, the load current needs to be eliminated from the compensation equipment so that the voltage across the AVR is a reflection of busbar voltage. This condition can be achieved for any load current, at a particular power factor, by suitably setting the compensation X and R settings.

If, with reference to Figure 6.9*b*, it is assumed that the AVC circuitry for each transformer is as in Figure 6.8*b*, except that the intertransformer wiring is removed, and if it is also assumed that the components of $I_L R$ and $I_L(-X)$ in phase with V are equal and opposite, then:

$$I_L(-X)\sin\alpha = I_L R\cos\alpha$$

where α as used in this expression is approximately equal to the power-factor angle, α, i.e.

$$\frac{R}{(-X)} = \tan\alpha \qquad (6.5)$$

Note that this approximation is only valid because the effect of the compensation equipment is small compared with the magnitude of V.

If in eqn. 6.5, α is assumed to be about 22° (the expected power factor angle), then

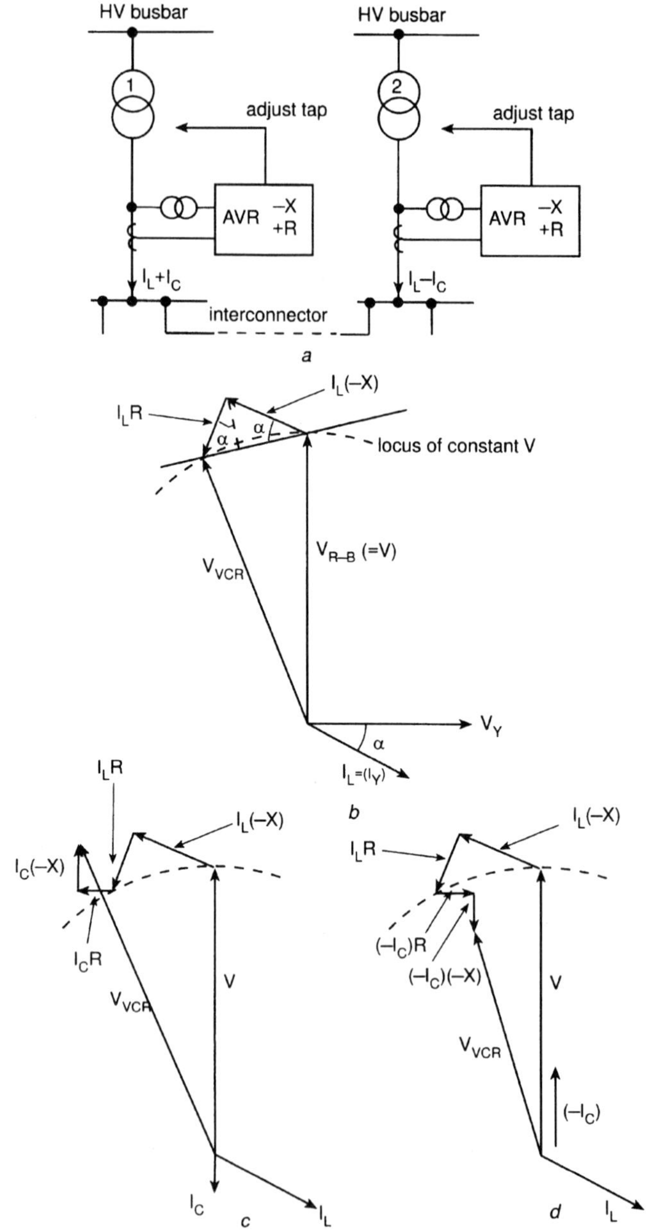

Figure 6.9 Application of compensation equipment to an interconnected power system

a Interconnected power system
b Effect of I_L on transformers 1 and 2
c Effect of I_C on transformer 1
d Effect of $(-I_C)$ on transformer 2

$$\tan \alpha = 0.4, \text{ and thus } \frac{R}{(-X)} = 0.4$$

Since in Section 6.13 X was calculated as being set to position 8 (i.e. equivalent to 12% compensation at full load, or 0.8% when one tap apart), then if $R = 0.4\ X$, R should be set to position 3 (i.e. approximately equivalent to 4.8% compensation at full load, or 0.34% when one tap apart). It is assumed that the quadrature components of $I_L R$ and $I_L(-X)$ have almost no effect on the magnitude of V_{VCR}, which is a reasonable approximation.

Consider now the situation in Figure 6.9*a* where transformer 1 is on a higher tap than transformer 2, so that circulating current flows in the direction from 1 to 2. The resulting phasor diagrams are shown in Figure 6.12*c* and *d*.

It can be seen that the effect of the circulating current on the compensation equipment is to cause transformer 1 AVR to see a higher voltage and transformer 2 AVR a lower voltage. Thus the tendency is for transformer 1 to tap down and transformer 2 to tap up, thereby keeping both transformers in step.

The above analysis has incorporated a number of approximations and thus in practice the interconnected transformers may run some taps apart. However, the effect of the compensation will create a tendency to pull the transformer taps together. Reverse-reactance compounding of this type is found on both distribution and transmission systems.

The question which usually arises at this point is why the AVC circuitry used for the interconnected power system cannot be used for the radial system, given that a standardised approach would result. The answer is that it could be. However, the intertransformer-wiring approach shown in Figure 6.8*b* results in much better control of the tap positions, because it embraces fewer approximations and inaccuracies.

6.15 Mixed systems

Figure 6.10 shows the application of AVC to mixed systems. In Figure 6.10*a* the two remote sites have their tap changers kept in step by use of reversed reactive compensation, and the two transformers on the same site by use of a master–follower system. This approach is more likely to be found on distribution systems.

Figure 6.10*b* shows all four transformers kept in step by use of reversed reactance compounding. Some investigation work may be required to obtain optimum X and R settings.

Figure 6.10*c* shows an AVC system which simultaneously provides negative-reactance compounding for local tap-change control, and line-drop compensation to provide a voltage level to a remote busbar. This approach is again more likely to be found on the lower-voltage systems.

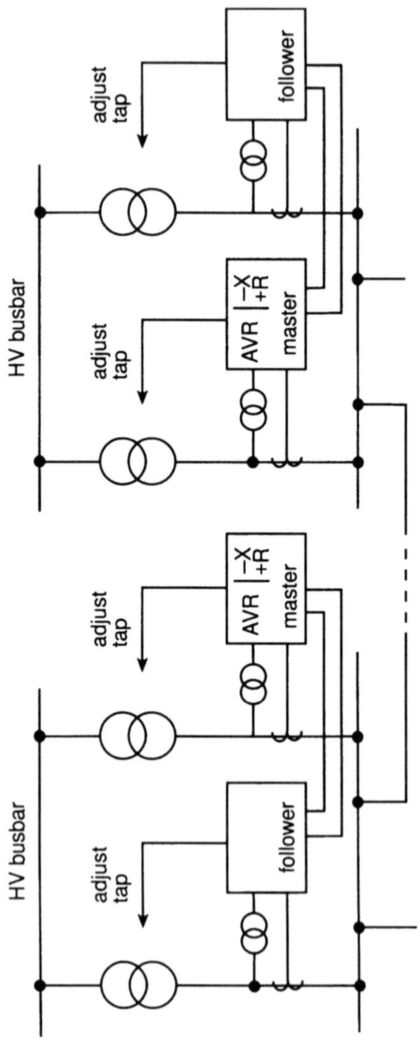

Figure 6.10 Mixed systems

a Tap changers kept in step through a combination of reversed reactance and master-follower systems

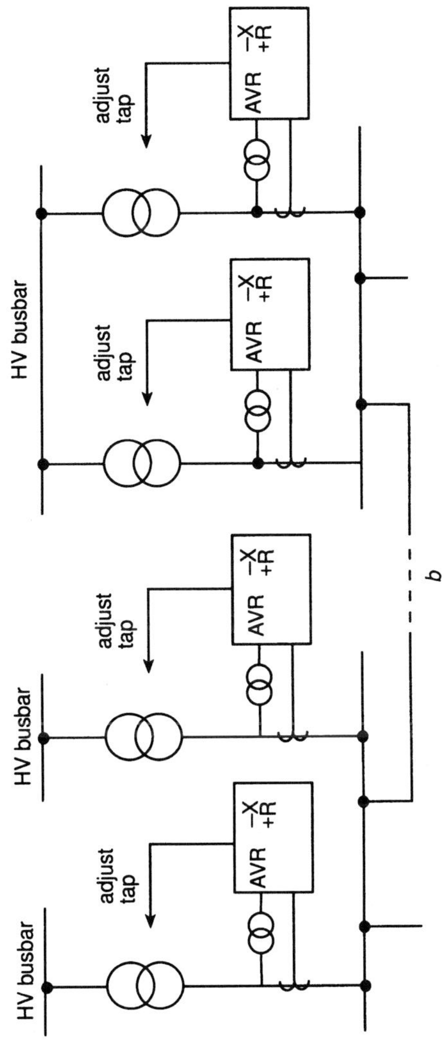

Figure 6.10 continued

b Tap changers kept in step through reversed reactance systems

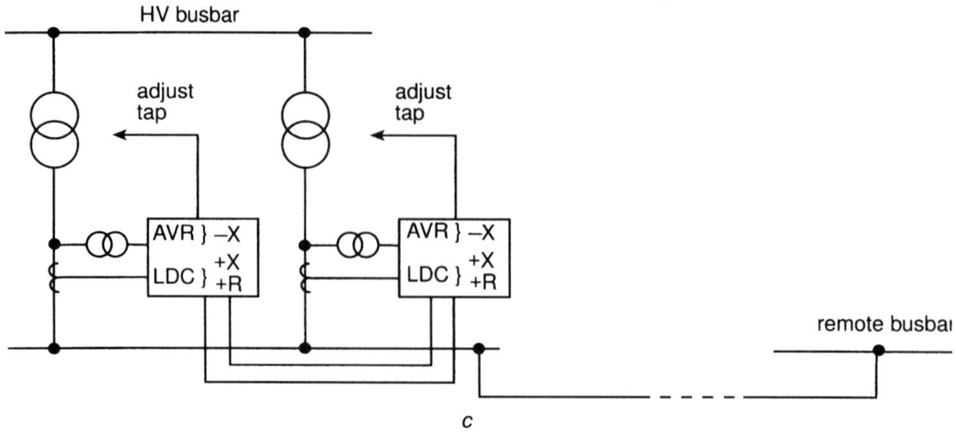

Figure 6.10 continued

c Tap changers kept in step and remote busbar voltage maintained through combination of reversed reactance and line drop compensation systems

COMMISSIONING

6.16 On-load tap changers

Tests should be carried out on on-load changers in accordance with the manufacturer's handbook. The following is a typical range of tests:

(*a*) functioning of the overload device as described in Section 6.5. Where single-phase motors are employed, the overload will need to be proved by injection of current;
(*b*) raising and lowering of taps from both the local and remote positions. A check should be made over the full tapping range, and the tap-position indicators checked at each position. Functioning of the limit switches should be proved by attempting to tap beyond the tapping range;
(*c*) a check of all circuitry and alarms, as typically shown in Figure 6.2;
(*d*) a manual operation of the tap changer, including verification of the mechanism limit stops at the end of the range;
(*e*) recording of the time for a complete tap change.

6.17 AVC

The AVC equipment is proved by injecting voltage and current into the equipment. In particular, the following off-load tests should be carried out:

(*a*) a check of the sensitivity of the relay by varying the injected voltage to obtain the dead band. The nominal voltage setting can usually be taken as the midpoint of the dead band (see Figure 6.5);

(*b*) verification of the polarity of any current compensation circuits. The commissioning engineer should first satisfy himself that the polarities shown in the schematic diagram are correct. Vector diagrams should then be constructed to predict the effect of in-phase and quadrature currents, flowing in the compensation circuitry, on the voltage-control relay (see Figure 6.8). These vector diagrams should be verified by secondary injecting such voltages and currents;

(*c*) a check of the effect of any voltage reduction resistors by injecting nominal voltage into the VT circuit, manually operating the voltage-reduction circuitry and measuring the voltage change experienced by the voltage-control relay (see Figure 6.8);

(*d*) verification of lock-out features and alarms.

Satisfactory completion of the above tests will result in confidence that the AVC has been correctly installed and connected. The ultimate proof however can only be through on-load commissioning tests. These should consist of the following:

(*a*) *Master–follower systems:* selecting permutations of master and follower, manually tapping the transformers up a number of taps (and then down) and observing the AVC system restore nominal voltage;

(*b*) *Reversed-reactance systems:* with reversed reactance schemes, manually tap the two transformers a number of taps apart, and then observe that the AVCs bring the taps either together or one tap apart.

MAINTENANCE

6.18 Maintenance

To large measure, the maintenance tests are a repeat of the commissioning tests.

Chapter 7

High-voltage pressure tests

7.1 Introduction

This chapter examines on-site high-voltage pressure testing. In the UK, the 1937 Electricity Supply Regulations demanded that certain pressure tests should be carried out, largely for reasons of safety, during both commissioning and maintenance. However the 1980 UK Electricity Supply Regulations have removed this requirement, so that pressure testing, if carried out at all, is primarily for reasons of quality assurance.

Generally speaking, industry practice is to carry out commissioning pressure tests on all high-voltage equipment at voltages of 33 kV and below. Above 33 kV the testing is more selective and tends to be carried out only on cables and GIS switchgear. The criteria for pressure testing are largely based on:

(*a*) the need to verify the integrity of equipment which has been subject to significant on-site assembly, particularly equipment whose conductors are also surrounded by an insulating material, such as cables and GIS switchgear;
(*b*) the practicality of providing pressure-testing equipment. Above 33 kV the physical size and power requirements of the test equipment tend to render the test impracticable; at the higher voltages, pressure tests are therefore confined to cables and GIS switchgear and trunking.

When high-voltage testing is carried out it is usual to apply an alternating-current test to all equipment except cables. The relatively high capacitance of a cable, and hence high charging current, results in it being more practical, in terms of test equipment size, to carry out a direct-current pressure test. The AC test has traditionally been applied for 1 min and the DC test for 15 min.

7.2 Preparation

If the item of equipment to be pressure tested is under safety rules, it will be necessary to issue a safety document to provide the necessary authority for carrying out the tests; otherwise a high standard of safety precautions will be required. The connecting wire from the source of the pressure test to the equipment being tested may frequently be noninsulated (since it is impractical in terms of level of insulation, and length of wire, to utilise an insulated conductor); this often requires some ingenuity in terms of positioning the test source and connecting wire, to avoid the latter coming too close to earth, hence reducing the insulation level.

Before commencing, the equipment to be tested should be subject to a 1000 V insulation resistance test (this may be difficult, or even impracticable, on long lengths of cable where significant charging current is required). The equipment should then be discharged to earth; where there is significant stored energy it may be necessary to discharge through a resistance (for reasons of safety and prevention of possible equipment damage).

7.3 Test method

The test should commence by increasing the applied voltage incrementally until the final value is obtained. At the same time the leakage current should be observed on an ammeter connected in series with the equipment under test. If the leakage current appears to be unusually high the test should be terminated. The leakage current should be recorded at the beginning and end of the test.

The test source should be fitted with a safety cutout, which terminates the test if the leakage current is unusually high. At the end of the test the voltage should be slowly reduced, and the equipment discharged through a resistor to remove all stored energy.

When outdoor equipment is tested, the weather conditions should be noted since damp weather can influence the test results.

7.4 Switchgear

It is usual to carry out the following commissioning tests on switchgear:

- pole–earth
- pole–pole
- across the open switching device

At higher voltages, where the phases are usually segregated, the pole–pole test is unnecessary.

The usual method of carrying out the complete set of tests is:

(*a*) With all switches under test closed:
  (i) Apply voltage to R with Y and B earthed.
  (ii) Apply voltage to Y with B and R earthed.
  (iii) Apply voltage to B with R and Y earthed.

(*b*) With all switches under test open: short-circuit R, Y and B to earth on one side of the open contacts, and short-circuit R, Y and B together on the other side of the open contacts. Apply the test voltage between the unearthed side of the contacts and earth.

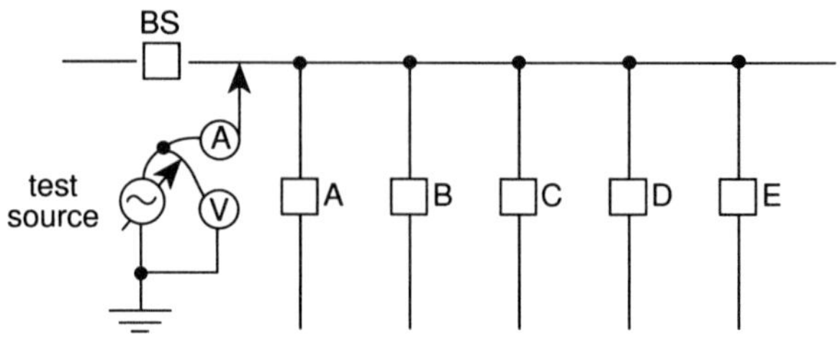

Figure 7.1 *Pressure test on an 11 kV switchboard*

Test sequence
1 Open bus section (BS)
2 Close circuit breakers A–E
3 Apply test source:
to R phase with Y and B to earth
to Y phase with B and R to earth
to B phase with R and Y to earth
4 Open circuit breakers A–E
5 Short-circuit feeder side of all circuit breakers R, Y and B to earth
6 Apply test source to R, Y and B busbars

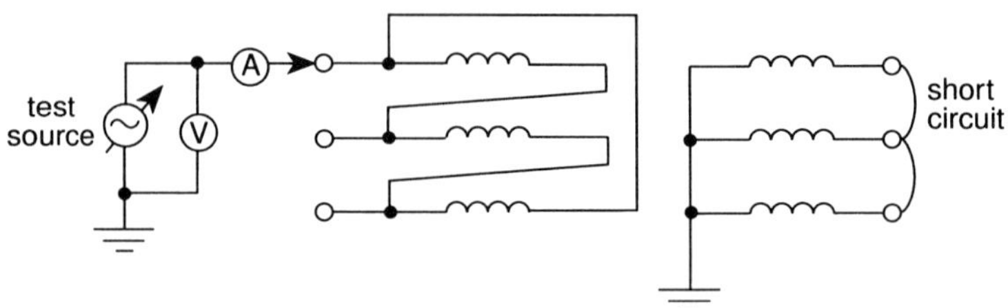

Figure 7.2 *Pressure test on a two-winding transformer*

Figure 7.1 illustrates a typical pressure test on an 11 kV metalclad circuit-breaker board, comprising a section of five bays and a bus section. A typical leakage current for each test, with a required test voltage of 24 kV, is about 30 mA.

The test voltage to be applied is specified in International Standard *IEC 684*.

7.5 Transformers

Figure 7.2 illustrates a pressure test on a typical 33/11 kV two-winding transformer. It is not usual to pressure test transformers having higher-voltage windings.

Table 7.1 Typical 11 kV cable tests

Phases under test	Earth connections
R, Y and B	Sheath
R to Y and B	Sheath
Y to B	R and Sheath

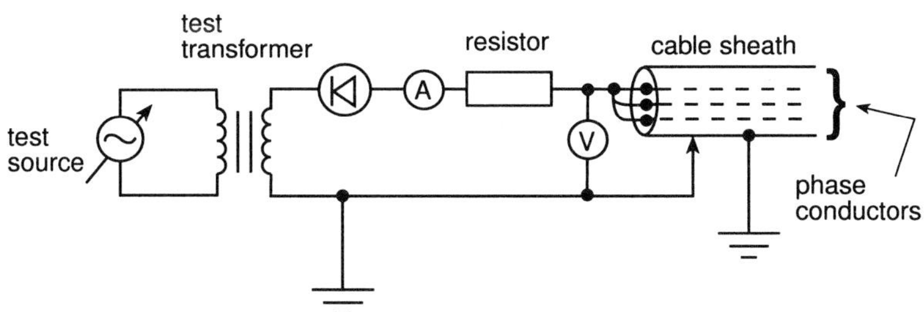

Figure 7.3 Cable DC phase-to-earth pressure-test arrangement

7.6 Cables

The voltage pressure test is the most important site test for a cable. It is carried out to ensure that no damage has occurred to the cable during laying, and that all joints and sealing ends are sound.

Belted cables, which predominate up to system voltages of 11 kV, have no earth screen around each individual phase conductor, and therefore it is usual to carry out both phase-to-earth, and phase-to-phase tests. The usual test format employed is described in Table 7.1.

At lower voltages the direct current is usually produced by use of rectifiers. Figure 7.3 illustrates the basic arrangements and method for a phase-to-earth test, and Figure 7.4*a* the method for a phase-to-phase test. With the phase-to-phase test the phases are charged positive and negative respectively with reference to earth; the phase-to-phase voltage is twice that shown on the voltmeter. The equivalent circuit in terms of capacitance is shown in Figure 7.4*b*. It is usual to charge (and discharge) the cable via a resistor, which controls and limits the charging current as the test voltage is increased.

At higher system voltages it is usual to employ a voltage-multiplying device to generate the test voltage, such as a Cockroft and Walton generator. This is capable of generating direct voltages up to 1000 kV. Single-core cables may be tested individually, or all three phases simultaneously, with respect to earth.

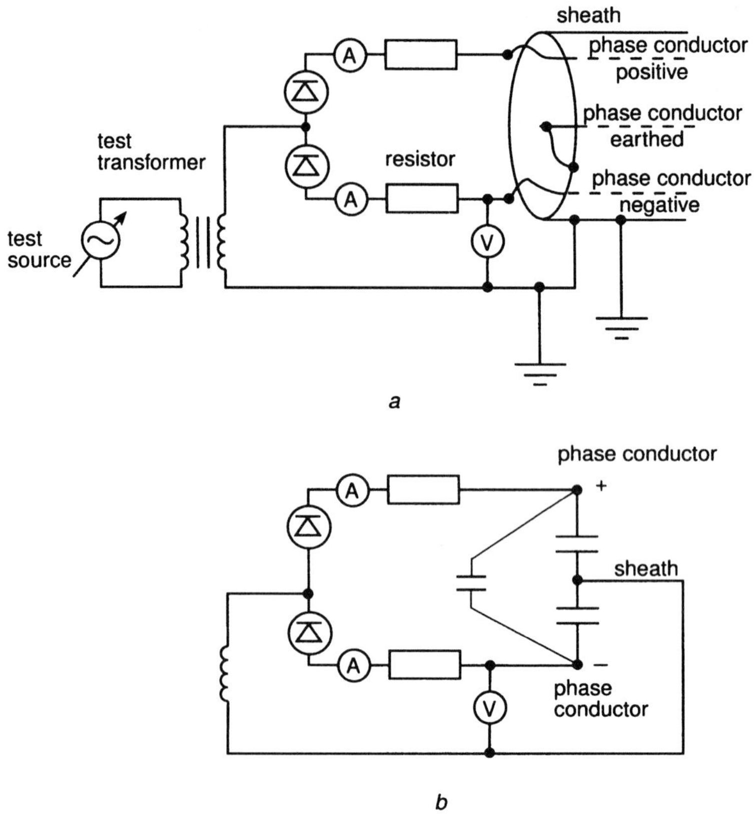

Figure 7.4 Cable DC phase-to-phase pressure-test arrangement

a Physical arrangement
b Equivalent circuit

On cross-bonded systems (see Chapter 8) it is usual practice to bond together the sheaths of each section at each joint position.

Recently there has been concern that HV direct-current tests may degrade the insulation of cross-linked polyethylene (XLPE) cables. As a result, a lower voltage test of 10 kV DC has been adopted with the objective of identifying gross defects only.

7.7 Maintenance

It is not usual to carry out pressure tests prior to return to service of equipment following maintenance. There are a number of exceptions to this general rule, such as cable-sheath-insulation testing (see Chapter 8), and HV-bushing discharge tests when bushings have had a history of failure or are suspect.

Pressure testing should also be utilised as an aid to fault diagnosis, where comparison should be made with the commissioning results.

Chapter 8

Underground cables and overhead lines

8.1 Introduction

The commissioning and maintenance requirements for cables and lines are generally less onerous than for other items of plant. However cables and lines are subject to, and are a source of, induced voltages arising from electric and magnetic fields, the effect of which needs to be contemplated before any work is carried out. This chapter therefore examines the practical effects and implications of electric and magnetic fields on the design, operation, commissioning and maintenance requirements of high-voltage cables, overhead lines and auxiliary cables and connected equipment. Reference should be made to Chapter 3 for discussion of the theoretical concepts.

CABLES

8.2 Mains cable

Mains cables for 415 V systems consisted, until the early 1970s, of the following typical specification:

- 4-core solid aluminium (or stranded copper) conductor
- impregnated-paper insulation—belted
- lead sheath
- steel-tape armouring
- hessian tape with bitumen compound and lime-wash finish

Figure 8.1 illustrates a typical arrangement. The electric field is contained entirely within the cable and is of relatively low intensity. The sum of the magnetic fluxes within the cable approximates to zero and thus the flux existing outside the cable is also virtually zero.

Induced voltages into the cable arising from external sources of flux (i.e. other cables or fault currents) are of low magnitude. This is because the armouring and lead sheath tend to be in direct contact with earth through the porous Hessian tape, throughout the cable length, thus acting as a low-impedance/short circuit to any induced voltages, and so screening the cable inner conductors.

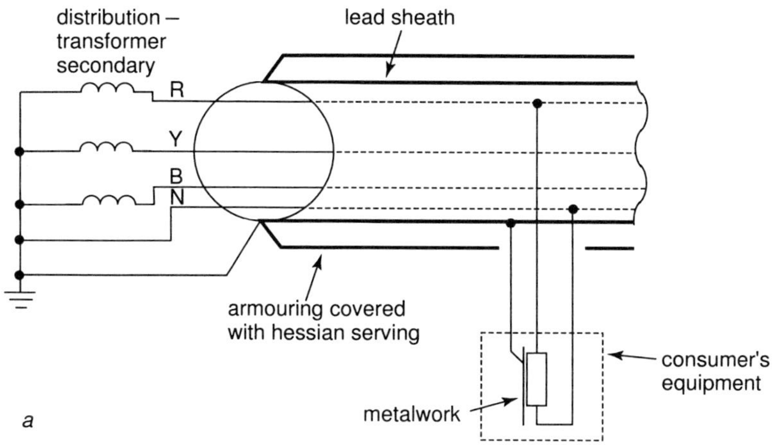

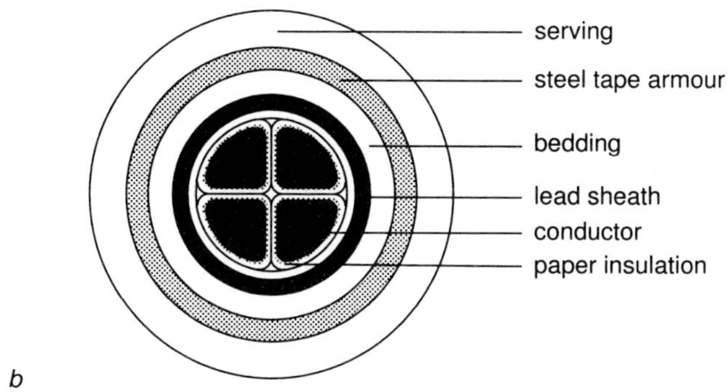

Figure 8.1 *Four-core medium-voltage mains cable*

a Schematic representation
b Cross-section

From about 1968 onwards there has been a move, for mainly financial reasons, towards the use of concentric neutral cables, as illustrated in Figure 8.2. A typical specification is:

- 3-core solid aluminium conductor
- impregnated-paper insulation—belted
- aluminium neutral sheath
- PVC insulating oversheath

The aluminium sheath acts as both the cable neutral and the return path for earth-fault currents (for both cable faults and consumers' apparatus faults). To ensure adequate return paths for consumers' fault current, the concentric neutral is usually earthed at various points along its length. This is known as protective multiple earthing (PME). An analysis of this type of cable in terms

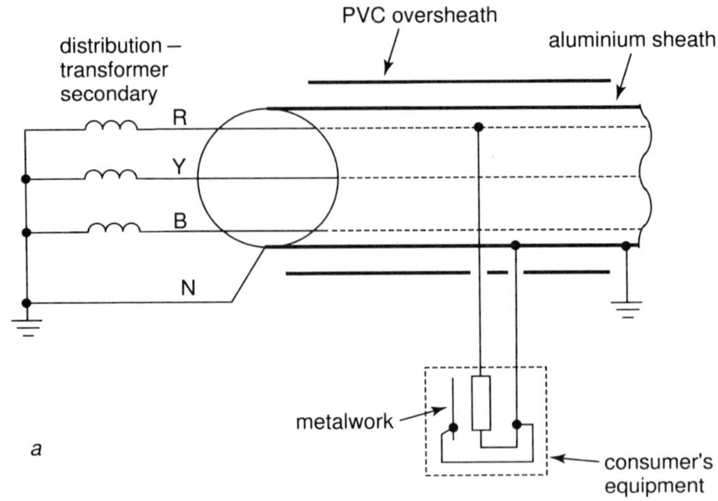

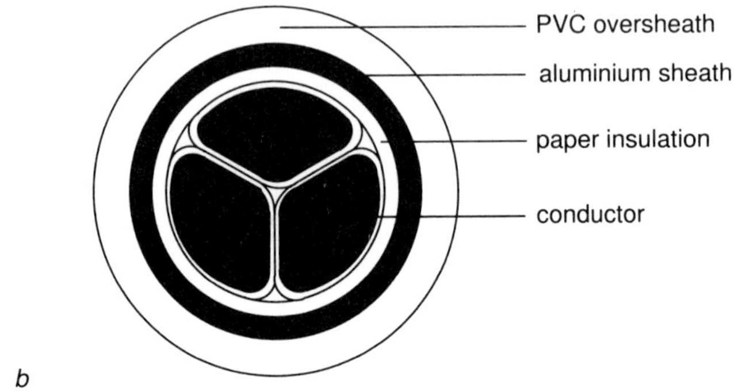

Figure 8.2 Three-core concentric-neutral medium-voltage mains cable

a Schematic representation
b Cross-section

of containment of the electric and magnetic fields and induced voltages is virtually identical to that of the four-core cable above.

When working on these types of cable no special precautions to limit the effects of induced voltages are usually necessary.

8.3 11 kV distribution system cable

As with medium-voltage cables, 11 kV distribution-system cables, until 1970 or thereabouts, consisted of a lead, as opposed to aluminium, sheath, with typically the following specification:

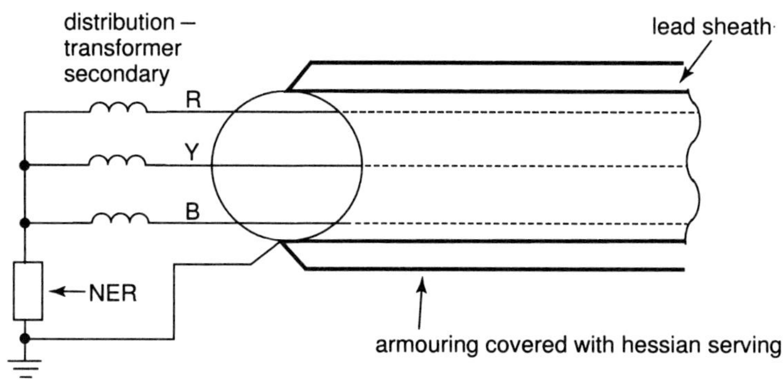

Figure 8.3 11 kV distribution-system cable with lead sheath

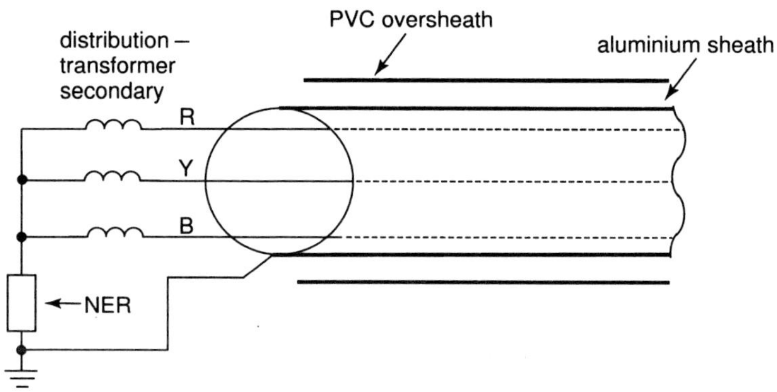

Figure 8.4 11 kV distribution-system cable with aluminium sheath

- 3-core stranded aluminium conductors
- impregnated-paper insulation—belted
- lead sheath
- wire armour
- hessian tape with lime-wash finish

Again, as with the medium-voltage cable, the electric field is contained within the cable, and the sum of the magnetic fluxes outside the cable approximates to zero. The lead sheath is also in continuous contact with the earth, thus short-circuiting any voltages induced into the cable arising from external sources of current (e.g. other cables). Thus once the cable is de-energised no specific precautions are required to ensure safety against induced voltages. Figure 8.3 illustrates the arrangement.

From about 1970 onwards there has been a move, for financial reasons, to the use of aluminium-sheathed cables (see Figure 8.4). A typical specification would consist of:

- 3-core stranded aluminium conductor
- impregnated-paper insulation—belted
- corrugated or smooth aluminium sheath
- PVC insulating oversheath

It would be normal practice to earth the cable sheath at each end of the cable. Aluminium, although lighter in weight, and mechanically stronger, is more prone to corrosion than lead, which necessitates an insulating PVC or polyethylene oversheath. This in turn results in a loss of the continuous earthing provided by the lead-sheathed cable. Significant voltages could then be induced into the aluminium sheath, from external sources such as other cables in close proximity, or from fault currents flowing through the earth. For this reason, safety precautions must be taken before working on the cable (i.e. following de-energisation of the cable). Such precautions are discussed in Section 8.16.

8.4 The Hochstadter cable

The cables discussed thus far have been belted cables, i.e. a belt of paper insulation is wrapped around all three (or four) conductors. However, belted type cables are only viable at voltages up to about 25 kV (although short lengths are in use at 33 kV). At higher voltages the electric field which emanates from each phase conductor, and flows to each other phase conductor and to earth, causes breakdown and failure of belted cables. This is because the electric field varies in intensity and direction, within the cable, as the conductor voltages vary sinusoidally with time. As a result, the paper laminations of a belted cable become subject to two components of electric field, a component tangential to the paper (i.e. along the paper lamination surface), and a component which is radial to the paper (i.e. through the paper laminations). Now the electric strength of paper along the lamination is only 20% of that through the lamination. This reduced electrical strength results in the tangential electric field causing leakage currents to flow along the surface of the paper. This causes local heating, carbonisation of the paper and eventual electrical breakdown.

This failure mechanism can be prevented by ensuring that the electric field is always radially directed through the paper laminations. This can be accomplished by surrounding each conductor with an earthed screen. Such a solution was first patented by Hochstadter in 1914, and the resulting cable is termed a 'Hochstadter' or simply 'H'-type cable.

Figure 8.5*a* illustrates an 'H'-type cable. Each phase conductor has a layer of earthed metallised paper or copper tape wound around the insulation of each core, and in contact with the screens on the other two phases. There is also usually an additional screen of copper woven fabric tape around all three per-phase screens.

An alternative to the 'H'-type cable is the Hochstadter Single Lead type, abbreviated to 'HSL' or just SL. This consists of a sheath around the

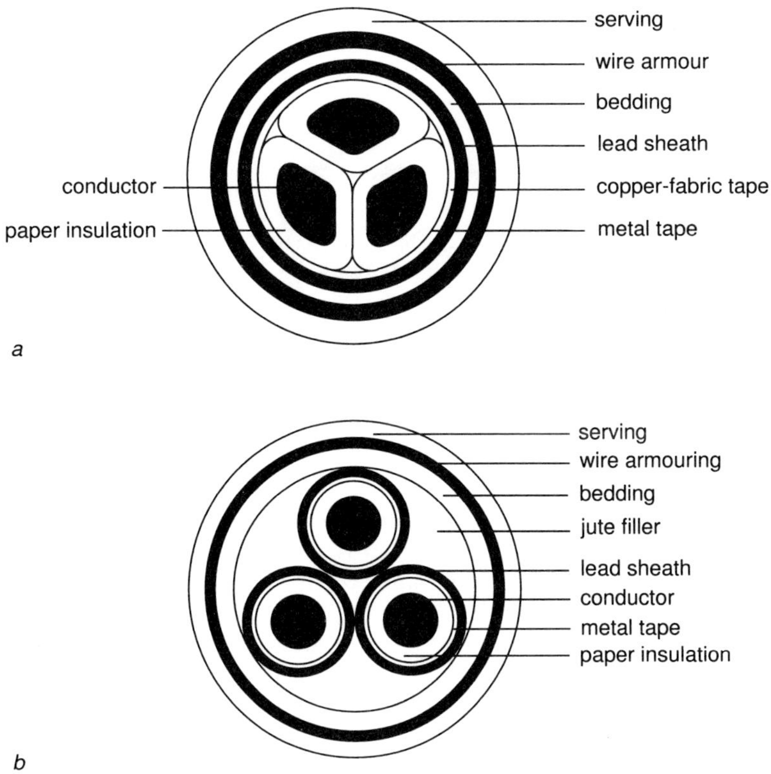

Figure 8.5 *Hochstadter cables*

a H-type cable
b HSL-type cable

insulation of each phase conductor, which is in electrical contact with the sheaths of the other two phases, with armouring around all three sheaths, as illustrated in Figure 8.5*b*. HSL cable overcomes terminating difficulties experienced with H cables.

8.5 Assisted cables

The H and HSL cables discussed above are known as 'solid-type' cables. They are suitable for voltages up to 33 kV. Their limitation is due to the detrimental effect of the electric field on cavities within the cable. Small cavities arise as a consequence of imperfect manufacture, or as a result of expansion/contraction effects as the cable load is varied. Cavities have a lower dielectric strength than solid insulation and are prone to electrical breakdown as a consequence of ion bombardment of the cavity walls at a rate of 100 times/s (i.e. twice per cycle). The insulating properties of the surrounding paper

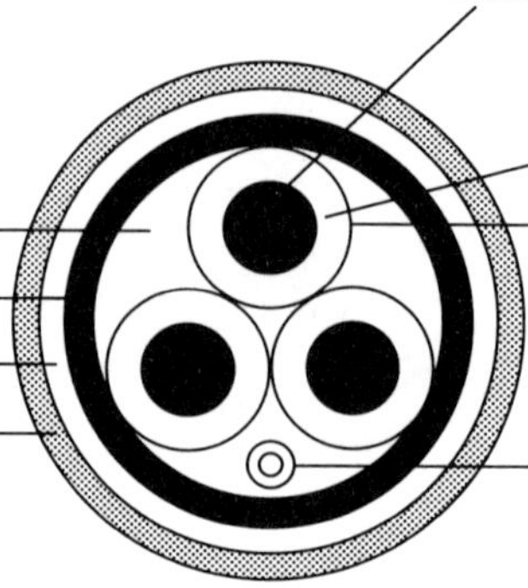

Figure 8.6 Gas-filled cable (three-core)

insulation are eventually destroyed and the cable subsequently fails. A test to determine whether the paper insulation has been subject to ionisation is to dip it in blue magenta dye. If the paper turns a waxy white colour then it is ionised.

Assisted cables, which are also Hochstadter-type cables, overcome the problem of cavities by either preventing them from being formed or filling them with an insulating medium of high dielectric strength. One of the following techniques is employed:

(*a*) The cable is filled with gas under pressure (gas-filled cable).
(*b*) The cable is subject to an external gas pressure (gas-compression cable).
(*c*) An insulating oil is maintained under pressure within the cable (oil-filled cable).

Gas-filled cables have a high-pressure inert gas (such as commercial nitrogen) contained within the metallic sheath and in contact with the paper dielectric. The gas percolates through the cable via a porous gas-channel pipe(s) within the cable or via an annular perforated gas pipe within the cable. Figure 8.6 illustrates a typical construction. The gas pressure, typically of the order of 200 psi, is maintained by cylinders of gas located at each end of the cable.

The gas pressure is monitored by gauges which give an alarm (typically 180 psi) and a trip (typically 150 psi). Gas-filled cables were popular at 33 kV and 132 kV, in particular, but now seem to have lost popularity in favour of oil-filled or XLPE cables.

The gas-compression cable consists simply of a cable located within a pipe which is subject to a gas under pressure which in turn compresses the cable to prevent the formation of voids.

In some countries, including the UK, the oil-filled cable has become, by far, the most common assisted cable. The oil is piped throughout single-core cables via a central oil duct within the conductor itself, as shown in Figure 8.7, or in three-core cables via oil ducts within the annular space between the

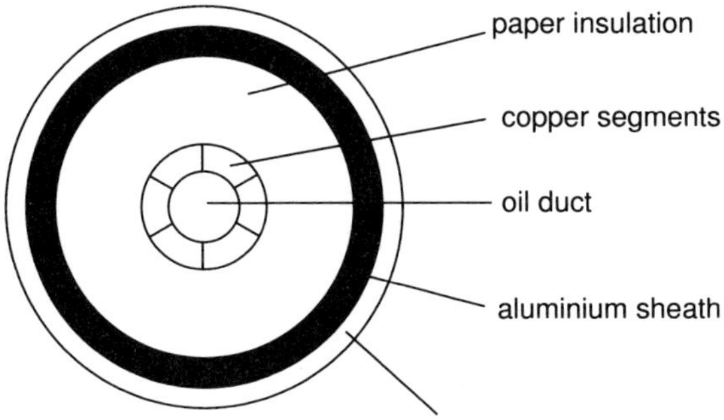

Figure 8.7 Oil-filled cable (single-core)

cable core and the sheath, as in the gas-filled cable. The oil pressure is maintained by means of pressure tanks of oil, which facilitate the expansion and contraction of the oil with changes in temperature. To maintain oil pressure on long cable runs, or in cables which are laid in an undulating ground profile, it is often necessary to install pressure tanks in intermediate positions along the cable. As with gas cables, the oil-cable pressure tanks have gauges which generate alarm and trip conditions if the oil pressure falls, for whatever reason.

8.6 Single-core versus three-core cable

The greater the system voltage the greater the insulation thickness of a cable. This, in turn, makes the cable more difficult to handle physically and causes heat dissipation from the cable to be reduced. At the higher voltages, the current rating of a cable tends to be higher than at lower voltages, so increasing cable heating. All these factors dictate the need for single-core cables at the higher voltages.

At 33 kV, cables in the main tend to be of three-core design, at 132 kV they are mainly single-core, and at 275 and 400 kV they are exclusively single-core.

Note that with three-core cables, the voltage induced into the cable sheath (and armouring) from balanced three-phase load current is virtually zero. This is because the induced voltages arising from three-phase positive-phase-sequence load currents cancel each other. Significant induced voltages can arise, however, from single-phase-to-earth faults, or from load currents which (unusually) contain significant zero-sequence currents. In contrast, the sheaths of single-core cables have voltages induced into them from balanced three-phase load currents. This in turn influences the method of earthing the cable sheath.

If a single-core-cable sheath is earthed at both ends of the cable, any induced voltage causes a circulating current around the earth loop, resulting in the generation of heat, which may require derating of the cable in terms of current-carrying capacity. The alternative of earthing the cable sheath at one end only would result in dangerously high voltages occurring at the unearthed end, particularly during fault conditions. For reasons of safety, the sheath voltages should ideally not be allowed to exceed about 60 V under maximum load conditions. The following sections identify compromise solutions to these mutually exclusive constraints. A further point to note is that single-core cables negate the use of armouring consisting of ferrous, and hence magnetic, materials, since such materials amplify the magnetic flux and hence the induced voltages and related losses.

8.7 33 kV distribution cables

In the main 33 kV distribution cables tend to be three-core H-type cables, either assisted or unassisted. It is usual to earth the cable sheath at both ends of the cable.

8.8 132 kV, 275 kV and 400 kV cables

132 kV cables tend to be mainly, but not exclusively, single-core assisted H-type cables. Three-core cables are usually terminated as single-core cables through use of a trifurcating joint. The sheaths of three-core cables would usually be earthed at each end of the circuit. The methods for earthing single-core cable are discussed in Sections 8.10–8.12.

275 and 400 kV cables invariably consist of single-core, insulated-sheath, assisted (usually oil) H-type cables. Sheath earthing is usually via solid bonding, single-point bonding or cross-bonding systems.

8.9 XLPE cables

The cables discussed so far have all been impregnated paper insulated. This technology has been the mainstay of power-cable installations. However since the early 1950s thermoplastic insulation has been increasing slowly in use. The most widely known thermoplastic material is probably PVC which is widely used for 415/240 V installation work. At higher voltages, cross-linked polyethylene or XLPE tends to be used because of its ability to work to temperatures as high as 90°C (some 15°C higher than paper).

XLPE takes its name from the chemical bonding, termed cross linking, of the polyethylene which describes its molecular structure, and which provides its tolerance to high working temperatures. Being a plastics material it may be thought that XLPE is unaffected by moisture, but moisture can be induced

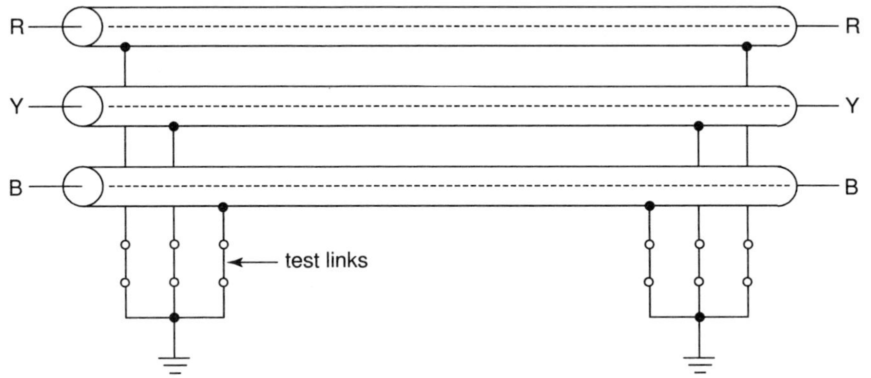

Figure 8.8 Solidly bonded single-core-cable system

into the XLPE and permeates it in the form of a 'water tree', resulting in eventual electrical breakdown.

The XLPE insulation is therefore encompassed in a conventional solid metallic sheath (lead, aluminium or, more recently, copper) to form a moisture barrier as well as an earth-fault-current return path.

XLPE-insulated cables require a greater insulation thickness than their paper equivalent. This results in the cables being more difficult to handle physically. XLPE, however, does not suffer to the same extent as paper from breakdown due to the direction of the electric field (see Section 8.4), but it is prone to the effects of cable breathing (i.e. cavities created through expansion and contraction, as described in Section 8.5). Discharges arising from these cavities can be nullified by the use of semiconducting layers around both the conductor and the insulation, which have the effect of reducing the concentration of field lines. XLPE cables also have the advantage of not requiring the additional equipment associated with paper-assisted cables, e.g. oil reservoirs, alarms etc.

XLPE cables are widely used in many countries (although the experience at 275 and 400 kV is limited). In the UK their rate of introduction has been slowed by the continuing satisfactory performance of paper-insulated cables. However, XLPE is now being increasingly used, and has been installed at most voltages.

8.10 Solidly bonded single-core cables

Solid bonding consists of simply earthing the cable sheath at both ends of the circuit. This would normally apply to short lengths of cable in the region of tens of metres long, where the sheath circulating currents would be small. The sheaths are normally earthed through test links, which are installed to facilitate cable-oversheath-integrity testing. Figure 8.8 illustrates the arrangement.

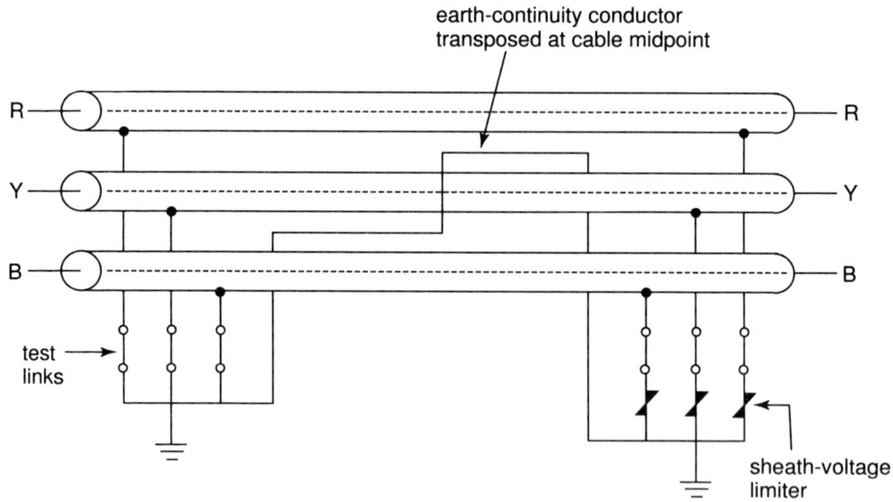

Figure 8.9 Single-point-bonding system

8.11 Single-point-bonding cable systems

Single-point bonding would normally be utilised on single-core-cable circuit lengths of typically 100–300 m. Figure 8.9 illustrates the arrangement. The sheath is solidly earthed at one end only, to prevent sheath circulating currents. If transient voltages experienced at the unearthed end are likely to be problematic, a sheath-voltage limiter (SVL, see Section 8.13) should be connected between sheath and earth.

If the cable sheath is unearthed at one end, there is no return path for earth faults which occur beyond the cable, other than via the earth itself. To avoid undue interference with communication circuits, it is desirable for the power system to provide its own earth-fault (zero-sequence) return paths. It is therefore usual for an earth-continuity conductor to be run parallel to the cable. To limit voltages induced into the earth-continuity conductor from the power cables, causing circulating currents around the earth loop, the earth-continuity conductor is transposed at the midpoint of the cable, thus resulting in partial cancellation of the induced voltages.

The sheath standing voltage at full load is usually limited in the UK to 65 V at 132 kV, or 150 V at 275 kV and 400 kV, which in turn usually limits the single-point-bonding system to cable lengths of about 300 m.

Cables with a typical circuit length of between 300 and 1000 m may have their sheaths earthed by a variation of the single-point-bonding system. This is illustrated in Figure 8.10. As shown, the cable sheath is electrically interrupted at the midpoint with each side of the interruption earthed via an SVL: this is termed a split-sheath single-point-bonding system.

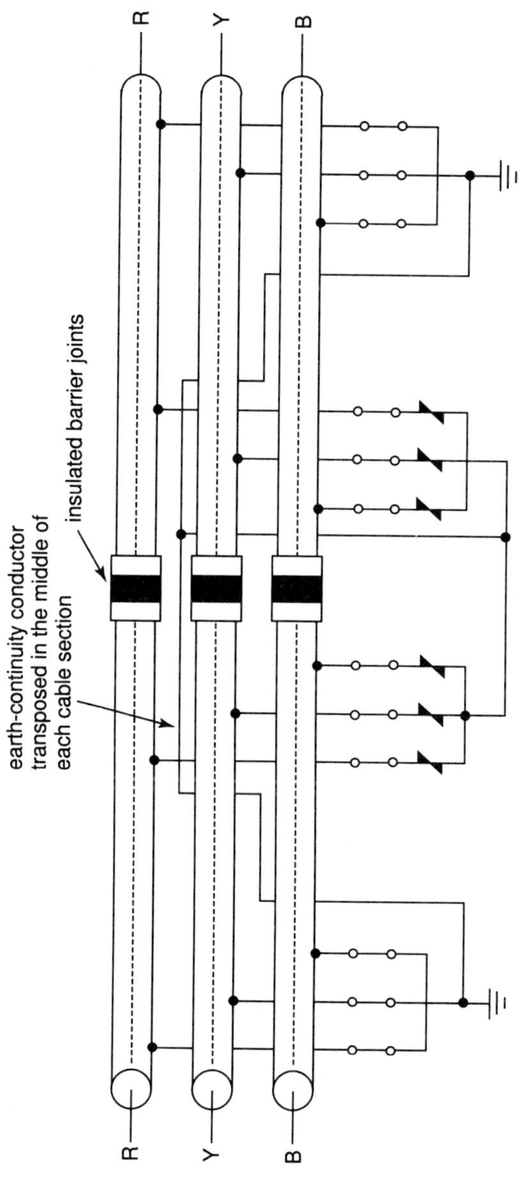

Figure 8.10 *Split-sheath single-point-bonding system*

8.12 Cross-bonding cable systems

Cross-bonding systems are used on the longer cable circuits. The system is both technically complex and financially expensive, but resolves the problems of excessive sheath circulating currents on the one hand, and dangerously high induced sheath voltages on the other. It additionally disposes of the need for a cable-continuity conductor. A typical system is illustrated in Figure 8.11.

When cables are laid in close trefoil formation the induced voltages arising from load current into each sheath are symmetrical. However, if, as happens frequently in practice, the cables are laid flat (to facilitate heat dissipation), the voltages induced into the sheaths of the two outer cables are greater than the voltages induced into the sheath of the inner cable. To overcome this, the three cables are transposed at every joint position. The joints also contain an insulating barrier to enable the cable-sheath sections to be connected R–B–Y in sequence. The summated induced sheath voltages over three sections arising from load current thus summate to zero (almost) as shown by Figure 8.11*c*. This enables the sheath to be solidly bonded to earth at each end of the circuit (continuous cross-bonding) or at every third joint position (sectionalised cross-bonding) in addition to the circuit end. Note that the number of sections of cable in the circuit will be a multiple of three.

At joint positions 1, 2, 4, 5, 7, 8 etc., there will be a standing induced voltage with respect to earth, which will be higher under through-fault conditions than under load conditions, and it is therefore usual to connect an SVL (see Section 8.13) at these positions. Positive and negative phase-sequence fault currents will cause induced voltages which, because of their symmetry, will cancel at every third joint position; however zero-sequence currents will cause in-phase induced voltages which will summate throughout the whole length of the cable, but whose magnitude will be limited by the SVLs.

8.13 Sheath-voltage limiters

Sheath-voltage limiters (SVLs), which used to be termed cable-covering-protection units (CCPU), are nonlinear resistors. They present a high resistance (approximating to an open circuit) to the normally low-magnitude standing voltages, and a low resistance under higher-voltage transient conditions. Their function is to limit sheath circulating currents under normal cable-loading conditions and to prevent damage to sheath insulation and insulating barrier joints from transient overvoltages which may occur during cable switching, through faults, lightning etc. They are located at unearthed joint positions on cross-bonded systems, and at the unearthed end of single-point-bonded systems. SVLs are usually housed in a lockable box, together with links to facilitate testing. Older SVLs were based on silicon carbide: these are rapidly being replaced by units utilising zinc oxide.

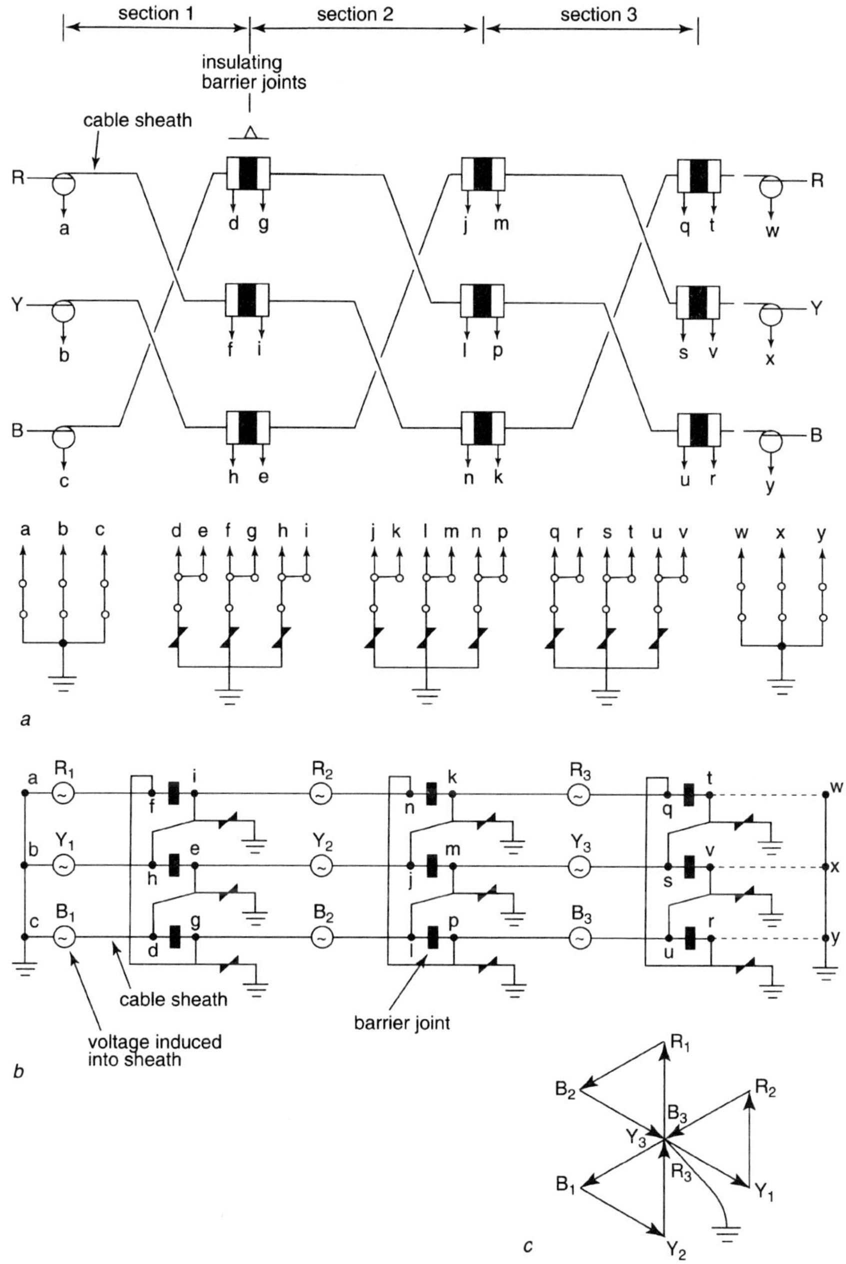

Figure 8.11 Crossbonded cable system

a Physical aspects of crossbonding
b Electrical connections
c Phasor relationship

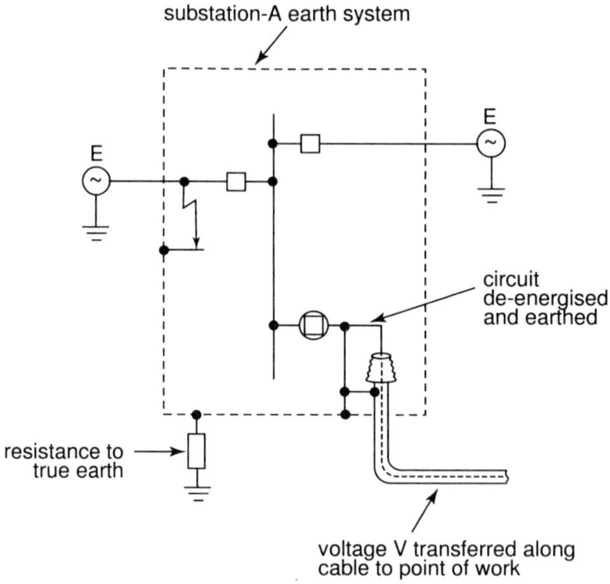

Figure 8.12 Transferred potential

8.14 Water-cooled cable systems

A small number of transmission-voltage cables are water cooled to enhance the current-carrying capacity. This usually consists of water pipes positioned immediately adjacent to the cable. Careful consideration needs to be given to how the water pipes are earthed, if they are metallic. More recent schemes tend to use plastics pipes. The case for water cooling is not as strong as it once was.

8.15 Transferred potential

A danger which needs to be noted when working on cables is that of transferred potential. This is illustrated in Figure 8.12. During a system fault, the potential of the earth mat at substation A rises above true earth potential by a voltage *V*, and this voltage is transferred along the cable conductors and sheath. It is therefore a source of danger when working on the cable. In the UK, telecommunications regulations require the rise in potential at a substation not to exceed 650 V for a period of 200 ms (i.e. cleared by main protections), or 430 V for periods greater than 200 ms (i.e. to be cleared by overcurrent protection). In practice, these voltage limits may be difficult to achieve.

8.16 Induced-voltage working

The need for induced-voltage working practices was briefly alluded to when discussing distribution-voltage cables in Section 8.3. At supergrid voltages the approach is more rigorous. The main dangers to personnel working on cables arise from the following sources:

(*a*) voltage transference from the points of earthing the cable conductor and sheath to the point of work; see Section 8.15;
(*b*) induced voltages from cables or overhead lines which run physically parallel to the cable to be worked on.

The relative complexity of having to resort to induced-voltage working is such that it need not be carried out if it can be proved by calculation that it is unnecessary. In general, induced-voltage working is unnecessary if it can be shown that the steady-state induced voltage is less than 60 V, or that the transient voltages due to either induction or transferred potential are less than those specified in Section 8.15.

Induced-voltage working consists of 'earthed working', where the part of the cable to be worked on, and the jointer, are at earth potential, and 'insulated working', where the jointer is insulated from earth. Section 8.17 describes a specific instance of induced-voltage working. The theme of this approach is to create an equipotential environment so that the jointer, cable and anything else with which the jointer may come into contact, while working, are at the same potential. Where possible, earthed working is adopted, since insulated working is a more complex activity.

8.17 Removing part of a cable sheath

Consider a cable which has been identified as requiring induced-voltage working, where it is necessary to remove part of the metallic sheath (which has a PVC oversheath) as shown in Figure 8.13.

While removing the PVC oversheath, the jointer could inadvertently make contact with the metallic sheath and thus 'insulated working' is necessary. This requires the jointer to work from an insulated platform. The quality of the insulation must be verified using an insulation-test meter. Once the metallic sheath is exposed, it is connected to earth at either side of the proposed cut using an insulated earthing pole. Earthed working should now be adopted. This consists of working from an earth mat, to which all metalwork capable of being touched by the jointer is connected. Given that jointer and sheath are now both at earth potential the sheath can be removed safely. A point worthy of note is that, to avoid 'transferred potential' to the point of work, it may be necessary to leave the ends of the cable circuit, which would normally be earthed, unearthed: this would normally require the cable to be classified as removed from the system and hence outside the constraints of the 'safety rules'.

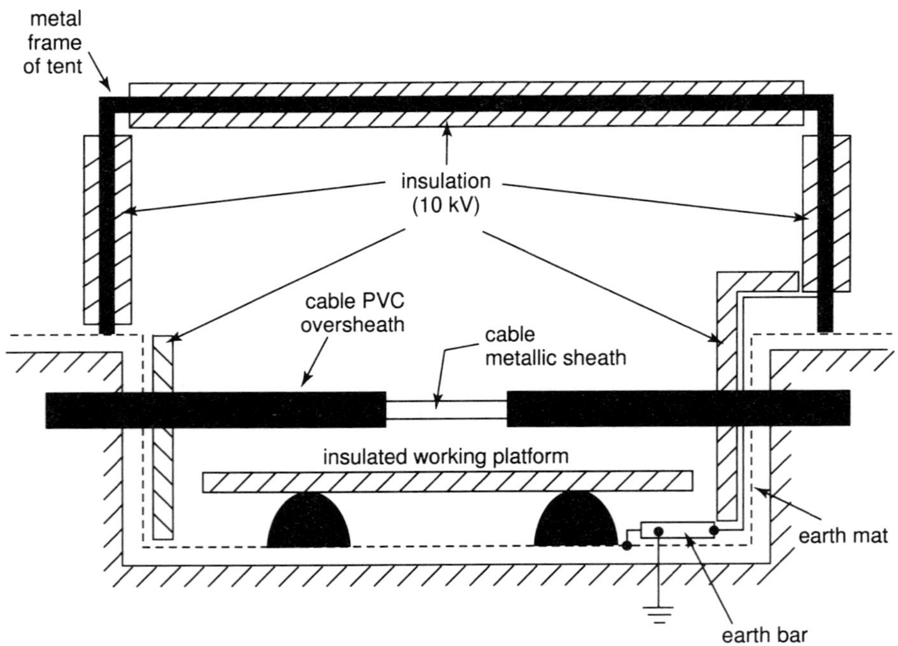

Figure 8.13 Induced-voltage working

Illustration shows insulated-working facilities. For earthed working, platform and insulation are removed

8.18 Cable-commissioning tests

As stated in Section 8.1 the commissioning requirements for a cable are few. Generally, all that is required is the following:

- pressure test (see Chapter 7)
- cable-loop resistance; however this may be problematic for single-core cables which may have a standing induced voltage from another parallel cable
- proving of gauges, valves and pressure alarms on assisted cables
- measurement of sequence impedances on cross-bonded cables
- oil tests on oil filled cables
- check of sheath insulation on insulated sheathed cables
- proving of cross-bonding on cross-bonded cables
- check of SVL characteristic
- collating of cable data

Sections 8.19–8.23 discuss these tests.

8.19 Sheath-insulation-commissioning test

To test the insulation of insulated sheathed cables it is necessary to disconnect the metallic sheath from earth. A test voltage of about 10 kV DC should then be applied between sheath and earth for about 1 min. Note that induced-voltage working may be required. Generally speaking, the leakage current should not exceed 1 mA. A current in excess of 10 mA would indicate a sheath fault.

For split and cross-bonded sheath systems, the sections not under test should be connected to earth. The section under test should be disconnected from the SVLs. The insulated barrier joints are thus also proved at the time of test. It is also usual to apply a voltage pressure test to prove the insulation, and to measure the conductor resistance of the bonding leads.

8.20 Cross-bonding-commissioning test

As part of the commissioning procedure, it is necessary to prove the sheath transposition and sheath-earthing system of cross-bonded systems. This can be accomplished by injecting a three-phase balanced current down the cable phase conductors, and measuring the current in the sheath-to-earth connections. The injected current induces a voltage into the sheath and for correct cross bonding there should be negligible current measured in the sheath-to-earth connection. The injected current should be at least 100 A. The injection source should be connected to one end of the cable and all three phases short-circuited to earth at the remote end to provide a low-impedance path. In the first instance, an incorrect cross bond should be made to prove that significant induced current will flow. The contact resistance between each joint and its bonding lead, and the resistance across each test link, should also be measured. The maximum resistance measured should not usually exceed 50 μΩ.

It is also usual to measure the positive- and zero-sequence cable impedances. This necessitates injecting three-phase (for positive-sequence) and three single-phase (for zero-sequence) currents down the cable, respectively, in the range 50–100 A while measuring the current, voltage and power, to enable the impedance, resistance and reactance to be calculated. Note that the reactive component of the zero-sequence impedance of a cable is, unlike that of an overhead line, usually a lower value than the resistive component. This arises because the sheath provides the return current, so other factors also influence the apparent resistance circuit for the zero-sequence (in-phase) resulting in a flux which opposes that arising from the cable conductor, hence reducing the overall reactance.

8.21 Oil filled cable commissioning tests

In general four tests are carried out:

(*a*) a test for residual gas pressure (mainly air) due to gases remaining in the cable after filling with oil. This is usually a specialist test carried out by the manufacturer;
(*b*) a test to obtain the coefficient of impregnation of the cable insulating paper. This again is usually a specialist manufacturer's test;
(*c*) a test of oil flow through the cable. This is done by timing how long it takes to fill a bucket connected to one end of the cable. Again, the manufacturer specifies the test;
(*d*) tests to check the insulating quality of the oil, generally as specified in Section 5.33.

8.22 SVL-characteristic-commissioning test

Modern testing of SVLs consists of injecting a direct voltage (in both directions in turn) across the SVL, measuring the corresponding current and comparing it with the manufacturer's recommended limits.

8.23 Cable data

It is important at the commissioning stage to acquire cable data from the manufacturer, both for record purposes and as a source of data for fault investigations. In particular, the following should be obtained:

- conductor resistance for the installed cable, based on cable sections
- cable capacitance
- dielectric power-factor/voltage characteristics
- a precise record of the route
- positive-, negative- and zero-sequence impedances.

8.24 Cable maintenance

If a cable is correctly installed and jointed, little maintenance is required. A typical list of maintenance requirements is as follows:

(*a*) visual inspection of terminations (this would range from, say, two-weekly for transmission cables to two-to-three-yearly for HV distribution cables;
(*b*) checking and recording of pressure gauge readings on assisted cables every 2–4 months;
(*c*) annual proving of alarms on assisted cables;
(*d*) sheath-insulation test on insulated-sheath systems. This is a repeat of the commissioning test in Section 8.19 except that the applied voltage may be reduced to 5 kV;
(*e*) SVL-characteristic test. This is a repeat of the commissioning test in Section 8.22;
(*f*) oil tests for oil-assisted cables, generally as specified in Section 5.33.

8.25 Cable-fault location

Cable faults may arise from two main sources: electrical breakdown (usually a function of age) or mechanical damage. With the latter it is usually evident that damage has been caused. However, in many instances it is necessary to locate the point of fault. Faults may be categorised as a high- or low-resistance, conductor-to-earth or conductor-to-conductor, or a conductor open-circuit fault. Fault location has historically been carried out by use of 'Murray loop'-type bridge devices. More recently, pulse-echo-type devices have been used. Contemporary literature more than adequately describes the various techniques and therefore there is little justification in duplicating such work here.

8.26 Cable spiking

Once the position of a cable fault has been located, it is necessary to ensure that, before it is worked on, the cable identified is the correct cable and that it is dead. This usually involves spiking the cable. Spiking consists of driving a metal spike into the cable at the point of fault. The spike is usually fired from a spiking gun which is clamped to the cable, and fired by pulling a lanyard from a position of safety some metres away.

Spiking is usually carried out in every instance on solid cables, but is used more discerningly on assisted cables, where the effect of spiking significantly increases the cost of repair. If the cable can be identified at the point of work without doubt, such as it being traced visually from a point on the cable already proved to be dead, spiking need not be carried out.

LINES

8.27 Overhead lines

The previous sections indicated that sources of induction arising from cables are due almost entirely to magnetic fields. This is because the electric field is contained between the live inner conductor and the earthed sheath. Overhead lines have no such constraints and both the electric and magnetic fields are sources of induction, and hence must be considered when working on overhead lines.

Distribution systems frequently consist of single-circuit wood-pole lines. When such a circuit is de-energised, the voltage remaining on the line due to induction is usually zero, except when the circuit runs parallel to another live circuit. This contrasts with a double-circuit transmission line, where the de-energised circuit is in close proximity to the live circuit on the same side of the tower. In such circumstances the induced voltages into the de-energised circuit, arising from electric and magnetic fields, can be very significant, rising in some instances to over 100 kV. The severity of such induced voltages is apparent to any engineer who has had cause to open an earth switch on a supposedly dead transmission line, where arcs of the order of 3 m or so can be experienced before they self extinguish.

Induction can also take place from transmission-line circuits to nearby distribution circuits (and vice-versa), or to other metallic objects such as fencing or pipelines. At the design stage, care must be taken to minimise the distance over which such circuits or other metallic objects run in parallel.

8.28 Overhead-line earthing (poles)

Medium-voltage overhead distribution systems are normally supported on wooden poles and have two main types of earthing.

The first uses an individual aerial earth wire, as illustrated in Figure 8.14*a*, and the second a protective-multiple-earthing system (PME) as shown in Figure 8.14*b*. The PME uses the neutral as the return wire for both load and fault currents. To ensure that there is a return path, should the neutral between consumer and supply transformer break, the neutral is earthed at a number of places, and all consumer's earthed metalwork (e.g. water pipes) is bonded to the neutral at the consumer's premises. This also ensures the

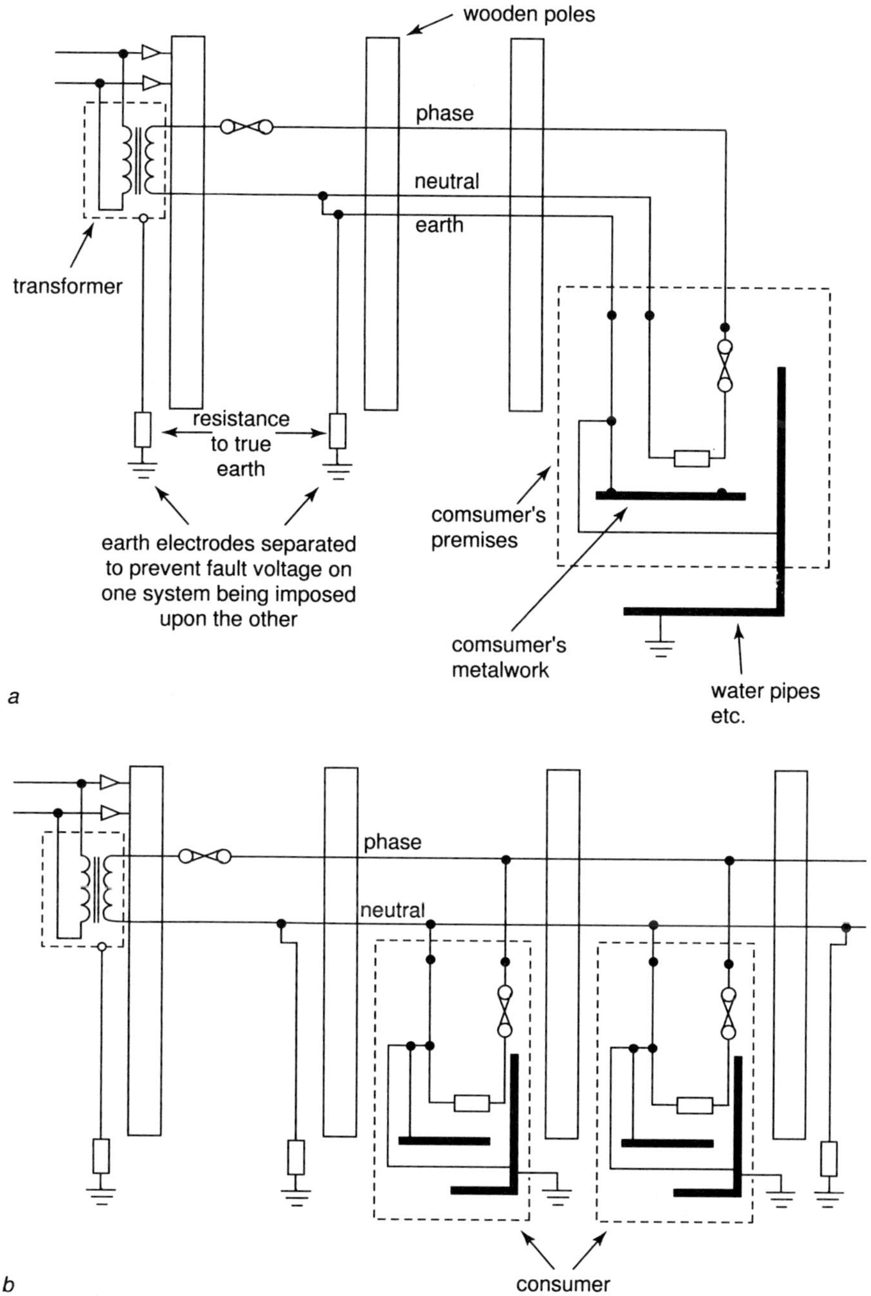

Figure 8.14 *Medium-voltage overhead-line earthing systems*

a MV system using aerial earth
b MV protective-multiple-earthing system

consumer's premises are an equipotential environment for reasons of safety. In the UK, the 1988 Electricity Supply Regulations specify the requirements for PME systems.

33 kV and 11 kV distribution systems are also usually supported on wooden poles. Earlier practice was to bond the pole-top metalwork to an earth connection, which was clipped to the pole and attached to an earth electrode. Experience, however, showed that the majority of overhead line faults were due to transient short circuits (caused by birds or wind-borne material) between phase conductors and pole-top metalwork. If the pole-top metalwork is left unearthed, the wood of the pole can be used as an insulator and faults of the type described will thus not develop into permanent faults; hence the circuit would not be tripped unnecessarily. This tends to be modern practice.

8.29 Overhead-line earthing (towers)

400 kV, 275 kV and 132 kV overhead lines are usually supported on steel lattice towers. An earthwire usually connects the top of each tower, and runs the whole circuit length to each terminal tower. The terminal towers are usually connected to the substation earth mat. The primary function of the earth wire is to shield the phase conductors from lightning strikes.

Each tower footing, being implanted into the ground, has a resistance to true earth. To conduct a lightning stroke successfully from the earthwire to ground (thus preventing a 'backflashover') the footing resistance should ideally be less than about 30 Ω for the 132 kV system, and less than 10 Ω for the 275 and 400 kV systems.

8.30 Overhead-line conductors

Aluminium has become the most common overhead line conductor material in several countries, although there is still significant use of cadmium copper on the lower voltage networks. The most common aluminium conductor is aluminium conductor, steel reinforced (ACSR). This was previously referred to as steel-cored aluminium (SCA). It consists of a central core of galvanised steel surrounded by aluminium conductors, as illustrated in Figure 8.15. The steel core is necessary to provide mechanical strength. Most of the current, but not all, is carried by the aluminium. Each wire on an ACSR conductor is greased during manufacture. This is necessary to:

(*a*) minimise electrochemical action (corrosion) between the aluminium and steel;

(*b*) minimise the formation of aluminium oxide, which will occur almost instantly if the aluminium conductor surface is removed due to, say, one conductor rubbing against another. Aluminium oxide is a poor con-

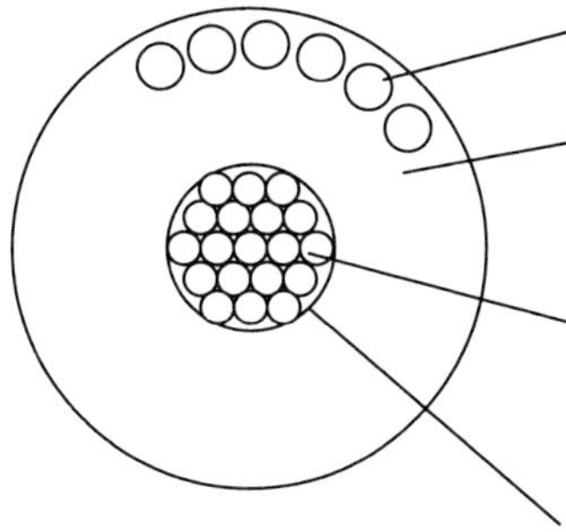

Figure 8.15 Aluminium conductor, steel reinforced

ductor and thus its formation must be avoided. In addition, it occupies a larger volume than the aluminium from which it is formed, so forcing the aluminium strands apart, which then allows the ingress of pollutants. The grease coating alleviates the formation of aluminium oxide;

(*c*) act as a lubricant between the wires of the conductor, thereby preventing grazing through rubbing.

More recently, there has been a move to aluminium-alloy conductors where materials such as magnesium and silicon have been added to the aluminium to provide mechanical strength.

At voltages of 132 kV and above, it is frequently necessary to run conductors in a twin or quad formation per phase (known as bundle), as opposed to using one large single conductor. This has the advantages of being mechanically easier to handle and reducing the corona loss. Recent installations have tended to move away from quad in favour of twin all-aluminium-alloy conductors which are less prone to conductor galloping.

8.31 Transposition

Figure 8.16*a* and *b* illustrates typical conductor configurations. As can be seen, in each instance, the central conductors will be subject to a larger induced voltage (from currents flowing in the other conductors), than the outer conductors. This inbalance will result in unwanted negative- and zero-sequence voltages being imposed on the power system. The imbalance can be mitigated by transposing the conductor positions physically, as shown in Figure 8.16*c*. Some account is taken of the problem on the UK transmission system, where overhead lines are relatively short, by transposing at the circuit ends (substations) only. This would result in one circuit having conductors positioned, for example, as RED (top), YELLOW (middle), BLUE (bottom) and the other circuit YELLOW (top), BLUE (middle), RED (bottom) etc., as shown in Figure 8.16*b*. Different circuits would have different transpositions

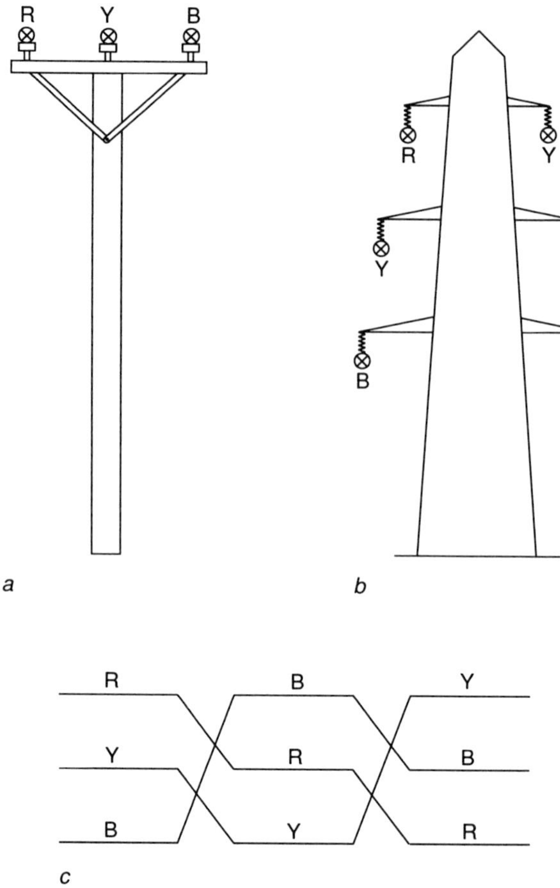

Figure 8.16 Reduction of voltage imbalance through transposition

a Pole-mounted configuration
b Lattice-steel-tower configuration
c Circuit transposition

with the objective of endeavouring to create a balance across the whole system.

8.32 Commissioning of overhead lines

The commissioning requirements for overhead lines are even fewer than those required for a cable. They largely consist of the following:

- measurement of joint resistances
- measurement of earthing resistance/tower-footing resistances
- line patrol

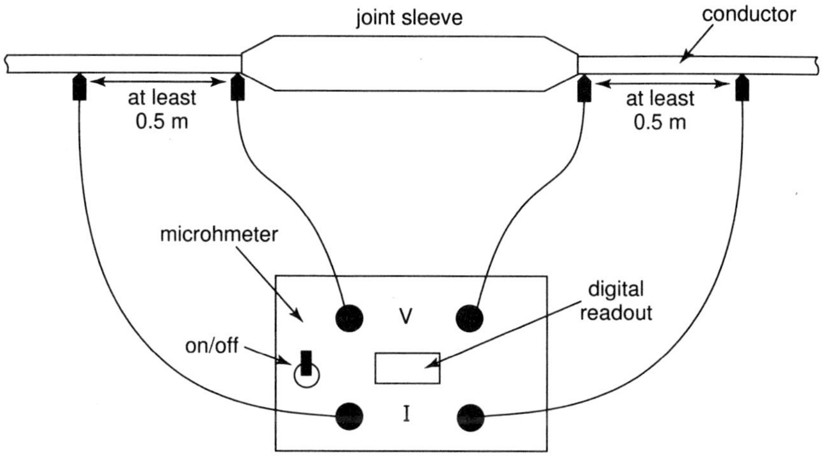

Figure 8.17 Use of microhmeter to measure joint resistance

- collation of line data
- phasing tests at the circuit ends

8.33 Measurement of joint resistance

At the higher voltages, jointing is used in preference to screwed-clamp terminations. The number of jumpers and terminations, and hence joints, on a transmission line is very considerable.

It is usual to measure the resistance of each joint with a microhmeter. Modern microhmeters typically inject about 10 A into the joint, and the voltage V across the joint is measured. Figure 8.17 illustrates the arrangement. It is important to place the current-injection leads I at least 0.5 m from the joint position to give the current the opportunity to spread uniformly throughout the conductor; otherwise an artificially high resistance could be measured.

Joints should only be accepted if the measured value is less than a specified figure. These figures generally fall into the range 10–100 $\mu\Omega$, depending on conductor size. A comprehensive record of all joint resistances, the location, the date, the jointer and the tester, should be kept.

8.34 Measurement of earthing-electrode resistance

The term earthing resistance, as used in this section, refers to the resistance between an earth electrode, or tower footing, and the general mass of earth. A specialist testing device is used for this purpose. The test method is shown in Figure 8.18. When this test is carried out it is important to ensure that the

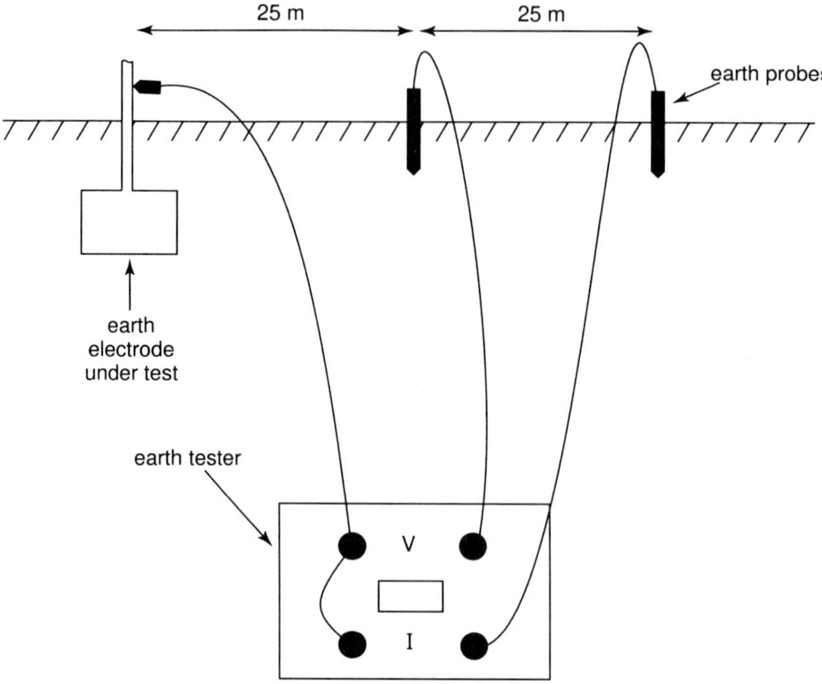

Figure 8.18 Measurement of earth-electrode resistance

I=injected current
V=measured voltage

electrode under test is not connected to other electrodes, e.g. the tower must not be connected to the aerial earth wire. The test connections are spaced some 25 m or so apart to ensure that the resistance measured is with respect to the general mass of earth.

8.35 Pre-energisation line patrol

Before the line is energised for the first time, it is important that a line patrol is carried out. The following points, in particular, should be verified:

- all portable earths have been removed;
- trees and foliage are not infringing the safety clearances;
- colour plates and flag brackets are fitted;
- where applicable, danger plates are fitted and anticlimbing guards are secured;
- the sag appears reasonable.

8.36 Overhead-line maintenance

In general, the only routine maintenance required is to carry out a visual inspection while walking (patrolling) the line route. All lines should ideally be patrolled once per year. Transmission lines, as a consequence of their greater importance to the power system, should be patrolled some three or four times a year. The patrol requirements should be in accordance with a standard report which should request commentary on the following:

- tower/pole number and type (suspension or tension)
- condition of any concrete muffs and tower/pole footings
- condition of towers for rust and loose bolts, or poles that are leaning, split or rotted
- broken or cracked insulators and condition of arcing horns
- broken strands or conductors
- condition of anticlimbing devices, danger plates, circuit-colour plates and flag brackets
- broken or slack stays
- condition of vibration dampers
- visual evaluation of the sag
- proximity of trees and buildings with respect to conductor clearances

To obtain the above information on towers, climbing will be necessary. The above is a qualitative examination, and the condition needs to be categorised in terms of 'good', 'poor' and 'needs action'.

8.37 Faults on overhead lines

Overhead lines are the most fault-prone items of plant on the power system. Typical faults are as follows:

(*a*) lightning flashover;
(*b*) conductor clashing as a result of galloping, i.e. high amplitude oscillation of the complete span resulting from high winds;
(*c*) conductor clashing as a result of large volumes of ice falling away and causing the conductor to rebound;
(*d*) contact with overgrown trees and foliage;
(*e*) contact with kites and hot-air balloons;
(*f*) stubble fires;
(*g*) unintentional contact by cranes, large vehicles etc.;
(*h*) bridging of distribution conductors by large birds.

If an overhead line trips and trips again on re-energisation, a line patrol should be carried out as soon as possible. Attention should be paid to failures such as broken conductors, shattered insulators, fallen trees, signs of flashover and dead animals. In bad weather conditions, or in times of emergency, a helicopter may be used to expedite the patrol.

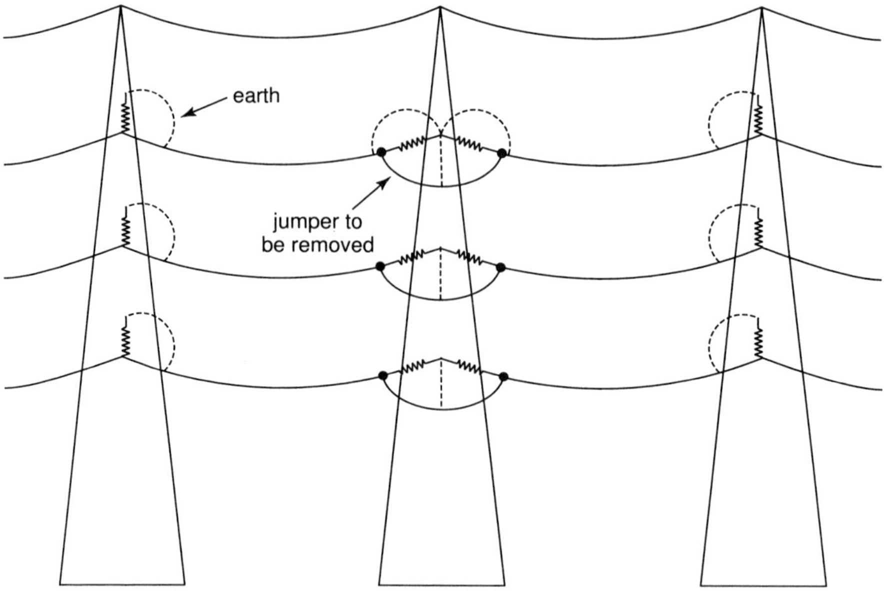

Figure 8.19 Earthing requirements for jumper removal

8.38 Overhead-line-earthing requirements

Significant attention needs to be paid to safety requirements when working on overhead lines to ensure safety from induced voltages. Figure 8.19 illustrates typical requirements for removing a jumper from one phase of a tension tower. In addition to earthing the circuit ends, 11 portable earths (shown dotted) need to be applied at the point of work. If the line was of quad conductor, each subconductor would require earthing, resulting in a total of 44 earths.

A point worthy of note, when considering induced voltages arising from magnetic fields, is that the highest induced voltages occur with fault, as opposed to load, currents, particularly those faults with significant zero-sequence current.

AUXILIARY CABLES

8.39 Types of auxiliary cable

Auxiliary cables serve two main purposes:

(*a*) the interconnection of protection, control and communication systems within a substation; and
(*b*) the interconnection of protection, intertripping and communications

Table 8.1 Auxiliary-cable features (information courtesy of the Electricity Association)

	PVC-insulated multicore cables	PVC-insulated multipair cables	Polythene-insulated multipair cables
Use	Electrical equipment and protection circuitry	Substation control, indication and alarm circuitry	Interconnection of protection and control systems between substations
Composition	7/067 (2.5 mm^2) copper conductor	1/0.8 (0.5 mm^2) copper conductor	1/0.8 (0.5 mm^2) copper conductor
	PVC-insulated, steel-wire-armoured, PVC sheathed	PVC-insulated, unarmoured or steel-wire-armoured, PVC sheathed	Polythene-insulated, steel-wire-armoured, polythene-sheathed
Max working voltage	600 V AC RMS 1000 V DC	110 V AC 150 V DC	Various
Core resistance/km, (Ω)	7.4	37	37
Capacitance/km, (nF)	440 core-to-armour	150 core-to-core	50 core-to-core
Induced voltage-insulation withstand (1 min)	5 kV AC RMS	5 kV AC RMS	Two types: 5 or 15 kV AC RMS
Core identification	Each core is numbered	Each pair has a unique colour combination	Each pair has a unique colour combination
Standard cable sizes	2,3,4,7,12,19,27 or 37 cores	2,5,10,20,40,60, 100, 200 pairs	4, 7, 19, 37 or 61 pairs

systems between substations. These are alternatively known as pilot cables.

Three main types of auxiliary cable are currently installed; these are detailed in Table 8.1.

Multicore cables are 'wormed' and follow a helical lay. They are generally used to connect equipment which is operated at a nominal voltage of 110 V

DC or 240 V AC. They are required when voltage-drop considerations are important. During commissioning, particular attention should be paid to voltage drop associated with trip-coil currents. For example a trip coil with a typical current of 12 A would drop 15 V (see Section 9.24) across a multicore cable with a loop resistance of 1.25 Ω, i.e. a cable run of only 80 m. Longer cable runs would require paralleling of cores, or alternatively larger-cored cables can be installed. Some older multicore cables are impregnated paper insulated.

PVC-insulated multipair cables are used for light-current control purposes within substations, and generally operate at a nominal voltage of 48 V.

Polythene-insulated multipair cables are used as substation-to-substation pilot cables. The 5 kV or 15 kV insulation level should be chosen to match the level of induced voltage; however, present practice at transmission voltages is to standardise at the higher insulation level of 15 kV. In this context, 15 kV is usually taken as being the maximum permissible induced EMF into auxiliary cables. Where calculations indicate that the maximum induced voltage will exceed 15 kV, aluminium-wired armouring or solid-aluminium armouring may be required to improve the shielding. The capacitance of polythene cables is about one-third of that of PVC-insulated cables, and is therefore superior in terms of loss minimisation for long-distance signalling.

The general practice with all auxiliary cables is to earth the armouring at both ends to screen the conductors from induced voltages, and to keep the armouring at a safe potential.

8.40 50 Hz induced voltages

The cores of auxiliary cables are reasonably well shielded from external electric fields by the earthed armouring, but 50 Hz induced voltages can arise from magnetic fields resulting from power-system currents. The magnitude of 50 Hz induced voltage depends on:

(*a*) the distance between the power cable, or line, and the auxiliary cable;
(*b*) the magnitude of the current in the power circuit. If the phase conductors of the power circuit are physically symmetrically positioned with respect to the auxiliary cable, only power-circuit zero-phase-sequence current contributes to the induced voltage;
(*c*) the length of parallelism;
(*d*) the resistivity of the earth. The higher the resistivity, the less the screening effect of the earth, the greater the induced voltage; and
(*e*) the screening effect of the auxiliary cable armouring.

Voltages which are induced along the length of a cable core are termed longitudinal. Figure 8.20 illustrates in very simplistic terms the impact of induced longitudinal voltages for the instance of the auxiliary cable being both unearthed and earthed.

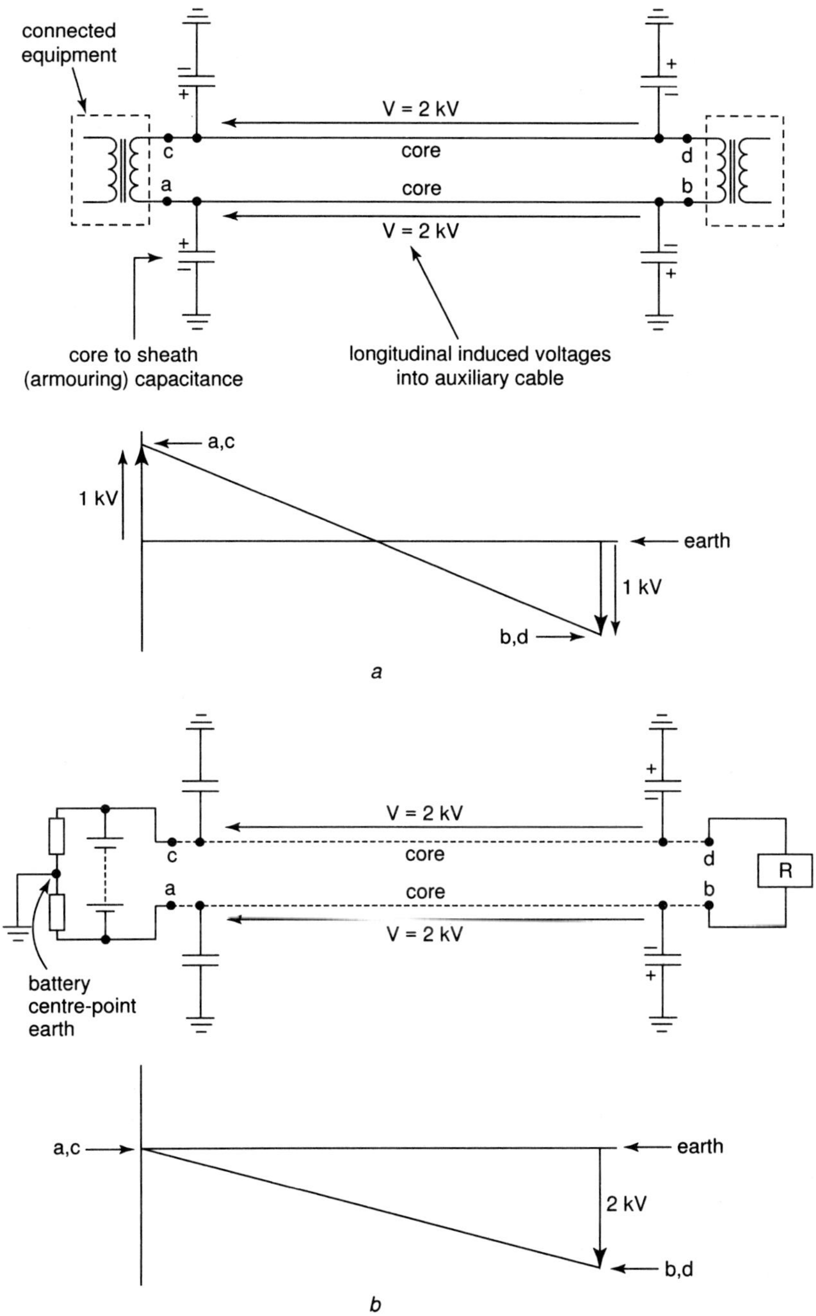

Figure 8.20 Effect of longitudinal voltages

a Unearthed auxiliary cable
b Earthed auxiliary cable

As can be seen, although the longitudinal voltage refers to the voltage difference between the cable ends, the effect of the core-to-armour capacitance is to create a voltage difference between core and armour.

The marginal differences which exist between core-to-armour capacitances of the various cable cores result in the longitudinal voltages, creating a core-to-core voltage difference, termed a 'transverse' voltage. This voltage also arises as a result of the very small differences between the longitudinal voltages in each core of multicore cables, caused by the marginal difference in separation from the power circuit. The design of these cables is such that transverse voltages are not expected to exceed 0.1% and 0.01%, of the longitudinal voltage in multicore and multipair cables, respectively.

8.41 Cables within substations

50 Hz induced voltages, arising from power-system load or fault currents, into either multicore or multipair cables which are totally within the boundary of a substation fence, are usually of such a small magnitude that they have negligible effect on equipment performance (multipair being much less susceptible than multicore cables).

This arises because the parallel runs between auxiliary cable and power circuits are short, and also because the significant number of auxiliary cables, with earthed armouring within the substation, results in multiple screening of any one cable by the remainder. The substation earth mat itself also has a screening effect.

50 Hz fault currents which flow in the earth mat also cause induced voltages into auxiliary cables but the earthed armouring again provides significant screening.

Induced voltages can also occur in multicore cables due to changes in the current in one core(s) inducing voltages into other cores, particularly during relay operations, or trip- or close-coil energisation. This should generally not be problematic if the length of multicore cable is less than 500 m.

Disconnector operation may also be a source of induced voltage when charging/discharging lengths of busbar connected to one side of an open circuit breaker. The resulting arc causes high-frequency radiated electromagnetic waves, the effect of which is to induce HF 'noise' into exposed wiring, and to a much lesser extent into auxiliary cables, which are less susceptible as a result of the armour screen. Note that at high frequencies the screen is less effective than when subject to 50 Hz, since loop impedance increases with frequency. In addition, the high-frequency arc current flows through insulator leakage capacitance to earth, and in turn may flow through the armour screen, so inducing voltages into the cable cores. The HF noise so produced has, in extreme cases, been known to produce relay maloperation, particularly if the relay is not suitably screened and earthed.

Some auxiliary cables, which are contained entirely within a building, are

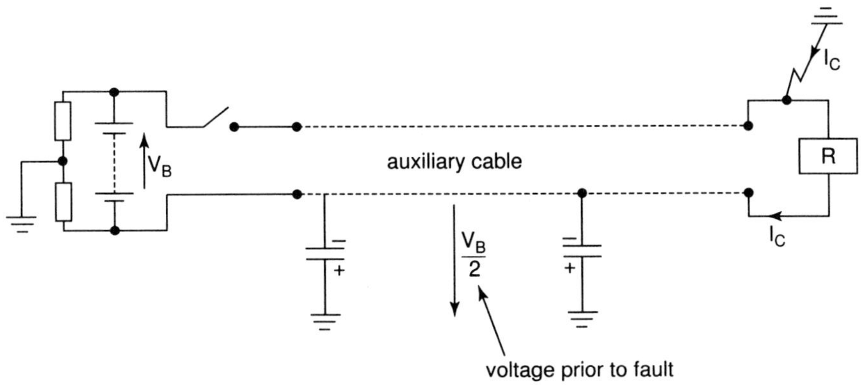

Figure 8.21 Relay maloperation arising from auxiliary-cable capacitance

unarmoured; this is acceptable because the risk of mechanical damage, and induced-voltage pickup, is low.

Multicore cables in particular, however, exhibit significant capacitance, the effect of which, coincident with a wiring earth fault, may be to cause relay maloperation, as shown in Figure 8.21. Thus relays which are not of trip relay standard, EB2 or ESI2 as described in Section 9.19, which are interconnected by auxiliary cables, should ideally be double-pole switched.

To obtain the necessary speed of intertripping which is required for transmission systems, it is necessary to interface intertrip-communication equipment with sensitive reed relays, having a nominal operating time of 2–3 ms. The interconnection of the intertrip-send circuitry with the reed relays, is usually accomplished via auxiliary cables. At dispersed-relay-room (blockhouse) sites, the length of such cables could be as much as 500 m; this can result, particularly under system fault conditions, in voltages being induced into the auxiliary cable to produce a transverse voltage sufficient to cause reed-relay operation. This, again, requires double-pole switching of the send circuitry, and the use of twisted-pair cable to minimise induced voltage pickup.

8.42 Pilot cables

Privately owned pilot cables are usually laid in the same trench as the associated high-voltage cable. They may, alternatively, be strung on poles or towers, or be laid in a trench which, for part of its distance, runs below the pole or line. In either instance, the cores of the pilot cable are subject to longitudinally induced voltages. The magnitude of induced voltage, for a pilot laid adjacent to a high-voltage cable, can be as high as 0.16 V/km/A of residual current (residual current is equal to three times the zero-sequence current). Thus a residual current of, say, 5000 A, flowing in a 16 km length of cable would produce an induced voltage of 12.8 kV, although this figure could be reduced with improved screening. Surprisingly, the induced voltage arising from overhead-lines is generally higher than with HV cables, up to a maximum of 0.19 V/km/A residual current. This arises because the screening effect of an overhead-line earth wire is less effective than a cable sheath.

Pilot-wire protection systems are usually designed to be relatively immune to circulating currents arising from transverse voltages. The limiting factor to the length of multicore pilot cable that can be used for such systems is usually pilot capacitance, which results in a maximum distance of about 25 km (400 Ω loop). To minimise induced voltage, the optimum cores as described in Section 8.43 should be employed. Multipair cable is both cheaper (core-for-core) and less susceptible to induced voltages than multicore, and the limiting factor in its application to pilot-wire protection is usually pilot resistance; maximum distance is about 33 km (approximately 2500 Ω loop). The pilots are usually electrically connected to the current transformers via a summation transformer, which should have an insulation strength which is sufficient to prevent breakdown across the windings, when subject to either longitudinal induced voltages in the pilots or substation rise in earth potential.

It is internationally accepted that no special insulation requirements or safe-working procedures are necessary if it can be shown that the induced voltages into pilots, or substation rise in earth potential, is less than 650 V RMS for the duration of fault-clearance time. This is usually achieved with distribution-voltage networks of 33 kV and below. However, at transmission voltages, the higher fault currents usually dictate the need for special insulation precautions. Thus at 132 kV there has been a tendency to standardise on a summation-transformer insulation level of 5 kV, which has been raised to 15 kV for the 400 and 275 kV networks. Wiring within the relay panel is also insulated to the 5 or 15 kV insulation level, respectively; see Figure 8.22.

Intertripping systems which use private pilots may be placed into three categories:

- direct
- surge-proof
- voice-frequency

110 V direct-wire intertripping consists of closing of a contact to energise a remotely positioned conventional trip relay, as shown in Figure 8.21, and it is usually suitable for pilot lengths up to 1 km. Voltage drop along the pilot will usually necessitate the use of multicore cable. To avoid the problem illustrated in Figure 8.21, double-pole switching should be employed (double-pole switching is now used in almost all new intertripping schemes).

Pilot lengths in excess of 1 km may be subject to induced voltages which could cause maloperation of the intertrip receive relay. This may be as a consequence of transverse voltages imposed across the relay, or due to a

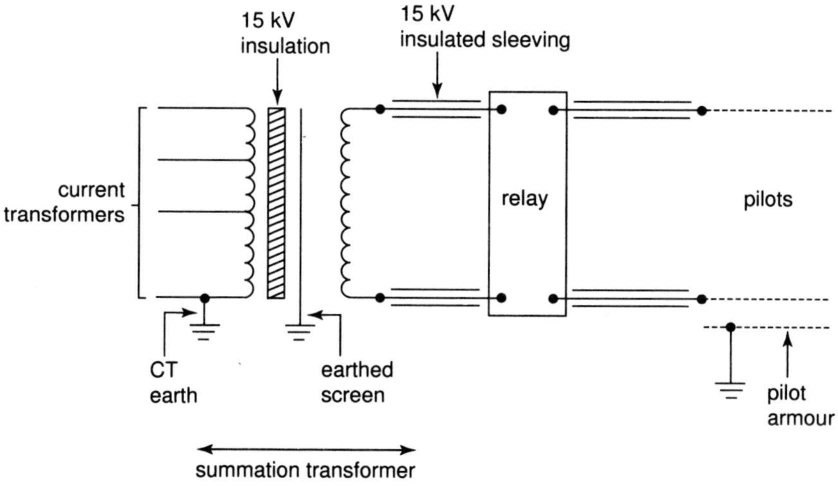

Figure 8.22 Pilot-wire-protection barrier equipment

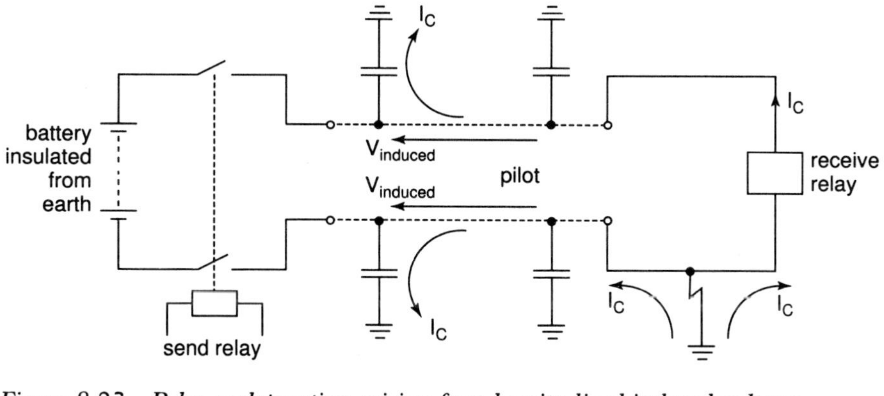

Figure 8.23 Relay maloperation arising from longitudinal induced voltages

combination of longitudinal voltages and a wiring fault to earth, so resulting in pilot capacitance current through the relay, as shown in Figure 8.23. Surge-proof relays may be employed to overcome these problems.

Figure 8.24 illustrates the basic features of a *surge-proof* relay. The objective is to maximise the transfer of direct voltages to the receive relay while rejecting the alternating interference voltages. Both multicore and polythene-insulated multipair cables may be used. Multicore cables have a lower resistance to accommodate voltage drop, but a higher capacitance. The limit of application of surge-proof relay systems is the ability of the relay to withstand alternating interference voltages, and the corresponding insulation level of the cable.

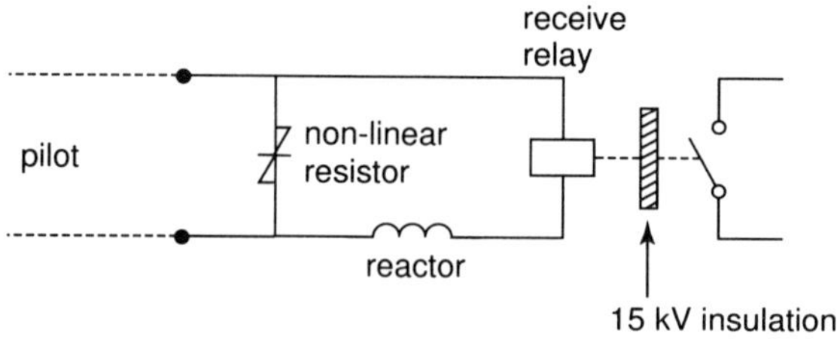

Figure 8.24 Basic surge-proof intertrip relay

Surge-proof relays are normally fed from a battery installed specially for the purpose. It is usually 15 kV insulated from earth to prevent pilot-induced voltages impacting on substation equipment, and substation rise of earth potential impacting on the pilot. Surge-proof relays have fastest operating times of the order of 80 ms and are thus too slow for application to transmission systems.

Voice-frequency intertripping usually employs a coded or frequency-shift system of communication, over polythene-insulated (15 kV) multipair pilot cables. These systems have a high security against maloperation arising from induced voltages. The pilot should be connected to the intertrip equipment via an interposing transformer, with insulation levels identical to those for protection, as described above. The limit of application is when the induced voltage exceeds the pilot-insulation level. The majority of VF intertripping systems are used with rented pilots.

8.43 Multicore-cable core selection

It is expected that almost all future private pilots used for protection purposes will be polythene-insulated multipair types, to minimise the effect of induced voltages. However, many existing installations use multicore cables, and no doubt situations will arise requiring the continued use of multicore cables in future (e.g. direct-wire intertripping in the same pilot).

To minimise the effect of transverse voltages on pilot-wire-protection equipment, the optimum cores must be selected. This requires a pair of cores with a high value of core-to-core capacitance (i.e. relatively low core-to-core impedance to minimise the induced voltage difference) and a low value of unbalance capacitance (i.e. pair-to-earth and pair-to-pair).

One method of evaluating the optimum pair of cores is to apply 50 Hz 240 V between each core and earth in turn, with all other cores unearthed, and measuring the voltages appearing between all the other cores, and all the other cores and earth. Table 8.2 shows an example of this for a 12-core multicore cable of some 10 km length with the voltage applied to core 3.

A table similar to Table 8.2 is produced for each core-to-earth injection. A high-impedance voltmeter must be used to avoid corruption of the test

Table 8.2 240 V applied between core 3 and earth

Core	1	2	3	4	5	6	7	8	9	10	11	12	E
1	—	0	0	60	60	60	60	60	60	60	60	60	100
2	—	—	0	0	60	60	60	60	60	60	60	60	100
3	—	—	—	—	—	—	—	—	—	—	—	—	240
4	—	—	—	—	0	0	0	0	0	0	0	0	50
5	—	—	—	—	—	0	0	0	0	0	0	0	50
6	—	—	—	—	—	—	0	0	0	0	0	0	50
7	—	—	—	—	—	—	—	0	0	0	0	0	50
8	—	—	—	—	—	—	—	—	0	0	0	0	50
9	—	—	—	—	—	—	—	—	—	0	0	0	50
10	—	—	—	—	—	—	—	—	—	—	0	0	50
11	—	—	—	—	—	—	—	—	—	—	—	0	50
12	—	—	—	—	—	—	—	—	—	—	—	—	50

Table 8.3 Collation of all voltage injections

Core	1	2	3	4	5	6	7	8	9	10	11	12
1	—	1	1	60	60	60	60	60	60	60	60	60
2	—	—	1	60	60	60	60	60	60	60	60	60
3	—	—	—	60	60	60	60	60	60	60	60	60
4	—	—	—	—	50	50	50	50	50	50	45	45
5	—	—	—	—	—	50	48	48	50	50	46	47
6	—	—	—	—	—	—	42	48	48	48	49	50
7	—	—	—	—	—	—	—	50	47	47	47	49
8	—	—	—	—	—	—	—	—	52	51	47	49
9	—	—	—	—	—	—	—	—	—	51	50	48
10	—	—	—	—	—	—	—	—	—	—	40	49
11	—	—	—	—	—	—	—	—	—	—	—	48
12	—	—	—	—	—	—	—	—	—	—	—	—

results. When all tests have been completed, the maximum recorded voltages between cores, taken from all tables, should be collated onto a summary sheet, as shown in Table 8.3.

From the above the best pairs are:

1 and 2, 2 and 3, or 3 and 1

followed by cores 10 and 11 and so on.

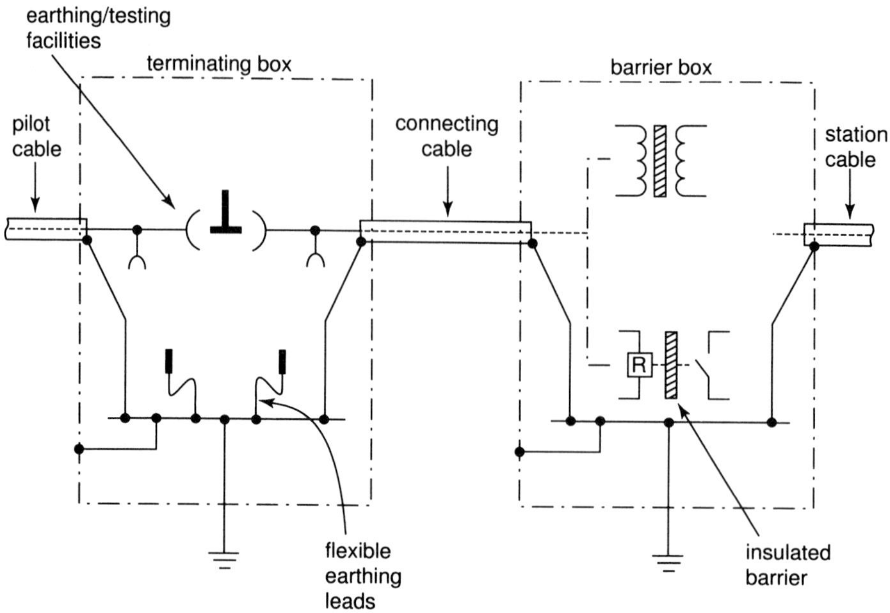

Figure 8.25 Typical pilot-cable-termination arrangements

8.44 Pilot-cable-termination equipment

If the high transient voltages induced into pilot cables are transferred to the wiring within a substation, damage to termination equipment could occur, and danger to personnel arises. To prevent this, for the instance of induced voltages exceeding 650 V RMS, the pilot should be terminated via barrier equipment located within the substation.

Figure 8.25 illustrates a typical arrangement of termination and barrier equipment. The two-box arrangement facilitates safe working on either the pilot or the substation-side equipment. The pilot-terminating box permits disconnection and earthing of the pilot using in-built facilities.

The relay arrangement shown in Figure 8.25 accommodates DC signals, such as alarm data sent from an intermediate position in a cable circuit (e.g. low oil pressure). The transformer accommodates AC signals such as voice-frequency data.

Protection equipment would not usually pass through this box, but has its own barrier equipment, located in the relay panel as described in Section 8.42.

8.45 Rented pilots

Many rented pilots consist of 5 kV insulated multipair cables. Rented pilots will not be laid in the same trench as power cables, nor usually follow the route of an overhead line. Thus induced voltages into a rented pilot from a power circuit are usually not the major concern. Of greater concern is a rise of earth potential being transferred onto the rented pilots. Such cables are usually unarmoured, but one core of a two-wire circuit would be earthed at the telecommunication exchange. In summary, requirements for connecting to rented pilots are:

(*a*) If the equipment to which the pilots are connected is connected directly to the HV system via CTs, VTs etc., such as with a pilot-wire or voice-frequency protection, 15 kV insulation level is required. This is usually provided by the protection system.
(*b*) If the equipment to which the pilots are connected is indirectly connected to the HV system, the required insulation level is dependent on the substation rise in earth potential (RIP). For an RIP less than 650 V RMS, no special insulation is required. For an RIP between 650 and 1000 V RMS, 2 kV insulation is required, and for an RIP in excess of 1000 V, 5 kV insulation is required. The insulation level is usually obtained by use of an interposing transformer on the substation side of the pilot cable.
(*c*) The peak voltage developed on the pilot cores under maximum fault-level conditions must not exceed 130 V, and the maximum transmitted current must not exceed 60 mA RMS. These values are only permissible during the short duration of the fault.

These conditions are very much concerned with safety to telecommunication-company personnel and equipment.

Chapter 9

Substation power supplies

9.1 Introduction

This chapter examines substation batteries and associated equipment, and the relationship between battery power supplies and the voltage-limitation requirements of trip relays and trip coils. In brief the design, commissioning and maintenance requirements of the following are discussed:

- substation batteries, chargers and earthing
- trip-relay types and performance limits
- trip-coil performance limits
- power-supply arrangements and voltage limits

BATTERY-SYSTEM FUNDAMENTALS

9.2 Battery systems

Although batteries are probably the most basic item of electrical equipment, a discussion with most site engineers will reveal that surprisingly little is known about them. This is perhaps not surprising since the deeper one investigates them, the more thought provoking they seem. A typical battery system would comprise a battery, a charger, a distribution board, a method of earthing and monitoring relays.

9.3 The lead-acid cell

The lead-acid Planté cell, or accumulator, was for many years almost the only type of cell utilised in substation batteries, but its dominant position is now being challenged by recombination cells. The lead-acid cell consists of a container, traditionally glass but more recently plastics, containing a solution of dilute sulphuric acid (the dielectric), into which is immersed a positive pole consisting of a plate of lead peroxide, and a negative pole consisting of a plate of lead. Figure 9.1 shows the arrangement. A modern cell would consist of a number of positive and negative plates, each held by a group bar.

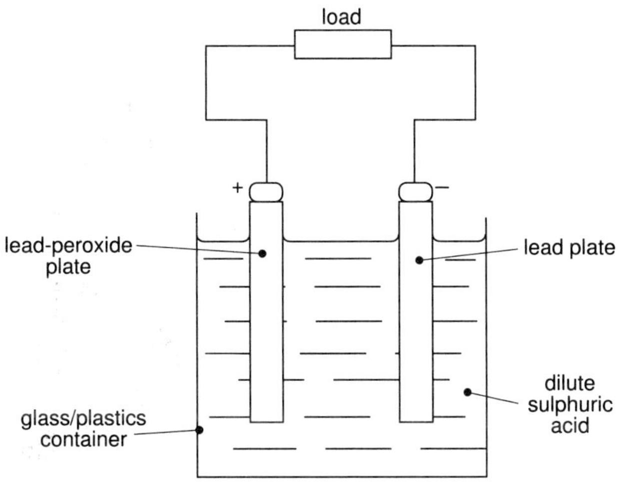

Figure 9.1 Basic design of a lead-acid cell

When a circuit is connected across the terminals, the cell commences to discharge and electrons flow from the negative to positive pole around the circuit. Inside the cell, the negative sulphate ions of the electrolyte move towards the lead, and the positive hydrogen ions drift towards the lead peroxide. This results in lead sulphate being formed at both plates, and the sulphuric acid in the cell being displaced by water. If the cell is subsequently recharged from an external voltage source, the reaction is reversed, and plates of lead and lead peroxide, and the sulphuric acid, are reformed.

Since the process of cell discharge converts sulphuric acid to water, the specific gravity (SG) of the electrolyte falls. The specific gravity can thus indicate the degree to which the cell is discharged. A fully charged cell has an SG of about 1215, and a fully discharged cell an SG of about 1150, at a temperature of 10°C.

9.4 Discharge characteristic

If a cell is discharged at very low rates of discharge current, for say 10 hours or longer, the cell terminal voltage initially settles at 2 V, and slowly falls over a long period of time, as illustrated in Figure 9.2. The nominal voltage of the cell is thus described as 2 V. In the UK, substation batteries are usually (but not exclusively) either 110 V nominal (55 cells) or 48 V nominal (24 cells). 110 V batteries are usually employed for circuit-breaker tripping, closing and protection purposes; 50 V batteries are employed for telecommunication and control purposes. In the following sections, the voltages associated with each battery type are depicted in the form (110 V/48 V).

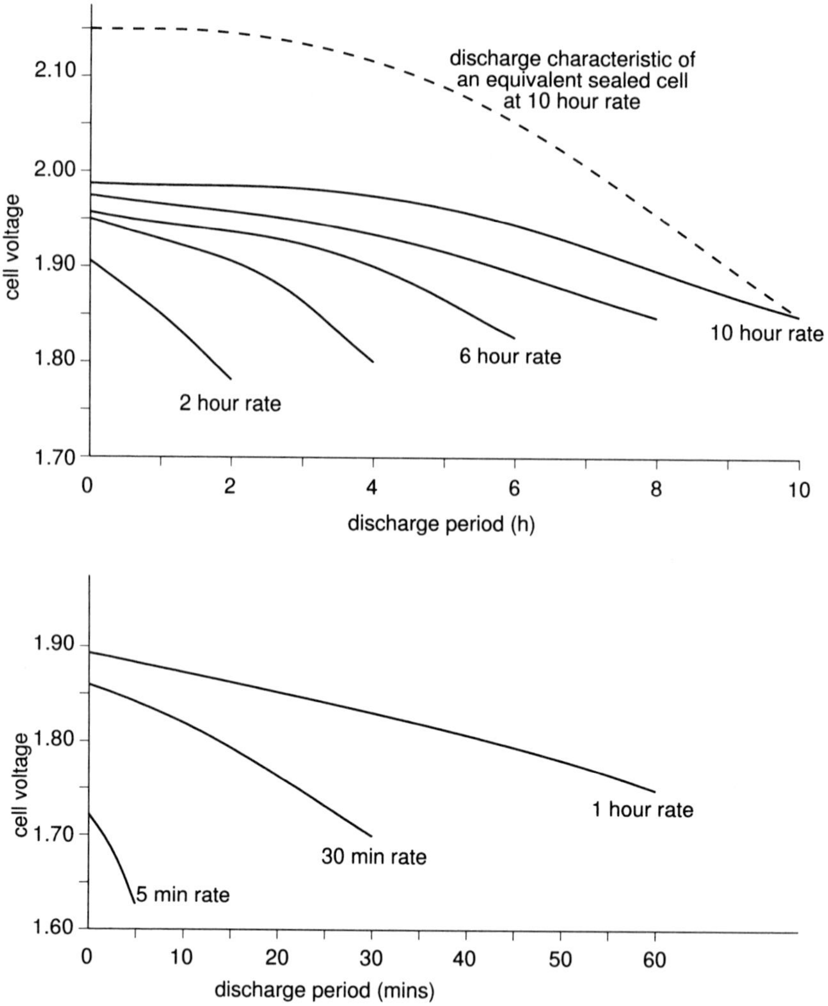

Figure 9.2 Discharge curves of a 250 Ah Plante battery

Discharge period (h)	10	8	6	4	2	1	0.5	5 min
Current (A)	25	30	37	54	92	150	240	580
Cell end voltage (V)	1.85	1.84	1.83	1.80	1.78	1.75	1.70	1.63
Battery end voltage								
55-cell (V)	102	101	100	99	98	96	93	90
24-cell (V)	44	44	44	43	43	42	41	39

The capacity of a battery or cell is usually expressed in ampere-hours (Ah), at the 10 h rate. For example, a 250 Ah battery or cell is capable of providing 25 A for 10 h without the cells suffering any deterioration. In order to take a steady current of 25 A from the battery, the load resistance must gradually decrease, since the battery terminal voltage falls with time. The concept of a varying resistance may seem inconsistent with practical loads, which are usually fixed (e.g. the fixed resistance of a relay). However, the fixed current-discharge rate is used as the basis of battery-performance discharge curves, as provided by manufacturers.

Figure 9.2 illustrates a typical range of discharge curves for a 250 Ah cell. Note that:

(*a*) The terminal voltage at the commencement of discharge decreases with increasing discharge current. This reflects the voltage drop across the cell internal resistance.
(*b*) As shown in Figure 9.2, the terminal voltage at the end of the discharge decreases with increasing discharge current. These figures are determined by manufacturers to ensure that the cell suffers no deterioration.
(*c*) Typical manufacturers' discharge curves range from the 10 h rate to the 5 min rate.

On termination of a discharge, the cell terminal voltage should slowly recover to about 2 V.

9.5 Charging requirements

If a cell which has been discharged is recharged, the voltage will rise to a maximum of about 2.7 V (unless the recharge is limited to a lower voltage). As the recharge takes place, the lead sulphate plates are reconverted to lead and lead peroxide, respectively. As recharge of the cell continues, electrolysis of the water, which is part of the sulphuric-acid solution, takes place, and hydrogen and oxygen gas are liberated. Excess charging would then lead to a loss of water from the cell. In practice, electrolysis commences before the plates are fully reconverted, as a result of localised reactions. The electrolysis process is known as 'gassing' and commences at a cell voltage of about 2.2 V (121/52.8 V). The ideal voltage to which a cell should be charged, to maintain it in a healthy condition with a minimal loss of electrolyte, is 2.25 V (123.75/54.0 V).

9.6 Battery chargers

In general, battery chargers for substation Planté batteries serve three purposes:

(*a*) to recharge a battery which has become discharged, as quickly as possible, without damaging the cells;

Table 9.1 Performance requirements for class EB1, EB2, ESI1 and ESI2 relays

	EB1	EB2	ESI1	ESI2
DC system nominal volts (V)	110	110	110	110
Relay rated voltage (V)	125	125	125	125
DC-system normal voltage (V)	125	125	125	125
Operative voltage range (V)	66–143	66–143	87.5–137.5	87.5–137.5
Minimum operating current (i.e. shall not operate below this value) (mA)	25	50	25	50
Able to withstand a capacitance discharge of 10 μF 150 V	No	Yes	No	Yes

(*b*) to provide a float charge to hold the cell voltage at 2.25 V (123.75/54.0 V) while supplying the standing load;
(*c*) to boost charge the battery from 2.25 V/cell up to the maximum voltage of 2.7 V/cell (in practice the maximum voltage is often limited to less than 2.7 V/cell; see Table 9.1). To avoid damage to the cell, the boost-charge finishing current should be limited to 7% of the 10 hour capacity (i.e. 7 A per 100 Ah battery rating); this is termed the finishing rate.

Specialist literature on battery chargers describes an almost bewildering variety. Many modern substation chargers, however, are characterised by float and boost limit adjustment for both voltage and current; this permits the setting of voltage and current limits, i.e. V_{FLOAT}, I_{FLOAT}, V_{BOOST}, I_{BOOST}.

Battery-charger rated capacity (output current) is usually specified as being equal to the maximum standing load plus the finishing rate. For example, a 200 Ah-capacity battery with a maximum standing load of 10 A would dictate a charger capacity of

$$10 + (7\% \times 200) = 24 \text{ A}$$

At the commissioning stage, the voltage and current limits should be set as follows:

V_{FLOAT} = 123.75/54.0 V (2.25 V/cell)
I_{FLOAT} = maximum standing-load capacity
V_{BOOST} = between 137.5 and 148.5 V or 60 and 64.8 V (2.5 and 2.7 V/cell) (see Section 9.9 for limitations of boost voltage)
I_{BOOST} = 7% of 10 hour capacity + maximum standing-load capacity

Recharging of a discharged battery should commence with the charger being selected to float. The start current will then usually commence as (and be limited to) the maximum standing-load capacity. Some literature recommends that there is little point in recharge current exceeding 14% of 10 hour capacity, e.g. 28 A for a 200 Ah battery, since higher currents are not 100% efficient in the plate-reformation process. As the cell voltage rises, the current will fall; when the cell voltage reaches 2.25 V/cell (123.75/54.0 V), the V_{FLOAT} voltage limit will limit the voltage to this value. At this point, boost charge can be selected; the current will then initially rise to the I_{BOOST} current-limit value, and then fall as the voltage rises to the V_{BOOST} limit setting.

A fully discharged cell would be brought to 2.25 V/cell in about 12 h, and raising the voltage to 2.7 V/cell, via boost charge, would take about another 8 h.

9.7 Undercharging

The result of undercharging the battery is sulphation of the plates. The term sulphation relates to the conversion of the plates to lead sulphate, which occurs during the discharge process. However, if the battery is left in a state of sulphation for sustained periods, it becomes difficult to reconvert the lead sulphate back to lead/lead peroxide, and the battery loses capacity. Sulphation commences at a cell voltage of 2.15 V (118.25/51.6 V); the lower the cell voltage the worse the sulphation. Sulphated plates tend to be gritty and harsh in appearance, eventually becoming a sulphury yellow in colour.

9.8 Overcharging

Overcharging occurs if cells are left in a sustained state of charge above 2.27 V(125/54.5 V). Symptoms are: excessive water consumption; distortion and expansion of positive plates; excessive deposits of flaky dark peroxide material; spongy grey lead deposits on the side of negative plates; and possible positive-plate-group bar and plate-edge scaling—which may possibly bridge the positive and negative plates, so causing partial short-circuits.

9.9 Boost charge

If a battery is float charged at 2.25 V/cell over a long period, there is a tendency towards electrolyte stratification, i.e. strata of electrolyte with differing specific gravities. This results in a loss of output when the battery is required to discharge into a load. Boost charging turbulates and mixes the electrolyte as a result of the gassing process.

A programme for boost charging may consist of:

(*a*) once per four years to counteract stratification;
(*b*) whenever the battery is discharged by more than 20% of ampere-hour capacity to counteract the effects of sulphation;

(*c*) whenever the battery has been left on open circuit for a period longer than two months. This is a conservative figure as the rate of self discharge on open-circuit is about 4% per month.

9.10 Safety

The liberation of hydrogen and oxygen gas during charging creates a potential source of danger. A hydrogen/oxygen gas mix, can, if ignited, be explosive. Before any work on batteries takes place, the battery room should be well ventilated. The access plugs should also be removed from each cell and a hydrometer used to puff in air, to expel any hydrogen or oxygen gas. No flame or spark should be employed close to the battery.

9.11 110 V nominal battery earthing

There are two main systems of battery earthing: battery-biased systems and battery-centre-point-earthed systems.

Figure 9.3*a* illustrates a battery-biased system; a DC quantity of about 30 V is derived from an AC source, via isolating transformers, and connected to the positive pole of the battery. This system has the disadvantage of possessing different sensitivities to earth faults occurring on the positive and negative pole of the battery, respectively. The negative pole also stands at some 155 V with respect to earth, and thus anyone inadvertently touching the negative pole wiring receives a significant shock. The case for battery biasing was based upon maintaining all equipment at a negative potential with respect to earth, to prevent corrosion of equipment due to electrolytic action. This case is, however, weak, since:

(*a*) equipment (relays etc.) is usually directly connected to the negative pole of the battery, and not the positive, and therefore less subject to corrosion;
(*b*) relay and equipment rooms are usually well heated and ventilated, thus minimising moisture—a necessary factor for electrolytic action;
(*c*) the capacitance to earth of the wiring connected to the negative pole (shown dotted in Figure 9.3*a*) stands at −155 V with respect to earth. An earth fault between a tripping contact and a trip relay (shown dotted in Figure 9.3*a*) thus results in the capacitance discharging through the trip relay, endeavouring to cause operation. This situation is further aggravated by virtue of the full 155 V of battery, and battery biasing voltage, circulating a continuous current around the faulted circuit;
(*d*) centre-point-earthed systems have not, in practice, exhibited unacceptable levels of corrosion.

Figure 9.3*b* shows a centre-point-earthed system. This is now applied to almost all new installations. This system is equally sensitive to faults on the positive

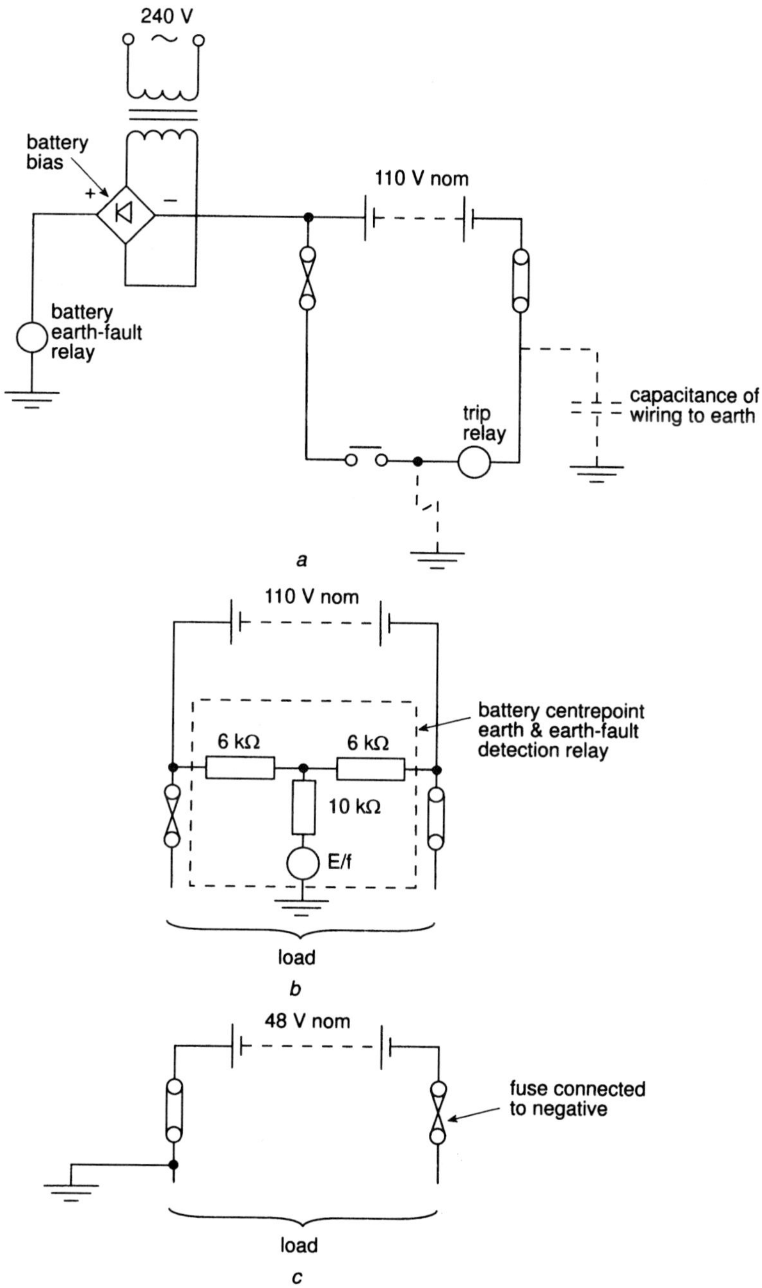

Figure 9.3 *Battery earthing*

a Battery-biased system
b Battery centre-point-earth system
c 48 V battery positive-pole earthing

and negative poles, and can typically detect earth faults on either pole down to a value of 50 kΩ. It removes virtually all the disadvantages of the battery-biased system.

9.12 48 V nominal battery earthing

This follows telecommunications practice of positive-pole earthing. As a consequence, the fuse is placed in the negative pole, as shown in Figure 9.3*c*.

9.13 Battery monitoring

In addition to a battery earth-fault-detection relay, a battery-monitoring system suitable for a transmission substation would comprise:

(*a*) low-voltage alarm relay, required, essentially, to monitor charger failure. This would typically have a setting of 2.15 V/cell (118/51.6 V), to detect the voltage at which sulphation commences;
(*b*) high-voltage alarm relay, which is also to monitor failure of the charger. This would typically be set to 2.32 V/cell (128/55.7 V), which is the voltage at which gassing, leading to significant loss of electrolyte, commences;
(*c*) battery-open-circuit alarm relay. This unit injects a pulse to measure battery internal resistance and outputs an alarm when the resistance exceeds about 1 Ω. Figure 9.4 illustrates a typical complete battery system.

9.14 Recombination cells

In the mid 1980s an alternative to the lead-acid Planté cell, the recombination 'cell', appeared on the market. *BS 6290* describes the cell as a 'lead-acid valve regulated sealed cell', more colloquially known as a 'sealed cell'.

As previously noted, the Planté cell liberates hydrogen and oxygen while the cell is being charged. The sealed cell recombines the hydrogen and oxygen, within the cell, to form water. Thus, unlike the Planté cell, there is no loss of electrolyte. The electrolyte within the sealed cell is held within a glass microfibre sponge-like separator, and thus, even if the cell container is fractured, there is no spillage of electrolyte. The sealed cell is fully sealed except for a pressure-relief valve, which vents to atmosphere if a predetermined internal pressure is exceeded.

In most types of sealed cell, several cells, usually of the order of six, are grouped together into one case, termed a monobloc. A monobloc of six cells would result in a nominal terminal voltage of $6 \times 2 = 12$ V. A consequence of this arrangement is that the equivalent of a 110 V nominal Planté battery is nine monoblocs of 12 V, a total of 108 V; with a 48 V nominal battery, four monoblocs would suffice. Although the sealed cell has a lower nominal

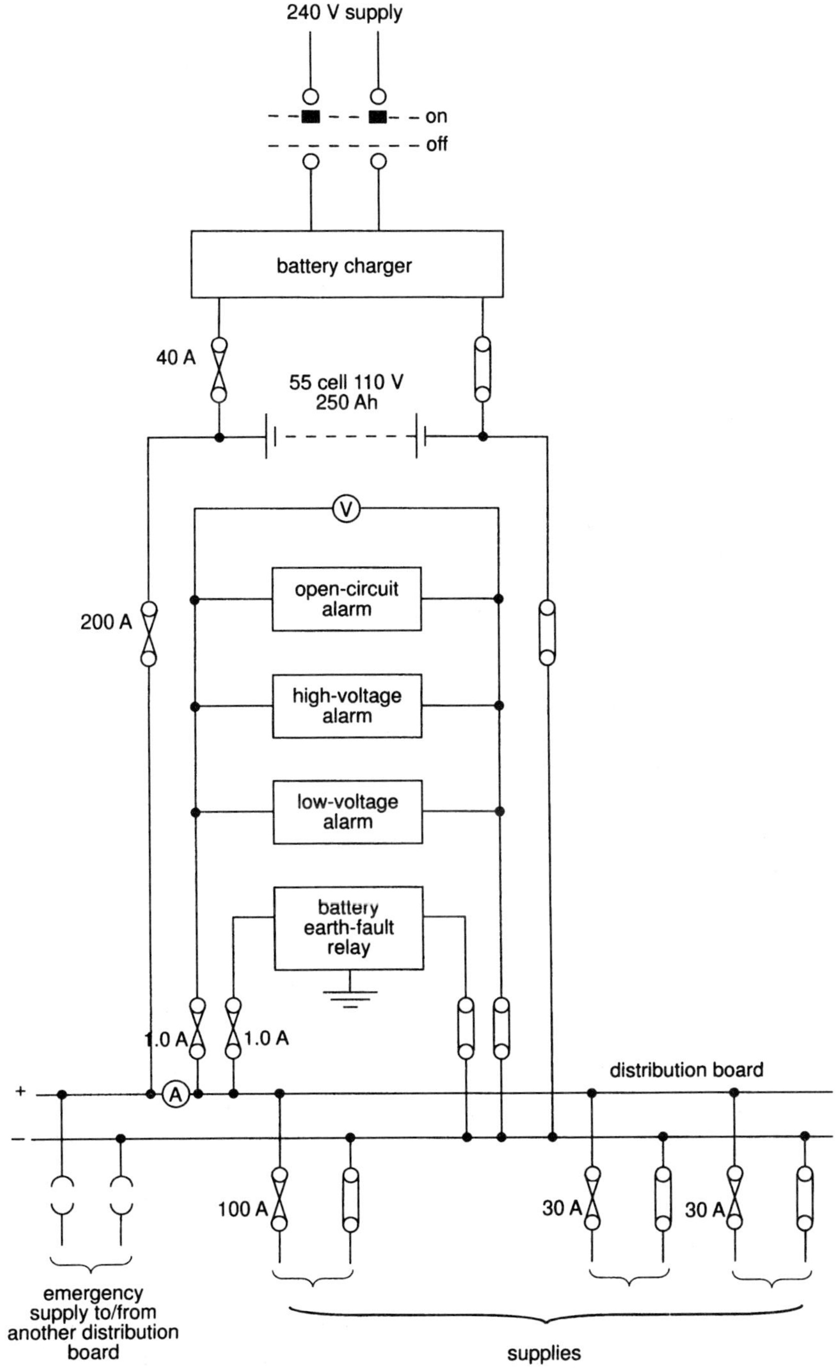

Figure 9.4 Typical transmission-substation battery-system arrangements

voltage than the Planté, it has a superior voltage profile for the same discharge current, as shown in Figure 9.2.

The recommended float voltage is 2.27 V per cell with an acceptable range of 2.24–2.48 V/cell. This results in the following battery voltages:

110 V battery (equivalant) nominal = $54 \times 2.27 = 122.6$ V float voltage
48 V battery nominal = $24 \times 2.27 = 54.48$ V float voltage
The low voltage alarm should be set at 2.13 V/cell (115/51.1 battery volts).

Unlike the Planté cell, there is no case for boost charging sealed cells.

The relative merits of sealed cells are:

(*a*) relatively maintenance free, no topping up of electrolyte;
(*b*) much smaller in size than the Planté equivalent;
(*c*) has an estimated life of between 10 and 15 years, compared to 20–30 years for Planté;
(*d*) cheaper in price than the Planté equivalent.

BATTERY COMMISSIONING

9.15 Battery commissioning (Planté)

Individual cells are usually delivered to site in a fully charged condition. They should be installed and connected to the charger as soon as possible. In particular, they should not be left in an uncharged state for periods longer than eight weeks. During erection, the pillars and intercell connections should be liberally coated with petroleum jelly. After erection, the electrolyte should be topped up to the recommended level using distilled or de-ionised water. Larger cells may be delivered to site devoid of the electrolyte; they should be filled as soon as possible. Note should be taken of manufacturers' recommendations. Following installation the following tests should be carried out.

(*a*) *Battery-charger tests*

Before connecting the charger to the battery, special care should be taken to ensure that the polarity is correct. It needs to be noted that the open-circuit voltage of some chargers (particularly older designs) can be very high, with a 'peaky' waveshape. When connected, the battery acts as a voltage-smoothing device. Care must be taken, however, to ensure that at no point during the tests does the charger become connected to the load, independent of the battery, since the charger voltage may damage the load equipment. Note that the load itself also acts as a partial smoothing device.

Once the charger is connected, it should be left to float charge the battery for at least 24 h. The charger should then be switched off and the battery discharged down to about 1.85 V. As the battery voltage falls, the low-voltage alarm should be observed. The charger should then be reconnected and a check made on the float-current-limit feature (which is usually factory set to the charger rated output current).

Once the recharge current has tapered down to a negligible value, the charger should be selected to boost and the boost-current limit checked. The over-voltage-alarm relay can also be checked at this time. The battery should then be fully boost charged, during which the specific gravity and voltage should be taken for each cell. Boost charging should continue until three consecutive hourly specific gravity and cell-voltage readings remain constant. The charger can then be selected back to float. A visual inspection of the cell plates should reveal colours of brown–black positive plates and light grey negative plates.

(b) Battery-discharge tests

The final stage is to carry out a discharge test. Some literature states that Planté batteries have proved to be so reliable that a discharge test is not needed. Other literature points to the fact that the battery is such an important component in the control of the power system that positive proof of the battery capability should be acquired, thus necessitating a discharge test. In the author's opinion the case is made for the discharge test.

Two types of discharge test can be carried out: a low-current discharge or a heavy-current discharge. The low-current discharge consists of switching the charger off, and discharging into a resistor bank at either the 6 h or 10 h rate (e.g. 25 A for 10 h for a 250 Ah battery). Cell voltage and specific gravity should be recorded periodically and should accord with the manufacturer's published data.

The heavy-current test consists of switching off the charger and allowing the voltage to settle at the nominal voltage of 2 V per cell. A heavy current, typically of magnitude equal to the 1 h rate (e.g. 150 A for a 250 Ah battery), should then be discharged into a resistor bank, for about 90 s. The voltage immediately prior to switching off the discharge should be recorded and compared against manufacturers' data. It is necessary to maintain the discharge for up to 90 s because the voltage profile tends to fall initially, and then rise again before levelling out, as illustrated in Figure 9.5.

Note that the requirements of the battery-discharge and battery-charger tests should be integrated to avoid duplication of effort.

(c) Earth-fault relay

The battery earth-fault relay should be proved by placing a variable resistance between each pole, in turn, and earth to find the earth-fault resistance which causes operation. After the load is connected, various points in the wiring should be short-circuited to earth to ensure that the relay is operating, and resetting when the earth is removed.

9.16 Battery commissioning (sealed cells)

This largely follows the same format as for the Planté cell except that no filling, topping up, boost charging or measurement of the specific gravity of

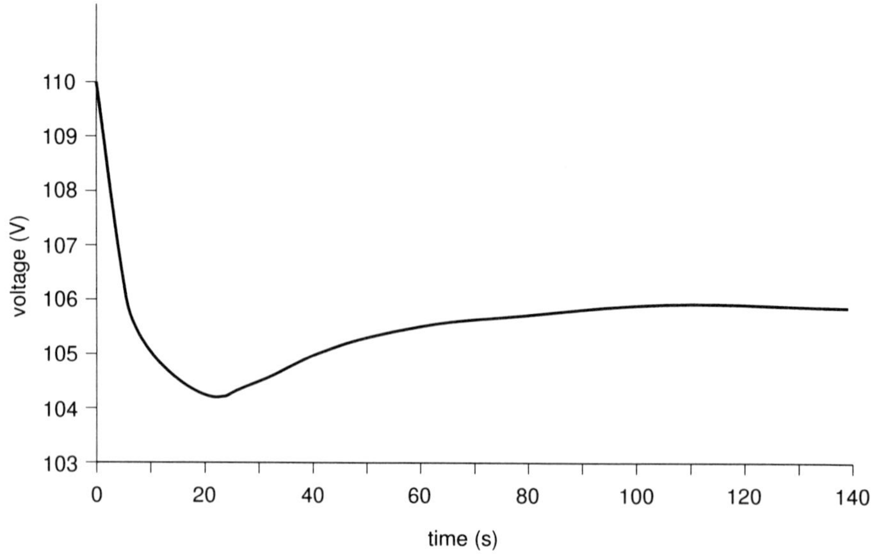

Figure 9.5 *Typical heavy-current discharge curve*

300 Ah, 110 V nominal battery, discharge = 180 A

the electrolyte are required, and there is no visual inspection of the plates since the cell container is usually opaque.

BATTERY MAINTENANCE

9.17 Battery maintenance (Planté)

Battery-maintenance requirements can be summarised as follows:

- visual inspection
- periodic check of cell voltage and specific gravity
- discharge test
- cell-termination regreasing and cleaning
- boost charge
- check of charger output current
- battery-monitoring relay test

These requirements are examined in greater detail below.

(a) Visual inspection

A visual inspection is a good indication of the condition of a battery. As stated earlier, a healthy cell has brown–black positive plates and light grey negative plates. There should be no debris covering the group bar, or attached to the cell walls (which may cause a short-circuit).

New Planté cells have a tendency to shed material, which looks like cigar ash, onto the cell floor; this is not a problem as long as it does not short-circuit the cell plates. Older cells frequently shed material from either plate onto the cell floor; this is a sign of aging.

(b) Cell voltage and specific gravity

Most manufacturers' literature recommends that once per month, the battery voltage, individual cell voltage, and specific gravity are measured and recorded. The cell voltage should be between 2.25 and 2.28 V per cell, and the specific gravity between 1.205 and 1.215 at a temperature of 15°C. The temperature at which the readings are taken should be recorded, and reference made to manufacturers' literature to interpret readings which may be taken at temperatures other than 15°C.

(c) Discharge test

It is usually not an option to discharge fully a battery which is in service, since it may be short of capacity if it is required in an emergency during the discharge. For this reason, the heavy-current 90 s discharge, as described in Section 9.15, is preferred. One of the advantages of the heavy-current discharge is that it proves that the battery will produce and stand up to a high current demand, synonymous with that required during circuit-breaker tripping. A reasonable timetable for the discharge test is two years after commissioning and four yearly thereafter.

(d) Termination regreasing/cleaning

Dirt and grit should be removed from the intercell connections as required. Peroxidisation, which is the blackening of the pillar terminal due to acid creep, should be removed, using a fine emery cloth, and the terminal wiped clean. Every four years or so, the old grease should be removed and replaced by a new coat of petroleum jelly. It is usually not necessary to break down and remake the connections.

(e) Boost charging

The case for regular boost charging is weak. An optimum maintenance timetable is described in Section 9.9. The four-yearly boost charge should follow the discharge test.

(f) Charger output current

The charger standing load current should be recorded at planned intervals, to check whether the connected load is approaching charger capacity. The recording interval will depend on the substation. Battery systems which may regularly be subject to additional load should be checked at least once per year.

(g) Battery-monitoring relays

The battery low- and high-voltage alarms should be checked during the discharge and subsequent boost-charge tests. The battery earth-fault relay should be checked, whenever a high-voltage circuit is out of use for maintenance, by momentarily connecting a short-circuit from wiring to earth.

9.18 Battery maintenance (sealed cells)

This can be summarised as follows:

- termination greasing and cleaning, as Planté cells;
- periodic check of the voltage of a block of cells;
- discharge test as for Planté cells. Given that a visual examination of the cells cannot be undertaken, nor specific-gravity readings taken, this test assumes a greater importance in verifying battery condition than for a Planté cell;
- check of charger output;
- battery-monitoring-relay tests as with Planté cells.

DC RELAYS

9.19 Trip relays

The function of a trip relay is to increase the number of contacts of a protection output relay. This is required because a protection relay may only have one output contact, but it may be required to carry out a number of functions (e.g. to trip more than one circuit breaker, to intertrip, to initiate fault recorders or to generate an alarm). In addition at least one trip-relay contact must be robust enough to make, carry, and in some instances break, trip-coil current: this is not always within the design capability of protection output relays. A modern trip relay will usually be specified to have an operating time of 10 ms or less; in practice this is often of the order of 6 ms. An essential requirement of a trip relay is its ability to operate within specified fluctuations of battery power supplies.

Trip relays may be categorised as:

(*a*) *Self reset:* i.e. the relay remains energised only as long as the initiating contact is closed (see Figure 9.6*a*);
(*b*) *Manually reset:* i.e. the relay is equipped with a cut-throat contact. Once operated it remains operated until manually reset (see Figure 9.6*b*);
(*c*) *Electrically reset:* This consists of a double-coil relay with a cut-throat in the energisation-coil circuit. When the relay is operated, a contact is made in the reset-coil circuit. The reset coil is then usually energised via a timer circuit to reset the relay (see Figure 9.6*c*). Such an arrangement is beneficial if (for example) a prolonged intertrip signal needs to be sent from the trip relay.

At least one contact of the trip relay is usually of the magnetic blow-out type,

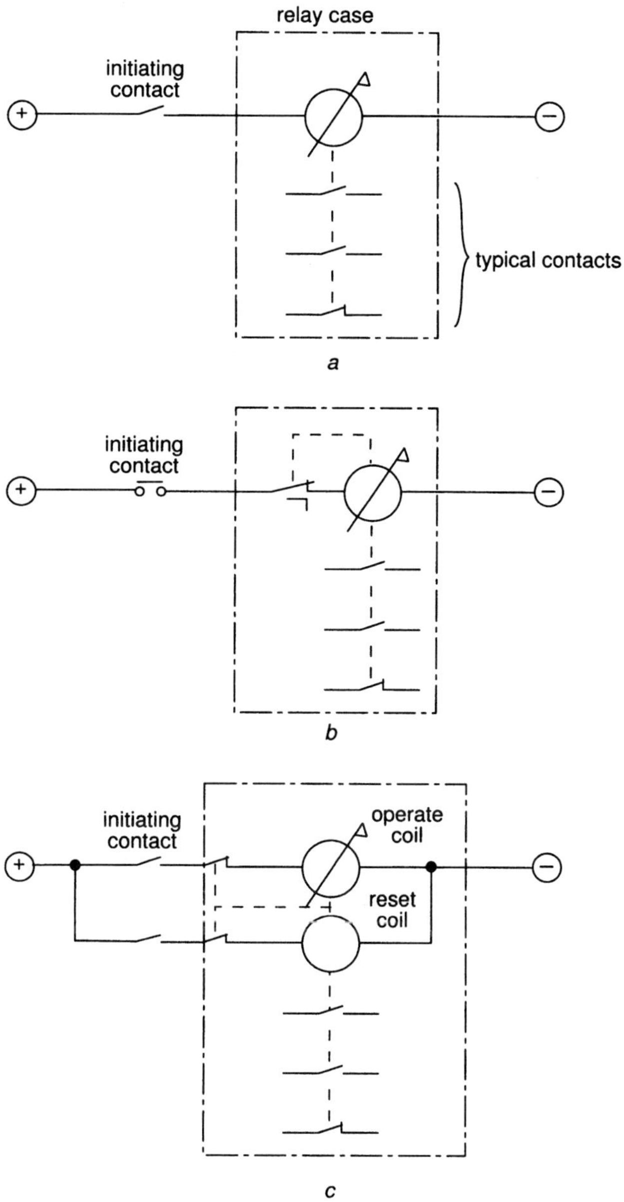

Figure 9.6 *Trip-relay types*

a Self-reset relay
b Manually reset relay
c Electrical reset

capable of interrupting circuit-breaker trip-coil current. In most instances, such a contact is not essential since, as shown in Figure 9.9, a circuit-breaker auxiliary contact usually breaks the trip-coil current when the circuit breaker opens.

In general, DC relays with low minimum operating currents and high operating speeds are liable to maloperation from capacitance currents, which may flow through the relay coil circuit if an earth fault occurs on the DC circuits associated with the relay. Relays are less liable to electrical maloperation if they:

(i) have both sides of their coils switched, i.e. are double-pole switched; and
(ii) do not have extensive external wiring associated with them; for example, if the relay wiring is confined to the relay room.

In recognition of this problem, UK industry standards define the design parameters of four classes of 110 V-nominal DC relays suitable for tripping purposes as shown in Table 9.1.

Type EB1 and ESI1 relays are not required to withstand the capacitance discharge. They are thus used where, from the nature of their coil connections, they are not likely to be liable to maloperation, i.e. they are double-pole switched or the coil is connected to wiring which is contained entirely in the relay room. Figure 9.7*a* illustrates the application of EB1 and ESI1 relays.

Type EB2 and ESI2 relays are, however, required to withstand the capacitive discharge. The 10 μF, 150 V criterion is taken as being a reasonable reflection of onerous site conditions, although it is acknowledged that it is not the theoretical worst condition. The 150 V corresponds to the negative-pole-to-earth voltage of a battery-biased system. Figure 9.6*b*, illustrates the application of EB2 and ESI2 relays.

Type EB1 and EB2 relays refer to an earlier specification than ESI1 and ESI2. All new installations should accord with ESI1 and ESI2. The operating voltage range, and hence permissible battery-voltage range, of EB1 and EB2 is 66-143 V (i.e. a cell voltage of 1.2–2.6 V). This determines the boost-charge voltage limit of the battery charger. Conversely, the ESI1 and ESI2 operating-voltage range is 87.5–137.5 V (i.e. a cell voltage of 1.59–2.5 V) and again this determines the boost-charge voltage limit of the battery charger as 137.5 V.

Table 9.1 identifies the requirements for 110 V relays, but similar specifications also exist for 30 and 48 V DC system nominal voltages.

In practice the relay operating currents tend to be higher than the minima quoted in Table 9.1 and more of the order of 0.25–1.0 A. The larger the trip relay current, the greater the immunity to maloperation, but the greater the wiring voltage drop.

At the commissioning stage, trip relays should be injected with increasing currents and the values of current and voltage at the point of operation should be recorded.

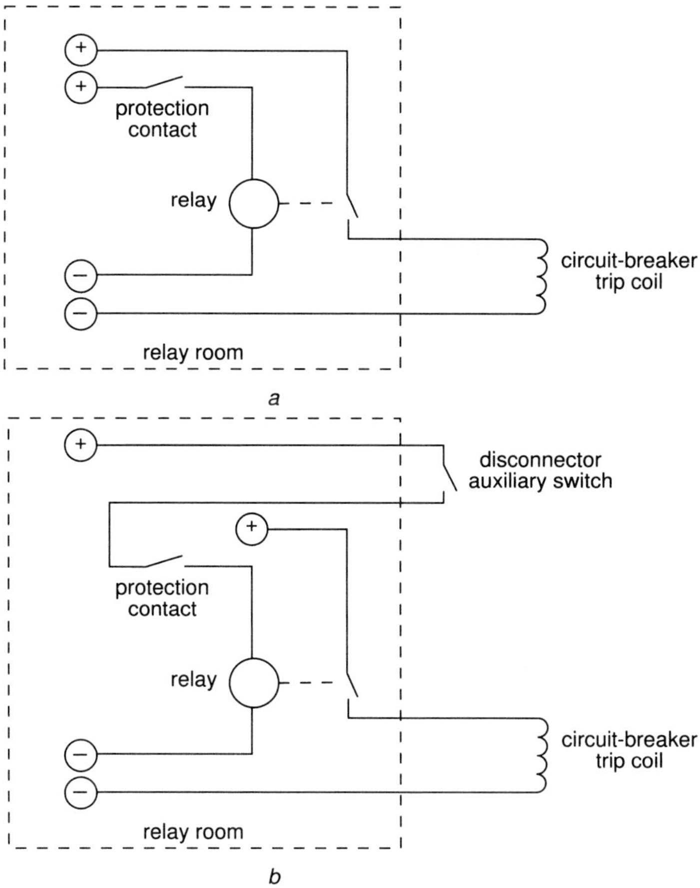

Figure 9.7 Trip-relay application

a Application of EB1 and ESI1 relays
b Application of EB2 and ESI2 relays

9.20 DC relays (general)

DC relays are required for a number of reasons other than trip-relay requirements. Typical applications are trip-circuit supervision (see Section 9.26), DC-supply supervision and general-purpose auxiliary relays. *BS 142* requires all such relays to have an operating range of 87.5–137.5 V, identical to trip relays.

The DC relays so far considered have been based upon 110 V nominal battery systems, since it is this voltage which is generally utilised for protection and tripping purposes. Circuit-breaker manual trip and close interpose relays, however, frequently utilise a 48 V battery system. Such relays are usually double-pole switched and are required to operate over a pilot resistance of up to 200 Ω. They are required to operate down to a minimum voltage of 32.5 V (UK supply industry requirements).

TRIP COILS

9.21 Trip-coil voltage limits

It is necessary to ensure that the voltage supplied to the trip coil terminals does not fall below the minimum required for operation during circuit-breaker tripping. Earlier circuit-breaker types were designed around the requirements of *BS 116* which required trip coils to work down to a minimum voltage of 88 V while the circuit breaker was carrying load, or 55 V when not carrying load. However, given that most circuit breakers operate via a trip-coil releasing a stored-energy mechanism, the distinction between load and no-load conditions is irrelevant. Thus most of these types of circuit breaker will operate with a trip-coil voltage down to 55 V under all conditions.

More recently, trip-coil design has been in accordance with *BS 5311* which specifies a minimum trip coil voltage of 87.5 V, and a maximum of 137.5 V (i.e. the same as for DC relays).

The voltage limits applied to trip coils usually also apply to closing coils.

9.22 Elements of voltage drop

The voltage appearing at the trip-coil terminals, when a circuit breaker is required to trip, is dependent on a number of factors, as follows:

(*a*) the drop in voltage which takes place at the battery terminals, due to the step increase in current supplied by the battery (caused by the trip coil current). The magnitude of this voltage drop is influenced by the battery capacity (i.e. the larger the battery capacity, the smaller the internal resistance, and the less the voltage drop), and the capacity of the associated battery charger;
(*b*) the voltage drop in the wiring between the battery terminals and the trip coil. This in turn is influenced by whether the feed from the distribution board is a radial or ring circuit;
(*c*) the number of circuit breakers required to be tripped simultaneously.

9.23 110 V battery capacity

The capacity required of a 110 V nominal battery is dependent on how the battery is utilised. These typical applications are:

(*a*) a battery located in a distribution-voltage substation, solely for the purpose of circuit-breaker tripping and protection;
(*b*) a general-purpose substation 'common' battery located in either a distribution-voltage or transmission-voltage substation, for the purpose of circuit-breaker tripping, closing, protection, disconnector control, interlocking etc.;
(*c*) a per-circuit 'blockhouse' battery used for tripping, closing and protection requirements solely on that circuit. These are found in transmission substations.

The criteria for determining battery capacity by the erstwhile CEGB in the UK for 110 V nominal batteries for transmission substations were:

(*a*) *Common batteries*

(i) With the simultaneous tripping load of a number of circuit breakers applied for 50 ms (i.e. resulting from a busbar-type fault), and the charger and standing load connected, the battery terminal voltage should not fall below 102 V; and
(ii) With the tripping load of one circuit breaker applied for 50 ms, and after the charger has been disconnected for 6 h, the battery terminal voltage should not fall below 102 V.

Note that the criteria for determining the capacity of a substation 48 V common battery, for telecommunications and control purposes, is simply the ability to supply the maximum standing load for 6 h in case of charger failure.

(*b*) *Blockhouse batteries*

The strategy adopted was as for common batteries, item (ii) above, but with a reduced terminal voltage of 93 V. Since a blockhouse battery supplies only one circuit, the busbar-type fault criterion is not applicable. The reduction to 93 V is in recognition that the wiring runs, and hence voltage drop, between battery and trip coil are usually less than with common batteries.

9.24 Wiring voltage drop

With the battery terminal voltages of 102 and 93 V identified in Section 9.23, the permitted wiring voltage drop, to ensure at least 87.5 V at the trip coil, would be 14.5 V for common batteries and 5.5 V for blockhouse batteries.

9.25 Trip-coil voltage tests (common batteries)

Figure 9.8 illustrates the three most common ways of feeding circuit-breaker trip coils from a substation common battery and distribution board (for clarity, only the positive pole is shown). At the commissioning stage, it is

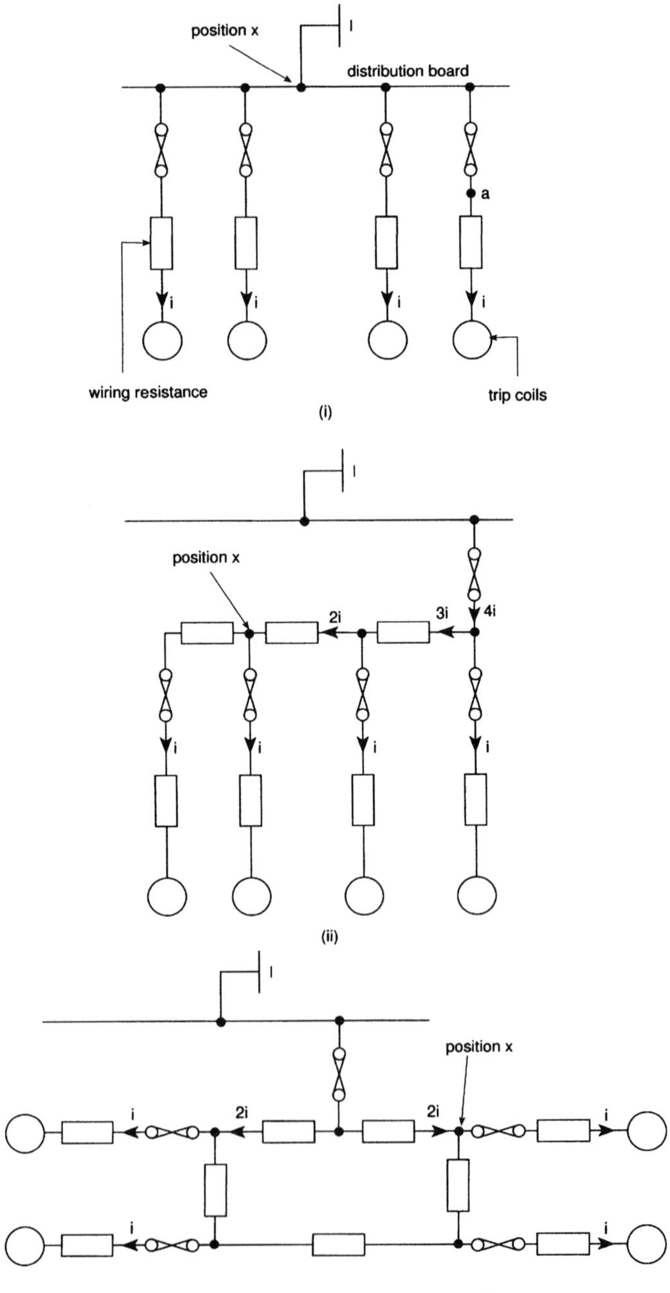

Figure 9.8 Trip-coil voltages

(i) Radial supply
(ii) Spur supply
(iii) Ring supply

Table 9.2 Short-duration discharge characteristics of a 200 Ah 110 V nominal Planté battery

	Time (ms)					
Current	0	5	10	20	50	100
(A)	(V)	(V)	(V)	(V)	(V)	(V)
20	125.0	123.5	123.3	123.0	122.8	121.1
40	125.0	122.7	122.4	121.9	120.8	119.0
60	125.0	122.0	121.5	120.7	118.9	116.5
85	125.0	120.5	120.0	119.2	117.0	114.8
125	125.0	119.1	118.5	117.3	114.5	112.7
160	125.0	118.2	117.2	115.8	112.5	111.1
180	125.0	117.1	116.1	114.4	111.1	110.3

necessary to verify that the voltages which appear at the trip coils exceed 87.5 V, and that the battery terminal voltage and wiring accord with the requirements of Sections 9.23 and 9.24 (or other similar supply authority requirements). The methodology is as follows.

(*a*) *Battery-capacity evaluation*
To evaluate battery suitability, it is necessary to have a plot of performance over typical periods of time in which circuit breakers trip (typically 50 ms). Table 9.2 illustrates such a plot, obtained by discharging a 200 Ah 110 V nominal battery, charged to 125 V, into a resistor bank, with the terminal voltage measured on an oscilloscope. Once acquired, this information would apply to any typical 200 Ah battery, and with such information the suitability of the battery can be determined without the need to carry out detailed tests. For example, suppose a 200 Ah battery is supplied from a 25 A charger and has to accommodate a maximum duty of tripping eight circuit breakers simultaneously, with each trip coil taking nominally 20 A, then the following approximations can be deduced:

Total current required = $8 \times 20 = 160$ A.
Current supplied by charger = 25 A.
Therefore current to be supplied by battery = 160–25 = 135 A.
From Table 9.2 the battery terminal voltage would be approximately 114.0 V after 50 ms.

Note that circuit-breaker trip-coil currents vary with the circuit-breaker type, and range typically between 8 and 20 A.

(*b*) *Single circuit-breaker trip*
When a new substation is being built, circuits are usually commissioned and entered into service as soon as they are constructed. In each instance, it is

necessary to prove the acceptability of the trip-coil voltage. This can be accomplished on a per-circuit basis by tapping the battery at the next tapping above 102 V (or as close as practicable), and with reference to Figure 9.8, applying the tapping lead via a switch to the trip-coil side of the fuse supplying the trip coil (with fuse removed). The trip-coil voltage should be monitored with a chart recorder or oscilloscope, and the switch and chart recorder should be initiated simultaneously to obtain the voltage profile across the trip coil, while the circuit breaker is tripping This voltage should exceed 87.5 V. Note that some trip coils are specified in terms of an operating current, and are fitted with an adjustable resistance in series with the trip coil to enable the current to be adjusted. Note also that the minimum charge should be taken from the tapped battery (i.e. the test, ideally, should only be carried out once), or the tapped cells will end up in a different state of charge from the remainder. Equalisation via boost charge will then be required.

(*c*) *Multiple circuit-breaker trip*
When all circuits of the substation are in service, a check should be made to evaluate performance under busbar fault conditions. The worst case should be checked, which is usually that of a busbar fault on a reserve section of busbar when reserve sections 1 and 2 are solidly connected. For example, this would be the situation when, say, main busbar 1 is out of use for maintenance, with the majority of circuits then being connected to the reserve busbars. For reasons of power-system security, it is usually not possible to trip all the circuit breakers attached to a busbar and thus the equivalent tripping load should be simulated, as follows (see Figure 9.8).

The circuit breaker with the longest wiring run should be identified. A resistance bank equal to the tripping load of the remaining breakers on the busbar should then be connected through a switch to position X (the point at which the breaker with the longest wiring run is connected). A wire should then be taken from the connection between the switch and the resistance bank to the trip-coil side of the fuse [e.g. position *a* in Figure 9.8(i)] supplying the trip coil with the longest wiring run (with the fuse removed). A chart recorder or oscilloscope should then be connected across the trip-coil terminals. The switch and chart recorder should be initiated simultaneously to obtain a record of voltage across the trip coil while the circuit breaker is tripping, the magnitude of which should exceed 87.5 V.

9.26 Trip-circuit supervision

Figure 9.9 illustrates a typical trip-circuit supervision relay plus associated wiring. When the circuit breaker is open, coils A and B and C are energised. When the circuit breaker is closed, coils A and C are energised, and when the circuit breaker has tripped, but the trip relay is still operated, coils B and C are energised. In the period between the trip relay being operated and circuit breaker opening, coils A and B are both short-circuited. The relay must thus

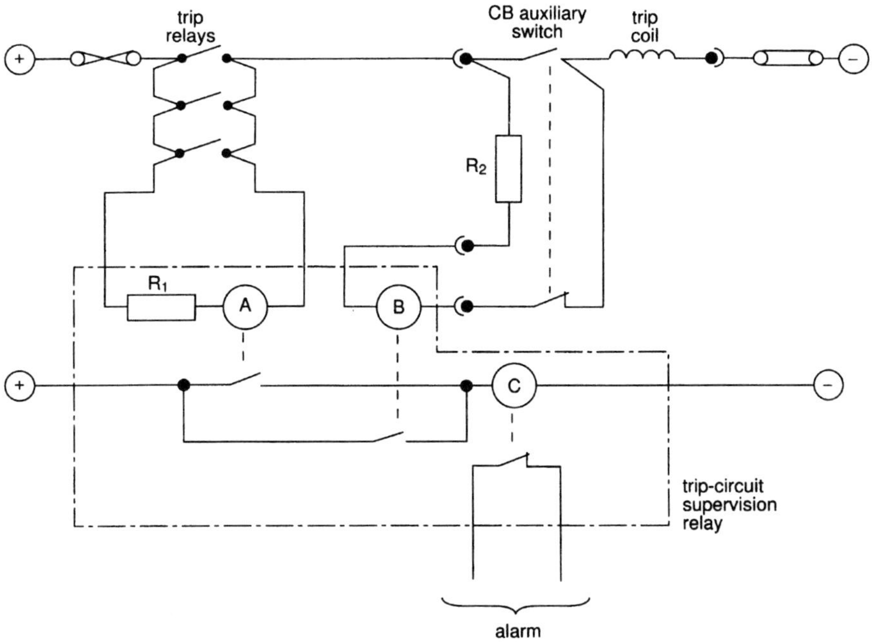

Figure 9.9 Trip-circuit-supervision relay

have a delayed-drop-off feature, typically of 500 ms, to prevent a spurious alarm being generated.

The supervision current through the trip coil is typically arranged via the resistors to vary between 10 and 20 mA, and not to exceed 60 mA if any one component of the supervision chain fails and short-circuits.

The trip wiring is arranged to loop from one trip-relay contact to another, with the supervision A coil at the end of the loop. This ensures maximum supervision of trip wiring. This looping should be proved during commissioning.

The supervision circuit shown provides complete supervision of the trip circuit. Some supervision schemes are simpler and may just involve a pushbutton-operated lamp which indicates continuity of the trip circuit only when the circuit breaker is closed. These simpler schemes are to be found in distribution substations which may not be equipped with extensive alarm facilities.

Chapter 10

Current and voltage transformers

10.1 Introduction

Current and voltage transformers provide the interface between the power-system high-voltage plant (primary plant) and the protection, control and metering equipment (secondary plant). The earlier sections of this chapter deal with current transformers, including primary injection techniques, and the later sections with voltage transformers.

CURRENT-TRANSFORMER THEORY

10.2 Current-transformer equivalent circuit

In the writer's experience many practising engineers have only a superficial understanding of the workings of current transformers (CT). In particular, the distinction between a current transformer and power transformer is frequently blurred. It is however important that the practising engineer is fully aware of these distinctions, which are best explained with reference to the transformer equivalent circuit.

Figure 10.1*a* illustrates, in schematic form, a typical 2000/1 current transformer such as may be found on a 400 kV network. For illustration purposes a current of 1000 A is taken to be flowing in the primary of the transformer and thus a secondary current of 0.5 A flows into a secondary load of 1 Ω. The secondary load of a CT is known as the 'burden'.

Figure 10.1*b* shows a simple equivalent circuit based on Figure 3.16*b*. In this instance, the magnetising impedance (as is usual) is shown on the secondary, and the magnitude given is a typical value for this type of CT magnetic-core *B*/*H* curve. The leakage reactance is small and is ignored. The power-system load imposed on the 400 kV network is the ratio of 400 kV-system voltage and current:

$$Z=\frac{V}{(\sqrt{3})\,I}=\frac{400\ 000}{1.73\times 1000}=230\ \Omega$$

Figure 10*c* shows all values referred to the CT primary side. In this instance, all secondary values are multiplied by

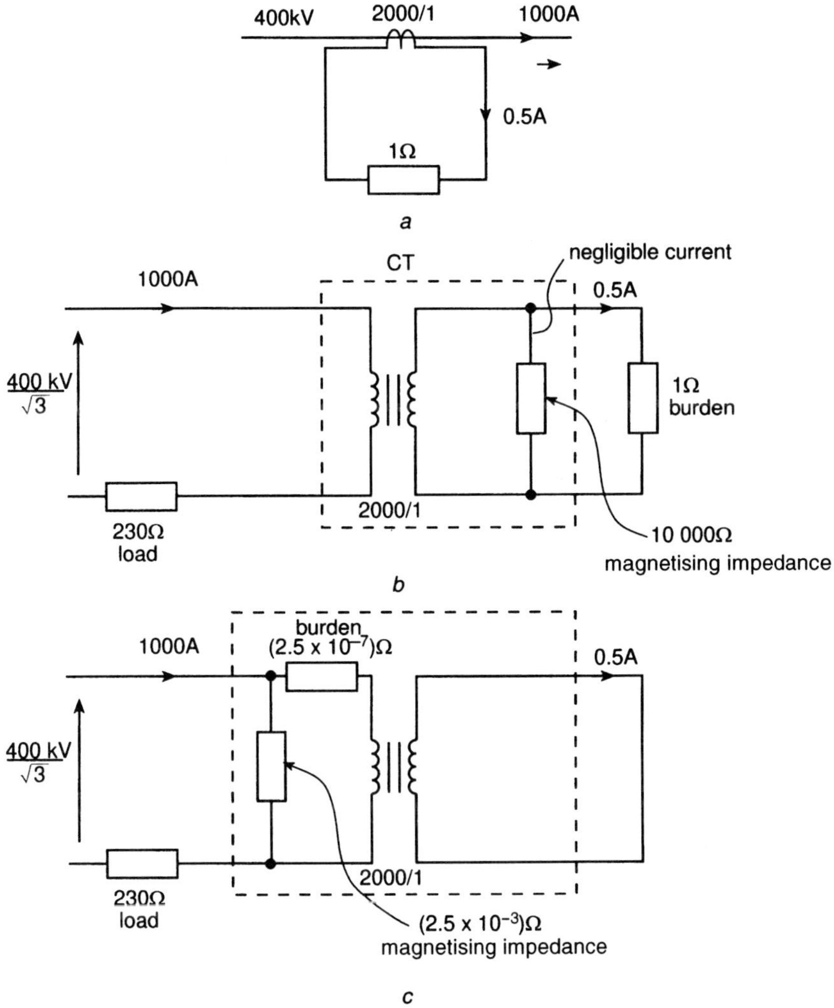

Figure 10.1 *The current transformer: a series-connected transformer*

a Circuit under consideration
b Simple equivalent circuit
c All values referred to primary side

$$\left(\frac{N_p}{N_s}\right)^2 = \left(\frac{1}{2000}\right)^2 = 2.5 \times 10^{-7}$$

As can be seen, the referred secondary impedances are so small compared with the load of 230 Ω as to be negligible. The current which flows in the CT primary (and secondary) is thus effectively determined by the primary

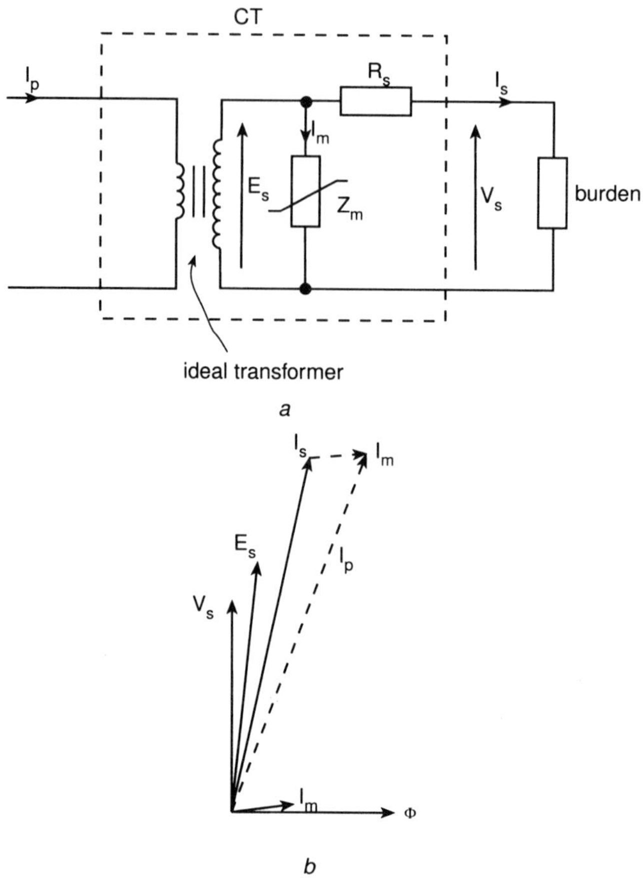

Figure 10.2 CT equivalent circuit

a Practical equivalent circuit
b Typical phasor diagram

(power-system) load. Further, the current flowing into the secondary is virtually independent of the burden on the secondary (or more precisely the parallel combination of the burden and magnetising impedance). This point will be examined in more detail shortly. The CT is a 'series-connected' transformer since the burden is effectively connected in series with the power-system load; this differs from a power, or voltage, transformer which is 'parallel connected' and has the full system voltage imposed across it.

Figure 10.2*a* shows the practical equivalent circuit. The resistance of the secondary winding R_s is usually much higher than the leakage reactance, which is usually ignored. The resistance and reactance of the primary winding are also relatively small and thus can be ignored. Figure 10.2*b* illustrates a typical phasor diagram.

10.3 Burden

The burden on a CT is a measurement of the load expressed in volt-amperes (VA) at the rated secondary current. For example if the rated secondary current was 5 A and the impedance was 2 Ω, the burden would be

$$\begin{array}{ccc} (5\times 2) & \times 5 & = 50\text{ VA} \\ (\text{V}) & \times (\text{A}) & \end{array}$$

However it is common practice, on site, to refer to the burden as the ohmic value of the connected load, although this usage is strictly incorrect.

Section 10.2 made the point that the current flowing into the burden is dependent upon the parallel combination of the magnetising impedance and burden. If, with reference to Figure 10.1*b*, the burden was increased to, say, a reactance of 1000 Ω, the current into the burden would be:

$$\frac{10\,000}{11\,000}\times 0.5 = 0.45\text{ A approx.}$$

i.e. the current would divide in proportion to magnitudes of the burden and magnetising impedance, thus leaving 0.05 A to flow into the magnetising impedance. Under such a scenario the voltage across the burden would be:

$$0.45\text{ A}\times 1000\ \Omega = 450\text{ V}$$

If we take the analysis one step further and replace the burden by an open circuit, all of the current would then flow through the magnetising impedance, and the voltage so produced would be

$$0.5\times 10\,000 = 5000\text{ V}$$

The current flowing in the power system has thus far been considered to be load current; if we now consider fault current of, say, 8000 A, the voltage developed across the magnetising impedance with the secondary on open circuit would be

$$\frac{8000}{2000}\times 10\,000 = 40\text{ kV}$$

At this stage two points are worthy of note. The first is that the voltage on the CT secondary increases with increasing burden, and rises to dangerously high levels if the CT secondary is open circuited. Such high voltages are a danger to life and can damage secondary wiring and equipment. The second point is that, when the above situations arise, the flux in the CT magnetic core rises to such a high level as to cause saturation. This significantly reduces, but does not eliminate, the high secondary voltages.

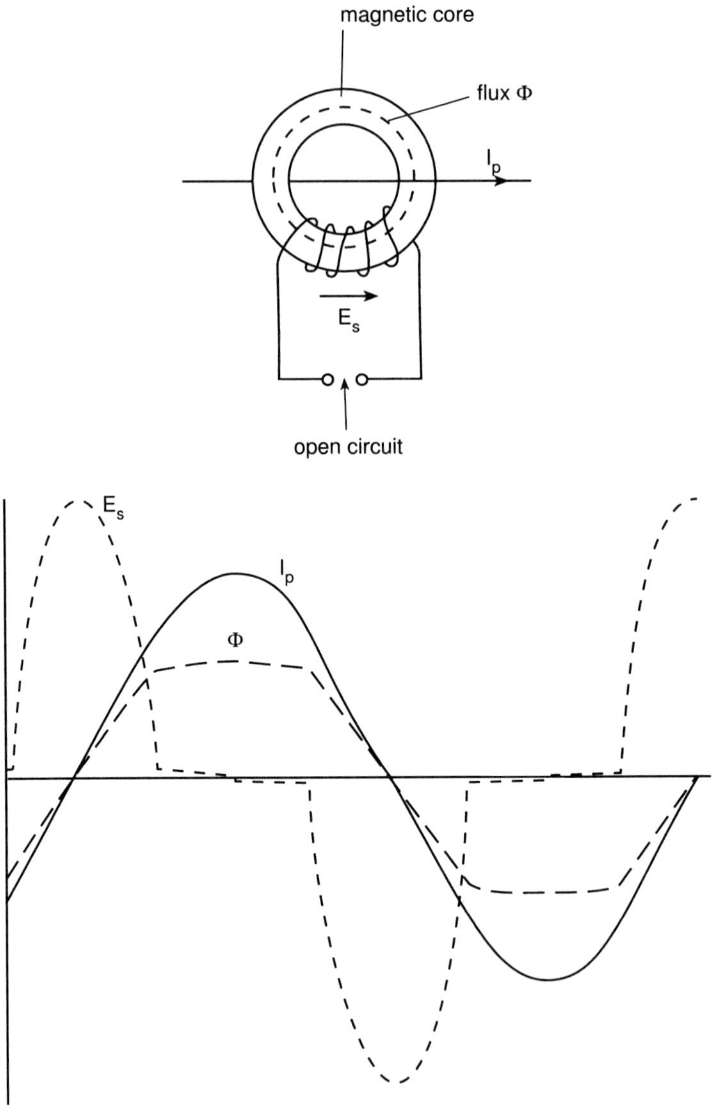

Figure 10.3 Effect of driving a CT into saturation

Figure 10.3 illustrates the conditions arising from a CT with an open-circuited secondary. The secondary voltage E_s is derived from the relationship

$$E_s = \frac{-Nd\Phi}{dt}$$

where Φ is the flux in the core and N is the number of primary turns. As can be seen, there is no voltage output from the CT secondary for large portions of the cycle: this occurs when there is little rate of change of flux due to core saturation. A relationship derived by Mathews to determine the magnitude of the peak voltage V_p developed by a CT under saturation is given by

$$V_p = 2\sqrt{\{2V_k(V_f - V_k)\}} \text{ volts}$$

where V_k is the CT knee point (see Section 10.15) and V_f is the secondary voltage which would be produced if the CT did not saturate.

For example, if V_k=2.5 kV and V_f=10 kV then V_p=12.2 kV.

Note that the above relationship is only valid when $V_f \geqslant 2V_k$. At lower values, the relationship loses accuracy.

For most applications the CT must not be driven into saturation and therefore a low burden or short circuit (if the CT is unused) should be connected across the secondary. Under such circumstances the following well known relationship holds:

$$I_p N_p = I_s N_s$$

where I_p=primary current
I_s=secondary current
N_p=primary turns
N_s=secondary turns.

10.4 CT performance

CT performance, in the first instance, is defined in terms of the class of accuracy. Table 10.1 summarises the situation. *BS 3938* specifies the accuracy

Table 10.1 CT-accuracy classification

Accuracy class	Application
0.1 0.2 0.5 1 3 5	Measurement purposes
5P 10P	General protection purposes
X	Special-purpose applications, usually protection

of each class in terms of current and angle errors. In general, the errors are so low that they are not pertinent to site testing.

10.5 Measurement CTs

Measurement CTs are required to maintain specified accuracy up to 120% of rated current, when the burden connected is equal to the rated output of the CT. For example if the rated current is 5 A and the rated output of the CT is 30 VA, the rated burden is:

$$\frac{(\mathrm{VA})}{I^2}=\frac{30}{25}=1.2\ \Omega$$

Thus the accuracy limits are maintained when up to $120\% \times 5\ \mathrm{A}=6\ \mathrm{A}$ flows into a burden of up to $1.2\ \Omega$. The rating plate should specify the rated primary and secondary currents, rated output and class, e.g. 10 VA class 1.

10.6 General-purpose protection CTs

Unlike measurement CTs, protection CTs are required to maintain their accuracy up to several times the rated primary current. Accuracy classes 5P and 10P are intended to cover simpler forms of protection such as IDMT, instantaneous overcurrent and earth-fault, biased differential etc. The primary current at which the limits of error of class 5P and 10P start to be exceeded is known as the 'accuracy-limit primary current', and the ratio of:

$$\frac{\text{accuracy-limit primary current}}{\text{rated primary current}} = \text{rated accuracy-limit factor}$$

The '5' and '10' referred to in the above CTs describes the percentage error between the ideal and actual secondary current when the accuracy-limit primary current flows.

BS 3938 has standardised the 'rated accuracy-limit factors' to 5, 10, 15, 20 or 30. The rating plate for classes 5P and 10P should show the rated output, the accuracy-limit factor and class, as given in the following example:

30 VA	Class 5P	10
rated output	class	accuracy-limit factor

The rating plate will additionally specify the rated primary and secondary currents. From the above somewhat complicated mix of information, the CT performance can now be evaluated. If we assume that the rated secondary current is 1 A, the rated burden connected to the CT in the above example is given by:

$$\frac{(\text{VA})}{I^2} \text{ (since } I^2 Z = \text{VA)}$$

$$\text{i.e. } \frac{30}{1} = 30\ \Omega$$

Now 10 A (1 A × accuracy-limit factor) can flow before accuracy is lost. This is equivalent to an output of:

$$10^2 \times 30 = 3000 \text{ VA}$$

Now the reason why accuracy begins to be lost is that the CT core (i.e. magnetising impedance) begins to saturate, which in turn implies that the voltage applied to the magnetising impedance has reached its maximum value consistent with maintaining the accuracy class. In this instance that voltage is

$$10 \text{ A} \times 30\ \Omega = 300 \text{ V}$$

If now the burden is reduced to, say, 10 Ω, the current which would flow before accuracy is lost would be

$$\frac{300 \text{ V}}{10\ \Omega} = 30 \text{ A}$$

The rated accuracy-limit factor has in effect now risen to 30. Thus accuracy-limit factor and load (burden) are interrelated (approximately inversely), and to determine what the CT is capable of supplying satisfactorily, as the ohmic value of the burden varies, the above process must be repeated. Suffice it to say that the upper limit of the accuracy-limit primary current (as burden decreases) is set by the CT's 'rated short-term thermal current'. Note also that CTs are designed, and frequently installed, to serve the dual function of metering and protection.

10.7 Class X CTs

BS 3938 defines class X CTs as being required for special-purpose applications. The performance specification is defined in terms of the following characteristics:

- rated primary current
- turns ratio (with an error not exceeding 1.25%)
- rated knee-point EMF at maximum secondary turns
- maximum exciting current at rated knee-point EMF
- resistance of the secondary winding at 75°C

Class X CTs are usually applied when a high knee point is required to avoid saturation of the core. In a protection context they are usually categorised into two classes: class A which is designed to transform accurately without saturation up to maximum fault current, with due regard to system X/R ratios (see Section 10.10); and class B, a less costly design which permits a degree of saturation under transient conditions. Class A is generally employed for feeder main protection and class B for high-impedance circulating-current schemes.

10.8 CT turns ratio and rating

A distinction needs to be drawn between the ratio of a CT and its current rating.

The term 'rated primary (or secondary) current' refers to the value marked on the rating plate, upon which the performance of the CT is based. For example, it is a factor in determining the rated accuracy-limit factor as identified in Section 10.6. Note that the 'rated secondary current' has no meaning in the context of class X CTs.

The term 'rated continuous thermal current' alternatively refers to the maximum continuous primary (and corresponding secondary) current which can be tolerated without damage. Usually, when rating is referred to on-site, it refers to the continuous thermal rating. As an example, the maximum continuous thermal rating of a 400 kV CT is 4000 A primary and 2 A secondary, frequently known as a 4000/2 CT. This differs from the CT turns ratio of 2000/1. It is important that these terms are not confused.

Note also that the CTs, like other items of plant, have short-time ratings. For example, the 400 kV CT referred to above would typically accommodate 57 500 A primary current for 1 s.

10.9 High reactance and low reactance

Protection current transformers are divided into two groups: high-reactance and low-reactance. The term reactance refers to the transformer leakage reactance; the lower the leakage reactance the greater the accuracy of the CT.

Classes 5P and 10P may be designed to be in either group. Class X transformers are always of the low-reactance type.

10.10 CT performance under system faults

When a fault occurs on the power system the resulting fault current comprises a decaying exponential current in addition to the sinusoidal steady-state current, as shown in Figure 10.4. The exponential current, alternatively

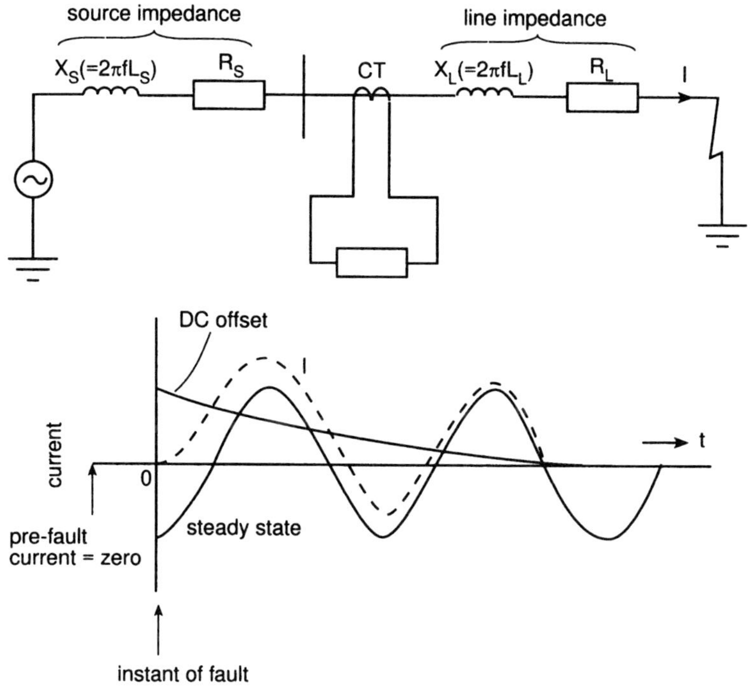

Figure 10.4 Typical fault current showing DC offset

$L_S + L_L = L$
$X_S + X_L = X$
$R_S + R_L = R$
At time t = 0 total post-fault current = pre-fault current

known as the DC offset, has a maximum value when the instant of fault coincides with peak steady-state current. The rate of decay of the DC offset is dependent on the *R*-to-*L* (resistance-to-inductance) ratio of the impedance to fault. The complete expression for the fault current is given by:

$$I_{total} = I_{max} \sin (\omega t + \phi) - I_{max} \sin\phi \, e^{-Rt/L} \qquad (10.1)$$

$$\text{i.e. } I_{total} = I_{steady\text{-}state} + I_{DC\,offset}$$

where ϕ=point on wave of fault.

Figure 10.5 shows the effect of a system fault current, with a DC offset, flowing through a current transformer. The resulting flux in the transformer core, as shown in Figure 10.5*b*, assumes no saturation of the core. However, the effect of the DC offset is to increase the flux in the core, and Figure 10.5*c* shows the resulting secondary quantities should the core be driven into saturation. To avoid saturation, and its effects, it is therefore necessary to take account of the DC offset when specifying the design of current transformers. Within this context it can be shown that, to avoid saturation, the CT knee point V_k is required to satisfy the following expression:

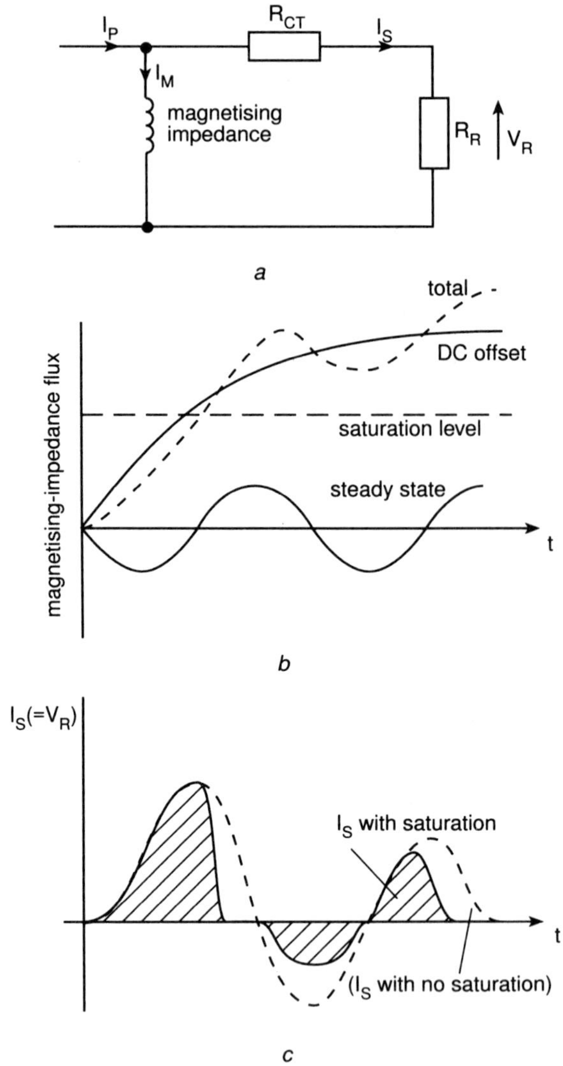

Figure 10.5 Effect of DC offset on CT performance

a Components of current
b Components of flux with no saturation
c Secondary current with saturation

$$V_k < \frac{I_p}{N_s}\left(\frac{X}{R}+1\right)(R_{CT}+R_R)$$

where I_p=maximum RMS steady-state fault current of the power system

Table 10.2 Typical X/R ratios

Plant item	X/R ratio
33 kV overhead line	1.4
132 kV overhead line	2.4
275 kV overhead line	8.5
400 kV overhead line	15.0
132/33 kV 90 MVA transformer	15.0
275/33 kV 240 MVA transformer	50.0
400/275 kV 1000 MVA transformer	130.0
660 MW generator	100.0

X = reactance of the power system between the voltage source and the fault ($X=2\pi fL$, where L is the value in eqn. 10.1)
R = resistance of the power system between the voltage source and the fault (i.e. the same R as in eqn. 10.1)
N_s = CT turns ratio
R_{CT} = resistance of CT
R_R = resistance of relay and associated wiring

The system X/R ratio thus has a significant impact on CT performance. Class X CTs, in particular, are designed around the X/R ratio. Table 10.2 illustrates some typical X/R ratios.

10.11 CT types and applications

In general, CTs can be categorised into two major types: the multiturn wound-primary and the single-turn bar-primary.

A wound-primary CT consists of a primary and a secondary arranged concentrically. The wound-primary CT tends to be more cost advantageous on the lower-voltage systems, particularly when low ratios are required with reasonably high accuracy. However, when large currents and/or long fault-clearance times are involved, wound primary CTs suffer from overheating and potential mechanical damage from electromechanical forces. In such instances, bar-primary CTs are required.

Current transformers found on low-voltage systems are frequently of the wound-primary type with butyl-rubber or epoxy-resin insulation.

For 11 and 33 kV metalclad switchgear, the tendency is more towards bar-primary CTs, again with butyl-rubber or epoxy-resin insulation. Figure 10.6*a* illustrates the typical location of CTs within metalclad switchgear. As can be seen, a number (usually no more than five) of CTs would normally be installed.

Outdoor dead-tank bulk-oil circuit breakers at 33 kV and above usually have CTs located within their bushings. This also applies to transformers which have bushing connections. Figure 5.13 illustrates such an arrangement. Some outdoor dead-tank bulk-oil circuit breakers at 33 kV have their connections enter the circuit breaker via glazed shed insulators. In such instances, the current transformers are located within the circuit-breaker tank, as shown in Figure 10.6*b*. In all the above instances, the primary would consist of a single turn.

Air-blast and live-tank sulphur-hexafluoride (SF_6) circuit breakers at 132 kV and above would normally be associated with separately mounted post-type CTs. Figure 10.6*c* illustrates a so-called 'hairpin' arrangement. The primary conductor within the porcelain housing would normally be insulated with oil-impregnated paper, and the complete unit would be filled with oil. A flexible oil reservoir within the top housing maintains an effective oil seal on the main chamber, while accommodating expansion or contraction of the main chamber oil.

Figure 10.6*d* illustrates an alternative arrangement. This type alleviates the high electromechanical forces on the primary conductor during system faults and minimises leakage fluxes. It has the disadvantage that the secondary conductors must be fully insulated from the live tank. Some post-type CTs have interleaved stress-control foils within the primary insulation (this is more likely to be associated with the type shown in Figure 10.6*c*), to which a tapping is connected to provide an output of 63.5 V, i.e. a capacitor divider. In the author's experience, such tappings are utilised during special tests but not as a matter of routine. A recent alternative to oil, as an insulator, for such CTs has been SF_6.

CT COMMISSIONING

10.12 CT-commissioning tests

The basic commissioning tests usually carried out on CTs consist of:

- check on CT position
- flick tests
- magnetisation curves (secondary injection)
- burden (resistance tests)
- insulation-resistance tests
- primary injections
- on-load tests (see Chapter 19)

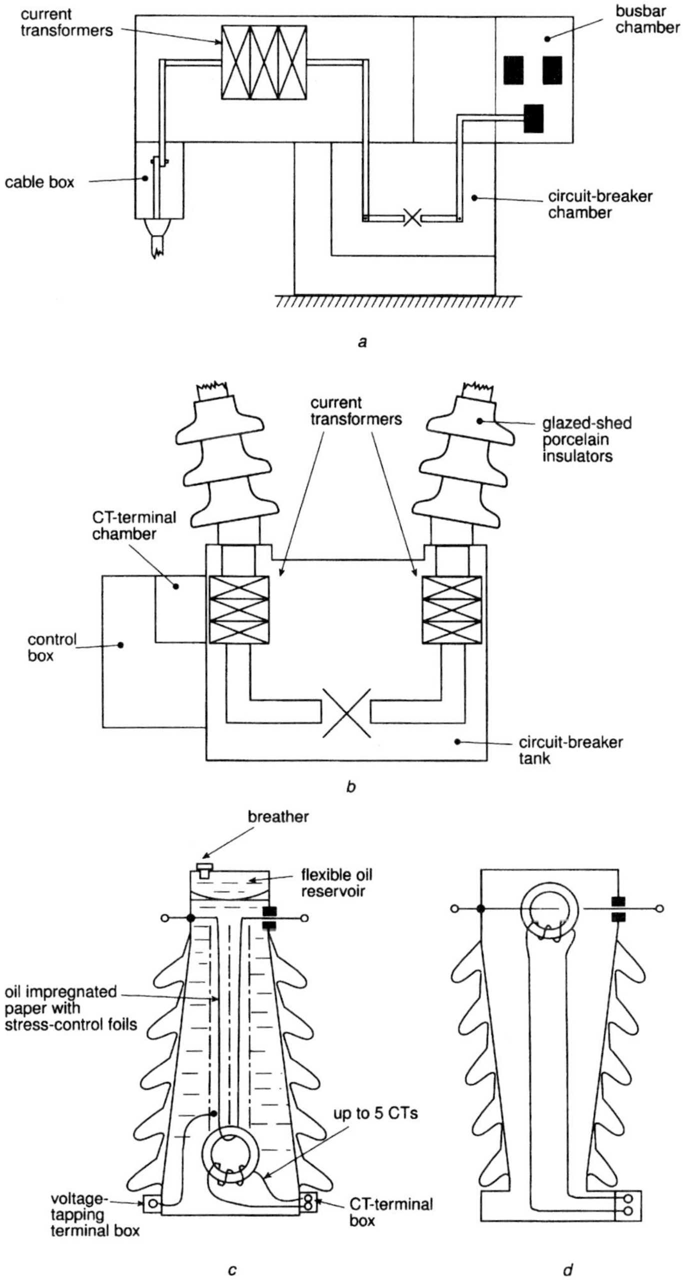

Figure 10.6 CT arrangements

a CT position with metalclad switchgear (distribution voltages)
b CT position, outdoor bulk-oil circuit breaker (33 kV)
c Post-type CT
d Alternative post-type CT

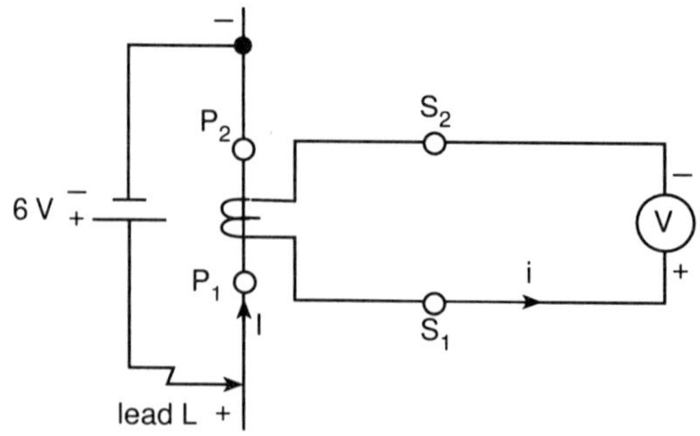

Figure 10.7 Flick test

Application of lead L causes kick of voltmeter in forward (positive) direction. Removal of lead causes kick in backward (negative) direction

10.13 Check on CT position

Prior to carrying out any tests on CTs, the commissioning engineer must satisfy himself that they are correctly positioned for purposes of protection and metering, through examination of the key, or circuit, diagrams.

10.14 Flick test

The flick test is carried out to prove that a CT, or group of CTs, has been manufactured and installed with the correct polarity, i.e. if the current flows into the primary in a certain direction (with respect to a reference voltage), the secondary current must flow out of the secondary winding in a corresponding and prescribed direction. Current transformers which accord with the requirements of *BS 3938* have their terminals labelled to enable polarity to be determined. In Figure 10.7, the primary-winding terminals are labelled as P1 and P2 and the secondary-winding terminals are labelled S1 and S2. The convention requires that if the primary current at a particular instant is flowing into terminal P1 (i.e. in the direction P1 to P2), then the secondary current at the same instant is required to flow out of terminal S1 (i.e. in the direction of the connected equipment).

To verify the polarity, a voltmeter selected to measure DC quantities should be connected with its positive pole to terminal S1 and its negative pole to terminal S2. A torch battery (typically up to 6 V) should then have its negative pole connected to terminal P2. The battery positive pole should then be applied to terminal P1 and, for correct polarity, at the instant of application the voltmeter should be observed to show a 'kick' voltage in the positive direction. Removal of the battery connection to terminal P1 should result in a kick voltage in the negative direction on the voltmeter. The magnitude of the kick voltage is usually in the millivolts range. In some instances, the magnitude of the kick is greater, and more easily observed, on a DC ammeter. The voltmeter or ammeter ideally needs to be an analogue device for ease of observing the 'kick'.

The polarity of a CT is only of concern when the equipment to which the CT is connected is polarity conscious, or when a number of CTs have to be interconnected on the secondary side. In such instances, it is preferable to carry out the flick test with the voltmeter connected at, and across, the equipment, thereby proving the correctness of the connections between the CT and the equipment. In practice, such a test frequently proves impracticable, either because it is necessary to prove the polarity of the CT before the wiring to the equipment is complete, or because the combined burden of the wiring and equipment, or complexity of the CT wiring, render the test impracticable. In such instances the polarity is not conclusively proved until the primary injections or, alternatively, the on-load commissioning tests are carried out.

Many CTs have a tapped winding (e.g. 1200/600/1) and all tappings should be flick tested at the time of commissioning.

Some CTs also have a test winding fitted to assist with future maintenance or recommissioning. The test winding should also be flick tested during commissioning. Figure 10.8 illustrates the terminal markings for these arrangements. Note that some older CTs may not have their terminals labelled in accordance with *BS 3938*, but use some alternative nomenclature which serves the same purpose.

A point worthy of note, and which occasionally causes a little confusion on site, is that with a tapped winding, such as 1000/500/1, the larger ratio, i.e. 1000/1, is obtained by connecting between S_1 and S_3, i.e. the whole of the winding, and the smaller ratio by connecting between S_1 and S_2. This follows directly from the relationship $I_p N_p = I_s N_s$.

10.15 Magnetisation curve

A CT-magnetisation curve is required to determine the secondary voltage and current at which the CT magnetic core begins to saturate. The results should then be compared with the manufacturer's curve to prove that the correct type of CT has been installed. The test to obtain the curve consists of injecting the CT secondary from a sinusoidal AC source and obtaining a plot of voltage against magnetising current up to the point of saturation. The point at which saturation is deemed to become significant is known as the 'knee point'. This is defined as that point on the curve when a 10% increase in voltage causes a 50% increase in magnetising current.

Figure 10.9*a* illustrates the arrangements required to obtain a magnetisation curve, and Figure 10.9*c* a resulting magnetisation curve. The following points need to be noted:

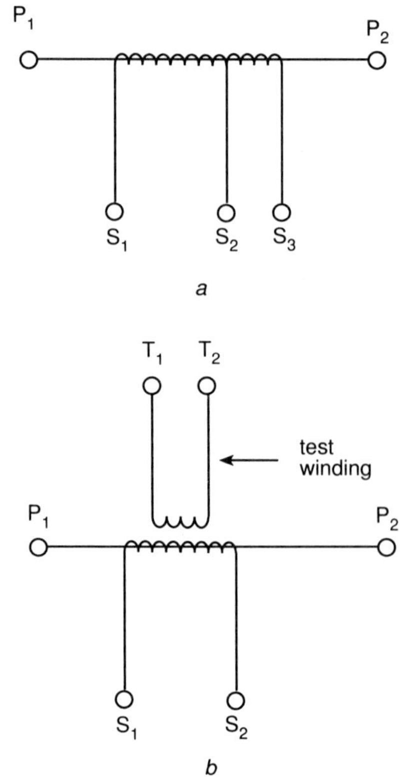

Figure 10.8 *CT nomenclature to BS 3938*

a CT with intermediate tappings
b CT with test winding

(*a*) When the test is carried out, the primary sides of the CT must not be earthed on both the P1 and P2 sides; otherwise a primary short-circuit will result and corrupt the test.
(*b*) The current which flows becomes nonsinusoidal as the CT saturates.

The test should commence by slowly increasing the voltage, and hence current, from zero, driving the CT into saturation, and then slowly reducing the voltage to zero. The purpose of this is to remove any remnant magnetism that might have been left in the core (either from manufacturer's tests, or following circuit de-energisation if it is tested during CT recommissioning). The remnant magnetisation arises from the hysteresis curve, and it otherwise causes the flux, at the start of the test, to commence from a value other than zero. Figure 10.10 shows a typical hysteresis curve. Once the remnant magnetism has been removed, the plot of voltage against magnetisation current can commence. After just passing the knee point the current should again be reduced slowly to zero to remove any remnant magnetism. Note that

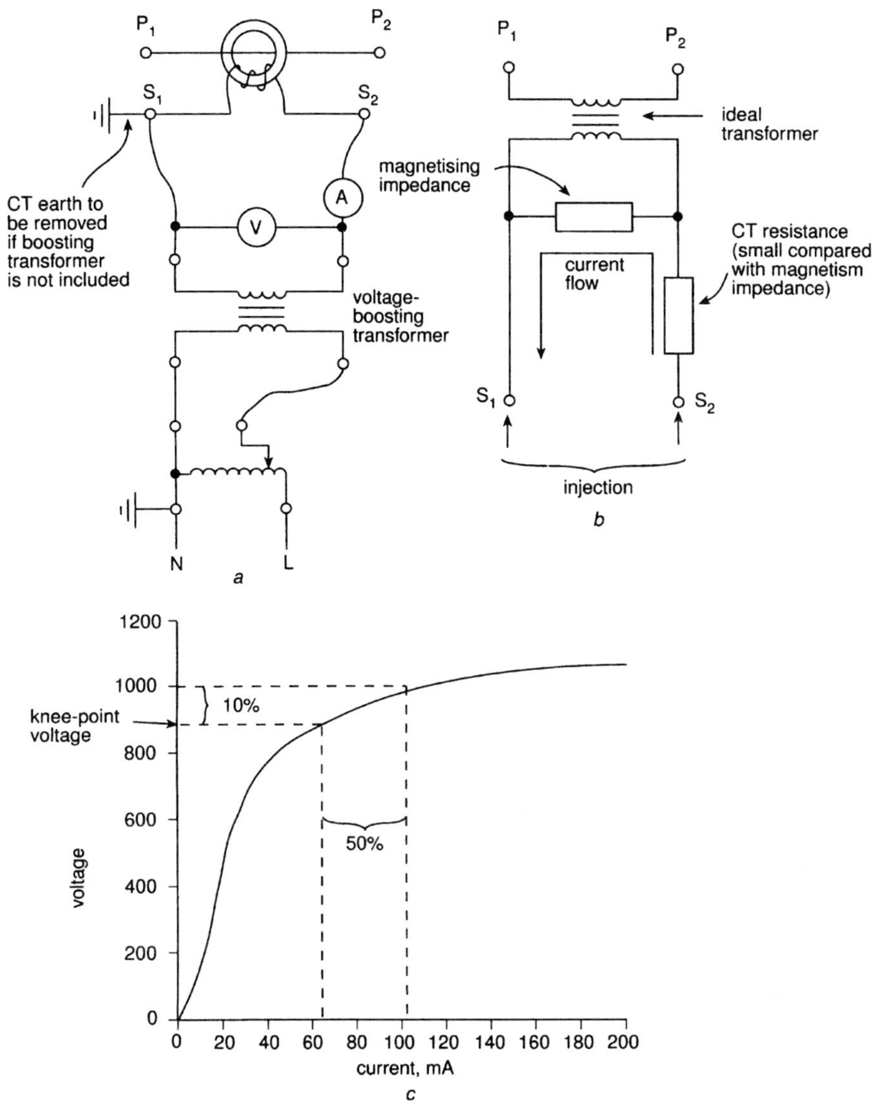

Figure 10.9 *Magnetisation curve*

a Test arrangements
b Equivalent circuit
c Typical magnetisation curve for class X CT with ratio of 600/1

there is a danger of damaging the CT insulation if the current is switched off at any value other than zero, because of the presence of induced voltages.

CT knee points tend to range from about 200 V for CTs on the 11 kV system to 3 kV for CTs on the 400 kV system. For CTs with a high knee point it is not essential to carry out a plot up to the knee-point voltage, but it is sufficient to

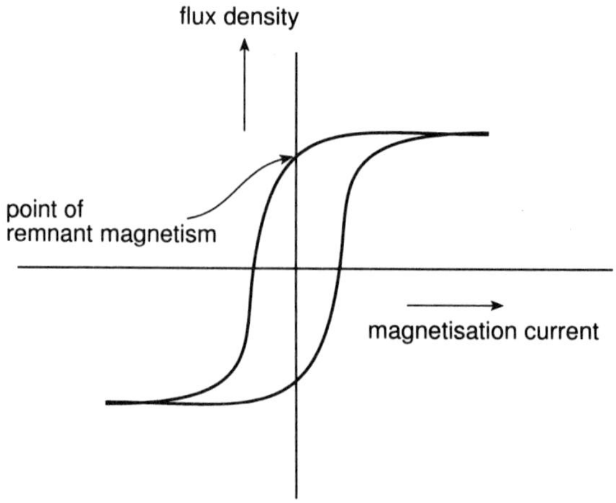

Figure 10.10 Typical hysteresis curve

compare a number of test results with the manufacturer's data. Furthermore, when injecting high voltages onto secondary wiring, thought should be given to the wiring insulation level, so that the wiring is not damaged.

10.16 Burden

This consists of measuring the CT-secondary resistance, with an ohmmeter capable of reasonable accuracy over the range 0.1–10 Ω. Measurements should be taken of all tappings. Most CT burdens tend to fall into the range of 1–5 Ω. This information is useful if the integrity of the CT needs to be verified during future fault investigations. It is essential information for calculating high-impedance circulating-current protection settings. The burden obtained should be cross checked against the manufacturer's factory test data.

10.17 Insulation-resistance tests

These should be carried out when the wiring between CT and the equipment to which the CT is to be connected is complete. It is usual in some countries to earth CTs on the star-point side via an earth link, as illustrated in Figure 10.11. The earth link should be opened and the insulation-resistance (IR) test device (1000 V megger) connected across it, and the test reading taken. A point worthy of note is that any connected equipment which is not designed to be subject to the test voltage should be short-circuited prior to the test. The earth link, when closed, is required to retain the secondary wiring at earth potential, for reasons of safety when the circuit is energised. Given that there

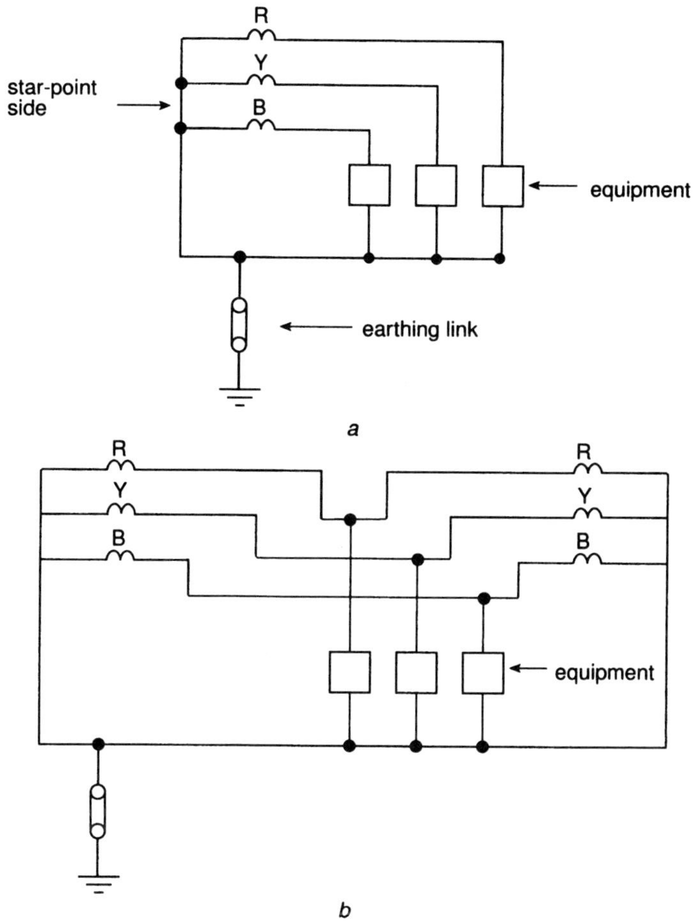

Figure 10.11 *Earthing of CTs*

a Single set of CTs
b Two (or more) sets of CTs

will be capacitance between the primary and secondary of the CT, the secondary wiring would stand at some intermediate voltage between system voltage and earth, without the earth-link connection.

PRIMARY INJECTIONS

10.18 Primary injection

Primary injection should be carried out when the wiring is complete, and should be more-or-less the last test to be carried out prior to energisation. The test consists of injecting a test current into the CT primary, to enable measurements to be made of the secondary quantities as verification of the correct installation of the CT, the connected wiring and the equipment. Before discussing the detail of the tests, it is worth noting some pertinent general principles, as follows:

(*a*) The equipment to be tested must be dead and isolated from the system.
(*b*) The tests will probably be carried out under 'sanction for test' documentation. This will permit the removal of safety earths under the direction of an authorised engineer, if the test so demands.
(*c*) The tests should be carried out in accordance with a programme. The programme should indicate the predicted result for each test. The test should be terminated the instant it is evident that the results are not as predicted.
(*d*) Consideration should be given to whether the test current will also flow in CTs other than those to be verified.
(*e*) When measurement instruments are inserted into CT wiring, every endeavour should be made to disturb the wiring as little as possible. Where possible, miniature clip-on instruments should be used.
(*f*) A typical primary injection set delivers up to 1000 A from a source of 15 V. With such a low driving voltage, the test connections must be very clean.
(*g*) Secondary currents will usually be in the milliampere range. Older ammeters possess a high burden on the lower ranges, and this can have an impact on the test results. Such ammeters should not be used.
(*h*) When the test is complete, ensure that all test connections are removed, and all CT link blocks are left in an appropriate state.

The tests themselves, which are often termed 'shots', are required to verify the following:

(i) *ratio shot:* a shot to prove the ratio and phase-to-neutral wiring of a CT;
(ii) *polarity shot:* shots to prove the polarity and phase-to-phase wiring of a group of three (R, Y, B) CTs (this may additionally also include a neutral CT). Note that this test only gives the relative polarity of one CT to another, whereas the flick test ascertains absolute polarity;
(iii) *matching shot:* a shot(s) to prove the polarity of one group of CTs with reference to another group;
(iv) *operation shot:* a shot to prove the wiring to, and the operation (setting) of, the connected device. This shot may be subdivided into operation from a single CT or operation from more than one CT. The magnitude of current is often insufficient to cause operation, owing to the limited output of the primary injection set, and thus indication that the current is in the operate direction will suffice;
(v) *stability shot:* a shot to prove that unit protection stabilises on through fault.

The number of primary injection tests carried out should be the minimum consistent with providing the required level of confidence that the equipment is correctly installed. Depending on the equipment circuitry being proved, a single injection often covers a number of the above tests. A point of debate is centred around the order in which the stability and operation shots should be carried out. Generally speaking the last shot should be with all connections normal, and should be an operation shot (as a reassurance that operation will occur, if required, on circuit energisation). However, this may prove impracticable with some installations.

A selection of typical primary injections will now be examined; note that in some instances the equipment could also be proved by alternative permutations of tests.

10.19 Primary injection of feeder protection CTs

Figure 10.12 shows a typical primary injection arrangement for feeder-protection CTs. The tests required are those of ratio, polarity and operation. The ratio and operation shots are achieved during the same test. The secondary currents are monitored via ammeters located adjacent to the protection equipment. Prior to the primary injection on the CT tap to be used in service, a ratio test should be carried out on the tap not to be used. Theoretically, it is only necessary to carry out R–E, R–Y and Y–B injection tests, but in practice it is more usual (and just as easy) to go for the full range of tests and carry out R–E, Y–E, B–E, R–Y, Y–B, B–R.

10.20 Primary injection of mesh corner CTs

Figure 10.13 illustrates the primary injection requirements for a unit protection, in this instance a mesh-corner protection. It is assumed that the scheme is of the high-impedance circulating-current type. Such schemes have their CTs connected on a per-phase basis, with a common neutral, and with a high-impedance relay connected to each phase. To simplify the illustration, the arrangements for only one phase are shown. The tests required are those of ratio, polarity, matching, operation and stability.

Figure 10.13*a* shows the requirements for ratio and polarity shots for location-A CTs. This test also serves as a single-CT-feed, operation shot. The test should be repeated on the B and C CTs by closing of circuit breakers and removing earths as required.

Figure 10.13*b* shows the test requirements for matching group A and C CTs. This test also serves as a stability shot. The test should be repeated on the A and B CTs.

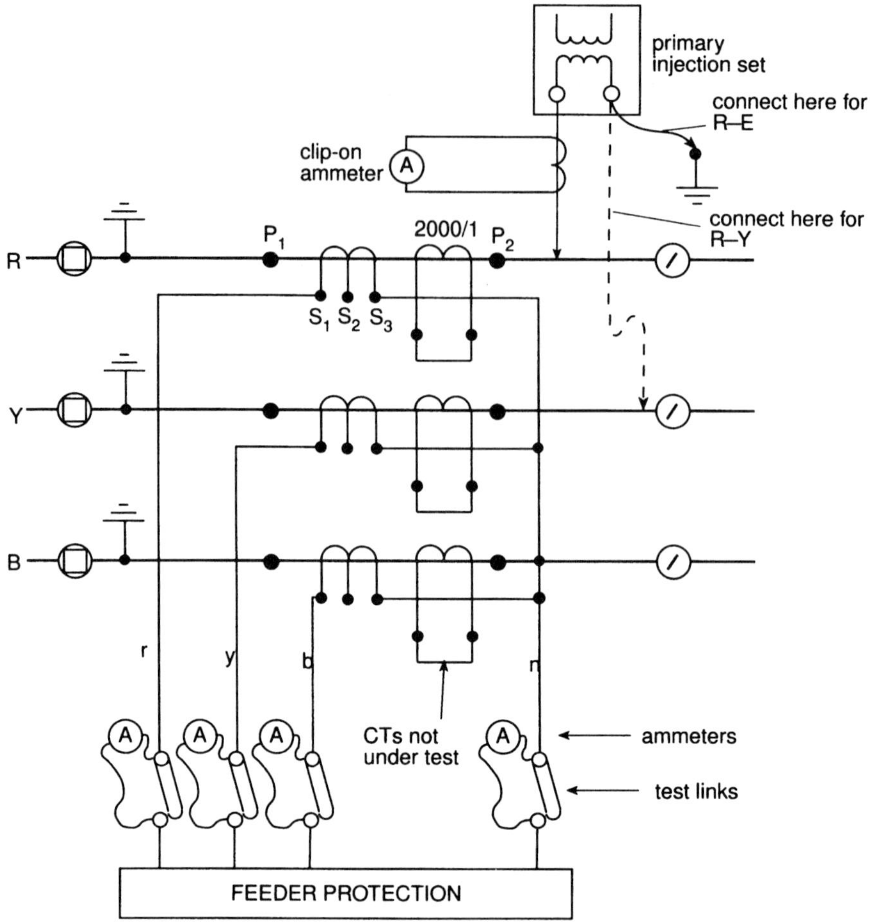

Figure 10.12 Primary injection of feeder CTs

	Injection	Primary current	Secondary current			
			r	y	b	n
		(A)	(mA)	(mA)	(mA)	(mA)
Ratio	R–E	200	100	—	—	100
&	Y–E	200	—	100	—	100
operation	B–E	200	—	—	100	100
Polarity	R–Y	200	100	100	—	—
&	Y–B	200	—	100	100	—
operation	B–R	200	100	—	100	—

Finally, Figure 10.13*c* shows the test requirements for a two-CT infeed operation shot for B and C CTs, which should be repeated for the A and B CTs.

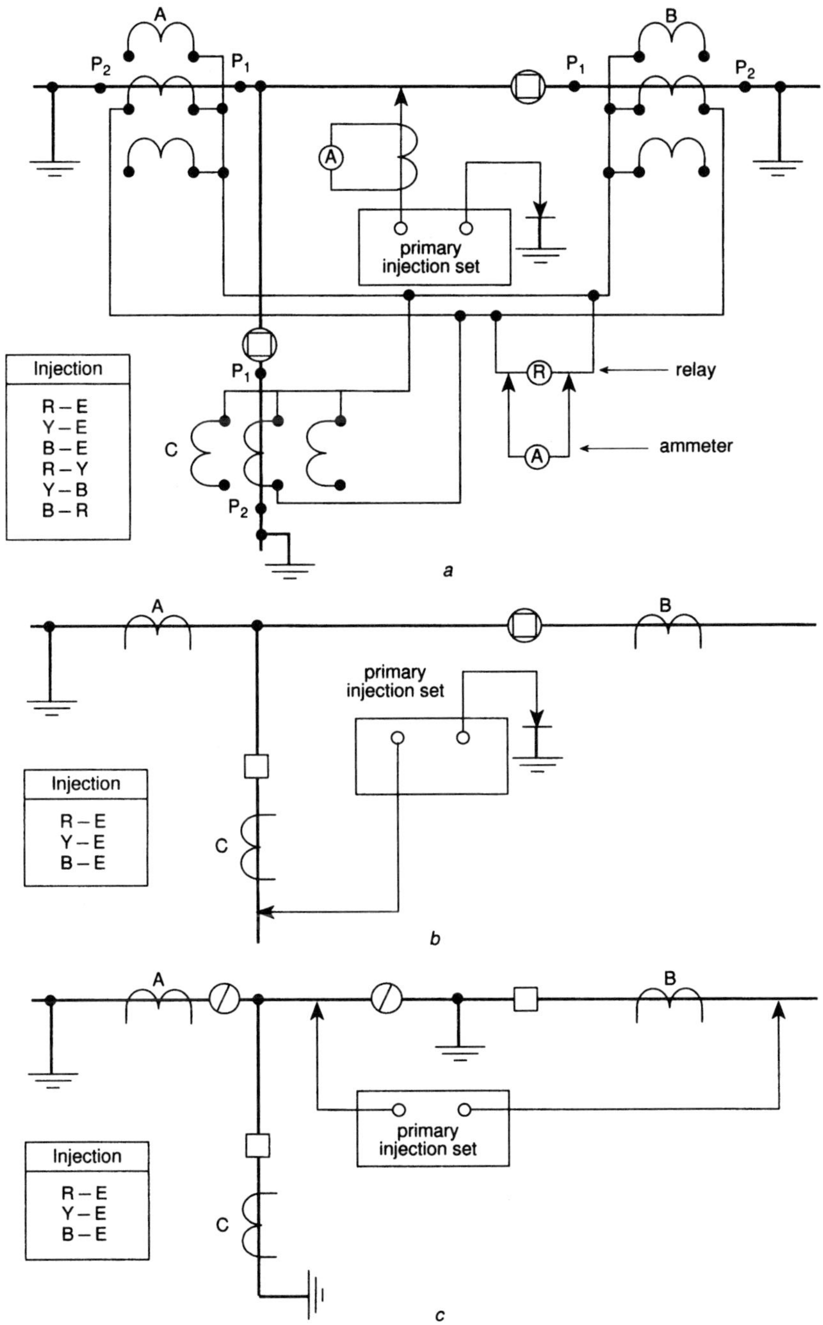

Figure 10.13 Primary injection of mesh-corner CTs

a Ratio and polarity of group 'A' CTs
b Matching of group 'A' and group 'C' CTs
c Operation shot of groups 'B' and 'C' CTs

10.21 Primary injection of autotransformer CTs

Figure 10.14*a* illustrates the essentials of high-impedance circulating-current protection as applied to an autotransformer. For ease of illustration only one phase of the protection circuitry has been fully illustrated, the other two phases being identical to that shown. An ammeter is connected across each high-impedance relay. The injections required are ratio, polarity, matching, operation and stability.

The tests usually commence with ratio and polarity shots on the transformer LV and neutral CTs in a similar fashion to those illustrated in Figure 10.13*a* (this assumes that the CT primary terminals are accessible). The currents are measured on the ammeters. These tests also serve as operation shots. It is, however, impossible to inject a high current into the CTs within the HV bushing, since the CTs are in series with the high impedance of the transformers. In such circumstances it is usual to employ a 415 V three-phase source (from a substation distribution board) to carry out the test.

Figure 10.14*a* illustrates the test arrangement for carrying out ratio and polarity shots on the HV-bushing CTs; the usual six tests should be carried out (R–E, Y–E, B–E, R–Y, Y–B, B–R). Given that the secondary currents are measured at the relay, it is necessary to short-circuit and disconnect the LV and neutral CTs, so that the only current measured emanates from the HV-bushing CTs. At first sight it may seem unusual to apply the three-phase supply to the LV side of the transformer, but this creates a higher current than if the three-phase supply is applied to the HV side.

Figure 10.15 shows the derivation of currents for the earth-fault injection from a sequence-components analysis. A similar analysis with the injection from the HV, i.e. 400 kV, side would give approximate currents as follows:

$$I\,(400\text{ kV side}) = 1.8\text{ A}$$

$$I\,(132\text{ kV side}) = 6.8\text{ A}$$

Note that in Figure 10.15 the earth on the three-phase supply effectively short-circuits the tertiary zero-sequence reactance, and thus there is little or no current in the tertiary. This arises because the source impedance of the 415 V supply is negligible compared with the transformer impedance.

After the LV and neutral CTs have been restored, a stability shot (which is also the matching shot) as shown in Figure 10.14*b* should be carried out. It is acceptable that the stability shot is carried out as a one-off three-phase shot. This is because, unlike the mesh-corner situation, an external fault on an autotransformer always results in current flow in all three groups of CTs, i.e. HV, LV and N (as opposed to any permutation of two), and secondly at this stage the wiring has been proven as pertaining to three individual phases.

Finally, a three-phase operation test involving all three groups of CTs, as shown in Figure 10.14*c*, should be carried out. The current flowing into the ammeter across the relay should be equal to:

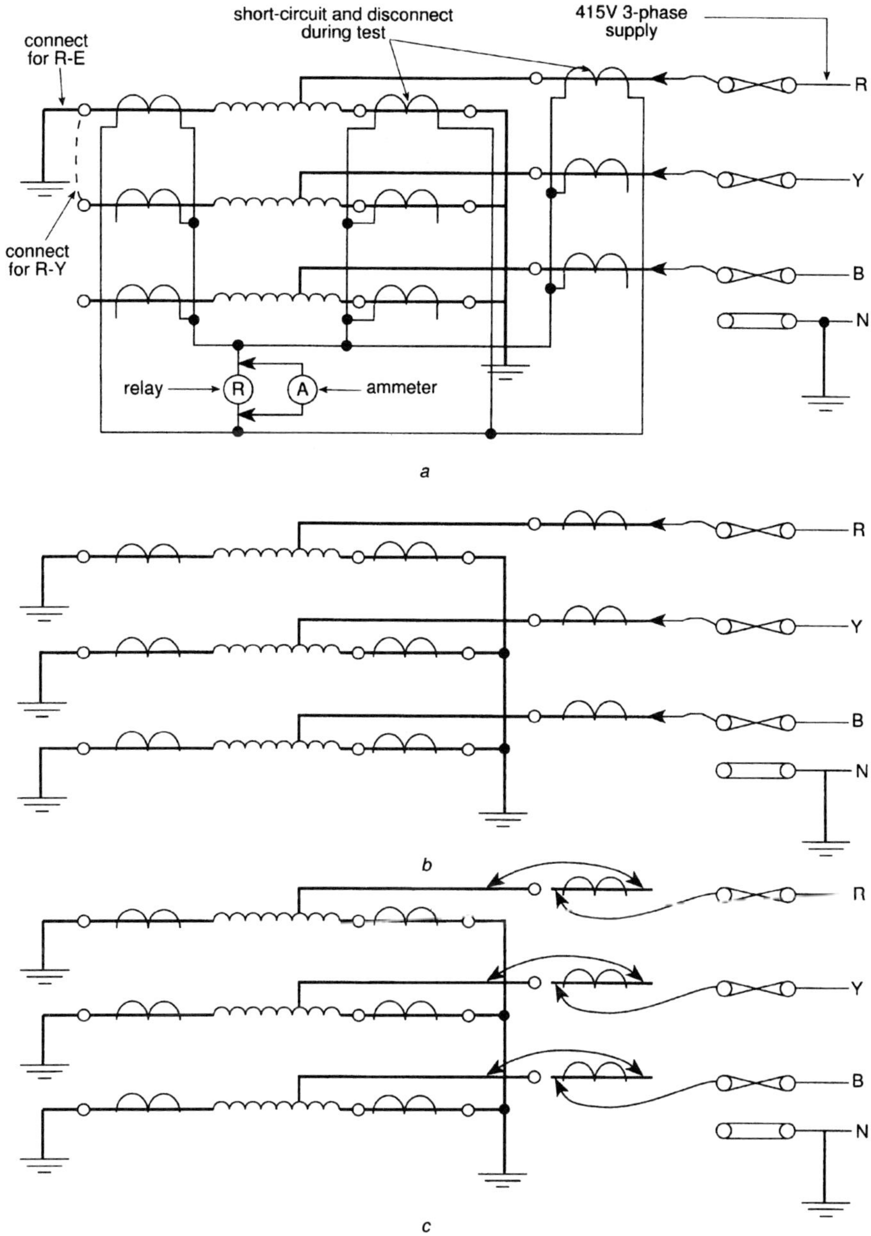

Figure 10.14 Primary injection of autotransformer-protection CTs

a Ratio and polarity shot on HV-bushing CT
b Stability shot
c Operation shot

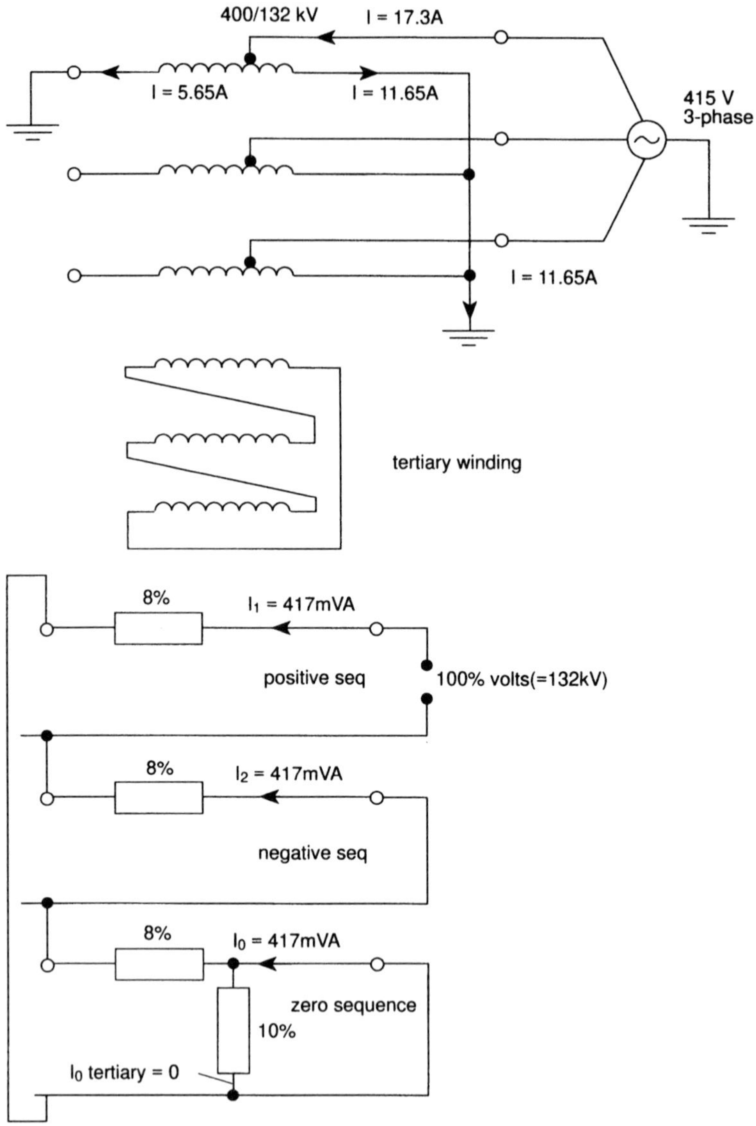

Figure 10.15 Sequence-component analysis of 415 V shot

With 132 kV applied, power = 100/24% = 417 MVA
= positive-sequence
= negative-sequence
= zero-sequence

Total power = 3 × 417 = 1250 MVA
Therefore I = 5500A (132 kV), 1800A (400 kV side)
With 415 V applied,
I (132 kV side) = (0.415/132) × 5500 = 17.3 A
I (400 kV side) = (0.415/132) × 1800 = 5.65 A

$$\frac{2 \times \text{current delivered from the three-phase supply}}{\text{CT ratio}}$$

Note that with new transformers there is a tendency to incorporate a primary test winding (piece of bar) through the bushing CT, to facilitate a high current ratio (and polarity) injection. The stability shot and overall operation shot should ideally still be carried out from a 415 V source since it provides a true simulation of in-service conditions. However, with some installations, particularly GIS substations, access restrictions may prevent a 415 V injection being carried out: under these circumstances only matching shots in conjunction with the primary test winding can be achieved and great care must be given to obtaining the correct polarity (see Figure 3.18).

10.22 Primary injection of double-wound transformer CTs

Figure 16.16 (see Chapter 16) illustrates one of the most complex examples of a double-wound transformer (without CT primary test windings), with associated CTs connected to biased-differential and restricted-earth-fault protections. The features to be proved are:

- ratio shots on HV and LV CTs (including the neutral CTs)
- polarity shots on HV and LV CTs (including the neutral CTs)
- stability and operation of HV and LV restricted earth fault relays (REF)
- stability and operation of biased-differential protection (B diff)

There are several possible test methods for achieving the above: this text will only describe one possible method. Before examining the detail two points are worth noting:

(*a*) It is usual to secondary inject the interposing transformers before commencing primary injections.
(*b*) Ammeters should be placed in the secondary wiring in each phase and neutral CT, as close to the CT as possible; across each (high-impedance) restricted-earth-fault relay; and in the operate and restraint coils of the biased differential relay. This should readily be achievable with the aid of modern miniature clip-on ammeters which can easily be repositioned.

The most thought-provoking tests to consider are those which prove the stability or operation of the relays. In particular, it needs to be verified that the protection systems are configured to accommodate the flow of sequence components of current across the transformer under all conditions. However, it is not particularly helpful to consider the tests from a sequence-component standpoint, except to be aware that zero-sequence currents cannot flow across this type of transformer. A more meaningful approach, in terms of test data, is to consider the usual permutations of faults (i.e. R–E, Y–E, B–E, R–Y, Y–B, B–R) with reference to in-zone and out-of-zone fault positions, on both the

HV and LV sides of the transformer. A further consideration is that, wherever possible, a high-current test source should be employed (i.e. injecting hundreds of amperes); however, this cannot be employed for overall stability shots owing to the high impedance of the transformer, or for any injection on the HV CTs since they are usually installed in the HV bushings, and the in-zone connections are therefore inaccessible. Under these latter circumstances it will be necessary to use a 415 V test supply, resulting in the injected currents being typically in the range of 5–20 A.

It is suggested that separate primary injections are carried out for the restricted-earth-fault and biased-differential schemes, and although this results in partial duplication of tests it facilitates clarity of purpose. A suggested test format is as follows:

(*a*) *LV restricted earth fault (REF)*
Using a high-current source (primary injection set):

(i) Inject across each LV CT in turn (r, y, b, n), as in Figure 10.12. This verifies the LV CT ratio and REF operation.
(ii) Inject through a combination of LV CTs in series (r-n, y-n, b-n), as in Figure 10.12, where 'n' refers to the LV neutral CT.

Note: as the polarity of each phase CT has been verified against the neutral CT, there is theoretically no need to carry out a phase-to-phase injection; however many engineers will carry out this injection for the sake of completeness. This verifies the LV CT polarity as a group and REF stability.

(*b*) *HV REF*
Remove the high-current source and apply a three-phase 415 V source to the LV terminals:
(i) Short-circuit and disconnect the HV neutral CT and apply an earth to each HV phase in turn (R–E, Y–E, B–E). This verifies the HV CT ratio and HV REF operation.
(ii) Reconnect the HV neutral CT, remove the 415 V source, and inject across the HV neutral CT with a high-current source (primary injection set). This verifies the HV neutral CT ratio and HV REF operation.
(iii) Remove the primary injection set and re-apply the 415 V source to the LV terminals, apply an earth to each HV phase in turn (R–E, Y–E, B–E). This verifies the HV CT polarity as a group and REF stability.

Note that if CT primary test windings are available, wiring disconnections are unnecessary since a high-current source (primary injection set) can be applied directly to each phase.

(*c*) *B diff operation and stability due to HV fault condition*
Remove ammeters from across the HV and LV restricted-earth-fault relays. Maintain the three-phase 415 V source to the LV terminals.

(i) Short-circuit and disconnect the R, Y, B HV CTs, apply a combination of short-circuits in turn to the feeder side of the HV CTs (R–E, Y–B, R–Y–B). This verifies B diff operation.

(ii) Reconnect all CTs, apply a combination of short-circuits in turn to the HV terminals (R–E, Y–B, R–Y–B). This verifies B diff stability.

(*d*) *B diff operation and stability due to LV fault conditions*

Apply a three-phase 415 V source to the HV terminals.

(i) Apply combinations of short-circuits in turn to the transformer side of the LV CTs (r–E, y–b, r–y–b). This verifies B diff operation.

(ii) Apply a combination of short-circuits in turn to the feeder side of the LV CTs (r–E, y–b, r–y–b). This verifies B diff stability.

Notes

(*a*) Throughout the tests, the LV neutral earthing resistor will require to be short circuited.

(*b*) All tests should be carried out at transformer nominal tap.

(*c*) Some modern GIS substations, with access restrictions, may prevent 415 V injections (usually arising from inaccesibility of earth connections). However, in such instances it is probable that the transformer will be fitted with CT-primary test windings. These will facilitate operation shots but not stability shots. Under these circumstances the only option may be to carry out a wide range of operation shots (injecting through both HV and LV CTs) and compare measured with calculated currents to provide an acceptable level of confidence. Some balance shots can be achieved with the use of two high-current primary-injection test sets. Final confirmation of balance will need to await the on-load tests.

10.23 Primary injection of busbar protection CTs

Figures 10.16 and 10.17 illustrate a typical busbar-protection arrangement for a transmission-voltage double-busbar substation. To minimise the complexity of the illustration, only one phase of the protection circuitry is shown and the neutral connection (as denoted by terminal C) of the CTs has been omitted. M1D, R1D, M2D and R2D represent the main and reserve bar discrimination-zone relays, and CK the check-zone relay. It is assumed that feeder F1 is a new circuit whose CTs are required to be connected into the existing busbar-protection scheme. The first decision to be made is whether the busbar-protection CT circuits should be proved with the existing busbar-protection trip circuitry in service, or out of service.

Given that, during the tests, earths will be applied, there is a danger of inadvertent application to live circuits, and thus the case is better made for the busbar protection remaining in service. Conversely, if the busbar protection is in service, it is imperative that it is not inadvertently operated. To avoid this the injected currents must not exceed 50% (say) of the setting current of any relay. The test should then proceed as follows:

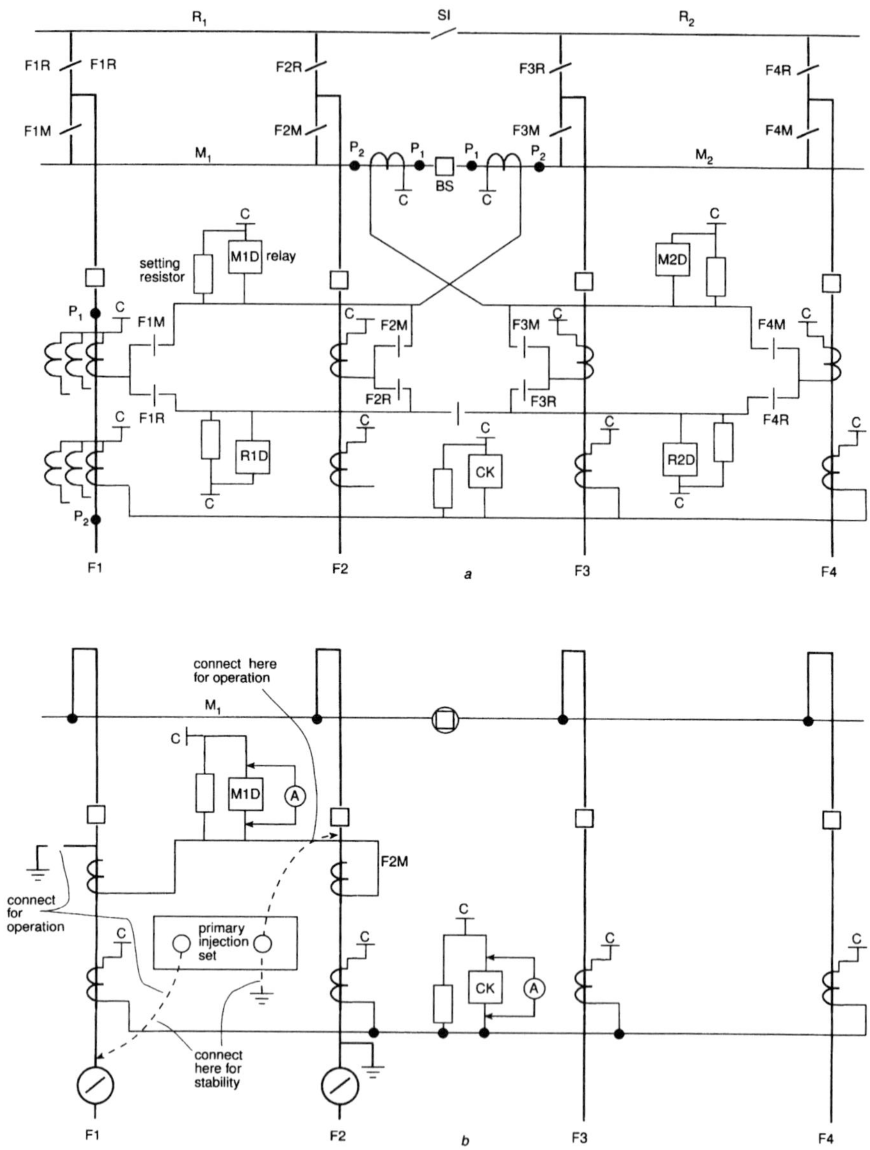

Figure 10.16 Primary injection of busbar protection

a Typical double-busbar-protection arrangement
b Operation and stability shots for M1 discrimination and check zones

(*a*) Short-circuit the trip contact of the check relays (Figure 10.17). This is necessary because the check CTs for the whole substation are permanently connected, irrespective of the busbar running arrangement. There is thus a

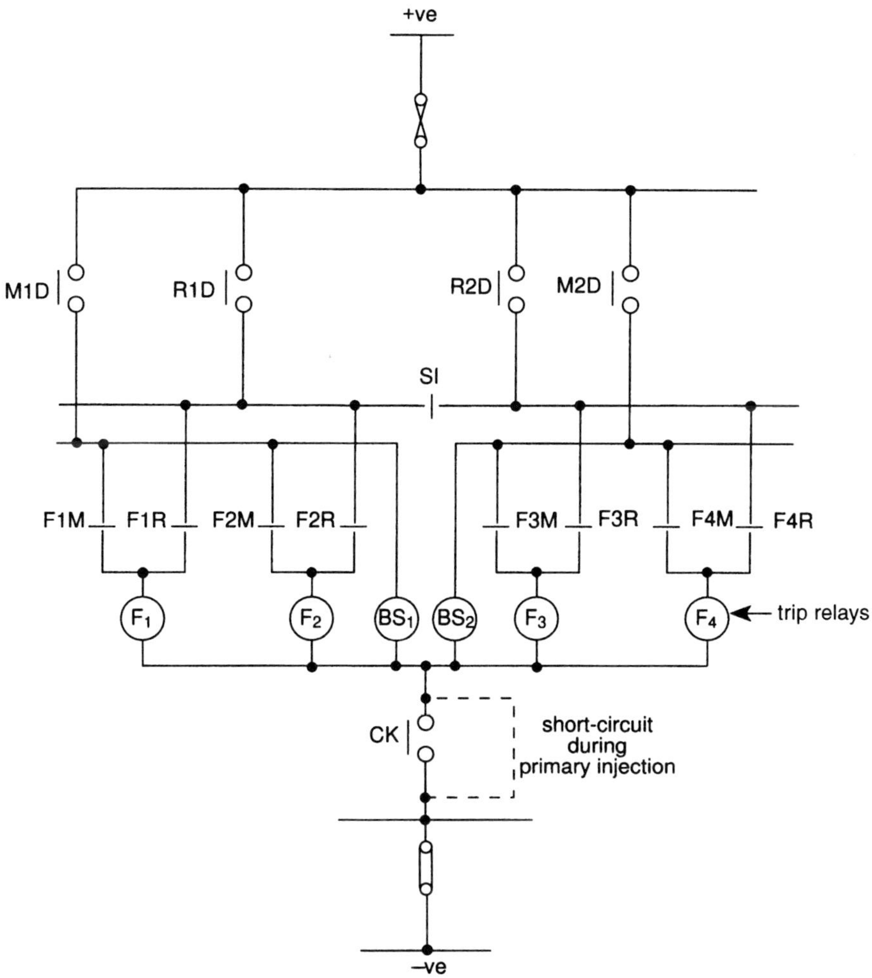

Figure 10.17 Trip circuitry associated with Figure 10.16

danger that, if an actual busbar fault arises during the tests, the test current may cancel the operating current in the in-service CTs. Thus during the tests the busbar protection is arranged to operate solely on the discrimination zones. This is illustrated in the tripping arrangements shown in Figure 10.17.

(*b*) De-energise F2 and busbar M1 and select F1 and F2 to M1. Primary inject F1 CTs only, in a similar fashion to Figure 10.13*a*, to prove ratio, group polarity and relay operation from single-CT infeed. Measure the secondary current on ammeters connected across the M1D and CK high-impedance relays.

Arrange the circuitry as in Figure 10.16*b* and carry out matching, stability and operation shots (two-CT infeed). Note that during the operation shot the circuit breakers should be open.
(*c*) The above tests should be repeated with F1 and F2 selected to busbar R1 to prove the R1D zone. On completion, the short-circuit on the check zone applied in (*a*) above should be removed.
(*d*) At this stage the primary injection is complete, but it is recommended that a secondary injection of the per-zone buswires is carried out, through a high-resistance test source, to ascertain the voltage and secondary current injected into the zone which will just operate the relays. With reference to the discrimination zones, each F1 busbar selector disconnector should be closed in turn (to insert the F1 CTs into the scheme), with the circuit breaker open, and all other circuits on that zone selected to that busbar (if operationally acceptable) while the injection is carried out. Consideration should be given to removing tripping for the duration of the injection. This information is necessary for comparison with future maintenance-injection results. It needs to be noted that the above tests illustrate the general approach to be adopted, and do not necessarily embrace all permutations of types of busbar-protection schemes (see also Section 16.16). Although a high-impedance busbar-protection scheme has been discussed, the general principles apply also to biased-differential schemes.

10.24 Primary injection of generator-protection CTs

This will be examined in Chapter 17.

10.25 Primary injection of metalclad switchgear

The previous sections have largely concentrated on primary injection of open-terminal substations. On metalclad switchgear the primary conductors are less accessible. On the traditional type of metalclad switchgear, to be found in distribution substations, a test bushing is usually supplied. The test bushing is inserted into the switchgear spouts (once the circuit-breaker truck has been removed), on either the busbar or feeder side, as required. Figure 10.18 illustrates a typical arrangement. At higher voltages, SF_6 switchgear usually has access ports at strategic points to enable primary injection to be carried out.

10.26 Cable slip-over CTs

It is frequently necessary to apply 'slip-over' CTs to cable circuits. In such instances, the sheath earth should either be located on the feeder side of the

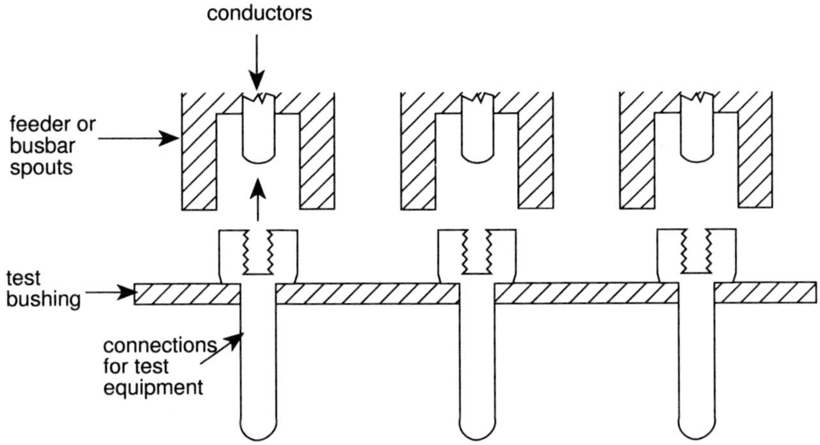

Figure 10.18 *Test-bushing arrangement*

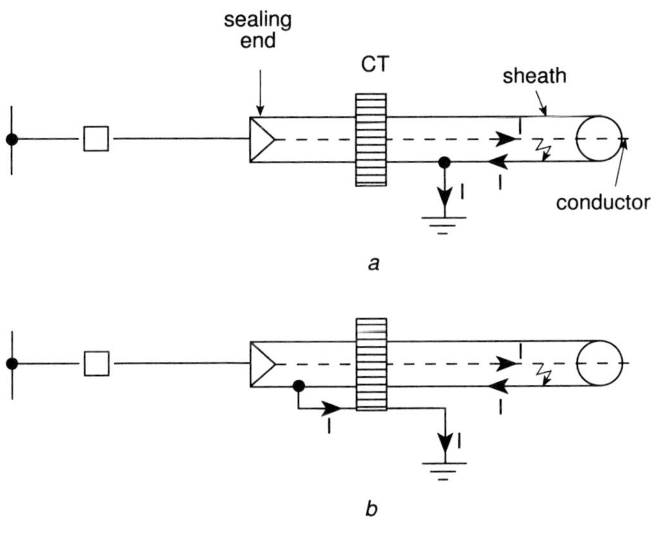

Figure 10.19 *Cable slip-over CTs*

a Cable earthing on feeder side of CT
b Cable earthing on busbar side of CT

CT or the earth should be fed through the CT. If this is not done, there is a danger of current cancellation between cable conductor and sheath. The point is illustrated in Figure 10.19. Special attention should be paid to cable earthing and its effect on protection systems, when carrying out primary injection.

10.27 The limits of primary injection

It is necessary to address the question of why significant effort is made to carry out primary injections within a substation but it is accepted that there is no primary injection to prove unit protections between different substations. The reality is that it is usually impossible both because there is too much impedance in the intersubstation circuit, and to ensure safety from induced voltages, to carry out such a primary injection, and thus proving of the protection is left to the on-load tests. However, as a point of principle, as much testing as possible should be carried out prior to placing the circuit on-load, given that if problems are experienced at this stage they have significant implications for the duration of the circuit outage (also see Section 19.17).

CT MAINTENANCE

10.28 Maintenance of CTs

Once commissioned, CTs require little maintenance. Almost the only maintenance required is a periodic test of insulation resistance. Some utilities measure magnetisation curves as part of the maintenance, but in the author's experience CTs give so little trouble that the case for this is weak.

10.29 Recommissioning CT circuits

Assuming that the CTs themselves have not been physically relocated, it is usually not necessary to carry out primary injections when CT circuitry is being recommissioned, following changes. All that is required is a secondary injection from a point in the wiring which has not been disturbed. Notwithstanding this, if the changes to the wiring are significant, most commissioning engineers prefer to err on the side of safety and carry out primary injection shots, particularly where unit protection schemes are involved.

10.30 Investigations relating to CTs

When carrying out investigations, especially where protections have either maloperated or failed to operate, it is worthwhile researching the following:

(*a*) Are the CTs on the correct tap?
(*b*) Has maintenance been carried out recently during which CT test links have inadvertently been left open, or a short-circuit has inadvertently been left on the CT?

Note that if a fault recorder is monitoring the event but is not connected to the same CTs as the protection, the above may not be apparent.

VOLTAGE-TRANSFORMER THEORY

10.31 The voltage-transformer equivalent circuit

The voltage-transformer (VT) equivalent circuit is essentially the same as the power transformer, as illustrated in Figure 3.16. This is because both power and voltage transformers are parallel connected, i.e. they have full system voltage imposed across their windings. The VT differs from the power transformer, however, in as much as the power-transformer leakage reactances are designed to have specific values, to limit the fault level under short-circuit conditions, whereas the VT leakage reactances are designed to be as small as possible, since the objective of the VT is to obtain a secondary voltage which is a true reflection of the primary voltage.

10.32 VT burden

The burden of a VT is defined in terms of the volt-amperes at the rated secondary voltage, but often on site, as with the CT, it is incorrectly referred to as the impedance of the secondary circuit expressed in ohms.

10.33 VT accuracy classes

The accuracy classes are defined in *BS 3941* as shown in Table 10.3.

The accuracy class defines, in the main, the voltage-ratio error, for example class 0.1 has a voltage-ratio error of ±0.1% and class 3P has a voltage-ratio error of ±3%. The accuracy class also specifies a phase displacement between primary and secondary.

Measurement VTs are required to maintain accuracy between 80% and 120% of rated HV voltage over a specified range of burdens. Protection VTs

Table 10.3 VT accuracy classes

Accuracy class	Application
0.1	Precision testing/standard VT
0.2	Precision indicating instruments
0.5	Precision metering
1.0	Commercial metering
3.0	Industrial measuring equipment
3P	Protection purposes
6P	Protection purposes

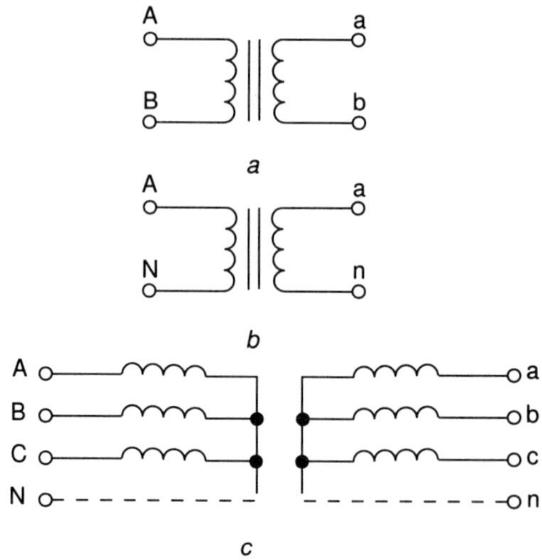

Figure 10.20 *VT terminal markings*

a Single-phase VT with fully insulated primary
b Single-phase VT with reduced insulation towards neutral
c Three-phase VT

are required to maintain accuracy between 5% and 150% (in some instances) of rated HV voltage, over a specified range of burdens. In the main, electrical power systems utilise protection VTs, although metering VTs are used when commercial metering is required. Privatisation of the supply industry in several countries has spawned a new generation of high-accuracy VTs (which are integrated with the CT into a single unit) specifically for metering purposes.

10.34 VT transformation ratios

The VT secondary windings usually have a nominal phase-to-phase voltage of 110 V, with a voltage of 63.5 V between phase and neutral.

10.35 VT terminal marking and polarity

Figure 10.20 illustrates the required terminal markings to *BS 3941*, for the electromagnetic windings of a VT. Letters A, B, C, N denote the primary winding and a, b, c, n the secondary winding. At any instant, the voltages A–N and a–n are in phase.

10.36 VT types

Voltage transformers fall into two main categories as follows:

- electromagnetic: single-phase or three-phase
- capacitor type (CVT): single-phase only

At distribution voltages the electromagnetic VT is more economic than the CVT. Greater accuracy can also be designed into electromagnetic VTs. At transmission voltages, the CVT predominates, but when greater accuracy is required for, say, metering purposes, recourse is made to electromagnetic VTs.

10.37 Capacitor voltage transformer (CVT)

Figure 10.21 illustrates the essential features of a capacitor voltage transformer. Consider first the load side of the inductance to be open-circuited as shown in Figure 10.21*b*. The voltage V_I in this case is equal to V_{C2}, which in turn is equal to

$$\frac{V_P\,X_{C2}}{X_{C1}+X_{C2}} \text{ where } X_{C1}, X_{C2}=\text{reactances of capacitors } C_1 \text{ and } C_2$$

$$\text{Now } X_{C1}=\frac{1}{2\,\pi f C_1} \text{ and } X_{C2}=\frac{1}{2\,\pi f C_2}$$

$$\text{Thus } V_I=\frac{V_P\left(\frac{1}{C_2}\right)}{\frac{1}{C_1}+\frac{1}{C_2}}=\frac{V_P}{\frac{C_2}{C_1}+1}$$

We can now adopt a Thevenin equivalent circuit to determine the effect of connecting the electromagnetic transformer and load Z_L to the inductance L.

This is shown in Figure 10.21*c*. The capacitance C is equal to C_1+C_2, the parallel combination of C_1 and C_2 as required by the Thevenin approach. If C and L are arranged to be resonant at 50 Hz, the voltages V_I and V_T are always the same value irrespective of transformer loading. By this means V_T is always in direct proportion to V_P. Figure 10.21*d* illustrates the vector diagram (the effect of the relatively small magnetising current has been ignored). A problem with the CVT is that, if the secondary of the transformer is subject to a short-circuit, the voltage V_{C2} can reach dangerously high levels. This is

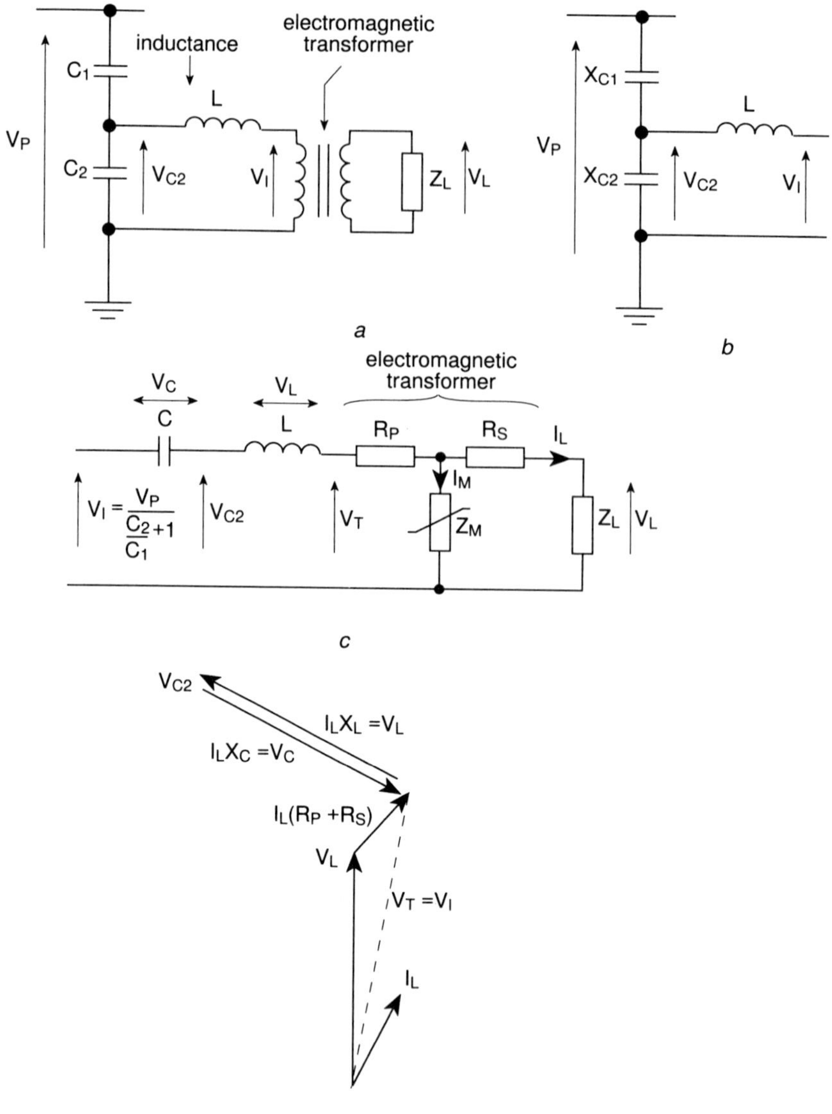

Figure 10.21 Principle of capacitor voltage transformer

a Essential elements
b Transformer disconnected
c Thevenin equivalent circuit
d Vector diagram

usually avoided by arranging L to saturate or by connecting a spark gap across C_2.

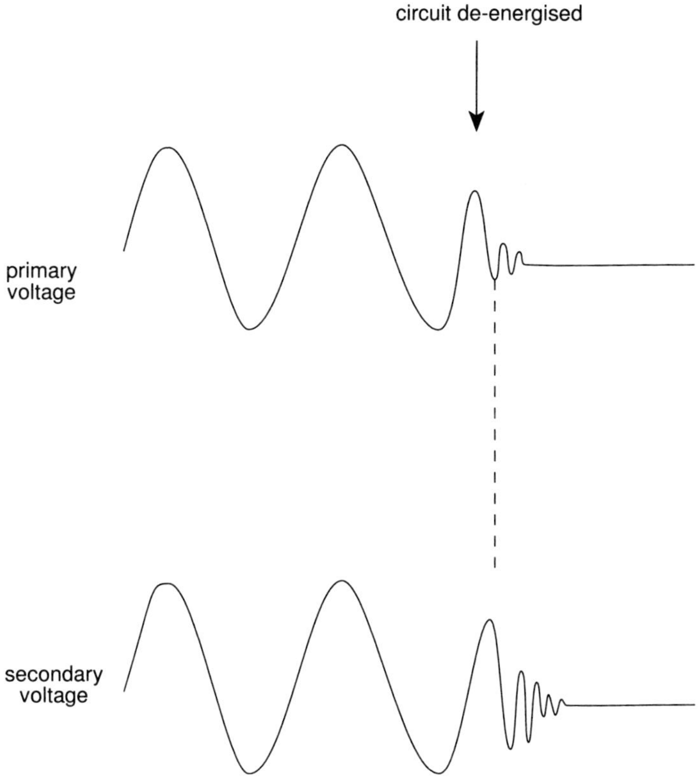

Figure 10.22 Illustration of CVT transient response

10.38 CVT transient response

Given that the CVT, unlike the electromagnetic VT, contains both inductive and capacitive components, it is prone to exhibiting oscillatory transient voltages, when subject to a step change in power-system voltage. *BS 3941* specifies CVT performance criteria, and requires oscillations to be limited to acceptable values. In particular, it requires that, in the event of a primary short-circuit to earth, the secondary voltage shall decay to a value less than 10% of the prefault peak value, within one cycle. Figure 10.22 illustrates this requirement.

10.39 CVT ferroresonance

The magnetising impedance Z_m, shown in Figure 10.21*c*, will saturate under certain circumstances. In particular, oscillatory transients can cause a state of ferroresonance in, and hence saturation of, the magnetising impedance

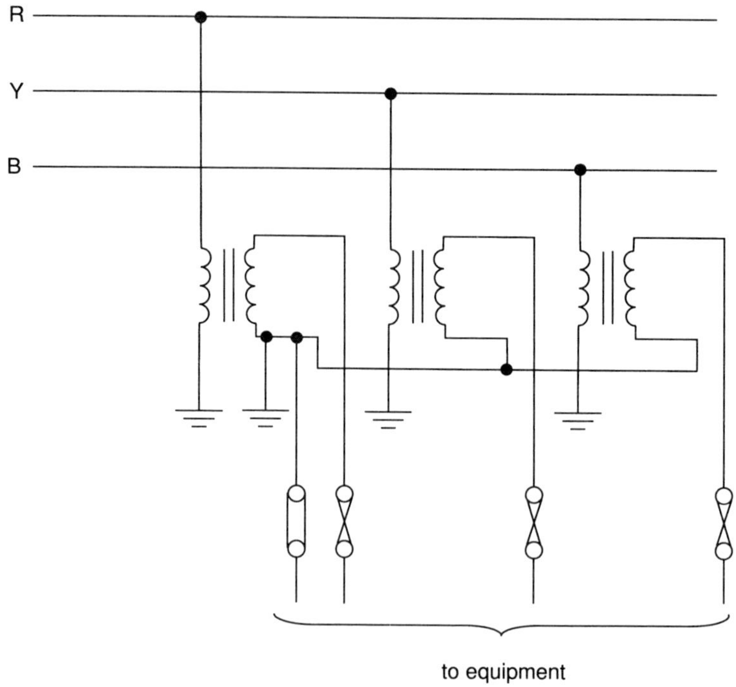

Figure 10.23 Single-point secondary earthing of three single-phase VTs

which, once established, can be sustained by the power system (see Chapter 5). The effect of ferroresonance can be an increase in secondary voltage of up to 150% nominal. *BS 3941* specifies CVT-performance criteria to minimise the effects of ferroresonance but, even so, it can be difficult to eliminate. The most effective measure for eliminating ferroresonance is to connect a suitable resistive load across the CVT secondary terminals to act in an oscillation-damping capacity. Electromagnetic VTs also exhibit ferroresonance, but this is much less common than in a CVT.

10.40 Transformation of zero-sequence voltages

Chapter 3 discussed the criteria for the transformation of zero-sequence quantities from the primary to the secondary of a transformer. This can be achieved via single-phase VTs, or a three-phase five-limbed core VT. Some protection devices require zero-sequence voltage information to function correctly, and in such instances it must be ensured that, if a three-phase VT is utilised, it is of the five-limbed design.

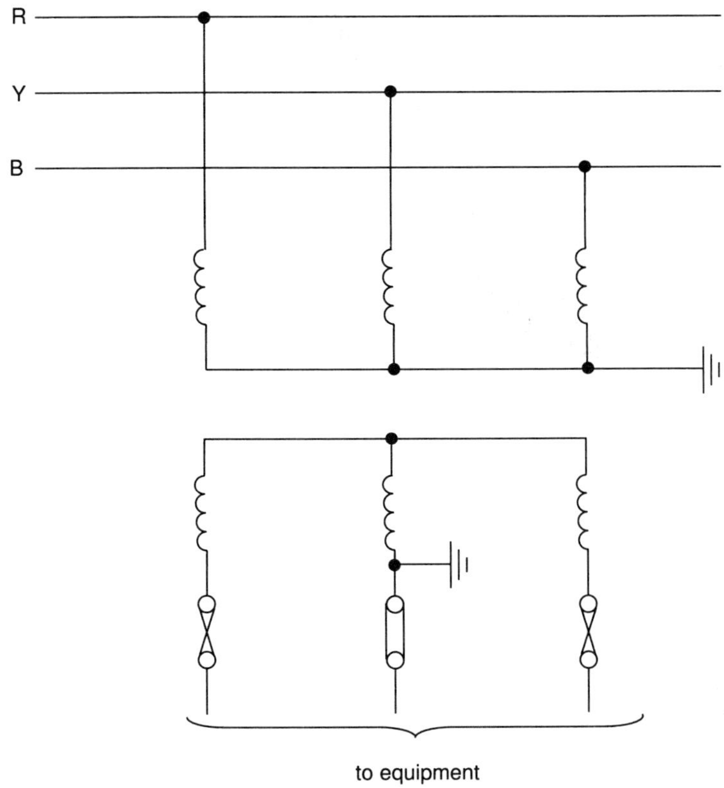

Figure 10.24 Earthing of VT via yellow phase

10.41 VT Buchholz

Electromagnetic VTs are usually oil filled. At voltages of 132 kV and above the volume of oil is significant and frequently necessitates a conservator, which in turn facilitates the use of a Buchholz relay (see Chapter 5). The Buchholz relay usually only consists of the float, and not the surge deflector plate. It may be connected into either trip or alarm circuitry depending on the position of the VT on the system.

10.42 VT fusing and earthing

Fusing of the primary winding of the VT, for protection purposes, is largely impracticable. For example, if we consider a typical VT secondary short-circuit current of say 100 A, the primary current for, say, the 132 kV system would only be about 80 mA. Notwithstanding this, primary fuses are sometimes applied to VTs connected to 11 kV and 33 kV systems.

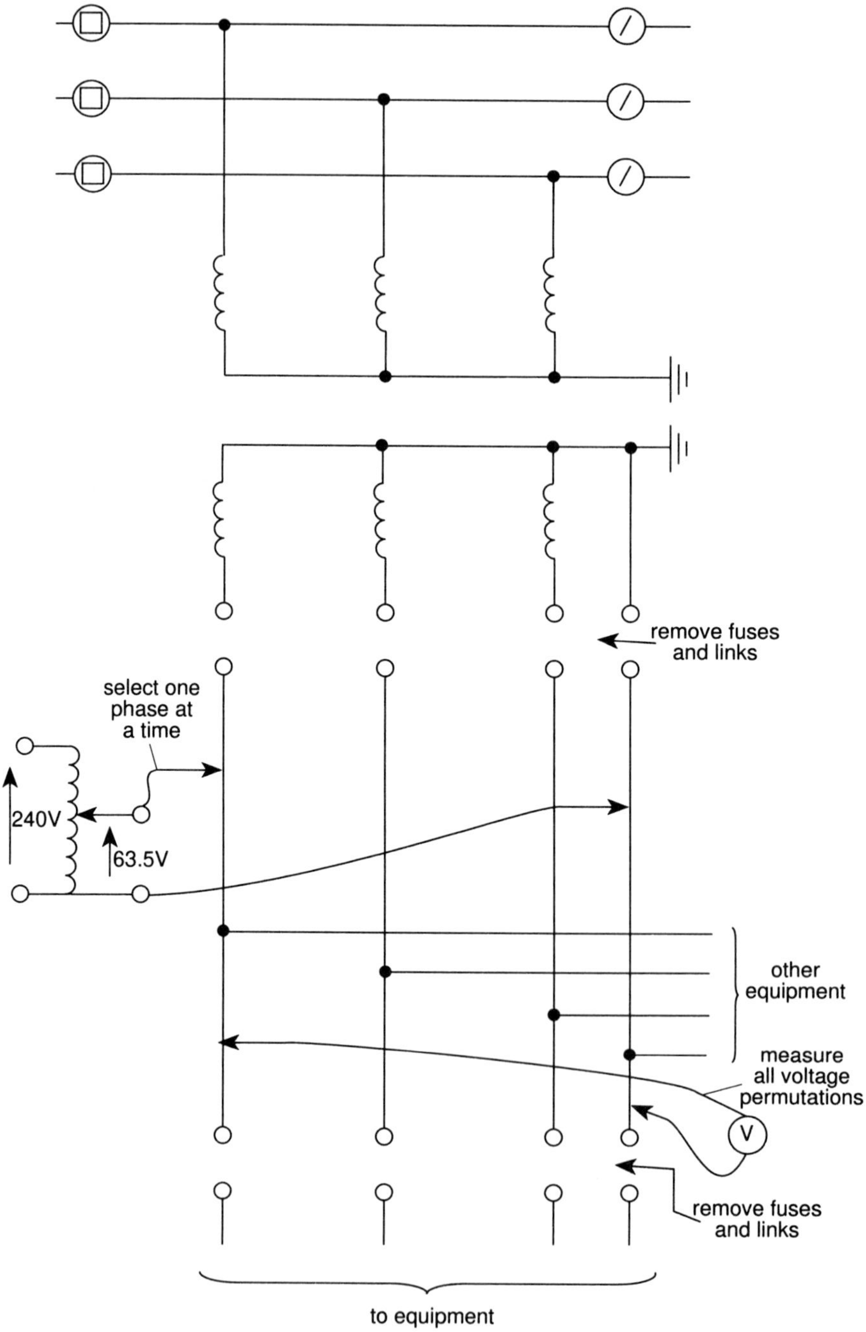

Figure 10.25 Secondary injection of VT wiring

The secondary of the VT is always fused. At transmission voltages it is usual to monitor the integrity of the fuses with a VT-supervision relay. Modern supervision relays usually measure the negative-phase-sequence voltage which arises with the loss of one or two fuses. System faults also result in negative-phase-sequence voltages. Thus, to prevent incorrect relay operation during system faults, the relay is fed from the circuit current transformers, to derive a negative-phase-sequence current interlock. Loss of all three fuses also causes an output from the relay. This output is usually interlocked with plant auxiliary switches to distinguish between a true loss of fuses and the circuit being switched out of service.

Three-phase electromagnetic VTs usually have the star point of the secondary earthed at the VT. Higher system voltages employing three single-phase VTs need to have the secondary earthed at a single point, as shown in Figure 10.23. The alternative of earthing each VT separately is not recommended since circulating currents arising from system faults could flow in the multiple earth-loop connections, so causing an unwanted voltage drop which in turn is imposed upon the equipment. It is usual to earth the secondary of some VTs via the yellow phase, and not the star point. This is shown in Figure 10.24.

VT COMMISSIONING

10.43 VT commissioning

The commissioning of a VT is not as involved a procedure as for a CT. In general it consists of the following:

- collection of data, i.e. serial number, type reference, ratio, rating etc.
- general inspection
- check on oil samples for electromagnetic VTs
- insulation-resistance test between windings and earth of an electromagnetic VT
- measurement of electromagnetic winding resistance
- check of Buchholz, where one is fitted
- flick tests
- secondary injection to prove secondary wiring (see Section 10.46)
- vector-group test (see Section 10.45)
- on-load tests (see Chapter 19)

10.44 VT flick tests

Flick tests can be carried out in a similar fashion to that for a CT, as described in Section 10.14. Many commissioning engineers, however, bypass this test and rely solely on the on-load tests.

10.45 VT vector-group test

A vector-group test can be carried out for three-phase electromagnetic VTs in a similar fashion to that described in Chapter 5, by applying a 415 V three-phase supply to the primary. However, the secondary voltages for such a test are small (e.g. 1.4 V for a 33 kV/110 V VT with 415 V applied) and make the construction of the vector diagram somewhat impracticable. For this reason this test is frequently bypassed, leaving reliance on the on-load phasing tests.

10.46 Secondary injection to prove VT secondary wiring

It is advisable to prove the secondary wiring between the VT and the equipment before the on-load tests. This can be done by injecting voltages at the VT, as shown in Figure 10.25, and measuring the voltages at the equipment fuses and links.

VT MAINTENANCE

10.47 VT maintenance

Apart from a periodic check of the VT oil (see Chapter 5), VTs are virtually maintenance free.

10.48 VT fault investigations

There are a number of points worthy of note in the context of VTs and fault investigations:

(*a*) Has a distance protection maloperated because the VT fuses have not been replaced following the last maintenance? Modern supervision relays should detect this at the point of circuit energisation, but older relays were not, in many instances, fitted with a three-phase VT-supervision feature.
(*b*) Has the CVT been ferroresonating (check on any connected fault recorder)?
(*c*) Is there evidence of any spark gap flashover?
(*d*) Is the VT properly earthed?
(*e*) Examine the voltages on any connected fault recorder.

Chapter 11

Switchgear control systems

11.1 Introduction

This chapter is concerned with the more technically complex switchgear control systems as applied to common substation configurations. General aspects of control systems are discusssed in Chapter 12. Note that the term 'control' refers to systems which are concerned with the operation of switchgear, as opposed to 'protection' systems which are concerned with the disconnection of faulted plant. The systems covered are:

- interlocking
- synchronising
- autoswitching

Voltage-control schemes are covered separately in Chapter 6.

INTERLOCKING

11.2 Purpose of interlocking

Interlocking is installed to protect both the equipment and the operator from the dangers of incorrect equipment operation. It can be categorised as follows:

(*a*) *Operational interlocking*: This is concerned with the operational sequencing of disconnectors (both on and off load) when system connections are being reconfigured. This is usually achieved by 'electrical-interlocking' systems and generally applies to open-terminal substations.
(*b*) *Safety, or maintenance, interlocking*: This is concerned with the safe operation of switchgear, the provision of safe access to plant for maintenance purposes and the application of fixed earthing devices. It is usually achieved by 'mechanical-interlocking' systems utilising keys or, for distribution-voltage metalclad switchgear, latches and bolts. Note that one school of thought terms the former as 'key interlocking' and the latter as 'mechanical interlocking'.

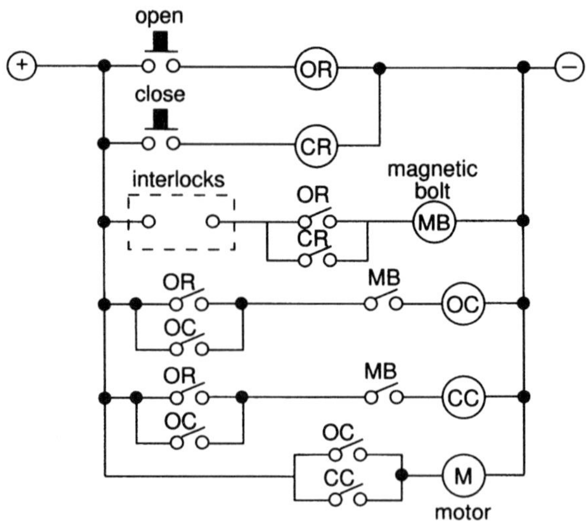

Figure 11.1 Typical disconnector-control circuitry

OR = open relay
CR = close relay
OC = open contactor
CC = close contactor

11.3 Operational interlocking

The operational interlocking of disconnectors is usually accomplished by installing an electromechanical bolt, termed a magnetic bolt or more simply 'mag bolt', in the disconnector control circuitry.

Figure 11.1 illustrates typical control circuitry for a motorised disconnector. The magnetic-bolt coil is interlocked via auxiliary switches from other disconnectors and circuit breakers to form an interlock chain.

Figure 11.2 builds upon Figure 11.1 to show the typical interlocking arrangement for busbar-selector disconnectors for a double-busbar substation. Figure 11.2*a* illustrates a simple example of a double-busbar substation with busbar-selector disconnectors 104 and 106 (also 204 and 206). Disconnector 104 can be opened and closed safely under two conditions, as follows:

(i) *No power path*: With circuit breaker 105 and disconnector 106 open, disconnector 104 can be opened and closed at will, since no power will flow through it (only busbar-charging current).
(ii) *Parallel path*: If disconnector 106 and circuit breaker 105 are both closed, disconnector 104 can be opened and closed at will, provided that it is short-circuited by another parallel path (e.g. 134, 130 and 136 are closed).

Figure 11.2*b* shows the circuitry required to achieve operation of the

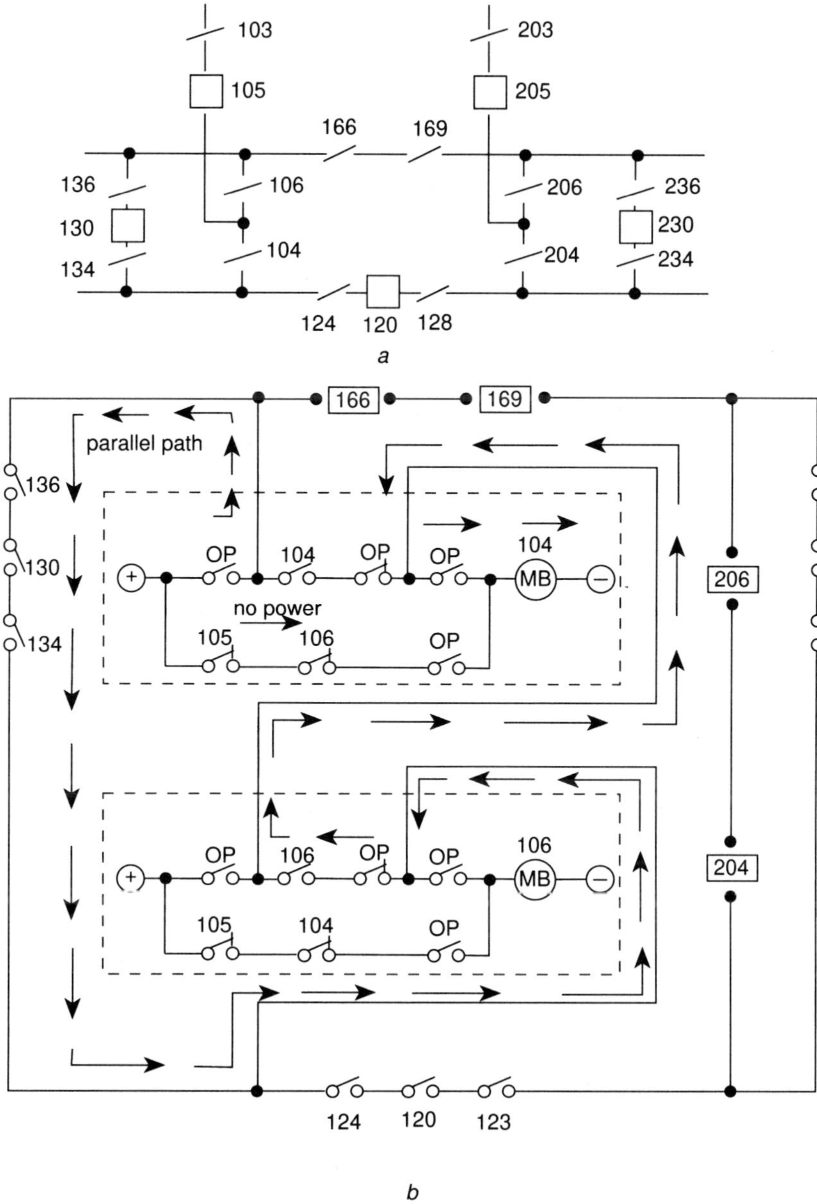

Figure 11.2 Electrical interlocking: busbar-selector disconnectors

a Double-busbar substation
b Circuitry required for operation of the magnetic bolt

OP is energised by either relay OR or CR (Figure 11.1)

magnetic bolt for disconnectors 104 and 106 via the 'no power' and 'parallel' paths. The boxes containing disconnectors 204, 206, 166 and 169 contain similar circuitry to the enclosed dotted areas for 104 and 106.

The operation of other disconnectors such as 134, 136, 103, 203, 124 and 128 is also dependent on magnetic bolts which have no-power-path interlocks. These are not shown in Figure 11.2.

To demonstrate the principles involved further, Figure 11.3 illustrates the magnetic-bolt-interlocking requirements for disconnectors 113 and 103, which are connected to one corner of a mesh substation. These disconnectors are not in parallel with other disconnector and circuit-breaker arrangements, and therefore the interlocking is based on no-power paths. Two interlocking routes exist for disconnector 103. One route occurs when 126 and 420 (or 424) are open, and the other route is when 113 (or 180), 120 (or 128) and 420 (or 424) are open. Note that the interlocking arrangements for disconnector 126 (not illustrated) would involve parallel-path interlocks, via the parallel path around the closed mesh.

The above illustrations refer to motorised disconnectors, but identical magnetic-bolt interlocking systems are also applied to manually operated disconnectors. In such instances, operation of the magnetic bolt is usually initiated by lifting the disconnector operating handle to the horizontal, which, provided that the interlock chain of contacts is made, frees the operating-handle mechanism to permit rotational movement of the disconnector.

11.4 Disconnector interlocking mechanism

Figure 11.4 illustrates a typical disconnector interlocking mechanism. When the position selector is selected to the free position, the magnetic bolt solenoid, when energised, can draw the plunger fully into the solenoid. This in turn enables the drive restrictor to be raised, allowing unimpeded movement of the rotating cam. Once the cam has rotated below the drive restrictor, it prevents the drive restrictor, and hence the whole magnetic-bolt assembly, from resetting, until the cam completes its travel at the end of the closing or opening sequence.

Safety interlocking (as described in Section 11.3) usually involves the repositioning of keys. Figure 11.4 illustrates such a key, inserted into its barrel. Selection of the position selector to the locked-open, or locked-closed, position, rotates the key-restrictor plate out of the jaws of the key-barrel assembly. This enables the key to be rotated through 90° and withdrawn. The disconnector is now locked in this position, since the orientation of the key-barrel assembly now prevents the position selector being reselected to the free position. If the magnetic-bolt solenoid is now energised, the position of the lock plate prevents the plunger from being drawn into the solenoid, which in

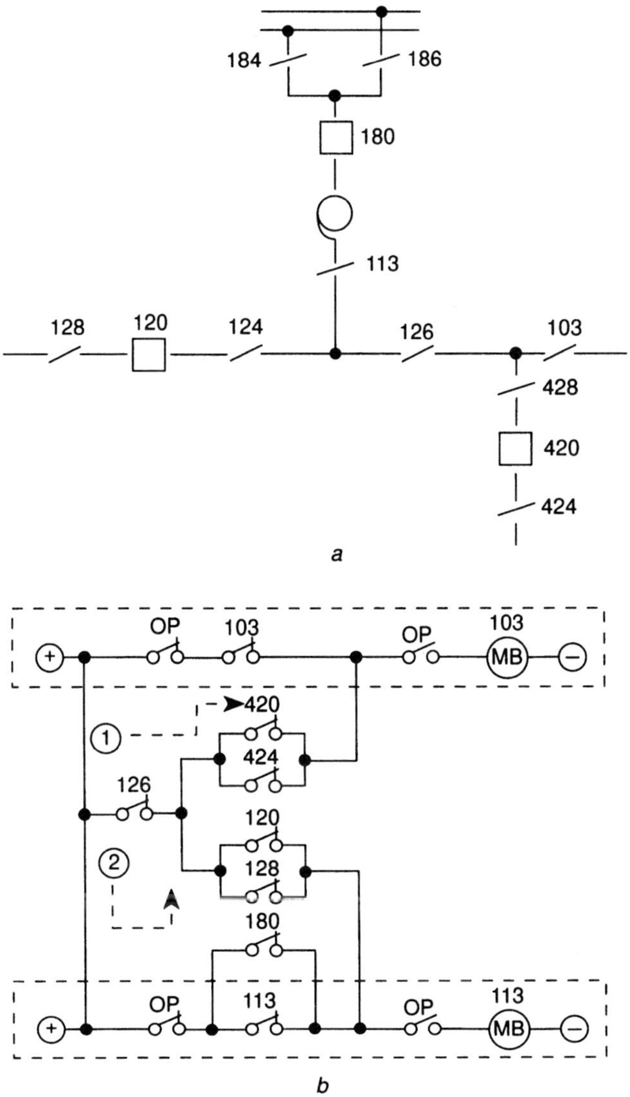

Figure 11.3 Electrical interlocking: mesh-corner disconnectors

a Substation configuration
b Circuitry required for operation

OP is energised by either relay OR or CR (Figure 11.1)

turn prevents the magnetic-bolt mechanism from being operated. Hence the disconnector is immobilised. Although only one key is shown, a disconnector would typically contain up to four keys.

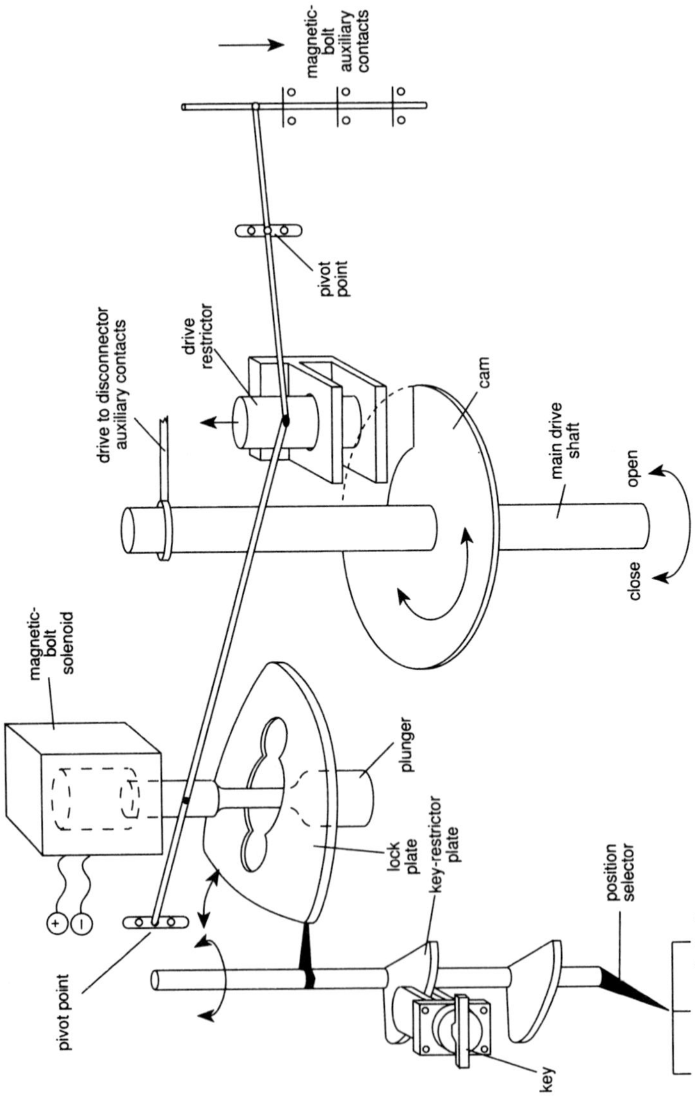

Figure 11.4 Typical mechanical and electrical interlocking

11.5 Safety interlocking

Interlocking associated with distribution voltage metalclad switchgear is usually a combination of electrical and mechanical and is usually concerned with the correct positioning of the circuit-breaker truck. Typical examples relating to a 33 kV metalclad circuit-breaker are:

(*a*) electrical interlocking to ensure that the circuit breaker cannot be raised or lowered unless its interrupter contacts are in the open position;
(*b*) electrical interlocking to ensure that the circuit breaker can only be closed when it is in the fully raised or lowered position;
(*c*) busbar and feeder fixed-contact spouts are automatically covered by shutters, operated by a mechanical interlock system, as the circuit-breaker is racked to the isolated position;
(*d*) mechanical interlocking to ensure the circuit breaker can only be closed in feeder or busbar earth position when the truck is physically located in the appropriate position.

Interlocking associated with the earthing of high-voltage switchgear other than distribution-voltage metalclad switchgear has traditionally been 'mechanical interlocking', involving the transfer of the type of key shown in Figure 11.4.

11.6 Mechanical-interlocking example

Figure 11.5 illustrates a relatively simple interlocking scheme utilising keys for earthing the high-voltage plant associated with the transformer end of a transformer feeder. The key symbols in the figure are interpreted as follows. A circle indicates a key which is released when the disconnector or earth switch is in the open position. A square indicates a key which is released when the disconnector or earth switch is in the closed position. A triangle indicates a key which is released when the circuit breaker is in the open position, or when the disconnector is in the open or closed position, but which must be inserted to permit operation of the disconnector.

The key positions which are shaded black represent the locations of keys which are in the locked position when the item of plant is in the in-service position, (e.g. circuit breaker and disconnector closed, and earth switch open).

To close earth switch 111, the line would have to be de-energised, all circuit breakers and disconnectors opened, and the following sequence carried out:

(*a*) A key transferred from 1T0 to 1T3. 1T3 can then be opened and a key from 1T3 transferred to E/S 1T1.
(*b*) A key transferred from 113 to E/S 1T1.
(*c*) E/S 1T1 can now be closed. This releases a key from E/S 1T1 which needs to be transferred to E/S 111.

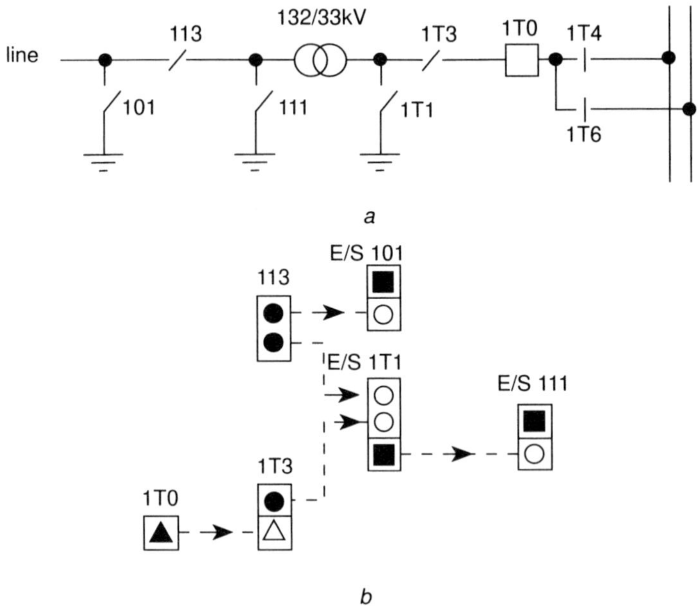

Figure 11.5 Mechanical interlocking: transformer feeder

a Substation configuration
b Key interlock arrangements

(*d*) E/S 111 can now be closed. An E/S 111 key can now be released and inserted into a safety lock out box. E/S 111 and 1T1 are now mechanically interlocked in the closed position.
The process needs to be reversed to put the circuit back into service.

Figure 11.6 illustrates a more complex mechanical interlocking system pertaining to the earthing of busbars for a transmission-voltage open-terminal double-busbar substation. The symbols are identical to those used in Figure 11.5. To facilitate the process, key-exchange boxes, which are usually located in a convenient central position within the substation, are required. Those key-exchange boxes in Figure 11.6 designated with an (OR) suffix can have their keys released by two routes. For example, to close E/S 121B and E/S 121C, the two options to release the keys from 'bus section 1 exchange box (OR)' are:

(*a*) via opening 124 and 128

or

(*b*) via the closure of E/S 121A (via MB1 E/S) and E/S 221A (via MB2 E/S). This latter option leaves disconnectors 124 and 128 available for operation during busbar maintenance.

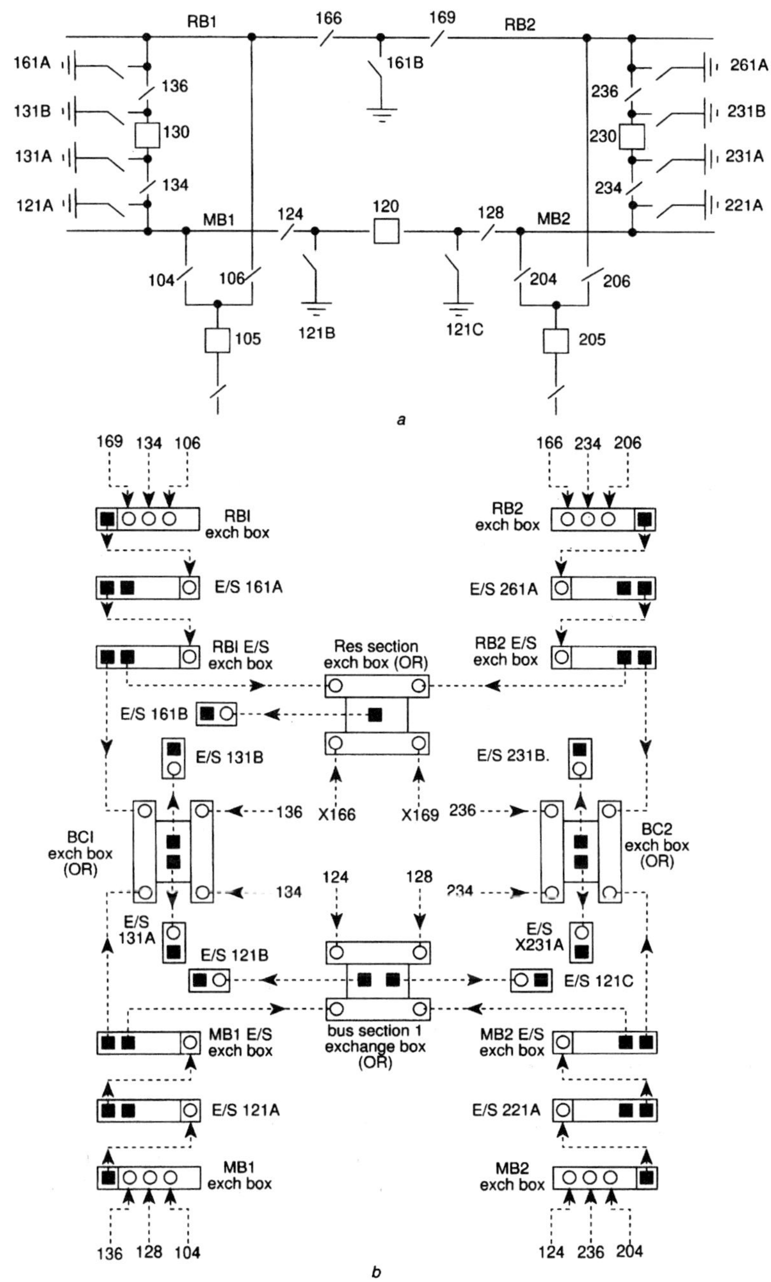

Figure 11.6 Mechanical interlocking: double busbar

a Substation configuration
b Key interlocking arrangements

As can be seen, the mechanical-interlocking process is a combination of logical AND or OR gates.

11.7 SF_6 metalclad switchgear

The modern tendency with transmission-voltage SF_6 metalclad switchgear is to utilise, for control purposes in general, a microprocessor-based co-ordinated control system. Such a system would contain all control facilities including both operational and safety interlocking circuitry; however, this interlocking would not be of the magnetic-bolt or mechanical type of design but a software interlock.

11.8 Commissioning of interlocking

As interlocking is concerned with security of plant and the safety of personnel, it is extremely important that the commissioning tests are rigorous; a written schedule of tests should therefore be produced.

For electrical-interlocking systems, each alternative chain of closed auxiliary switches which permits operation of the magnetic bolt must be verified by operating the plant physically. In addition, it must be proved that the opening of any one contact in the chain prevents operation. As circuits are being added to a substation, operational checks on existing plant will be required.

The commissioning of mechanical interlocks requires checks to ensure that keys can only be released when the item of plant is in the appropriate fully open or closed position, and that removal of an appropriate key prevents operation of the item of plant. With reference to Figure 11.4, it should be verified that the magnetic-bolt mechanism can only be operated when all keys are inserted. It is important to ensure that not more than the correct number of keys is available and that all keys are correctly inscribed.

11.9 Maintenance of interlocking

Day-to-day operation of the plant itself usually verifies the integrity of interlocking systems, and therefore almost no maintenance is required. Mechanical-interlocking systems will probably require periodic lubrication.

SYNCHRONISING

11.10 Criteria for circuit-breaker closing

Synchronising systems ensure that circuit breakers can only be closed under certain predefined conditions. This is to ensure the retention of system stability, the prevention of damage to plant, particularly that arising from generator pole slipping, and to facilitate the reparalleling of a split power system. Such systems are also routinely used to synchronise generators to the power system. Since synchronising systems are associated with generation, they tend to be installed at transmission substations and distribution systems with connected generation.

To achieve a synchronisation closure of a circuit breaker, voltages from both sides of the open circuit breaker are applied, via voltage transformers, to a synchronising relay; see Figure 11.8*a*. If the applied voltages satisfy the relay settings criterion, a contact within the relay closes to permit energisation of the circuit-breaker close coil, and thus circuit-breaker closure. The relay settings usually require the voltages on both sides of the circuit breaker to satisfy, within limits, three criteria. They must:

- have the same frequency
- have a zero phase-angle difference
- have nominal voltage amplitude

There are three basic types of synchronisation closure:

(*a*) *Check synchronisation*: With reference to Figure 11.7*a*, this occurs when circuit breaker A, which is required to be closed, is within a solidly connected power system.
(*b*) *Power-system synchronisation*: With reference to Figure 11.7*b*, this occurs when circuit breaker A closes to parallel two independent (split) power systems.
(*c*) *Generator synchronisation*: This is a special case of power-system synchronisation, whereby a single generator is connected to the rest of the power system as shown in Figure 11.7*c*.

A fourth method of closing a circuit breaker exists which does not involve synchronising systems, as shown in Figure 11.7*d*. This is known as over-ride (or dead-line) synchronisation, whereby circuit breaker A energises a de-energised (dead) line.

11.11 Check-synchronisation relay

Check-synchronisation relays are usually designed with three settings:

- phase angle
- voltage amplitude
- time delay (this is in effect a measure of frequency differences)

From Figure 11.7*a*, if current is flowing in the circuit in service, there will be a phase-angle (and amplitude) difference between the two substations X and Y (due to voltage drop across the circuit impedance), which will exist across the open circuit breaker A. To enable the check-synchronisation relay to

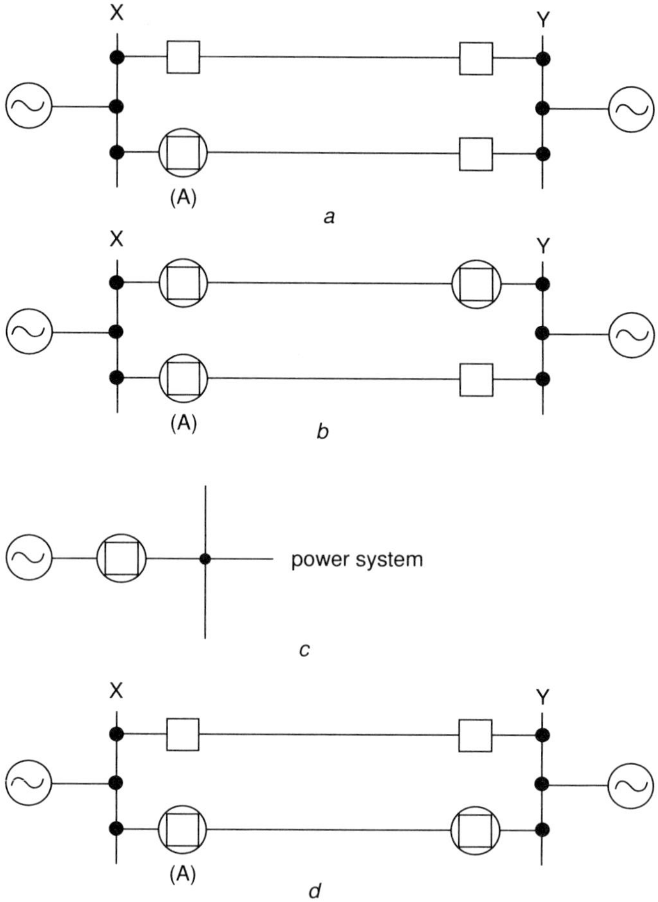

Figure 11.7 Circuit-breaker-closing conditions

a Check synchronise
b Power-system synchronise
c Generator synchronisation
d Dead-line

operate, it therefore requires an angle-setting window. This is usually arranged to be of the order of ±25°. If the angle is made too large, closure of the circuit breaker can result in a significant shock load to nearby generation. This is due to a sudden change in power-system transfer impedance, creating a flow of synchronising power.

The voltage-amplitude setting is usually that of an undervoltage limit, with a setting of 80–90% nominal rated voltage. Voltage inputs to the relay of amplitude less than the setting would inhibit the relay from operating. This setting prevents a system with a serious reactive-power deficit (resulting in low voltage) being connected to a healthy system and disrupting the latter.

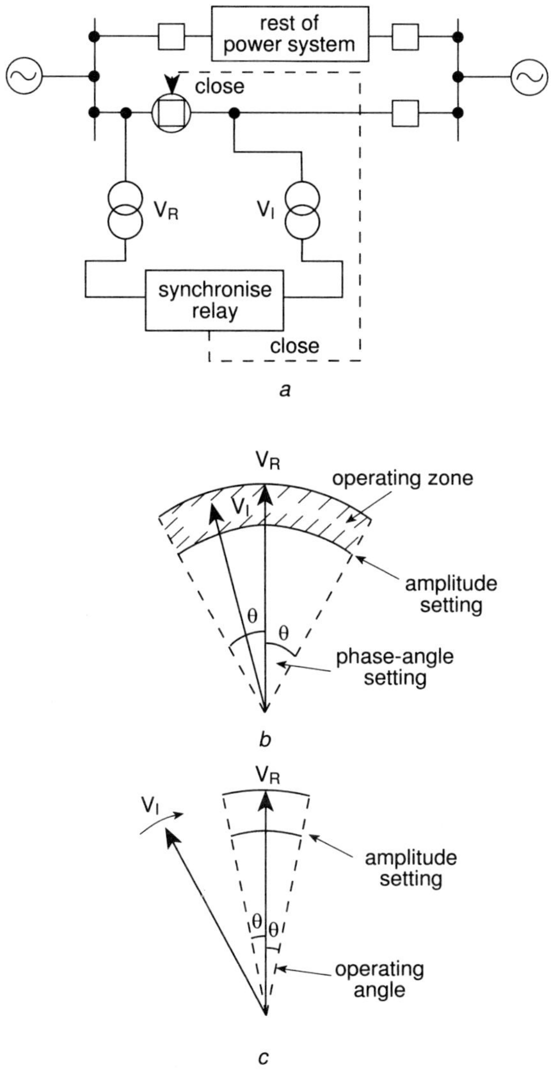

Figure 11.8 Synchronise-relay operating conditions

a Synchronising-relay connections
b Check-synchronise relay
c System-synchronise relay

Figure 11.8*a* illustrates a synchronising relay fed with voltage V_R and V_I via two voltage transformers. V_R is termed the running voltage and V_I the incoming voltage. Figure 11.8*b* shows the operating characteristics of a check-synchronisation relay; V_R is taken as the reference voltage, and the operating zone defines the possible relative position of V_I to permit operation. Because a parallel path exists through the rest of the power system, V_I will be stationary in relation to V_R, the precise phasor relationship depending on the voltage drop across the rest of the power system.

The time delay is usually of the order of 2–3 s. This is required for the situation of a power system being split and therefore no parallel path across the rest of the power system. In this instance V_I would rotate with respect to V_R, the speed of rotation (slip) being dependent on the relative frequencies of the split systems. Without a time delay, a check synchronisation relay could synchronise a split system (under autoreclose conditions) at the limit of the angle setting and at a very high slip, so causing a significant and undesirable shock loading to the power system. The time delay is arranged to ensure that the two voltages must be in the operating zone for a period in excess of that which is likely to occur with a split system.

11.12 Power-system-synchronisation relay

Power-system-synchronisation relays are usually equipped with four settings:

- phase angle
- slip frequency
- output pulse timer
- voltage amplitude

If, in Figure 11.8*a*, the connection through the rest of the power system is broken, the power system is split, and voltage V_I rotates with respect to voltage V_R, the speed of rotation depending on the slip between the split-power system frequencies. Figure 11.8*c* illustrates voltage phasor V_I rotating towards V_R. The angle setting θ of the relay, is typically 10°, and the slip-frequency setting is typically 0.125 Hz. When V_I becomes coincident with the angle setting and the slip is less than 0.125 Hz an output pulse, typically of 500 ms duration, is sent from the relay to the circuit-breaker-close coil, to close the circuit breaker. The objective is to ensure that the circuit-breaker contacts close when V_I and V_R are approximately in phase. Thus if, for example, the circuit-breaker-close time is 150 ms (i.e. from close-coil energisation to main contacts close) the angle of closure occurs at:

$$10° - \left(\frac{150\ \text{ms} \times 360°}{\text{slip period}}\right)$$

$$= 10° - (0.15 \times 0.125 \times 360°)$$

$$= 3.25°$$

Note: slip period = 1/slip frequency, as determined in Section 11.18.

The voltage-amplitude setting is identical to that described for the check-synchronisation relay.

11.13 Generator synchronising

A modern relay for synchronising a generator (often termed an autosynchroniser) is in effect a system-synchronisation relay with two additional control features:

(i) control of the turbine governor to effect generator frequency control;
(ii) control of generator excitation through the AVR, to effect generator-terminal voltage control.

This is a much more sophisticated relay than those described in Sections 11.11 and 11.12, and the output pulse is usually arranged to ensure that the circuit breaker closes when V_I and V_R are exactly in phase.

11.14 Busbar substation-synchronising scheme

Figure 11.9*a* illustrates the basic circuitry and principles associated with synchronising a circuit breaker connected to a double-busbar substation. Voltages are taken from feeder voltage transformers through tapped interposing voltage transformers (nominal ratio = 1:1) through plant auxiliary switches (e.g. CB1, M1 or R1) to voltage buswires (MBB1, RBB1, MBB2, RBB2). These buswires, termed the 'voltage-selection buswires', mimic the voltages on the busbars. A second set of buswires, to which the incoming and running voltages are selected during synchronising, and to which a synchroscope (older schemes) and system-synchronisation relay are also selected, are known as the synchronising buswires.

To illustrate how the circuitry functions, consider main busbar 2 (MBB2), energised via disconnector M2 and circuit breaker CB2, which in turn energises main busbar MBB1 via closed circuit breaker BS1, and it is required to carry out a check synchronisation of circuit breaker CB1 (which is open), which is connected to MBB1 via disconnector M1. The following steps are required to close the circuit breaker via check synchronisation:

(*a*) In Figure 11.9*b*, select the switch to the 'check-synch' position to energise the SY relay.
(*b*) A contact from relay SY in Figure 11.9*a* will connect the MBB1 buswire voltage via auxiliary contact M1 to the check synchronise relay S and to the running-voltage-synchronising buswires. Another SY contact will connect the voltage experienced on the feeder side of CB1 to the check-synchronisation relay and to the incoming voltage-synchronising buswires.
(*c*) If the incoming and running voltages are within the settings of the check-synchronisation relay, it will operate. The precise phase angle will be shown on any connected synchroscope.
(*d*) The operator can now initiate the close button in Figure 11.9*b* to energise the circuit-breaker-close coil CC via the check-synchronisation relay S contact.

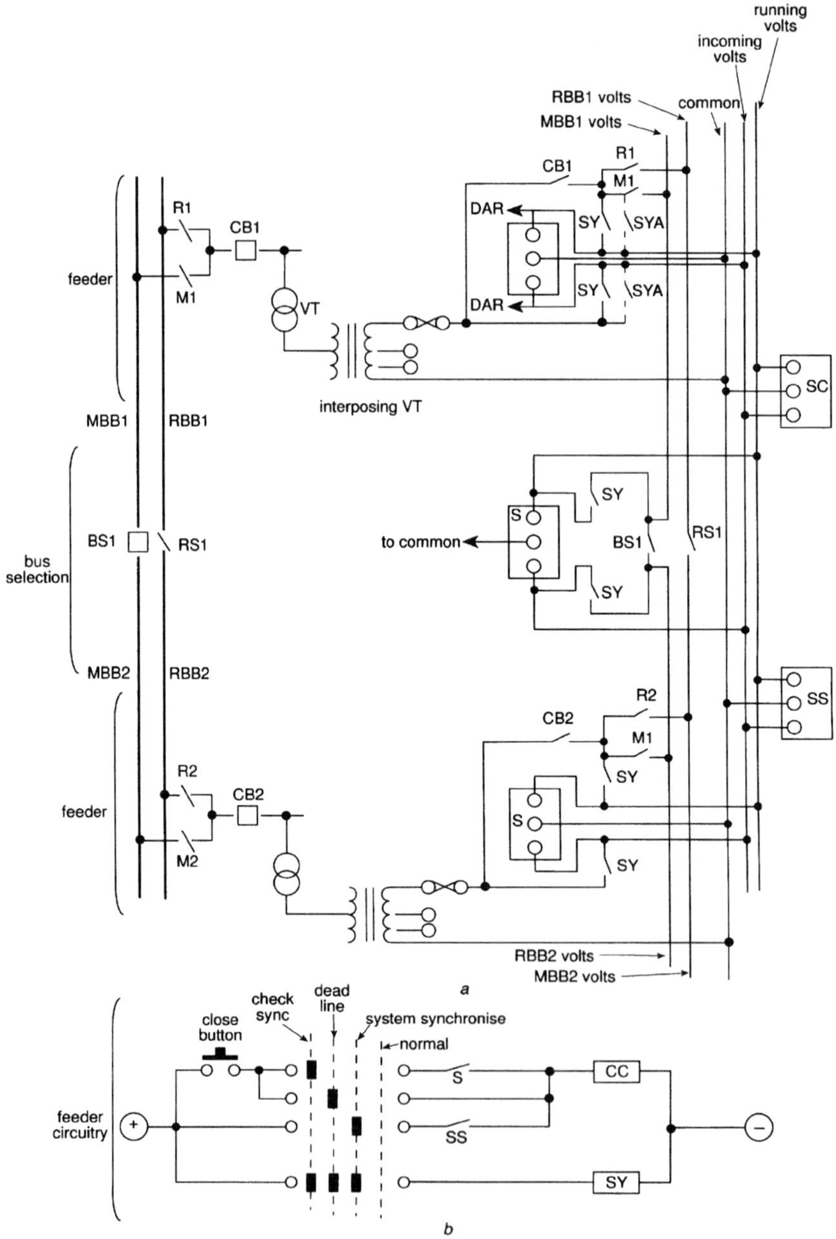

Figure 11.9 Synchronising facility: double-busbar substation

S = check-synchronise relay
SS = system-synchronise relay
SC = synchroscope
CC = CB-close coil
SY = synchronisation voltage-switching relay

When circuit breaker CB1 closes, the feeder voltage now becomes connected, via the CB1 auxiliary contact, to the MBB1 voltages buswires, thus mimicking the voltage on the busbars and connected feeders.

The system-synchronisation relay has traditionally been a common relay for the substation, whose contacts are routed (by circuitry not shown) to the close circuitry of the circuit-breaker selected for synchronisation. Synchronisation close via the system-synchronisation relay only requires the operator to select the switch in Figure 11.9*b* to the 'system synch' (SS) position. When the voltages which are applied to the relay satisfy the setting, the SS-relay contacts close to energise the circuit-breaker-close coil CC directly. Observation of the synchroscope during a system-synchronisation operation would show rotation of the phase-angle indicator, the speed of which will accord with the system slip frequency.

More recent designs have incorporated both check and system synchronisation facilities into the one relay. The effect of this on Figure 11.9*a* would be to remove the system-synchronisation relay SS, and to incorporate an SS feature into each of the check-synchronisation relays S.

The above is termed 'manual synchronisation' (as opposed to synchronisation via autoreclose), since manual selection of the synchronisation selection switch is required.

The interconnection of the secondaries of numerous voltage transformers can result in circulating currents around the scheme circuitry. The interposing VT is thus designed with a relatively high impedance to limit circulating current, and is equipped with tappings for equalisation of voltages during commissioning.

11.15 Mesh substation-synchronising scheme

The basic circuitry for a mesh synchronisation scheme is shown in Figure 11.10. For simplicity only half of the mesh is shown. The circuitry shown in Figure 11.9*b* also applies to mesh circuit breakers. As can be seen, the voltage-selection bus-wiring circuitry is more complex than that for the double-busbar substation. This circuitry comprises a clockwise voltage buswire (sometimes termed the A buswire) and an anti-clockwise voltage buswire (sometimes termed the B buswire), each segregated by contacts off voltage relays V_1, V_2 etc. The need for this circuitry is to facilitate delayed autoreclose (DAR) of a mesh corner. Mesh DAR usually has a facility to enable (or disable) dead-line reclosure in either a clockwise or anticlockwise direction around the mesh, independently. Note that dead-line DAR will usually not be allowed unless one side of the circuit breaker is energised. This, in turn, necessitates a voltage-selection scheme which provides voltages to the DAR in both clockwise and anticlockwise directions.

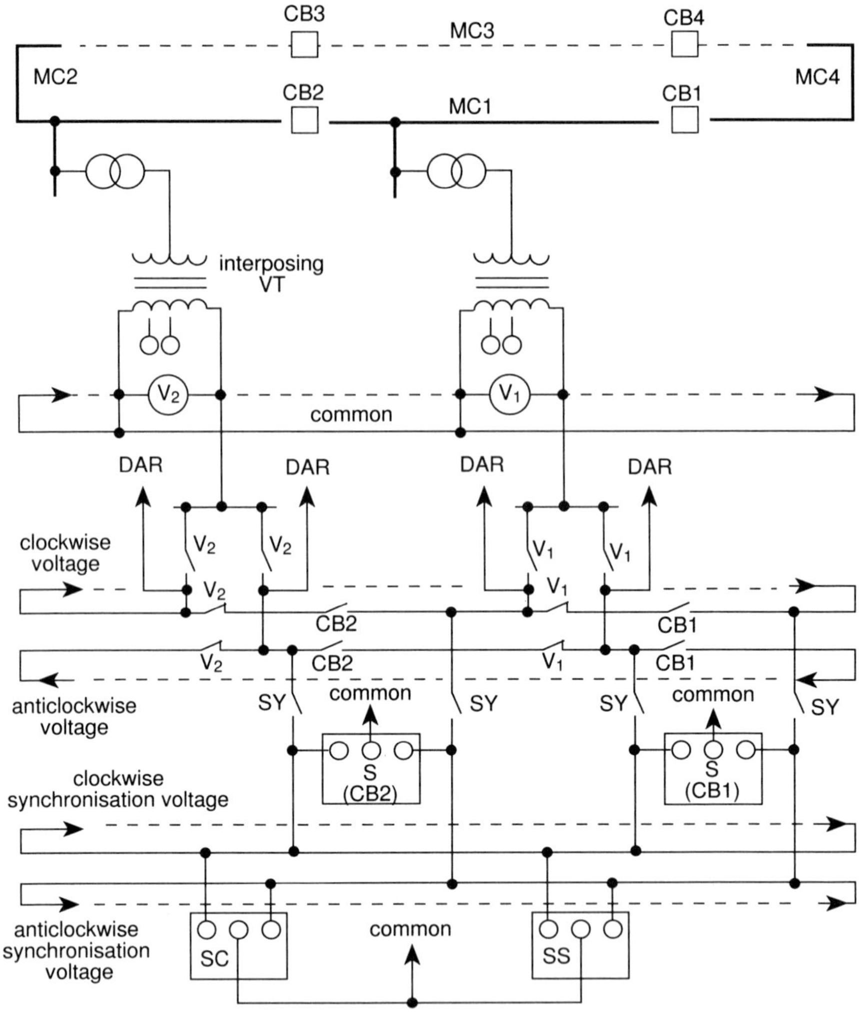

Figure 11.10 Synchronising-facility: mesh substation

V1, V2 = voltage-monitoring relays

11.16 Past and future schemes

Sections 11.9 and 11.10 have described the types of synchronising scheme which have been dominant in the UK for the last 25 years. There are, however, many schemes still in service of an earlier vintage, the most common of which are frequently known as 'plug and socket'. This type of scheme, which does not function via voltage-selection buswires (i.e. a mimic of the busbars), requires the positioning of plugs into sockets to apply voltages from the incoming circuit, and a circuit to which the circuit breaker in question is to be synchronised, to the synchronising relay.

Some modern microprocessor-based co-ordinated control schemes fully incorporate the synchronising scheme; they have limited application to date, but they point the way to the future. However, the engineer who understands the principles behind the discrete-component technology of Figures 11.9 and 11.10 will be well placed to understand the microprocessor equivalent.

11.17 Synchronising-scheme commissioning

Commissioning of a synchronising scheme would consist of the following:

(*a*) secondary injection of check-synchronisation relay and system-synchronisation relays, interposing VTs, and synchroscope;
(*b*) proving of the contacts within the AC and DC circuitry of the scheme;
(*c*) switching into service and on-load tests.

11.18 Synchronising-relay secondary injection

Secondary injection of the check-synchronisation relay typically consists of:

(*a*) determination of the phase-angle setting by injecting nominal volts into the incoming and running voltage terminals, and varying the phase angle of one voltage via a phase shifter. Operation is verified by observing closure of the output contacts with an ohmmeter. The process should consist of incremental increases in phase angle, interspersed with a pause for a period longer than the relay time delay, until the phase angle is reached at which the relay operates. Results should be obtained for both positive and negative phase angle.
(*b*) determination of the time delay by switching in-phase voltages (or voltages which are within the angle setting of the relay) onto the incoming and running voltage terminals. One pole of the switch should start a timer, which should be stopped by the operation of the relay output contact.
(*c*) obtaining the amplitude setting by injecting incoming and running voltages to operate the relay and then reducing the amplitude of one voltage until the relay resets.

Secondary injection of the system-synchronisation relay is illustrated in Figure 11.11. The phase-angle setting, slip-frequency setting and output-pulse duration can all be determined by the one test. Two variable-frequency oscillators, with a nominal frequency of 50 Hz, are connected to the relay. The effect of the frequency difference of the voltages between the two oscillators is observed on an analogue voltmeter V, which also is connected across a channel of a chart recorder. The output contact of the relay is connected to the chart-recorder via a battery (typically 6 V), which provides a deflection of

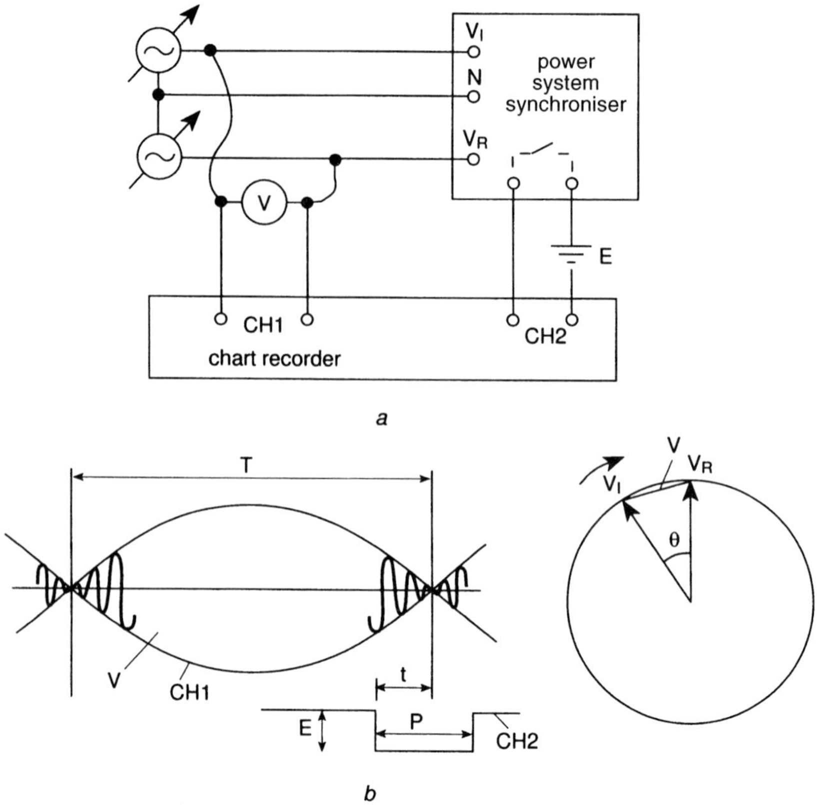

Figure 11.11 *System-synchronisation tests*

a Equipment
b Chart-recorder trace

the recorder needle when the relay operates. The difference voltage between V_R and V_I produces a beat envelope, as shown in Figure 11.11*b*. The relationship between the period T of the beat and the frequencies employed is briefly explained as follows. In the phasor diagram in Figure 11.11*b*:

Let phasor V_R be the reference phasor, with a frequency f_R
Let phasor V_I have a frequency f_I

In $1/f_R$ seconds V_R passes through one cycle of 360°
In $1/f_I$ seconds V_I also passes through 360°

but in $1/f_R$ seconds, V_I will pass through $\dfrac{1/f_R}{1/f_I} \times 360°$ and V_I will stand at an angle θ with respect to V_R where θ is given by

$$360 - \frac{1/f_R}{1/f_I} \times 360 = 360\left(\frac{f_R - f_I}{f_R}\right) \quad (11.1)$$

Now the period of time T (slip period) before V_R and V_I become in phase again will be given by

$$T = \frac{360}{\theta} \times \frac{1}{f_R} \quad (11.2)$$

Inserting eqn. 11.1 into eqn. 11.2 we obtain

$$T = \frac{1}{f_R - f_I} \quad (11.3)$$

where $f_R - f_I$ is known as the slip frequency.

To carry out the test, the voltage of both oscillators should be set to a nominal 63.5 V and a frequency of 50.000 Hz. The frequency f_I of one oscillator shall then be raised marginally above 50.000 Hz; the relay will then produce an output pulse once per slip cycle. The frequency f_I should then be increased incrementally until it is just outside the slip setting (relay fails to produce an output pulse) and then very slowly decreased until it is just within the slip setting. Relay operation is indicated by observing a flick of the chart recorder channel 2 needle when the pulsed output occurs. The chart recorder should then be initiated and traces, as in Figure 11.11*b*, obtained. The frequency f_I of the oscillator whose frequency has been varied should be recorded.

If the slip setting is, say, 0.125 Hz, and $f_R = 50.000$ Hz, then f_I should read 50.125 Hz, so T, the beat periodicity,

$$= \frac{1}{f_R - f_I} = \frac{1}{0.125} = 8 \ s$$

The angle setting of the relay (i.e. the angle between the two voltages at which the relay closes its output contacts), is given by

$$\text{angle} = \frac{t}{T} \times 360° \quad (11.4)$$

Thus, if t was measured on the chart recording to be 0.2 s, the measured angle

setting $= \frac{0.2}{8} \times 360 = 9°$.

Note that eqn. 11.4 is based on the beat period T elapsing once every revolution (360°) of the rotating phasor V_I (with respect to the reference phasor V_R).

The output pulse duration p is measured from the chart recording. It is typically of 500 ms duration. The voltage-amplitude setting can be obtained by reducing the slip until the relay operates and then reducing the voltage amplitude of one input until the relay fails to operate.

The above test should be repeated with frequency f_I now reduced below 50 Hz.

11.19 Commissioning of synchronising scheme

With reference to Figure 11.9, assume the circuit containing CB1 is to be commissioned. The following should typically be carried out:

(i) With M1 and R1 disconnectors in the open position and the VT fuses removed, a 50.0 Hz 63.5 V supply should be injected on to the buswire side of the VT fuses. Relay SY and circuit breaker CB1 should be operated and voltages measured at each point in the synchronising circuitry at which voltage should appear (e.g. at the check-synchronisation relay).

(ii) Using an accurate voltmeter, voltages should be measured at the VT (the source of injection), and the check and system-synchronise relays, and the interposing VT tapping adjusted to raise the lowest measured relay voltage to the same magnitude as that at the VT — thus compensating for voltage drop in the wiring.

(iii) Disconnectors R1 and M1 should be operated and the voltages fed from MBB1 and RBB1 buswires measured at appropriate points in the circuitry (e.g. at the check-synchronisation relay).

11.20 On-load synchronising tests

With the circuit ready for service, the on-load tests would usually consist of closing disconnector M1 or R1, energising the circuit from the remote end (with CB1 open) and closing CB1 via the check synchronisation relay as described in Section 11.14.

It would not normally be the practice to carry out a comprehensive on-load test on a system-synchronisation relay when adding a new circuit to the substation. However, such a test should ideally be carried out (although this is not always practicable) when the system-synchronisation relay is first installed. This would require a repeat of the test described in Section 11.18 but utilising the power system in place of test voltages. In addition, a circuit-breaker auxiliary contact should be connected via a battery to the chart recorder, to indicate when the circuit-breaker contacts close. The circuit breaker should initially be inhibited from closing until satisfactory traces have been obtained. This test would require the circuit in question to be selected to a busbar whose only other infeed would be a bus section or coupler. The incoming voltage supply would normally be from a single generator allocated specifically to the test. Difficulties can be experienced in bringing the generator frequency within the slip setting, and maintaining a constant generator frequency during the beat-envelope period. The test can therefore be time consuming.

When a new generator is synchronised, on-load tests are always required. In general, this test follows the same format as that of the system synchroniser, but additionally involves checking the performance of the governor and AVR.

11.21 Synchronisation-scheme maintenance

Synchronisation schemes are utilised on a regular basis for the day-to-day control of substations and thus prove their integrity as they are used. There is thus no case for maintenance of the scheme circuitry. A periodic test should, however, be carried out on the check and system-synchronisation relays (say two-yearly), although modern relays tend to have in-built duplication to minimise maloperations, thus lessening the case for regular testing.

With reference to Figure 11.9, there have been instances of the interposing VT secondary fuses of synchronisation schemes blowing. This situation can arise when a power-system fault is cleared, resulting in the depressed system voltage returning to normal, and transient charging currents in the synchronising circuitry causing the blowing of fuses. Such fuses are now required to be antisurge.

AUTOSWITCHING

11.22 Autoswitching

The term autoswitching embraces three activities:

(*a*) *Autoreclose*: This refers to the automatic reclosure of a circuit breaker following a trip caused by protection operation.

(*b*) *Autoisolation*: This refers to the automatic isolation of faulted plant, following tripping of all infeeding circuit breakers by protection operation. This is usually followed by re-energisation, via delayed autoreclose, of healthy plant.

(*c*) *Autoclose*: This is the switching into service of standby plant, which was out of service and on standby prior to the fault, e.g. a bus-section circuit breaker which may have been open prior to the fault for system fault-level reasons.

11.23 Types of autoreclose

The majority of faults on overhead lines are transient, due to lightning flash-over, conductor clashing in high winds, contact with overgrown folliage, pollution etc. Once the circuit has been tripped by protection, it can therefore quickly be returned to service. This is the role of an autoreclose system.

Autoreclose may be categorised into the following types:

- distribution-voltage pole-mounted autoreclose
- distribution-voltage ground-mounted circuit-breaker autoreclose
- high-speed autoreclose (HSAR)
- delayed autoreclose (DAR)

Many rural distribution systems are equipped with pole-mounted autoreclose equipments with a typical reclose time of about 1 s. In addition, from the mid 1960s onwards there has been an increasing application of autoreclose ('repetitive single shot') to ground-mounted distribution-voltage circuit breakers. The design and application of such equipment is well described in the 'Power system protection' series publications (see Bibliography).

High-speed autoreclose was once widely applied to the UK transmission system. After a circuit breaker had been tripped by protection, it was reclosed in a time of typically 400 ms. The advantage of high-speed autoreclose is that the power system is reconnected as soon as possible, thus aiding system stability. It also, however, has a number of disadvantages:

(*a*) If reclose takes place onto a persistent fault, the power system is subject to two faults in quick succession, so increasing the risk of system instability.
(*b*) No check is made of system synchronism before reclosure. Thus reclose of two systems with a wide angle difference could take place, with a risk of system instability.
(*c*) It does not give time for faulted plant to be isolated before reclosure takes place, i.e. it cannot facilitate an autoisolation feature.

The above disadvantages are overcome by delayed autoreclose. This system typically has a minimum reclose time of 10 s (sometimes longer). A 10 s delay gives time for generator oscillations to dampen following the first fault clearance. Almost all high-speed autoreclose systems have been replaced by DAR on the UK transmission system.

Three-phase tripping and reclosure is currently employed in the UK, although single-phase trip and high-speed reclose is used in many countries. The closely interconnected nature of the UK power system, however, makes the case for single-pole trip and high-speed reclose, with its added equipment complexity, weak.

The following sections concentrate on DAR systems as applied to 132 kV and higher-voltages substations, i.e. the more complex systems. In many ways, DAR systems are the focal point in the control of a circuit, since they receive information from, and give information to, nearly every other item of equipment associated with the circuit. Thus, if DAR performs correctly during commissioning, it provides a reasonable level of confidence that the other systems and items of equipment are performing correctly. Figure 11.12

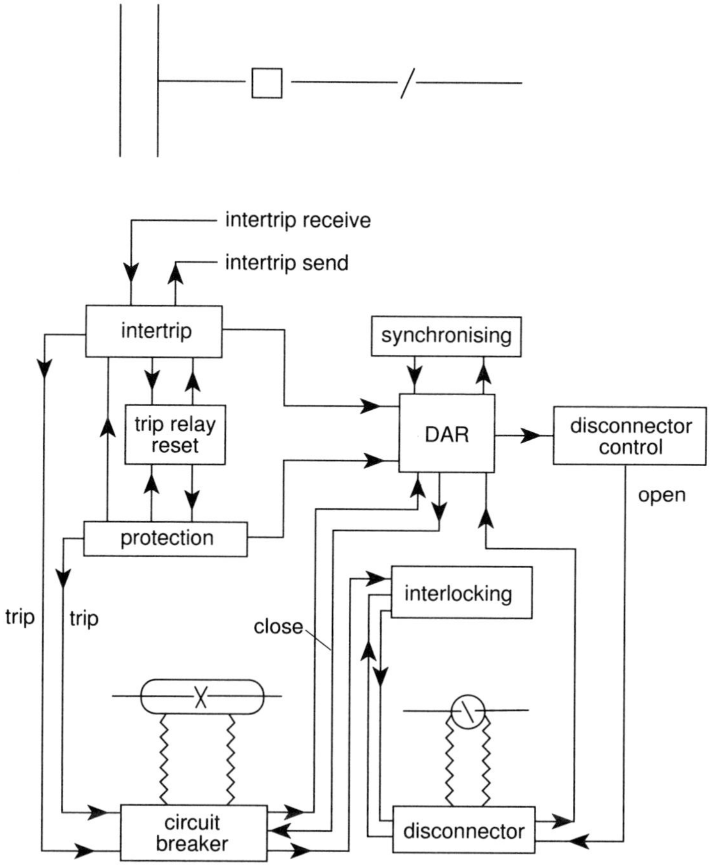

Figure 11.12 DAR system

shows in block diagram form how DAR interfaces with other items of equipment, for a feeder connected to a busbar substation.

11.24 Plain-feeder DAR

Figure 11.13*a* illustrates some of the essential features of circuitry required for DAR on a plain feeder. The design of the trip-relay reset and intertripping schemes is heavily influenced by the requirements of the DAR scheme, and must be understood to explain how the DAR functions.

Figure 11.13*b* shows some of the essential features of a typical trip-relay reset scheme. Operation of a trip relay energises two timers. TD1 usually has a setting of about 10 s and resets the trip relays during normal DAR operation. TD2 has a setting of about 70 s and is used to reset the protection-trip relay PTR, when the line disconnector (LI) opens following a persistent

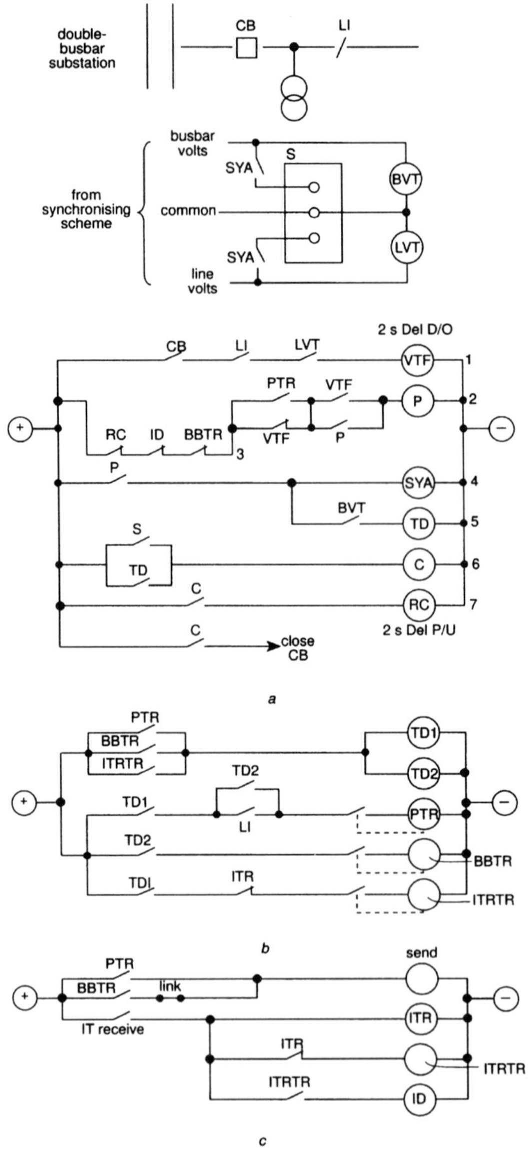

Figure 11.13 Plain-feeder DAR and associated circuitry

a Basic DAR circuitry
b Trip-relay reset
c Intertripping
P = reclose start
TD = dead-line time delay
RC = reclose cancel
TD2 = 70 s timer
BBTR = busbar TR
ITR = intertrip-receive relay
BVT = busbar VT relay
LVT = line VT relay
VTF = memory circuit
SYA = synchronise select
C = close relay
TD1 = 10 s timer
PTR = protection TR
ITRTR = intertrip-receive TR
ID = persistent-intertrip relay

line fault, and also to reset the busbar-trip relay BBTR. This long reset-time delay results in a long intertrip send to the remote end of the feeder, which in turn causes operation of the ID relay in the intertripping circuitry, as shown in Figure 11.13*c*. Relay ID, known as the persistent-intertrip timer, is usually set to about 60 s (i.e. shorter than TD2) and operation results in DAR cancellation.

The modern arrangement for intertripping is to double-pole switch both send and receive circuitry, but for clarity only single pole is shown. Traditionally, intertripping was only used for the remote-end clearance of transformer faults, but there is a modern tendency to install it on all transmission feeders to ensure remote-end clearance of feeder faults. Figure 11.13*c* shows the intertrip receive contact energising two relays, ITR and ITRTR. ITR is a self-reset relay which is used to prevent the reset of ITRTR in the trip-relay reset circuit until the receive signal actually ceases. ITR also forms part of the intertrip communication-channel end-to-end test facility (not shown in Figure 11.13), as discussed in Chapter 18.

In Figure 11.13*a* the legs of the DAR circuitry are numbered 1,2,3 etc. to facilitate the following description.

Leg 1: This circuitry energises relay VTF, which has a 2 s delayed drop off. It is a memory circuit which verifies that the feeder was in service and healthy for at least 2 s prior to the fault.
Leg 2: This is the reclose start circuit. Relay P is energised and seals itself if VTF was energised before trip relay PTR is operated, i.e. if the circuit was alive before the fault.
Leg 3: P is de-energised and DAR cancelled if there is a busbar fault (BBTR), a persistent intertrip received (ID) or after a successful DAR via the reclose-cancel relay (RC).
Leg 4: SYA selects volts to the check-synchronisation relay (S). Contacts from SYA are shown dotted in Figure 11.9.
Leg 5: The dead-line timer TD is energised, provided that the busbar (BVT) is alive.
Leg 6: The close relay C is energised by either the check-synchronisation relay (S), if the feeder has already been reclosed from the remote end, or via the dead-line timer TD, if the remote end has not reclosed.
Leg 7: RC resets the DAR circuitry after a 2 s delay.

Note that the circuitry described above is a simplification of circuitry found on-site, but embodies the basic principles.

11.25 Mesh substation autoswitching

Mesh substations are usually equiped to carry out DAR and autoisolation. The circuitry for mesh substation autoswitching is arguably the most complex and extensive of any, and it is therefore not possible to describe it in any detail in this book. Before about 1977, almost all relays were electromechanical in design. Each mesh corner was controlled by a composite relay termed an A unit, which was largely concerned with the detection of a persistent fault on a feeder connected to that mesh corner, and the opening of associated disconnectors to achieve autoisolation of (*a*) faulted transformers, or (*b*) feeders with a persistent feeder fault. Each circuit breaker was equipped with a DAR unit termed a B unit; this has many of the features of Figure 11.13*a*. There is a need to define a priority of reclosure around the mesh and this is usually in the order: circuit breaker 1, 2, 3, 4. A third unit, termed a C unit, controlled the DAR on the transformer LV circuit breaker, and a number of other units controlled DAR in/out switching, indications etc. These autoswitching equipments usually had facilities to switch out the DAR while leaving the autoisolation feature in services. All these units were interconnected by a web of wiring and many are still in service today.

From about 1977 onwards, most new mesh substation and complex plant configurations have been controlled by microprocessor-based autoswitching equipments. This has created significant savings in wiring and drawings, but, owing to the complexity of the software, it has necessitated exhaustive recommissioning tests whenever modifications have been necessary. In addition there has been a diminished understanding of the equipment, and how to modify it, by site engineers. To help remedy these limitations, there is currently a move to introduce modular microprocessor-based equipments, which present the scheme in a conventional contact-logic format based on the electromechanical design.

11.26 Autoswitching-equipment commissioning

There are four areas of testing pertaining to the commissioning of autoswitching schemes:

- simulation tests to prove the relay(s)
- scheme input-contact-logic and output-relay tests
- off-load site-simulation tests
- on-load tests

(*a*) *Relay-simulation tests:* These largely relate to tests carried out on microprocessor-based equipment to prove the software. Inputs representing the substation plant, protection relays etc. need to be simulated (probably requiring a computer-based simulator) with both inputs to, and outputs from, the equipment recorded on an event logger, the latter being to aid performance analysis and to provide a test record. A schedule of tests and expected results, similar to those described in Section 11.27, should be produced in advance of the tests and compared with the results.

For complex equipment the number of permutations of tests can be very large. In order of priority, the following should be carried out:

(i) common faults, e.g. line transient, line persistent, transformer;

(ii) common faults with conditions imposed, e.g. DAR switched out, low gas pressure etc.;
(iii) faults with sections of the power system out of service, e.g. a transformer or a feeder switched out of service prior to the fault;
(iv) multiple faults, e.g. faults on two different parts of the power system simultaneously.

Generally speaking, the number of permutations of tests would be in the order of 50–100, depending on substation size and complexity.

(*b*) *Contact-logic tests*: This refers to a check of scheme contact logic, in the form of inputs to a microprocessor-based scheme or chains of contacts in an electromechanical-relay scheme. It should be carried out on site by operating the plant and protection relays manually in sequence. Similarly, the output relays should be operated to prove that they control the appropriate item of plant, generate an alarm etc.
(*c*) *Off-load site-simulation tests*: For complex schemes, it is well worth while carrying out site-simulation tests. These should be done when all the equipment associated with the autoswitching inputs and outputs has been installed, connected and tested, and just prior to the on-load tests. They consist of simulating in-service conditions by closing the (de-energised) plant and simulating voltage-transformer inputs to the scheme by use of variacs, supplied from a 240 V source with a 63.5 V output, connected to the VT secondary wiring. A test schedule should be prepared in advance, with a format similar to that described in Section 11.27. The test method consists of operating protection relays manually (to trip the circuit breakers), simultaneously switching the supplies to the variacs, (to de-energise the circuit), and observing visually that the autoswitching equipment carries out DAR or autoisolation of the plant. A rough check of the sequence timing should be carried out with a stop watch. By this means, the scheme can be proved to a high level of confidence before it enters service.
(*d*) *On-load tests*: This refers to operating protection trip relays manually to trip the circuit breakers in accordance with a schedule similar to Section 11.27, when the circuit is on-load, and allowing the autoswitching equipment to prove itself. If off-load simulation tests have been carried out, the number of on-load tests can be reduced to simulating a small number of the most common fault conditions.

11.27 Examples of autoswitching

The three examples of autoswitching described below relate to the system shown in Figure 11.14. The synchronising facilities are based on Figures 11.9 and 11.10, and the trip-relay reset and intertripping follow the principles illustrated in Figure 11.13. The following points should be noted.

(*a*) The protection shown in Figure 11.14 illustrates only the feeder

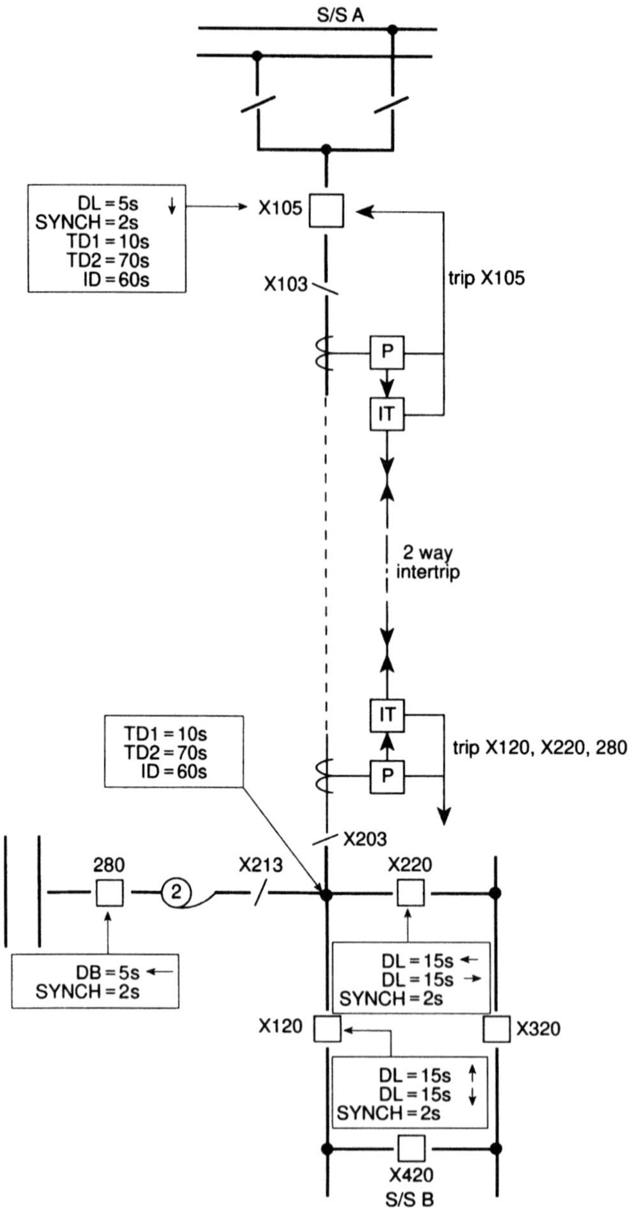

Figure 11.14 System for analysing autoswitching performance

DL = dead-line-close timer
SYNCH = synchronise-check time relay
TD1, TD2 = trip-relay reset timers
ID = persistent-intertrip timer
DB = dead-bar-close timer
P = protection
IT = intertrip

protection. To increase clarity, transformer and other protections are not shown.

(*b*) There is a need, in some instances, for the autoswitching equipment to output a contact to override the interlocking. This is not described in the event listing.

(*c*) The various times illustrated are typical only.

(*d*) The arrows in the boxes indicate the direction of dead-line reclosure of the associated circuit breaker.

(*e*) In predicting how an autoswitching equipment might function, it should be noted that it should ideally replicate the action of a manual operator faced with the same set of conditions.

(*f*) It is important that the commissioning engineer has enough knowledge to predict how the autoswitching system will perform; understanding how the software functions is of lesser importance.

Fault 1: Type of fault = line transient

Start conditions = system solid

Stage	*Time*	*Event*
1	0	Line fault inception
2	40 ms	S/S A + S/S B protection operate and IT send
3	80 ms	S/S A X105 trip and IT receive S/S B X120, X220, 280 trip and IT receive
4	10 s	S/S A + S/S B trip relay reset
5	15 s	S/S A X105 recloses (dead line)
6	17 s	S/S B X120 recloses (check synch)
7	19 s	S/S B X220 recloses (check synch)
8	21 s	S/S B 280 recloses (check synch)

Thus the circuit is fully restored to service.

Fault 2: Type of fault = line persistent

Start condition = system solid

Stage	*Time*	*Event*
1 to 5	15 s	As fault 1
6	15.04 s	S/S A protection operate and IT send
7	15.08 s	S/S A X105 trip
8	16 s	S/S A X103 commence to open
9	26 s	S/S A X103 fully open
10	75 s	S/S B timer ID operated
11	76 s	S/S B X203 commence to open
12	85 s	S/S A trip relays reset
13	85 + s	S/S B IT trip relays reset
14	86 s	S/S B X203 open
15	101 s	S/S B X120 reclose (dead line)
16	103 s	S/S B X220 reclose (synch-check)
17	105 s	S/S B 280 reclose (synch-check)

Thus disconnectors 103 and 203 are opened to isolate the line, and X120, X220 and 280 are reclosed and SGT2 re-energised.

Fault 3: Type of fault=SGT2
Start conditions=system solid

Stage	*Time*	*Event*
1	0	S/S B SGT1 fault inception
2	40 ms	S/S B SGT1 protection operate and IT send
3	80 ms	S/S B X120, X220, 280 trip
		S/S A intertrip received
4	120 ms	S/S A X105 trip
5	1 s	S/S B X213 commence to open
6	11 s	S/S B X213 open
		S/S B SGT prot trip relays reset
		S/S B IT send removed
7	11 + s	S/S A trip relays reset
8	16 s	S/S A X105 reclose (dead line)
9	18 s	S/S B X120 reclose (check-synch)
10	19 s	S/S B X220 reclose (check-synch)

Thus the transformer is isolated and healthy plant is restored to service.

11.28 DAR maintenance

Immediately before a circuit is switched out for maintenance, a number of on-load tests pertaining to the more common system-fault conditions should be carried out. However, if the DAR equipment has recently been subject to system faults and the equipment has functioned satisfactorily, the 'before-maintenance' tests can be dispensed with. Generally speaking, no off-load maintenance tests need to be carried out. A number of on-load tests should, however, be carried out when the circuit is returned to service.

Chapter 12
Protection and control: general

12.1 Introduction

Protection and control systems consume much more engineering time during the commissioning of a circuit or substation, or during maintenance, than high-voltage plant. A sound knowledge of protection and control technology is, therefore, a pre-requisite to managing the commissioning and maintenance process. This chapter examines some general issues associated with protection and control systems. Chapter 11 has already examined some of the more complex control systems and later chapters examine particular types of protection.

CHARACTERISTICS AND DRAWINGS

12.2 Purpose of protection systems

The purpose of a protection system is to detect a fault on the power system, and then to instruct the appropriate circuit breakers (or steam isolating valve) to trip and isolate the faulty item of plant. This process needs to be accomplished as fast as possible to prevent:

(*a*) power-system instability. Transmission-system faults which are not cleared rapidly can lead to loss of generator synchronism, with widespread loss of supplies;
(*b*) unnecessary damage to faulted plant. The quicker that faulted plant is disconnected, the less damage it will sustain;
(*c*) damage to healthy plant. Power-system plant is rated to carry fault current for only a short period;
(*d*) loss of supply to consumers, particularly industrial consumers with synchronous motors which, if the motors fall out of synchronism and trip, could lead to loss of production.

Note that protection systems cannot always be made responsive enough to guarantee safety of personnel.

12.3 Purpose of control systems

The purpose of a control system is to facilitate the manual or automatic reconfiguration of the power system, together with the acquisition of information on the status and loading of the plant to facilitate operational decisions.

The term 'control' is often used as a blanket term to cover plant controls, plant indications, metering and alarm systems as follows:

(*a*) *Plant control:* This embraces:

(i) Manual control from both local and remote control boards of:

- circuit breakers and disconnectors (embracing synchronising and interlocking);
- tap changers, voltage-reduction and voltage-compensation schemes;
- protection and autoreclose scheme in/out switching.

(ii) Automatic control, which covers:

- autoreclose, autoisolation and autoswitching;
- automatic-voltage-control and reactive-compensation systems.

(*b*) *Indications:* This refers to the transfer of information from circuit breakers, disconnectors, earth switches, tap changers, winding-temperature instruments, switchable relays etc. to local and remote control boards to indicate whether the item is open/closed, or in/out. A further in-between state can also exist, known as DBI ('don't believe it', implying neither open nor closed).

(*c*) *Metering:* This refers to the measurement of voltage, current, active power and reactive power and the conveying of such information to local and remote control boards. It also embraces special-purpose monitoring equipment.

(*d*) *Alarms:* This is the transfer of alarm information from high-voltage plant, protection and control systems to local and remote control boards.

(*e*) *Fault recorders:* These record voltages, amps and protection operations.

12.4 Protection characteristics

There are a number of commonly used terms which describe the characteristics of protection systems:

(*a*) *Protection equipment:* a measuring device which detects the existence of a power system fault, e.g. distance protection;

(*b*) *Protection system:* the CTs, VTs, protection equipment, trip relays, fuses and links etc. which need to be connected as a system to detect a specific type of fault, and trip the appropriate circuit breakers, e.g. busbar protection;

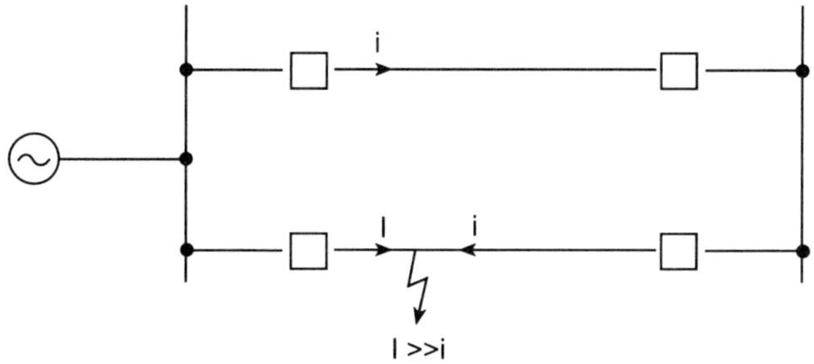

Figure 12.1 Fault resulting in sequential tripping

(*c*) *Unit protection:* a system which will only operate when a fault exists on the specific item of plant which is being protected, e.g. biased differential transformer protection. Conversely a nonunit protection is a system which may trip for a fault anywhere which it is capable of detecting, e.g. an overcurrent relay;
(*d*) *Discrimination:* the ability of a protection equipment to disconnect only the faulty section of the power system. A unit protection has good discriminative properties;
(*e*) *Sensitivity:* the ability of a protection to operate with low levels of fault currents;
(*f*) *Stability:* the ability of a protection to remain inoperative when faults occur outside the protected zone;
(*g*) *Reliability:* the protection's ability to operate correctly or stabilise as required;
(*h*) *Back-up protection:* at the higher voltages the power system is usually equipped with a main protection and a back-up protection. The main protection is invariably a high-speed unit type of protection. Back-up protection is usually slower in operation and of a nonunit type, and provides additional protection in two possible ways: to clear a fault on remote equipment if the protection or circuit breaker on that equipment fails (this is termed 'remote back-up'); or to clear a fault on the item of equipment with which the CTs are associated, if the main protection fails (this is termed 'local back-up'). Typical back-up protections are zones 2 and 3 of distance protection and IDMT relays. An example of back-up application consists of feeder back-up protection which is required to operate on the failure of busbar protection at the substation located at the remote end of the feeder (i.e. to provide remote back-up).
(*i*) *Sequential tripping:* as a consequence of circuit configuration, a feeder fault may result in almost all the fault current entering the feeder from one end, resulting in protection operation and circuit-breaker trip at the one end. This immediately results in an increase in current at the other end with corresponding protection operation and tripping of that circuit breaker. This is termed 'sequential tripping'. Figure 12.1 illustrates a typical situation.

12.5 Fault-clearance times

A guide to the target fault-clearance times for the UK power system is as follows:

(*a*) *400 and 275 kV transmission system:*
400 kV target clearance = 80 ms
275 kV target clearance = 100 ms

These times relate to new installations, and are measured from fault inception to circuit-breaker arc interuption. In general both the protection system and the circuit-breaker should have an operating time of about 40 ms, but slow operation by one may be compensated by faster operation of the other. Many existing installations, however, have protection and circuit-breaker operating times of the order of 60 ms, resulting in an overall clearance time of 120 ms. Short clearance times are necessary to guarantee the stability of connected generation, particularly the larger 500/660 MW machines.

(*b*) *132 kV and below*

Faults must be cleared as quickly as possible to minimise disruption to consumers; they must also be cleared within the short-circuit rating time of plant to avoid damage. At these voltages, financial considerations have a greater bearing on the protection employed. Many distribution system protections, however, are capable of operating within the range 40–100 ms with total fault-clearance times in the range 100–160 ms.

Distribution systems with directly connected generation will require high-speed fault clearance, typically in the range 100–200 ms, to ensure power-system stability.

12.6 Diagrams

Most substation diagrams are concerned with protection and control systems. The following summarises the various types.

(*a*) *Key diagram:* The purpose of a key diagram is to provide a clear overview of the protection systems. It consists of a single-line diagram of the primary plant, to which is added the CTs, VTs, protection equipments and tripping of the appropriate circuit breakers. An example is shown in Figure 12.2. Key diagrams seem to have lost some popularity as protection systems have become more complex but, in the author's experience, they are an invaluable aid to understanding the intended protection system design.

(*b*) *Functional diagrams:* Over many years, electricity supply companies have specified their scheme requirements to contractors via standard schematic diagrams which specify the complete design. These have ensured commonality of installation from substation to substation. However, European directives now require diagrams to specify the functionality of operation only. Figure 12.3 illustrates a simplified example of this type of diagram: at first

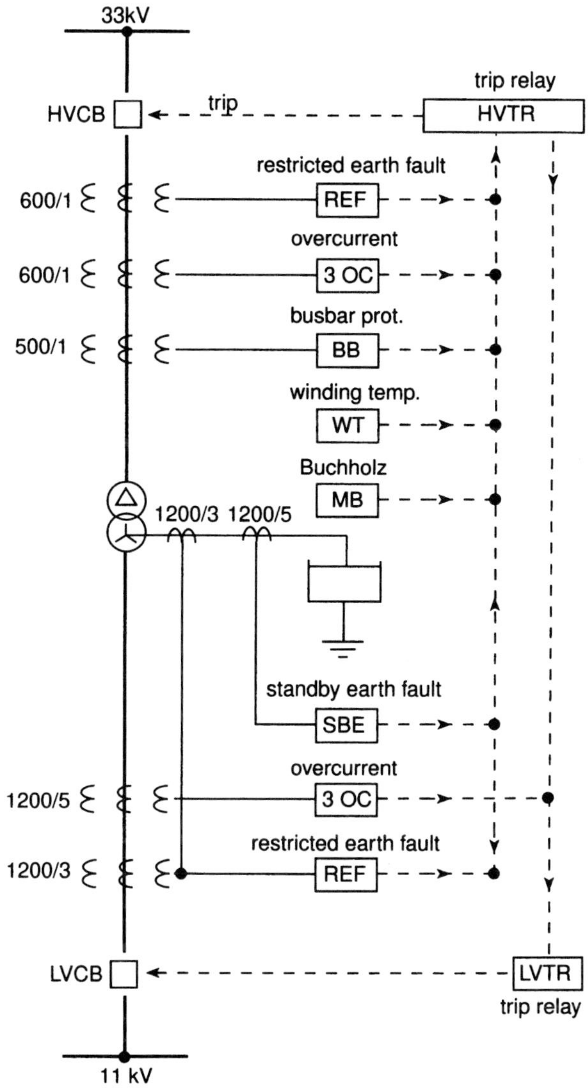

Figure 12.2 Typical key diagram

REF=restricted earth fault	MB=Buchholtz
OC=overcurrent	SBE=standby-earth fault
BB=busbar protection	HVTR=HV trip relay
WT=winding temperature	LVTR=LV trip relay

sight it may appear similar to a key diagram, but the functional diagram also includes power supplies and control systems to describe the functions of all circuitry.

(*c*) *Schematic (or circuit) diagrams:* These are electrical design diagrams of

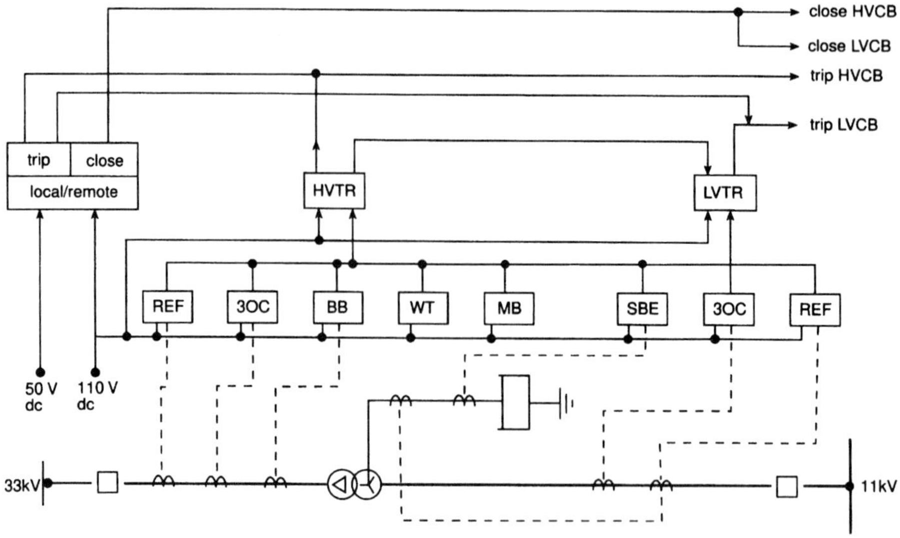

Figure 12.3 Functional diagram

REF = restricted earth fault
OC = overcurrent
BB = busbar protection
WT = winding temperature
MB = Buchholtz

SBE = standby-earth fault
HVTR = HV trip relay
LVTR = LV trip relay
HVCB = HV circuit breaker
LVCB = LV circuit breaker

protection and control systems, drawn as simply as possible with a minimum of crossovers, to provide an explanation of how the circuitry functions. Some manufacturers produce part-physical schematics to enable physical arrangements of equipment or point-to-point multicore connections to be shown. Commissioning tests are carried out to verify that the equipment is installed and performs in accordance with the schematic diagram. They are an essential tool in fault investigations. Figure 12.4 illustrates an example based on the key diagram in Figure 12.2: the schematic diagram would usually include a key defining the relay and fuse/link description and nomenclature, but this is omitted for brevity.

(*d*) *Wiring diagrams:* These show the physical position of wiring terminations and routing. An example is a wiring diagram for a relay-marshalling panel. These diagrams are required to facilitate installation.

(*e*) *Multicore diagrams:* These are a schedule of multicore cables required to facilitate installation. They provide:

- cable identity and termination locations
- the number of cores and their usage (ferrule numbers)

(*f*) *Panel-layout diagrams:* These show the physical disposition of relays, fuses, links and labels on a particular panel. They are required to facilitate installation.

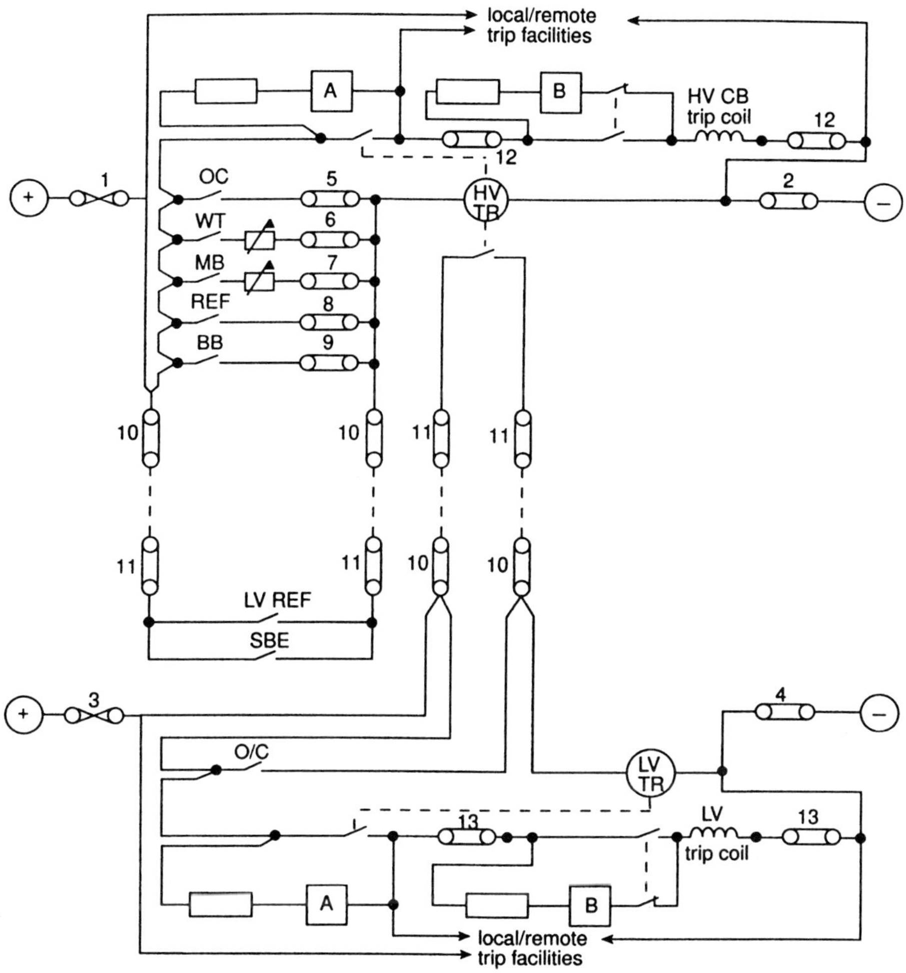

Figure 12.4 *Typical schematic diagram*

A, B = trip-circuit-supervision relays

(*g*) *Extracted diagrams:* These are schematic diagrams produced on site by the commissioning engineer, either to simplify a less-than-ideal manufacturer's diagram, or to bring a number of schematics onto a single diagram.

(*h*) *Logic diagrams:* The advent of microprocessor-based equipment has made the understanding of how equipment, or a system, functions much more difficult. A logic diagram is a simple representation of the logical process carried out by the equipment. Figure 12.5 illustrates a typical example for a distance protection.

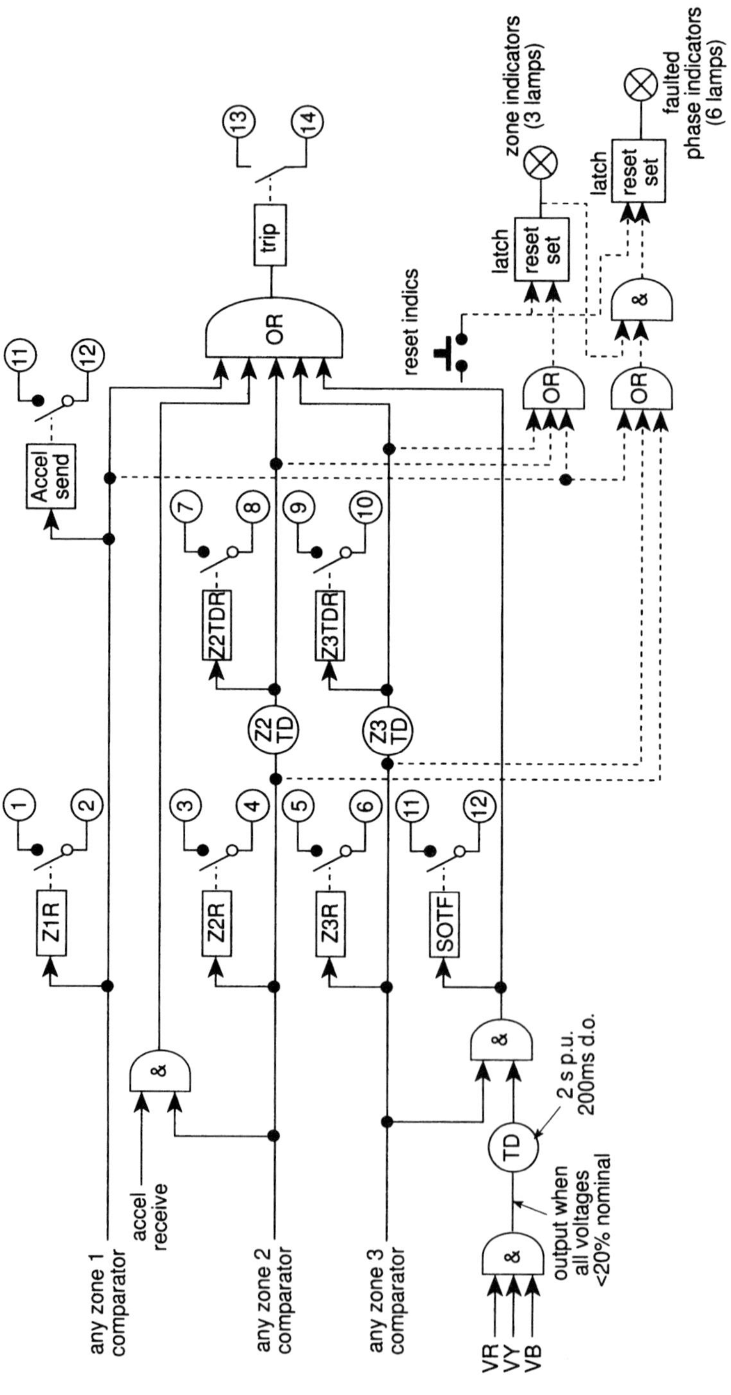

Figure 12.5 *Distance-protection logic diagram*

Z1R = zone-1 repeat relay
Z2R = zone-2 repeat relay
Z3R = zone-3 repeat relay
SOTF = switch-onto-fault-repeat relay
Accel send = accelerate-signal-send relay

COMMISSIONING

12.7 Commissioning of protection and control systems

The following specifies the typical range of commissioning tests required for protection and control systems:

- diagram examination
- data collection
- insulation-resistance tests
- CT flick tests
- CT magnetisation curves
- CT burdens
- relay inspection
- relay secondary injection
- DC (logic) operation
- primary injection
- off-load tripping tests
- pre-energisation checks and inspections
- on-load tests

Each of the above will be examined briefly.

(*a*) *Diagram examination:* This consists of two activities:

- a check to ensure that appropriate diagrams have been received for every system. They should then be numbered, entered into an index and filed;
- a thorough examination of the key and schematic diagrams to gain a complete understanding of the systems design.

(*b*) *Data collection:* Relay data (type, serial numbers, settings ranges etc.) should be collated for both relay-setting purposes and for entering into the commissioning log.

(*c*) *Insulation resistance:* These tests should be carried out in accordance with a test schedule, with a 500 V 'megger', after the installation is complete, and (where possible) with all links and fuses in position. Tests should be carried out as follows:

- CT circuits to earth; the earth link should be lifted during the test
- VT circuits to earth; with the earth connection removed
- DC circuits; from each pole and any isolated section of wiring to earth with the supply fuse and link removed
- between each of the CT, VT and DC circuits

Some equipment is not rated to withstand 500 V, and should, therefore, have the appropriate terminals short-circuited. Generally, all new wiring should have an IR well above 100 MΩ, although on a wet day long runs of outdoor wiring may be as low as 5 MΩ. A 'megger' will usually discharge the wiring on

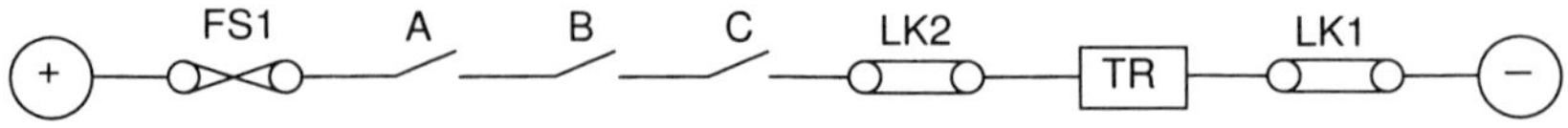

Figure 12.6 DC operation

cessation of the test but this is not necessarily true of all IR devices, and thus care should be taken not to touch the wiring immediately after the test.
(*d*) *CT tests:* The requirements for flick tests, magnetisation curves and burdens are covered in Chapter 10.
(*e*) *Equipment inspection:* All equipment should be checked for damage arising during transit or on-site installation, correct labelling, existence of CT short-circuit or DC isolation contacts on relays etc.
(*f*) *Secondary injection:* This refers to the injection of current into a relay from the secondary side of the CTs, usually from a panel-mounted test block. The purpose of the test is to verify that the relay has not been damaged during transit and to provide a set of figures which will form the basis for comparison during future maintenance and investigation tests. The relay should have been tested accurately in the manufacturer's works, and thus secondary injection need only prove that the relay is within tolerances.

Simpler relays are usually injected at a number of settings, but more complex relays with numerous settings are usually injected at the setting to be used.
(*g*) *DC operations:* This consists of proving that the circuit logic of each protection and control system accords with the associated schematic diagram. Before the fuses and links are inserted, the polarity of the DC supplies should be verified. The circuitry should then be proved by manually operating, or secondary injecting, each relay to prove each contact chain. For example, in Figure 12.6, it must be proved that relay TR can only be operated when all three contacts A, B and C are made. It must also be proved that relay TR cannot be operated when fuse FS1 or links LK1 or LK2 is withdrawn (negative checks). As each contact and link is proved, it should be ticked on the schematic diagram.

It must also be proved that all circuit breakers and motorised disconnectors can be operated from all control positions; that indication circuitry reflects the true position of the plant; and that all alarms are communicated to the appropriate control point. This will require operation of the plant.

Note that, when recommissioning an existing circuit, both secondary injections and DC operations should always be carried out from a point which is beyond the point of disturbance.

(*h*) *Primary injection:* This refers to an injection of current into the primary of a current transformer to prove the CT ratio, the polarity of CTs as a group, the polarity of intergroup CTs, the wiring between the CTs and relays, relay stability, relay operation and, where practicable, the current setting of the relay. Chapter 10 describes this subject in greater detail. Primary injection should ideally be carried out when installation is complete, thereby minimising the risk of inadvertent tampering with CT wiring after it has been proven.

(*i*) *Off-load tripping:* Before a circuit is energised, each trip relay should be operated manually to trip the circuit breaker. A full intertrip test should also be carried out to prove the circuitry from the initiating contact to the remote-end receive relay.

(*j*) *Pre-energisation check and inspection:* This consists of checking that each termination is tight; crosschecking the ferruling in each panel against the wiring diagram; a final check of labels and relay settings; checking that temporary connections and 'test short-circuits' have been removed; and that CT test links are in the in-service position.

(*k*) *On-load tests:* These will be considered in Chapter 19.

12.8 Relay settings

Relays fall into two categories:

(*a*) those which operate when a measured quantity exceeds a setting, e.g. an overcurrent or time delay relay; and

(*b*) those which change state on the application, or removal, of an electrical supply to their terminals, e.g. trip relay [an all or nothing (logic) relay].

For relays with settings, it is essential to have a setting-management procedure which clearly records the settings required, identifies responsibility for application and ensures promulgation of information to all who need to know (e.g. substation, office, control centre). Salient stages in a settings procedure are as follows:

(*c*) settings data collection
(*d*) settings calculations
(*e*) entry of data onto a settings sheet
(*f*) application of settings and confirmation that they have been applied
(*g*) filing and distribution of settings sheets
(*h*) emergency settings procedure

A good settings sheet should clearly state the following: substation; circuit; relay function; relay type; relay serial number; relay-settings range; setting to be applied to the relay; and the CT ratio to be used. When settings are applied the whole process should be double checked, and the settings sheet then signed and dated.

MAINTENANCE

12.9 Maintenance of protection and control systems

The maintenance of protection and control systems typically consists of repeating the following commissioning tests:

- insulation-resistance tests
- relay secondary injection
- DC operations, all that is usually necessary is to prove each contact chain — negative checks are usually not required
- off-load tripping tests

The other tests identified in Section 12.6 are not carried out. This is because they are either associated with installation-orientated activities (e.g. verification of the correct type of relay, labels etc.) or proving polarity (e.g. flick tests and primary injection), or they confirm the position of fuses and links which, if the wiring has not been disconnected/reconnected, cannot change. Some utilities carry out magnetisation curves and on-load tests as part of routine maintenance, but in the author's experience the case for this is not strong. It is essential to produce a record of any disconnections, and subsequent confirmation of reconnection. An inspection to verify that all equipment has been returned to normal should be carried out before the circuit is re-energised.

Note that it is essential to identify clearly by barriers, screens or labels the equipment which is to be worked on, to prevent inadvertent interference with in-service equipment.

The higher the system voltage, the greater the frequency of maintenance. For example, at supergrid voltages, tripping tests have for many years been carried out annually, with insulation resistance, secondary injection and DC operations carried out two-yearly (the current trend is for these tests to be carried out four-yearly). On the 11 kV system such tests might only be carried out five/six yearly or longer.

12.10 Investigations

Protection systems can fail in one of two ways:

- operating when they are not required to operate
- not operating when they are required to operate

The starting point for any investigations is thus to deduce what should have operated and then determine what has actually operated. Chapter 2 covers general principles of fault investigation; particular points to examine in the context of protection and control systems are:

(i) What currents were flowing at the time of the incident, particularly in a sequence-components context?

(ii) Are the CTs on the correct ratio, or have any CTs inadvertently been left with a short-circuit applied?

(iii) What was the state of the battery, what currents may it have been supplying transiently and what effect would that have on battery voltage?

(iv) Is the installation properly earthed?
(v) Have the settings been correctly applied?

12.11 Test equipment

A wide range of equipment is required for testing protection and control systems. Examples include a secondary injection set (a multipurpose equipment with variable current and voltage control), variacs, resistors, oscilloscopes, timers, frequency meters, chart recorders, voltmeters, ammeters, wattmeters, oscillators etc. A number of points are worthy of note during testing as follows:

(*a*) Most tests require the injection of sinusoidal quantities. If unacceptable test results are being obtained, it is recommended that the waveform is examined. Some test equipment, variacs in particular, are known to give distorted waveforms. Resistors can often be used to smooth the waveform.
(*b*) Before testing begins, the earthing arrangements should be checked. The earthy end of the test equipment should be connected to the earth of the device to be tested. Particular care should be exercised when using oscilloscopes; ideally, a differential amplifier input should be employed to overcome earthing problems. If such an oscilloscope is not available, it may be necessary to run the oscilloscope unearthed, which in turn has safety implications.
(*c*) When using a 415 V 3-phase-and-neutral test supply, do not assume that it is perfectly balanced. In particular, do not assume that the neutral is at earth potential; supply currents can cause the neutral to stand at a voltage with respect to earth (the author has measured as much as 20 V).
(*d*) If the test equipment has a high-input impedance, it may record induced pickup voltages from the power system. This can usually be negated by connecting a resistor across the terminals of the test equipment. High-impedance voltmeters are particularly susceptible to recording pickup voltages.
(*e*) Take care when handling solid-state modules and cards; it is recommended that an earthing bracelet connecting the wrist to the equipment earth be worn. When packaging such equipment, antistatic bags should be used. Manufacturers' recommendations should be adhered to.
(*f*) Be aware that some meters, when operating at the low end of their range, can exhibit a comparatively high impedance.

Until 1985, almost all tests were carried out manually. Since then, there has been an increasing dependence on autotesting, whereby a microprocessor-based equipment automatically runs through a complete sequence of relay tests and prints out the results. Ideally, an autotest equipment should only be connected to the relay terminals, but in some instances it may be necessary to connect it directly to the circuitry within the relay, to block certain events during the autotesting process. Automatic testing points the way to the future, but it has the disadvantage of the operator not fully understanding the tests carried out. An incorrect result during automatic testing also provides the investigating engineer with the dilemma as to whether the problem is with the autotesting equipment or with the relay. The autotest equipment may be more complex than the relay being tested. Any engineer regularly involved in relay testing, should carry a well equipped box of test equipment which should include wires, terminations, short-circuits, a variety of resistors, capacitors, switches, miniature time-delayed relays, miniature clip-on ammeter, single-phase and 415 V 3-phase supply leads, box spanners, termination screwdrivers, terminal blocks and a supply of fuses and fuse wire. In addition, at least two multimeters should be carried, one of which should be an analogue device.

12.12 Commissioning tips

(i) Before commencing work, consider whether there are any safety implications in terms of electrical or mechanical stored energy.
(ii) Before injecting current, consider whether the equipment is short-time rated, and limit the duration of the injection accordingly.
(iii) Before switching on test supplies, ensure that all meters are selected to the appropriate ranges.
(iv) Before carrying out insulation resistance tests, consider whether some equipments could suffer damage, and therefore require to be short-circuited.
(v) Care should be taken when using test equipment which has an inherent earth connection which may be damaged by inadvertently connecting the live rail to earth. This applies particularly to oscilloscopes.
(vi) If an equipment is not performing as expected when under test, first consider the earthing arrangements.
(vii) Keep a record of all temporary arrangements and connections, particularly 'short-circuits'.
(viii) When carrying out phasing tests, establish the voltage relationship between phase and earth before measuring across phases.

Chapter 13

Overcurrent and earth-fault protection

13.1 Introduction

Overcurrent and earth-fault protection are the oldest and most commonly used types of protection. Until the late 1980s almost all of these relays were electromagnetic in design, whereas new installations are predominantly solid state; even so, many of the simple, robust and reliable electromagnetic relays already installed are likely to remain on the system for many years to come. This chapter examines these relays from an electromagnetic design standpoint, since it provides a clearer understanding of how this type of relay functions. The design, application, commissioning and maintenance of the following relays will be considered:

- IDMTL overcurrent and earth fault
- directional overcurrent and earth fault
- instantaneous overcurrent

IDMTL-RELAY BASIC PRINCIPLES

13.2 IDMTL-relay principle of operation

Figure 13.1 illustrates the essential components of an inverse-definite minimum-time-lag relay (IDMTL). The upper electromagnet has a primary winding tapped at intervals and connected via a plug bridge to the current transformer secondary circuit. The plug bridge is used to vary the tappings to give the required current setting. The secondary winding on the upper electromagnet is energised by magnetic induction from the primary and is connected to the winding of the lower electromagnet. The fluxes produced by the upper and lower electromagnets each induce eddy currents into the metal disc. These, in turn, flow through the flux path of the other electromagnet. Fleming's left-hand rule indicates that a torque is produced on the metal disc in accordance with the relationship:

$$\text{torque} = I_1 I_2 \sin\theta \qquad (13.1)$$

where I_1 is the RMS current in the upper electromagnet
I_2 is the RMS current in the lower electromagnet
θ is the angle between I_1 and I_2

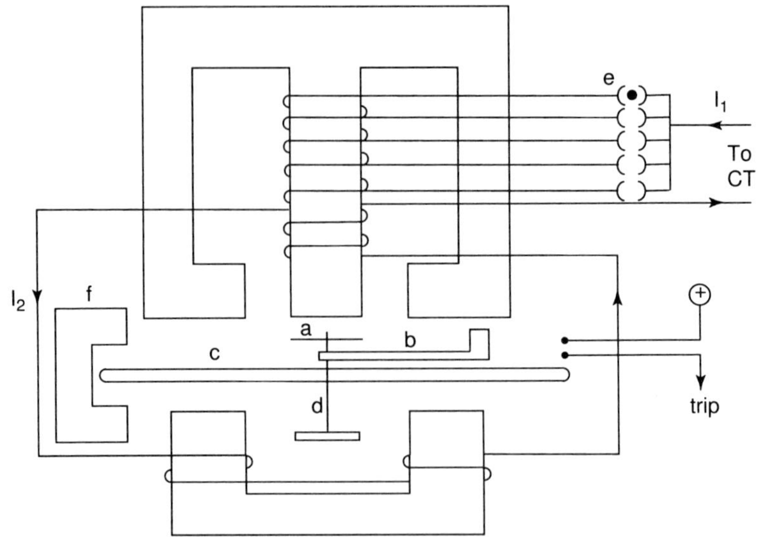

Figure 13.1 IDMTL-relay principle of operation

a Control spring
b Short-circuiting contact
c Metal disc
d Spindle
e Plug bridge
f Braking magnet

The spindle of the metal disc carries the control spring and moving contacts. At low currents the operating torque on the disc is proportional to the angle of travel. At high currents the magnetic circuit saturates and the torque is approximately constant.

A time-setting adjustment connected to the spindle is provided, which carries the back stop for the disc. By rotating the adjustment, the angle of travel of the disc can be altered to give a range of time/current characteristics. This is known as the time multiplier (TM).

The braking magnet ensures that when the operating current is removed, the disc is brought to a halt rapidly, to minimise the degree of overshoot.

13.3 Standard IDMTL-relay characteristic

The term IDMTL is often abbreviated to IDMT. This is the most commonly used type of relay. The characteristic of the relay is that the higher the current, the shorter the operating time; hence 'Inverse'. Above a certain current, the operating time is constant; hence 'definite minimum time'. The relay is known as a (3/10) relay and has a characteristic in accordance with the requirements of *BS 142* (which is gradually being superseded by *BS EM*

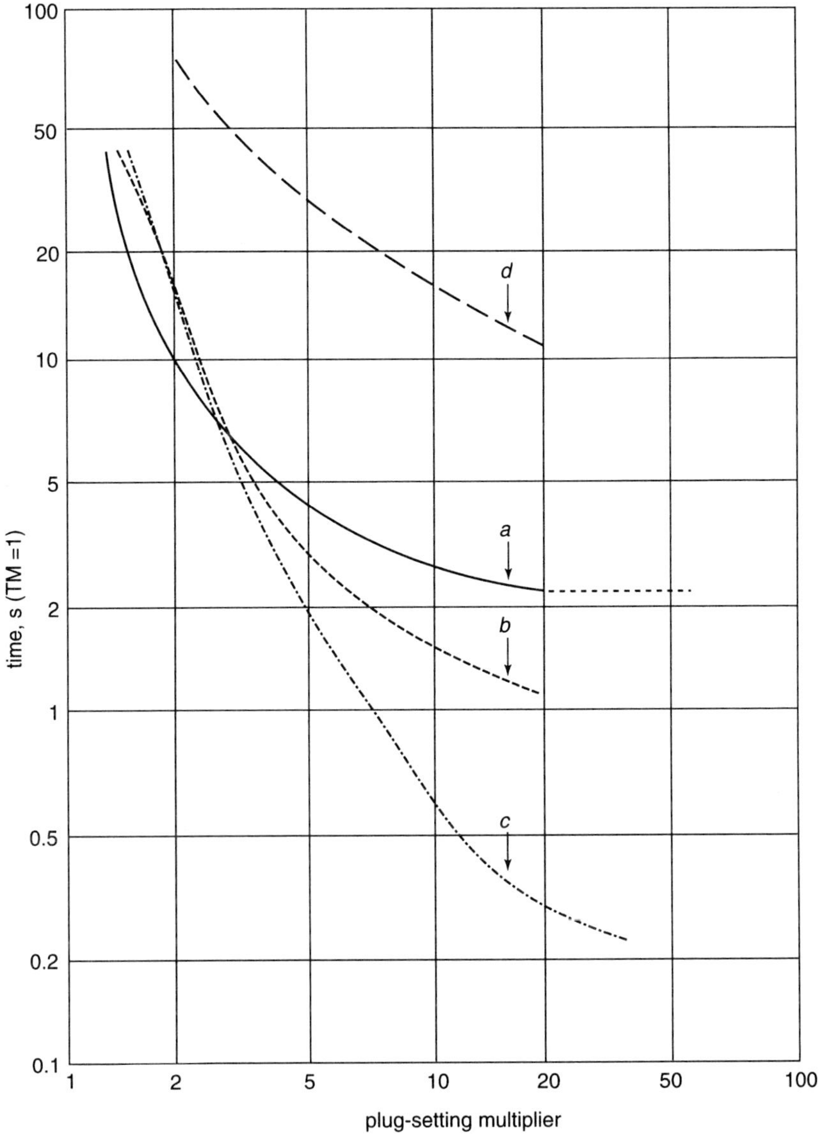

Figure 13.2 *Characteristic curves*

a Standard IDMT (3/10)
b Very inverse (1.5/10)
c Extremely inverse (0.6/10)
d Long time delay (30/5)

60255). This is illustrated in Figure 13.2. To facilitate greater accuracy at lower currents, it is usual to plot the characteristic to a logarithmic scale. The need

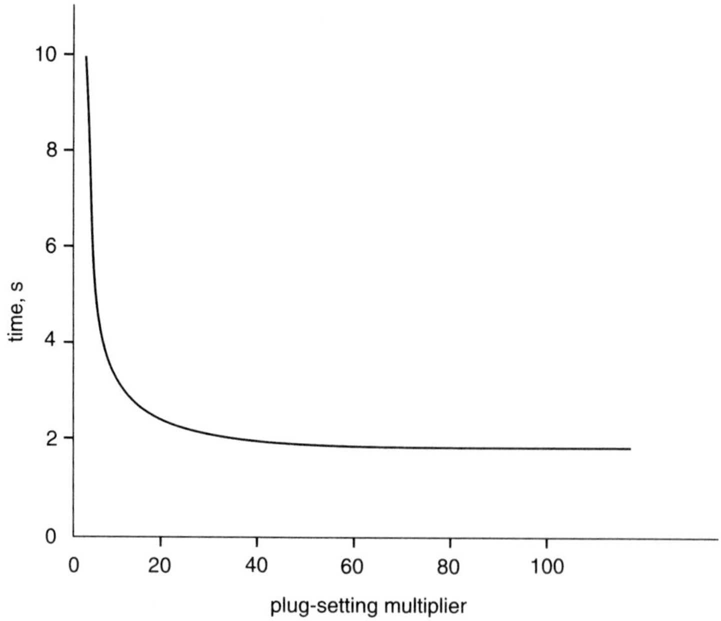

Figure 13.3 Standard IDMTL curve with linear scale

for this is evident if the characteristic is plotted on a linear scale as shown in Figure 13.3.

The relay is defined in *BS 142* as having a minimum pick-up or 'creep' current of 100% of the setting and a maximum pick-up current of 130% of the setting. The reset current must not be less than 70% of the pick-up current; this is important in ensuring that the relay will reset following clearance of a through fault, i.e. the relay must not remain partially operated. The term (3/10) defines a point on the characteristic at which a current of ten times the setting results in an operating time of 3 s, when the time-multiplier adjustment is at maximum travel (i.e. time multiplier = 1).

Figure 13.2 defines the horizontal axis as the 'plug-setting multiplier'. This term is defined as

$$\frac{\text{relay current}}{\text{relay setting}}$$

For example, suppose a relay with standard plug tappings of 25%, 50%, 75%, 100%, 125%, 150%, 175% and 200% of 1 A is set to 150% (ie 1.5 A). If the current into the relay from the CT is 15 A, then

$$\text{plug-setting multiplier (PSM)} = \frac{15}{1.5} = 10$$

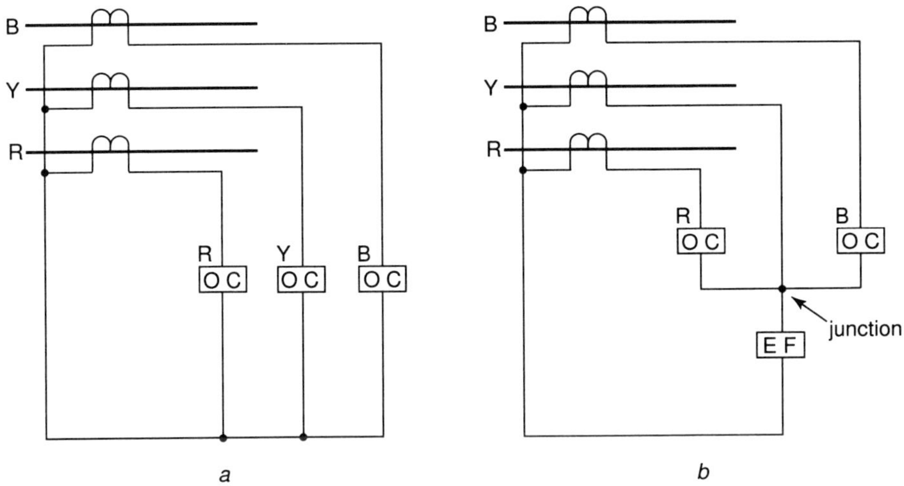

Figure 13.4 *Relay and CT connections*

a Overcurrent
b Overcurrent and earth fault

13.4 Relay and CT connections

Figure 13.4*a* shows the relay and CT connections usually employed for a three-phase overcurrent relay of the IDMTL type. The current in each phase of the line passes through the CT and flows into the corresponding phase overcurrent relay. The current setting of the relay must exceed the highest possible load current. Note that the usual application of this relay is to operate and trip the circuit breaker as a result of fault current and not to prevent marginal circuit overloading (i.e. it is not an overload relay).

Figure 13.4*b* illustrates the more usual CT and relay arrangement which includes an earth-fault relay. This arrangement dispenses with a surplus overcurrent relay, usually the yellow phase. A three-phase or phase-to-phase fault contains only positive-phase-sequence, or positive- and negative-phase-sequence, currents, respectively. These balanced symmetrical currents sum to zero at the junction of the relays and thus there is no current flow into the earth-fault relay. An earth fault additionally contains zero-sequence currents, which sum at the relay junction to result in a current into the earth-fault relay of three times the zero-sequence current. Since load current contains very little zero-sequence current (a small proportion due to the natural unbalance of the power system), earth-fault relays can have a more sensitive setting than overcurrent relays. It is worth noting that:

(i) overcurrent relays are responsive to positive-, negative- and zero-sequence currents;
(ii) earth-fault relays are responsive to zero-sequence current only.

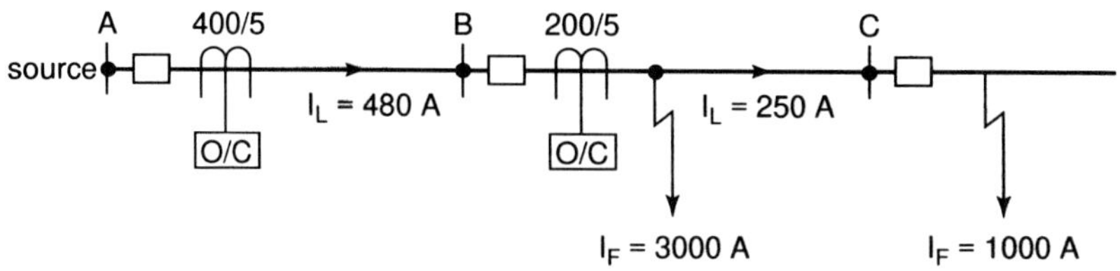

a

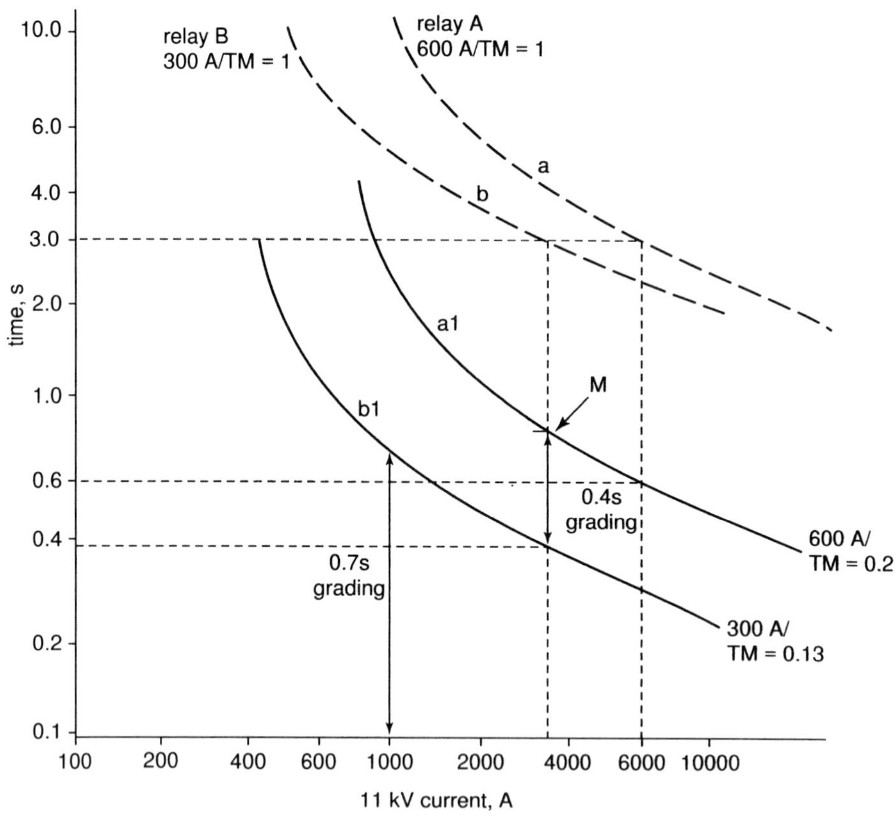

b

Figure 13.5 *Overcurrent grading*

a 11 kV radial network
b Required grading

13.5 Relay-grading criterion

The term 'relay grading' refers to relay settings which enable one IDMTL relay to discriminate with another. For example, with reference to Figure 13.5, which shows a simple distribution system, relay grading is required to ensure that, for the fault close to substation B, on feeder B–C, the overcurrent relay at substation B operates and trips the circuit breaker at B before the overcurrent relay at substation A operates. The approximate minimum operating-time interval between relays A and B to satisfy this requirement is made up of the following:

circuit B breaker opening time	=0.1 s
overshoot of relay A disc	=0.05 s
relay errors	=0.15 s
safety margin	=0.1 s
Total time	=0.4 s

With older relays, a time of 0.5 s was considered necessary. The theoretical relay error time, however, is more complex than the above simple approach would suggest, and is described in more detail in *BS 142*. With PSMs above 4, a 0.4 s margin can usually be accepted; otherwise *BS 142* should be consulted.

13.6 Relay-grading example

With reference to Figure 13.5, the settings of the relays at substations A and B for the 11 kV radial network are required. Note that I_L represents the maximum load current and I_F the maximum fault current at the points shown.

The settings are normally plotted on log-log graph paper, utilising a template of the (3/10) characteristic. From Figure 13.5*b*, any increase/decrease in relay-plug setting will move the characteristic horizontally right/left, respectively, and any increase/decrease in the time multiplier will move the characteristic vertically up/down, respectively.

Setting criteria are based upon the following:

(i) The plug setting should be higher than the circuit maximum load current.
(ii) The time-multiplier setting is based on achieving time grading (i.e. 0.4 s) when a current equal to the circuit maximum fault level passes through the relay.

Table 13.1 identifies the steps required to determine the plug setting. The steps required to obtain the grading curves, and hence the time multiplier with reference to Figure 13.5, are as follows:

(*a*) Utilising the plug setting determined in Table 13.1 and the IDMT template, plot the 3/10 characteristics, i.e. 3 s at 10 × plug setting. This results in curves *a* and *b*. Note that it is more convenient to work in 11 kV amps than PSMs.

Table 13.1 Plug-setting determination

	Load current (A)	CT ratio	Secondary load current (A)	Secondary setting current* (A)	Primary setting current (A)
Relay A	480	400/5	6	7.5	600
Relay B	250	200/5	6.25	7.5	300

* = next highest relay-tap position above load current

Table 13.2 Calculated settings

	Relay plug setting (A)	Primary current setting (A)	Time multiplier
Relay A	7.5	600	0.2
Relay B	7.5	300	0.15

(*b*) Assume that the relay at B is required to operate in 0.7 s to grade for the fault of 1000 A at substation C. The template is then lowered from curve *b* until it intercepts an operating time of 0.7 s at 1000 A; curve *b*1 is then drawn. The operating time of *b*1 at 10 × plug setting (i.e. 3000 A) is now 0.39 s. Thus

$$\mathrm{TM} = \frac{0.39}{3.0} = 0.13$$

In practice, this would probably be rounded up to a TM of 0.15.
(*c*) Relay A is required to grade with relay B with a fault current of 3000 A at substation B. Thus at 3000 A on Figure 13.5*b*, a discrimination margin of 0.4 s is drawn from curve *b*1, to give point M. The template is then lowered from curve *a* until it intercepts point M, and curve *a*1 is drawn. The operating time of *a*1 at 10 × plug setting (i.e. 6000 A) is now 0.6 s, and thus

$$\mathrm{TM} = \frac{0.6}{3.0} = 0.2$$

(*d*) The required settings are summarised in Table 13.2.

13.7 IDMTL relay family

The standard (3/10) relay is by far the most commonly applied IDMTL relay. However, a family of IDMTL relays with different characteristics exists. These are shown in Figure 13.2, and are summarised as follows:

standard relay (3/10)
very inverse relay (1.5/10)
extremely inverse relay (0.6/10) (see Section 13.23)
long-time-delay relay (30/5) (see Sections 13.18 and 13.20)

DIRECTIONAL RELAYS

13.8 Directional overcurrent

Figure 13.6 illustrates the basic design of a directional overcurrent relay. It comprises two elements: a directional element and an IDMTL element identical to that shown in Figure 13.1. The IDMTL element is inhibited from operating until the directional element has operated.

Torque on the directional element disc is given (see Figure 13.6) by

$$\text{torque} = II_v \sin \theta$$

Now the voltage V derived from a voltage transformer is applied to an inductive circuit; therefore I_v is both proportional to and lags V by 90°. Therefore

$$\text{torque} = kIV \sin (90\text{-}\Phi)$$

where k=constant of proportionality
and $90\text{-}\Phi = \theta$
So torque $= kIV \cos \Phi$

The directional element is thus a watt-metric device which measures the direction of power flow. Maximum torque occurs when V and I are in phase, reducing to zero torque when V and I are at 90° to each other. Figure 13.7 illustrates the resulting characteristic: current phasors I, I_1 and I_2 fall within the operate zone while I_3 is in the restraint zone.

The term maximum torque angle (MTA) denotes that angle between the relay applied voltage and current which results in maximum relay torque. For example, when V and I are required to be in phase to achieve maximum torque, the MTA of the relay would need to be 0°. During system faults the fault voltage and current are usually phase displaced, and thus it is usual to feed voltage and current quantities to the relay which result in maximum relay torque during fault conditions.

13.9 The 30° connection/relay MTA=0°

Figure 13.8 shows the voltage and current connections to a directional overcurrent relay with an MTA of 0°, which facilitates achieving maximum

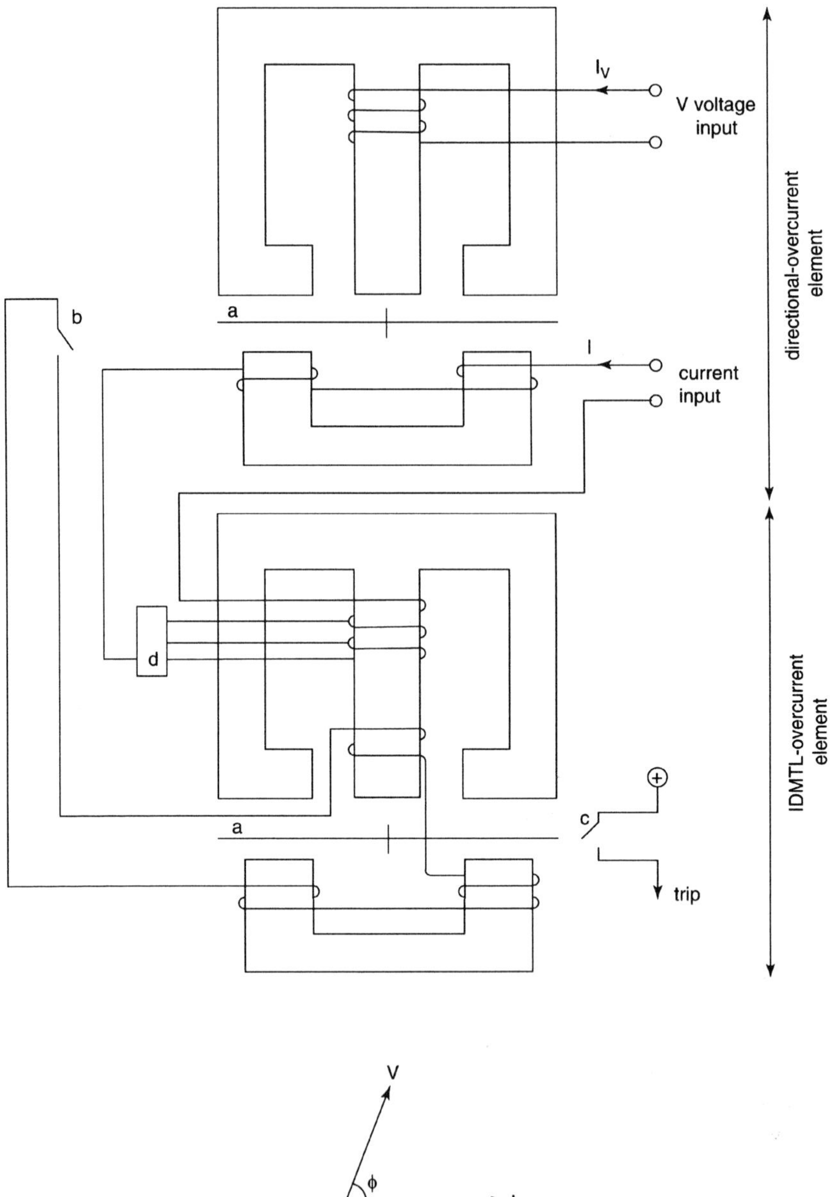

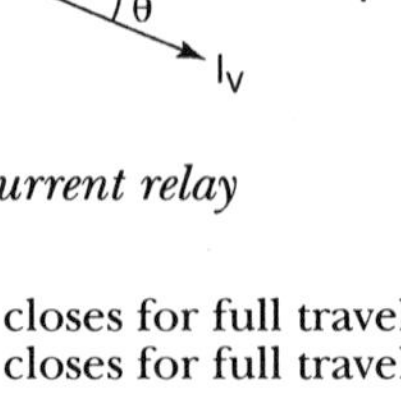

Figure 13.6 Directional overcurrent relay

a Relay discs
b Contact which closes for full travel of directional disc
c Contact which closes for full travel of overcurrent disc
d Plug bridge

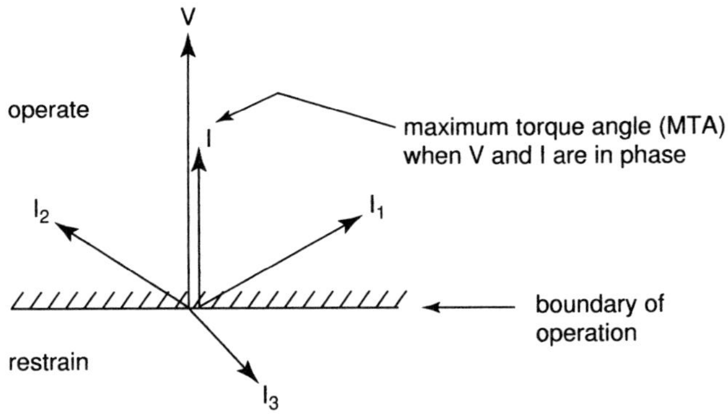

Figure 13.7 Basic directional characteristic

torque when the power-system phase current lags the phase-to-neutral voltage by 30°. Table 13.3 summarises the connections. This connection is rarely used in practice.

13.10 The 90° connection/relay MTA=45°

This relay has current and voltage connections which have a 90° phase angle. For example, the red element is fed with the red current and the yellow–blue voltage. Within the relay, the voltage applied to the voltage coil is arranged to lead the input voltage by 45° to achieve an MTA of 45°. This is accomplished by use of a resistor in series with the relay voltage coil, as shown in Figure 13.9*a*. The phasor relationship for the red-phase element is shown in Figure 13.9*c* and *d*. This results in relay maximum torque when the phase current lags the phase-to-neutral voltage by 45°. Table 13.4 summarises the connection.

When a system fault occurs, the voltages applied to the directional relays change in both magnitude and phase angle, but the phase-angle consideration is usually the most critical of the two in achieving satisfactory performance. The change in phase angle is usually not more than 30°; for example, with reference to Tables 13.3 and 13.4, the complete depression of any phase voltage would result in a 30° shift in phase angle (e.g. the angle difference between the R–N voltage and the R–B voltage). However, to maximise the torque available, careful consideration needs to be given to the application of directional overcurrent relays.

When directional overcurrent relays are also used to detect earth faults, the 90°/45° relay is superior, since the voltages applied to the relay are not from the faulted phase. In addition, the 90°/45° relay is preferred for three-phase faults because the phase angle of the fault current tends to be nearer 45° than 30°. A point worthy of note is that, for the instance of a three-phase fault close to the relay position, all voltages fall towards zero and the relay will not operate.

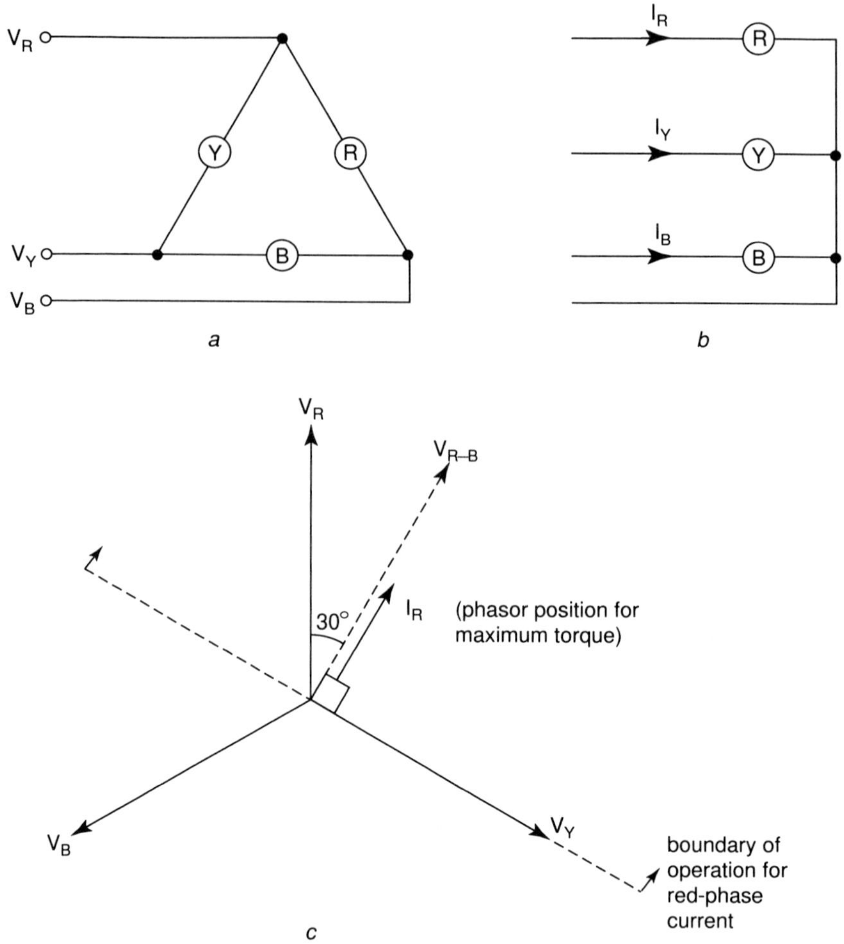

Figure 13.8 Directional-overcurrent 30° connection, relay MTA = 0°

a Voltage connections
b Current connections
c Phasor diagram for red relay

13.11 Directional earth-fault relay

When greater sensitivity is required for earth-fault protection, a directional earth-fault relay with an MTA of 0° can be used. In this instance, the voltage and current connections to the relay are fed from residual (zero-sequence) quantities. Figure 13.10*a* illustrates the required connections: the open delta of the VT secondary (which may alternatively be that of an interposing VT) results in the positive- and negative-sequence voltages summing to zero, thus leaving only the zero-sequence voltage. For an earth fault, the resulting

Table 13.3 Connections for a 30° connection/relay MTA = 0°

Relay	Current input	Voltage input
Red	R	R–B
Yellow	Y	Y–R
Blue	B	B–Y

residual voltage V_{RES}, as shown in Figure 13.10*b*, becomes antiphase to the fault voltage (since $V_{RES} = V_R + V_Y + V_B$). By use of a capacitor and resistor in series with the relay voltage coil, the voltage V_{REL} across the coil can be arranged to be approximately in phase with the residual current I_{RES}, thus achieving maximum torque under system fault conditions. This arrangement is applicable to directly earthed systems; resistance-earthed systems result in the angle between fault voltage and current being closer to zero, and thus there is not the need for the phase-angle-compensation circuitry shown in Figure 13.10*a*.

Note that to transform zero-sequence voltages accurately, the VT shown in Figure 13.10*a* must be of the five limbed variety as noted in Section 3.9.

INSTANTANEOUS OVERCURRENT

13.12 Instantaneous overcurrent

An instantaneous-overcurrent (or earth-fault) relay is one which operates instantaneously if the current exceeds the setting. The term 'high-set overcurrent' describes an instantaneous overcurrent relay which has a current setting above the circuit maximum loading, such that it will only operate for fault currents.

Modern instantaneous-overcurrent relays have operating times in the order of 5–15 ms. When applying instantaneous overcurrent relays, care must be taken to ensure that the design of the relay is such that it is immune to transient direct currents, particularly those associated with transformer-magnetising inrush currents.

APPLICATION OF IDMTL RELAYS

13.13 Grading of earth-fault relays across delta–star transformers

Section 13.4 pointed out that earth-fault relays are responsive only to zero-sequence currents. Figure 16.10 illustrates an earth fault on one side of a

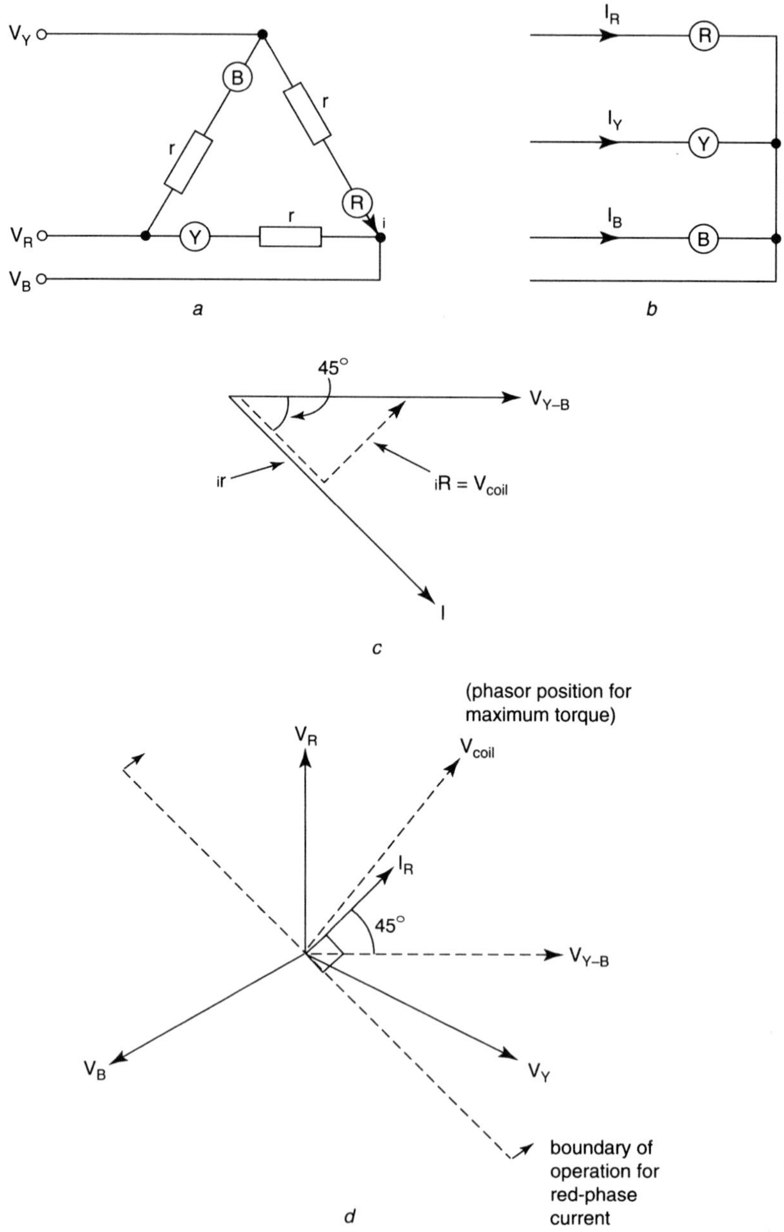

Figure 13.9 Directional-overcurrent 90° connection, relay MTA = 45°

a Voltage connections
b Current connections
c Red-phase voltage coil
d Phasor diagram

Table 13.4 Connections for 90° connection/relay MTA = 45°

Relay	Current input	Voltage input
Red	R	Y–B
Yellow	Y	B–R
Blue	B	R–Y

two-winding delta–star transformer. The accompanying phase-sequence network shows that zero-sequence current cannot flow through the transformer. Thus earth faults on one side of a delta–star (or star–delta) transformer do not cause current flow into earth-fault relays on the other side. Thus there is no requirement to grade HV and LV earth-fault relays across two-winding transformers when at least one winding is connected in delta.

13.14 Grading of overcurrent relays across star–delta transformers

Figure 13.11 shows the current distribution arising from a phase-to-phase fault on the LV side of a star–delta transformer, which has a turns ratio of *N*. Table 13.5 compares the current for a three-phase fault, with that for the phase-to-phase fault shown in Figure 13.11, assuming that the fault location is the same in both instances.

The implication of Table 13.5 is that grading of transformer HV and LV overcurrent should be based on fault currents arising from a phase-to-phase fault, since for the same HV current the phase-to-phase fault results in a smaller LV current, and hence a slower operating time. This compromises relay time grading since the LV relay is required to operate faster than the HV relay.

An identical conclusion can be derived if the HV/LV winding configuration is delta–star.

13.15 Earth-fault currents in interconnected systems

Figure 13.12 is a simple illustration of a red-phase-to-earth fault on a feeder circuit which is part of an interconnected system. The sequence network shows the current distribution for the red reference phase. Now if

$$I_{A1} = I_{A2} = I_{A0}$$

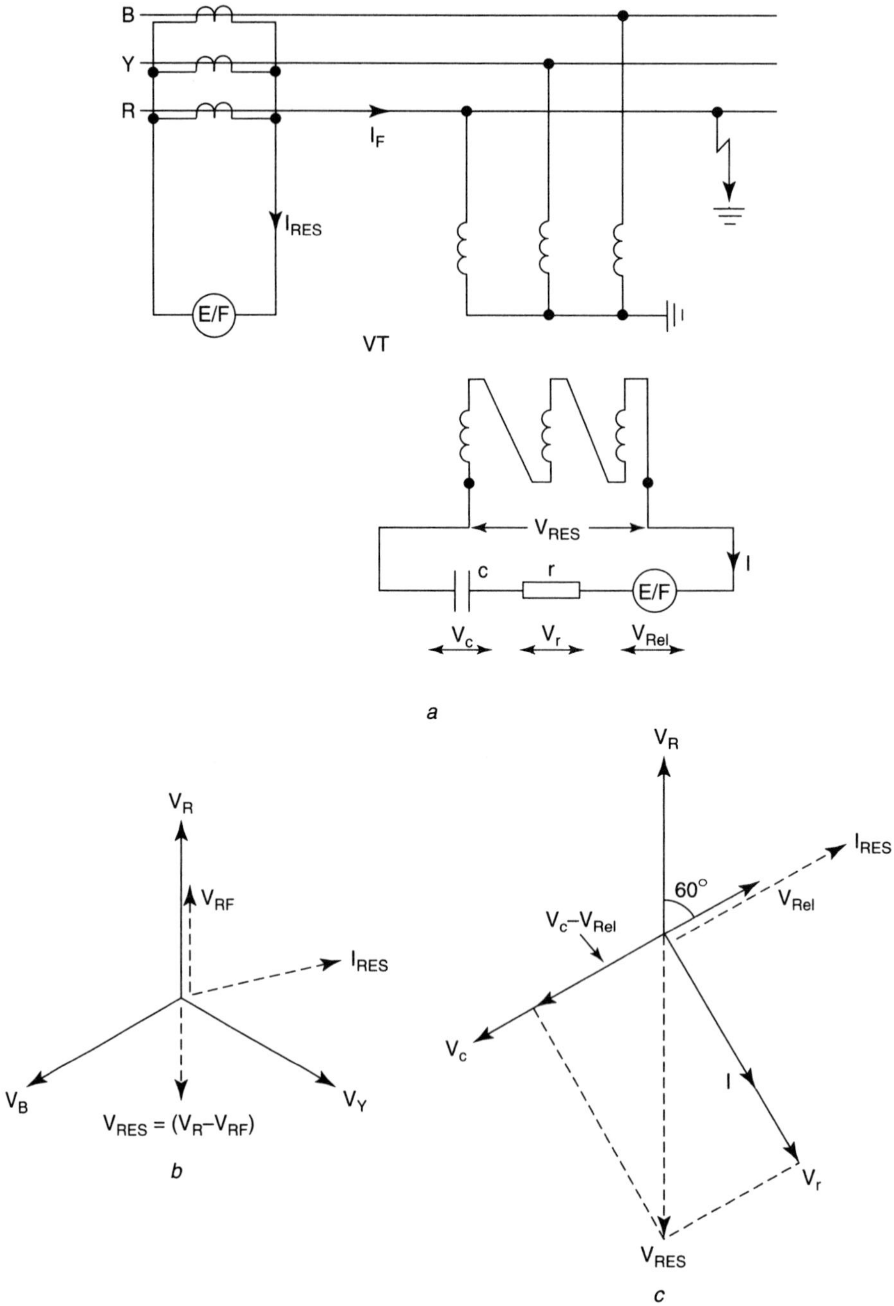

Figure 13.10 Directional earth-fault relay

a Connections
b Determination of V_{RES}
c Determination of relay voltage

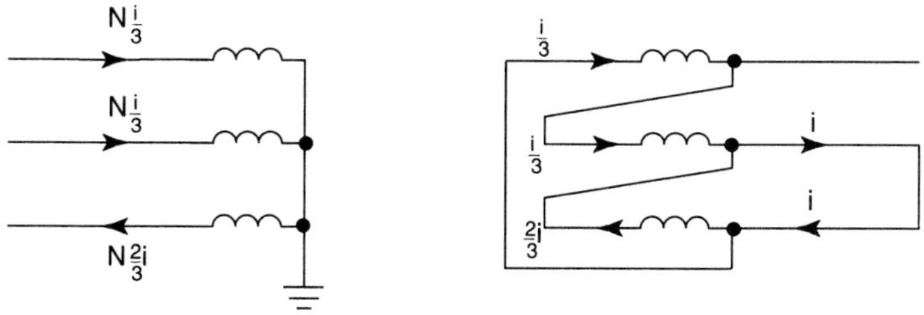

Figure 13.11 Passage of phase-fault current through a star–delta transformer

Table 13.5 Comparison of three-phase and phase-to-phase fault currents

	3-phase fault	Phase–phase fault
Maximum LV current	$2i/\sqrt{3}$	i
Maximum HV current	$2Ni/3$	$2Ni/3$

then

$$I_A = I_{A1} + I_{A2} + I_{A0} = 3I_{A1}$$

Then the total current supplied by end A is 3 I_{A1} which is also the current flowing through the earth fault relay. In addition, the current in the yellow and blue phases, fed from end A, is

$$I_Y = a^2\ I_{A1} + a\ I_{A2} + I_{A0} = 0$$

and

$$I_B = aI_{A1} + a^2\ I_{A2} + I_{A0} = 0$$

Now in practice the distribution of the positive- and negative-sequence currents will always be identical, but may differ from the zero-sequence currents (since zero-sequence impedances may differ from positive- and negative-sequence impedances). Thus $I_{A1} + I_{B1} = I_{A0} + I_{B0}$ but I_{A1} may not equal I_{A0}. For the sake of illustration, suppose that $I_{A0} = 0.5I_{A1}$, and $I_{A1} = I_{A2}$. Then the current in each phase fed from A consists of

$$I_R = I_{A1} + I_{A2} + I_{A0} = 2.5I_{A1}$$

$$I_Y = a^2 I_{A1} + aI_{A1} + I_{A0} = \text{-}0.5I_{A1}$$

$$I_B = aI_{A1} + a^2 I_{A2} + I_{A0} = \text{-}0.5I_{A1}$$

When such an earth fault is examined on a fault recorder, it can be puzzling as to why current is observed to flow in the healthy phases. The above provides the explanation.

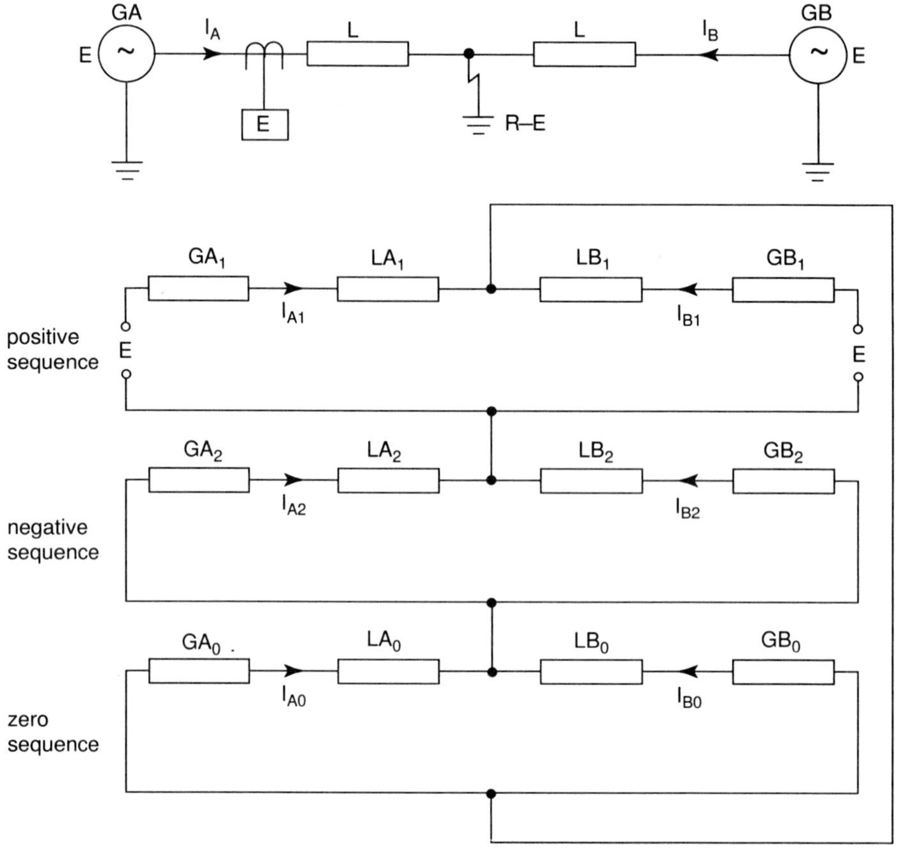

Figure 13.12 Earth fault on an interconnected system

The current in end-A earth-fault relay is, again, the summation of currents at the junction shown in Figure 13.4*b*, i.e.

$$I_R + I_Y + I_B = 1.5 I_{A1}$$

which is the same as the summation of the zero-sequence currents in each phase.

13.16 11 kV feeders

11 kV feeders are usually operated as radial circuits or as normally open rings. The feeder is usually protected by an overcurrent and earth-fault arrangement as shown in Figure 13.4*b*, with a relay-grading arrangement as shown in Figure 13.5. The plug setting of the overcurrent relay needs to be in excess of the highest load current to which the feeder will be subject. The earth-fault-

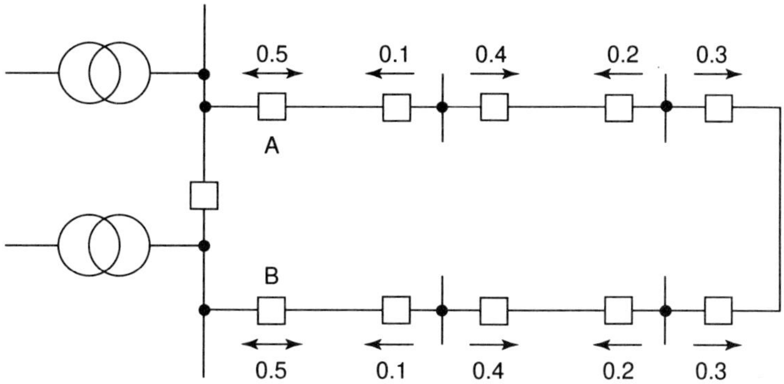

Figure 13.13 *Protection of a 33 kV closed ring*

⟷ nondirectional relay
⟵ directional relay
All times are in seconds

relay-plug setting should be set as low as possible consistent with achieving grading with other relays or any fuses on the system.

13.17 33 kV feeders

Figure 13.13 illustrates the typical application of directional and non-directional (i.e. standard 3/10) relays to a 33 kV closed-ring circuit. The illustration only defines discrimination in terms of time, and as can be seen a fault on any feeder results in the discriminative clearance of that fault by the relays connected to that feeder. The grading pattern is achieved by considering the fault current which would flow for a feeder fault close up to a source circuit breaker (e.g. circuit breaker A) when that source circuit breaker is open, i.e., circuit breakers A and B are considered open in turn and each circuit is considered as a radial circuit. The relay-grading exercise may be quite onerous, and is simplified in this illustration.

13.18 33/11 kV transformer

Figure 13.14 shows a typical protection arrangement for a 33/11 kV distribution transformer. Main protection of the transformer is provided by the restricted earth-fault protections (REF). The LV overcurrent is a standard 3/10 relay and provides backup protection for faults on the 11 kV network; operation of the relay would result in tripping of the 11 kV circuit breaker. The standby earth fault (SBE) provides backup protection for any earth faults on the 11 kV network; the characteristic used is usually the long time delay (30/5) as shown in Figure 13.2. This characteristic is usually designed, to

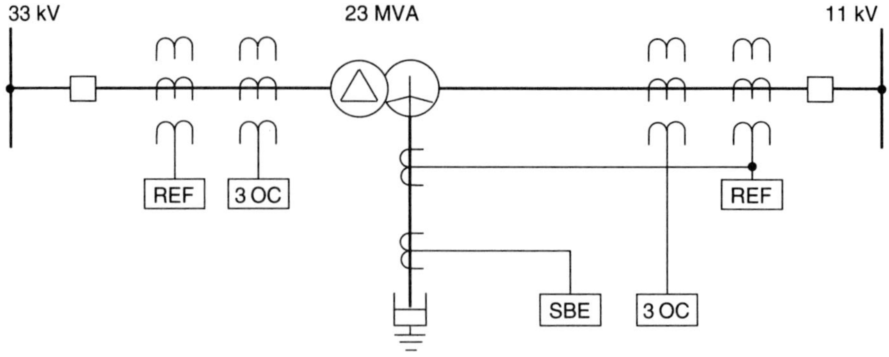

Figure 13.14 Protection of a 33/11 kV transformer

match the thermal rating of the liquid earthing resistor. Once operated, it is arranged to trip the 11 kV circuit breaker. The HV overcurrent provides backup protection for the zone between the transformer HV and LV overcurrent relays; its setting needs to be graded with the LV overcurrent relay. Once operated, it is arranged to trip both the HV and LV circuit breakers. Note that it would be usual for these transformers to be fitted with Buchholz and winding-temperature protection (not shown).

13.19 132 kV feeder

132 kV feeders are usually fitted with two-pole overcurrent and earth-fault relays, as shown in Figure 13.4*b*, as a backup to the feeder main protection (e.g. unit or distance protection).

13.20 132/33 kV transformers

Figure 13.15 illustrates the general protection arrangements for a 132/33 kV transformer. Salient points are as follows:

(*a*) *Main protection* of the transformer is accomplished by the biased differential (B dif) and restricted-earth-fault protections (REF). Buchholz and winding-temperature protection are fitted as standard, but are not shown.
(*b*) *The two-stage overcurrent relay* (2 STOC) arrangement is used, for both technical and economic reasons, in preference to separate HV and LV overcurrent relays. The AC circuitry for one phase is shown in Figure 13.15*b*, the other two phases being identical. Figure 13.15*c* shows the tripping arrangements; the first stage, initiated directly from the IDMTL overcurrent relay, trips the LV circuit breaker, and is used as backup to uncleared 33 kV system faults.

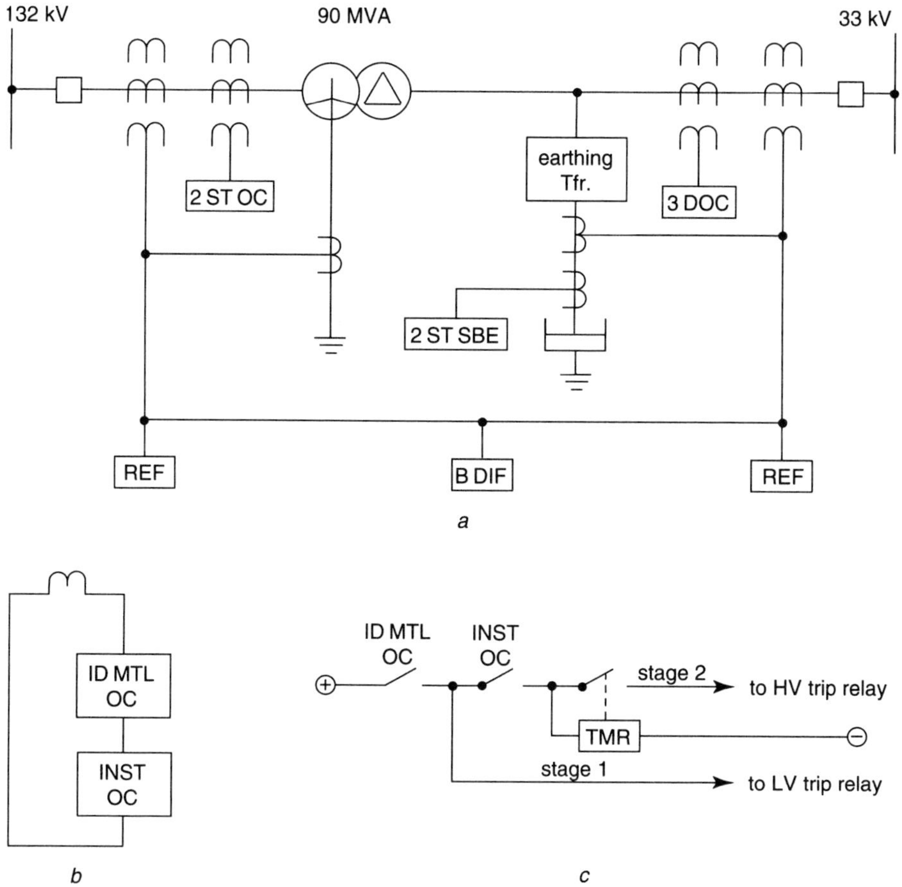

Figure 13.15 Protection of a 132/33 kV transformer

a General protection arrangement
b Two-stage overcurrent
c Two-stage-overcurrent trip circuitry

The second-stage trip is via an instantaneous overcurrent relay and a timer. The instantaneous overcurrent relay ensures that the timer is de-energised instantaneously, when current stops flowing as a result of the LV circuit breaker opening. This is preferred to the longer disengagement time of an IDMTL overcurrent relay. The timer has a typical time delay of 200 ms, which provides a better time discrimination than 400 ms (typically) which would be required for a separate LV overcurrent relay. Operation of the timer results in energisation of the HV trip relay, which trips the HV circuit breaker.

The stage-2 trip essentially protects against a fault on the transformer HV or LV connections which has not been cleared by the main protection. Figure 13.15 illustrates the case of an interbus transformer (i.e. HV and LV circuit breaker on the same site), and in such instances the instantaneous overcurrent and stage-2-timer contacts are frequently short-circuited to create simultaneous tripping of both LV and HV circuit breakers. However, for banked transformers or transformer feeders, particularly where there is a teed circuit connected to the feeder, the arrangement as shown in Figure 13.15 is almost always employed.

(*c*) *The two-stage standby earth-fault relay* (2ST SBE) employs similar circuitry to Figure 13.15*c* except that the SBE replaces the IDMTL OC and there is no instantaneous relay element. The first stage provides backup for uncleared 33 kV-system earth faults and trips the LV circuit breaker; the second stage provides backup protection for faults on the transformer LV connections and trips the HV circuit breaker. The relay characteristic is usually a long time delay (30/5), since this characteristic closely matches the thermal rating of the liquid earthing register.

(*d*) *The LV directional overcurrent relay* (DOC) is applied only in the following circumstances:

(i) where there is more than one interbus transformer between the same busbars. In this instance it ensures discriminative clearance of an uncleared fault on the transformer LV connections, thereby leaving the healthy transformers in service;

(ii) where the transformer is connected as a transformer feeder on the HV side. In this instance the DOC would detect faults on the HV feeder, resulting in tripping of the LV circuit breaker and initiation of autoreclose.

Figure 13.16 shows the vector relationships for the instance of a phase-to-phase fault on the HV feeder. Salient points are as follows:

(i) Figure 13.16*a* shows the physical arrangements of the transformer feeder.

(ii) Figure 13.16*b* shows the sequence network, with the position of the DOC CTs. Note that the relatively low impedance of L1 (part of the faulted feeder) shunts current away from the DOC relay. Opening of the HV circuit breaker would result in the L1 leg becoming open circuit, thus causing an increase in current through the DOC relay, i.e. a sequential tripping situation. Many modern installations have intertripping from the feeder HV circuit to the transformer LV circuit breaker, initiated by the feeder main protection. In such circumstances, the DOC relay would be acting as a backup to intertripping failure.

(iii) Figure 13.16*c* illustrates the voltages at the HV busbars.

(iv) Figure 13.16*d* and *e* shows the voltages and currents at the transformer T1 HV and LV connections, respectively. The phase angle between LV vectors i_y and $V_{b\text{-}r}$ shows that a DOC with a 90° connection and 45° MTA characteristic would be suitable for this application.

Figure 13.17 shows a similar illustration for an HV-feeder red-phase-to-earth fault. Note that the LV delta removes all zero-sequence currents from the DOC relay. The phase relationship between transformer T1 I_r and $V_{y\text{-}b}$ again

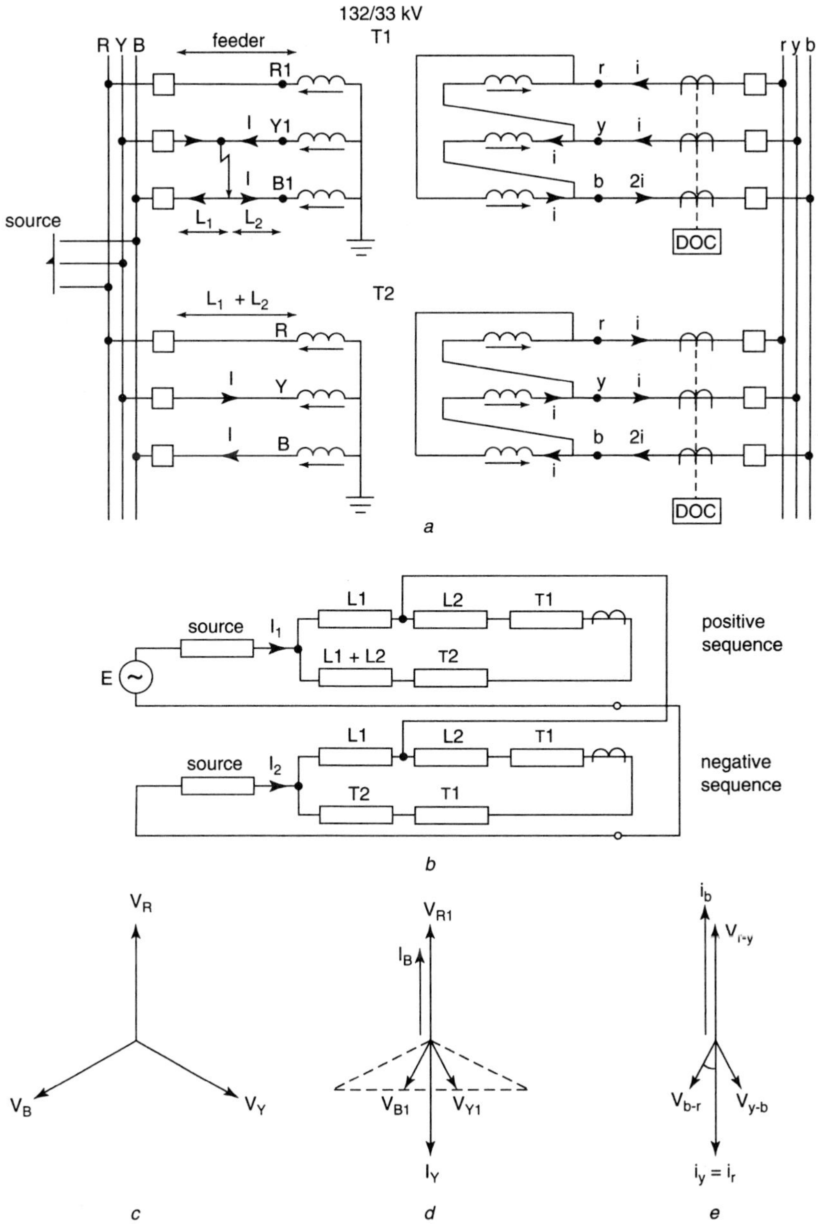

Figure 13.16 Performance of a DOC relay for a phase-to-phase fault

a Physical arrangement
b Sequence network
c Voltages at the HV breakers
d Voltages and currents at transformer T1 HV terminals
e Voltages and currents at transformer T1 LV terminals

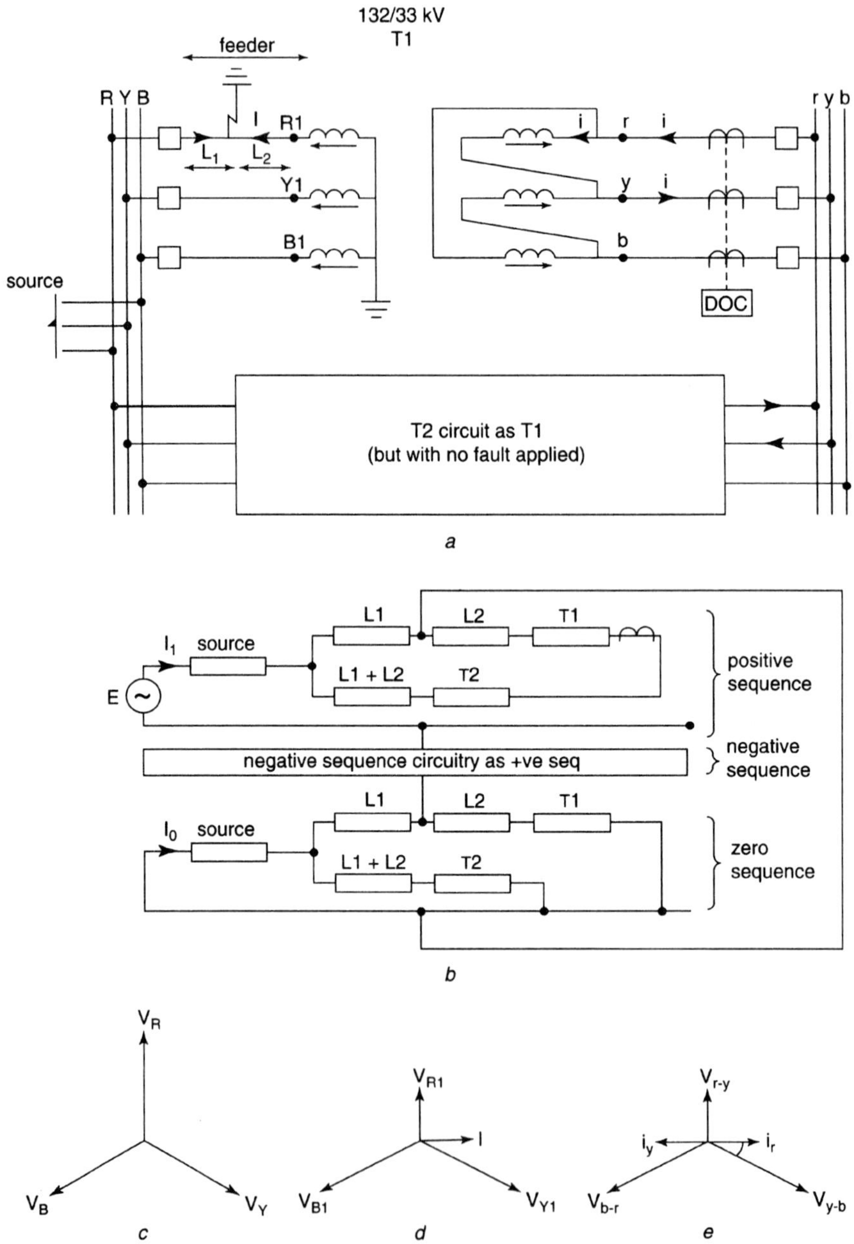

Figure 13.17 Performance of a DOC relay for a single-phase-to-earth fault

a Physical arrangement
b Sequence network
c Voltages at the HV breakers
d Voltages and currents at transformer T_1 HV terminals
e Voltages and currents at transformer T_1 LV terminals

shows the suitability of the 90°-connection 45°-MTA-characteristic relay. Both these Figures illustrate the relative complexity of the application of the DOC relay to transformer arrangements.

Note that it may not be possible to apply the DOC relay if there is significant LV generation (i.e. reverse power flow).

13.21 400 and 275 kV feeders

400 and 275 kV feeders are usually protected as follows:

(*a*) where the circuit is equipped with a distance protection, single-pole IDMTL earth-fault relays only are installed;
(*b*) where the circuit is not equipped with a distance protection, IDMTL 2-pole overcurrent and single-pole earth-fault relays are installed.

In (*b*) above, the overcurrent relays are required for the instance of an uncleared busbar fault on the busbars at the remote end of the feeder. The earth-fault relay is required to clear a high-resistance earth fault (arising, for example, from stubble burning) with resistances as high as 100 Ω. The operating times of these relays are required to be in the order of 1 s to provide grading with other backup protections such as zone-2 or distance protection.

In (*a*) above, zone-2 or distance protection provides the backup to an uncleared busbar fault and hence the overcurrent relays are not required.

13.22 400/132 and 275/132 kV transformers

The protection arrangements for 400/132 and 275/132 kV transformers is shown in Figure 13.18, salient points being as follows:

(*a*) *Main protection* of the transformer is accomplished by the circulating-current (CIRC CURR) protection. Buchholz and winding-temperature protection are fitted as standard, but are not shown.
(*b*) *The two-stage overcurrent relay* (2 ST OC) serves the same purpose as that described in Section 13.20. The first stage acts as backup protection to faults on the 132 kV system, and when operated trips the 132 kV circuit breaker. The second stage provides backup protection to the transformer HV and LV connections, and when operated trips the 400 kV circuit breaker.
(*c*) *The LV IDMTL* earth-fault relay (E) provides back-up protection for uncleared earth faults on the 132 kV system. This relay is positioned as shown, in preference to being connected to CTs on the HV side of the transformer, because any delta tertiary winding connected to the autotransformer shunts zero-sequence fault current away from the HV CTs, thus making an HV-connected earth-fault relay less responsive to 132 kV earth faults.
(*d*) *The HV-connected high-set overcurrent relay* (HSOC) is an instantaneous relay which is required to clear the faults on the transformer HV winding quickly.

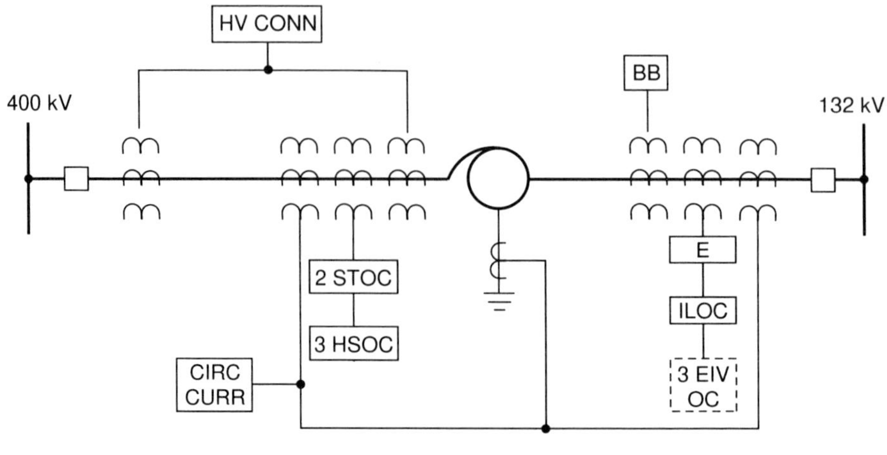

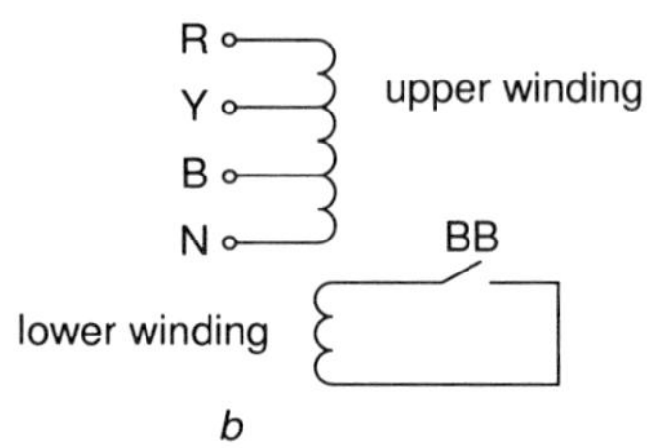

Figure 13.18 *Protection of a 400/132 kV transformer*

a Circuit
b Winding arrangements for interlocked overcurrent relay

This relay must be set such that it is unresponsive to faults on the 132 kV system.

(*e*) It is usual to install an *interlocked overcurrent relay* (ILOC) on the LV side of the transformer. Operation of the 132 kV busbar-protection scheme would usually result in tripping of the LV circuit breaker only. However, if the fault occurs between the LV CT and LV circuit breaker (i.e. within the busbar-protection zone), opening of the LV circuit breaker will not remove the fault and, without an interlocked overcurrent relay, relatively slow clearance via the two-stage overcurrent relay would result. The interlocked overcurrent relay consists of an induction-disc type of relay similar to that shown in Figure 13.1 with a sensitive setting (see Figure 13.18*b*). The upper winding is fed from CTs connected to each phase. The lower winding is conditioned by operation of the busbar protection (BB). Thus if the busbar protection operates and fault current continues to flow (implying a fault between the CT and circuit breaker), the interlocked overcurrent relay operates quickly to open the HV circuit breaker.

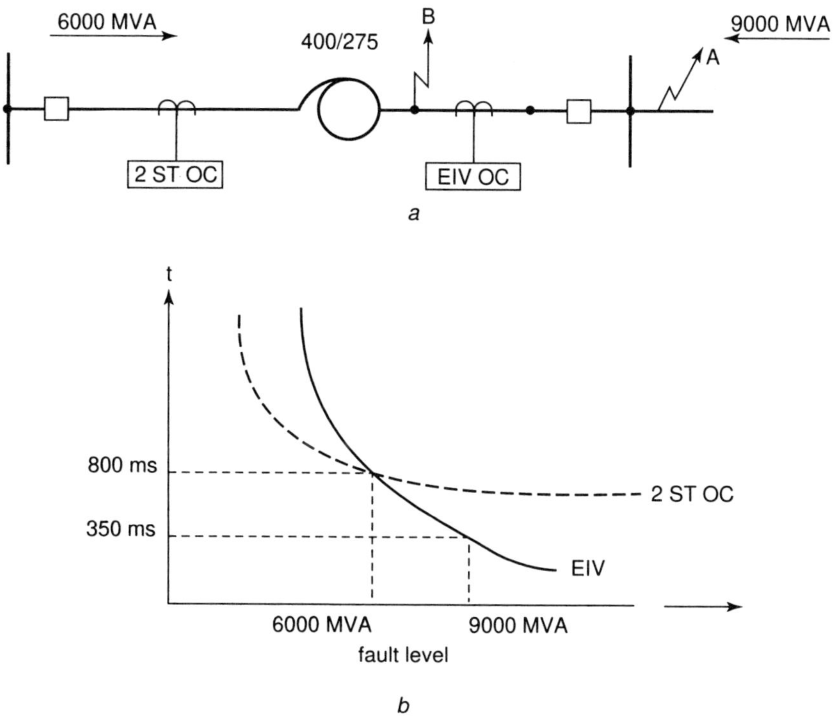

Figure 13.19 *400/275 kV-transformer winding*

a Circuit
b Fault level

13.23 400/275 kV transformers

The protection arrangements for a 400/275 kV transformer are identical to those shown in Figure 13.18, but with the addition of a three-pole extremely inverse overcurrent relay (3 EIV OC) connected to the 275 kV CTs. The EIV relay is a backup protection and serves the following purposes:

(*a*) In Figure 13.19*a* the 400/275 kV transformer is shown as having a fault at one of two possible positions, A or B. For the fault at position A, a current equivalent to about 6000 MVA will, in practice, flow through both the 2ST OC and EIV OC, and as shown by Figure 13.19*b*, both operate in a time of about 800 ms. For the fault at position B, the EIV OC is now subject to a 275 kV-system infeed of about 9000 MVA, with a reduction in operating time to about 350 ms, thus greatly assisting power-system stability
(*b*) Figure 13.20 illustrates the instance of a three-transformer infeed to a 275 kV system. For the fault shown, all three 2 ST OCs would, in the absence of an EIV relay, operate at the same time, with the loss of all three

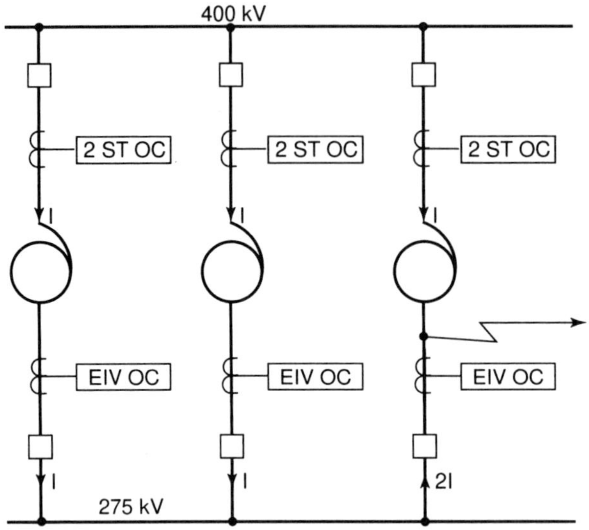

Figure 13.20 Discriminative clearance of an LV-connections fault

transformers. However the EIV relay connected to the faulted circuit, if fitted, would be subject to a current equal to $2I$ (about 10 000 MVA), and from Figure 13.19*b* it can be seen that the faulted transformer would be discriminatively cleared.

13.24 Generator transformers

The application of overcurrent and earth-fault relays to generator transformers is discussed in Chapter 17.

COMMISSIONING

13.25 IDMTL relay

Typical secondary-injection commissioning tests for this type of relay consist of the following.

(*a*) A check should be made of the minimum operating current with the plug inserted into the nominal setting (i.e. 100% plug setting) and the time multiplier set to 1. The injected current is incrementally increased, and the disc must not start creeping at currents less than 100% of setting; and full travel of the disc must occur, and the relay flag must have dropped, with an injected current less than 130% of setting. The current should then be

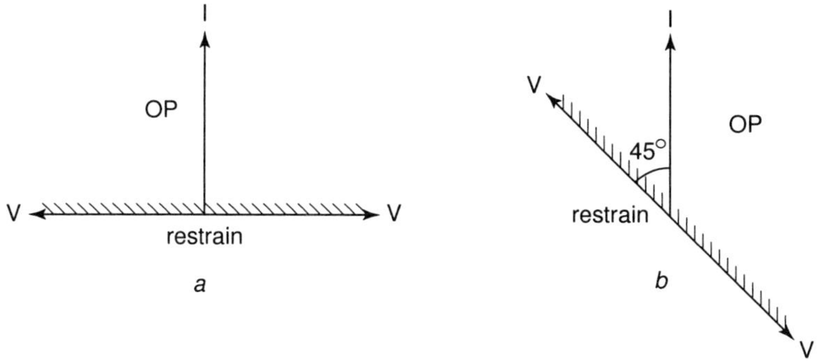

Figure 13.21 *Directional-overcurrent secondary injection*

a 30° characteristic
b 45° characteristic

V=injected voltage
I=injected current

incrementally reduced to enable the disc to reset, which must occur at a current not less than 70% of the pickup current.

(*b*) With the plug inserted into each position in turn, current should be injected to ensure that the disc rotates. The plug should then be withdrawn and current again injected to rotate the disc (with the plug withdrawn the plug bridge defaults to maximum plug position).

(*c*) A timing test is required with the plug at 100% setting and the time multiplier equal to 1. This is usually carried out with a stopwatch. With a (3/10) relay, the injected current is usually 2×plug-bridge setting (nominal operating time equal to 10 s) or at 4×plug bridge setting (nominal operating time equal to 5 s). An error of 10% is usually acceptable.

(*d*) A disc-reset time is required with the disc fully operated, and the current switched off.

(*e*) The on-load test usually consists of measuring the current in each relay (see Chapter 19). The measured current flowing into an earth-fault relay should be virtually zero.

13.26 Directional overcurrent relay

Tests on the directional elements may be carried out by injecting nominal current, while varying the voltage phase angle by use of a phase shifter. This should result in 180° of operation and restraint, respectively. The resultant characteristics for a relay with 30° connection/MTA=0° and a 90° connection/MTA=45° are shown in Figure 13.21. Proof that the voltage and current connections to the relay are correct should be obtained by primary injection of the CTs and injection of the VTs, as described in Section 10.46.

The minimum sensitivity of the relay should be obtained by injecting nominal voltage at the maximum torque angle (i.e. 90° from the boundary of operation) and slowly increasing the current until the relay operates. The volt-amperes to operate the relay can then be calculated and compared with the manufacturer's literature. Similarly a voltage of 10 V (representing a typical minimum that might be applied to the relay under fault conditions) should be applied at MTA and the current increased until the relay operates. Again the volt-amperes to cause operation should be calculated.

The IDMTL element of the relay should be tested as described in Section 13.25. It will be necessary to operate the directional element manually to achieve this.

The on-load test requires the directional feature of the relay to be proved. This can be accomplished by first checking the circuit metering to ascertain the direction of load current through the circuit. The relay should then be observed to operate or restrain, as appropriate.

The current-coil connection should be reversed, or voltage connections rotated (as described in Section 14.34 for distance protection) to prove the other mode of the relay. Note that when directional overcurrent relays are connected to transformers as in Figure 13.15, current through the transformer can be influenced by varying the taps (where there is more than one transformer on the same site) to cause circulating current as shown in Figure 6.4. This current has a quadrature relationship with respect to the voltage, but should be sufficient to prove a relay with a 45° characteristic.

With directional earth-fault relays it is necessary to short circuit one CT and remove its connections from the relay, to create a residual current into the relay. Similarly, the corresponding phase voltage should be disconnected to produce residual voltage.

13.27 High-set overcurrent

The secondary injection of this type of relay consists of slowly injecting current until operation occurs and then slowly reducing current until it resets. The drop-off-to-pickup ratio is generally required to be higher than 0.8; for many modern relays the result is usually higher than 0.9. A timing test measuring the time from the instant of current injection to contacts closed should be carried out. Typical operating times are of the order of 10 ms.

13.28 Solid-state relays

The preceding sections have discussed electromagnetic relays, but almost identical commissioning tests may be utilised on modern solid-state relays.

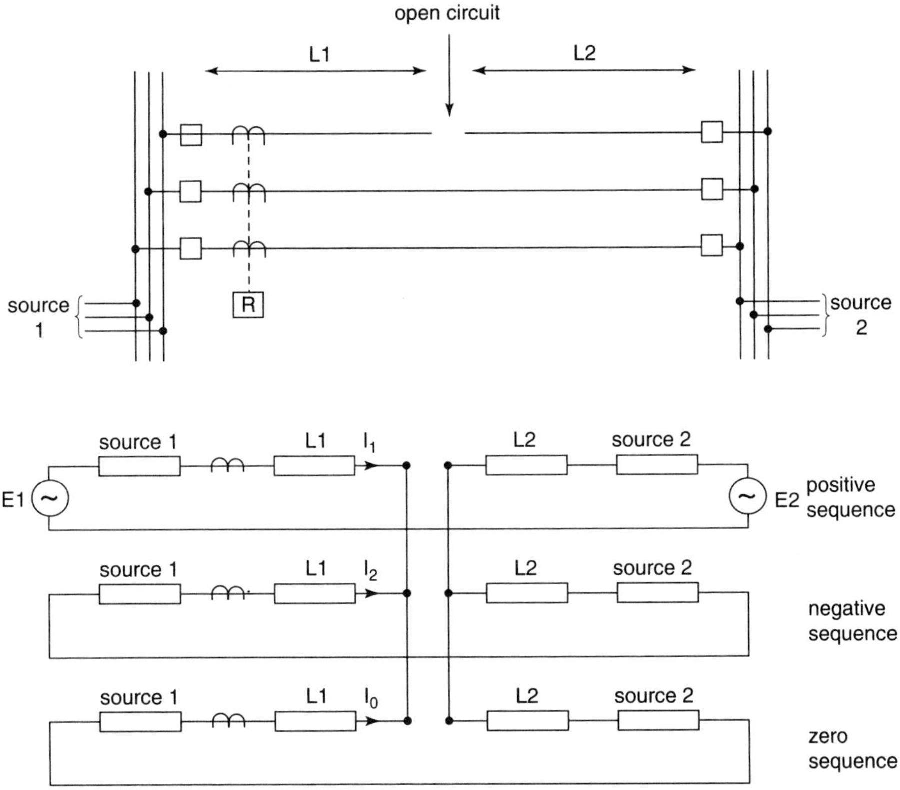

Figure 13.22 Current distribution arising from an open circuit

MAINTENANCE

13.29 Maintenance tests

These would generally be a repeat of the commissioning secondary-injection tests.

FAULT INVESTIGATIONS

13.30 Open circuit

An open-circuit phase conductor on an interconnected system may result in operation of earth-fault relays. Figure 13.22 shows (without proof) the resulting sequence network; as can be observed, the amplitude of zero-sequence current is influenced by the shunt effect of the negative-sequence network. Following tripping of a circuit breaker by such a fault, examination of a fault-recorder trace will tend to show no significant fault-current amplitude, thereby implying the advent of an open-circuit condition. It is worth noting that the open circuit may be due to one phase of a circuit breaker which failed to close when last switched, the effect of which is not apparent until the load current becomes of sufficient magnitude.

Chapter 14
Distance protection

14.1 Introduction

Distance protection was first applied to electrical power systems in the early 1920s and ever since has been an important design of feeder protection. It is a comparatively expensive protection, since it requires inputs from both current and voltage transformers, but it has the advantage of both flexibility of application and the provision of a backup-protection feature. It is used very extensively to protect 400, 275 and 132 kV system feeders, and numerous installations also exist on the 33 kV system. This chapter concentrates on basic principles, application, commissioning and maintenance requirements.

THEORY

14.2 Basic principles

The name 'distance protection' is derived from the fact that the impedance of the line is measured, and this in turn is directly proportional to line length. Thus the relay, in effect, measures the distance to a fault. Figure 14.1 illustrates the basic circuitry. The relay has inputs of three-phase voltage and current (only single phase is shown, by way of example) and compares the fault impedance Z_L with the relay-setting impedance Z_R.

14.3 Comparators

The device within the distance relay which measures the feeder impedance and compares it with the setting impedance is termed the comparator. Two basic types exist: the amplitude comparator and the phase-angle comparator. With reference to Figure 14.2, the amplitude comparator compares the amplitude of two phasors

$$S_o \text{ and } S_r$$

$$\text{and when } |S_o| > |S_r|$$

the comparator produces an output. Amplitude comparators were used in the early types of distance protection, when electromagnetic-relay designs predominated. With the advent of solid-state designs, dating from the early 1970s, the phase-angle comparator has predominated; and as shown in Figure 14.3, it functions as follows.

Let phasors S_1 and S_2 be the comparator inputs. Let θ=the angle by which S_1 leads S_2. The comparator operates when S_1 leads S_2 by less than ±90°, i.e. operation when

$$-\frac{\pi}{2} \leqslant \theta \leqslant \frac{\pi}{2}$$

The following sections will assume the use of phase-angle comparators to determine the various distance-relay characteristics.

14.4 Complex diagram

The impedance of an electrical power circuit is a complex quantity, expressed in the form $Z=R+jX$, where Z=impedance, R=resistance and jX=inductive reactance. These phasor quantities are best illustrated on a resistance–reactance diagram, e.g. Figure 14.5*a*. Such a diagram facilitates the illustration of the various characteristics which define the boundary of operation of distance-protection comparators.

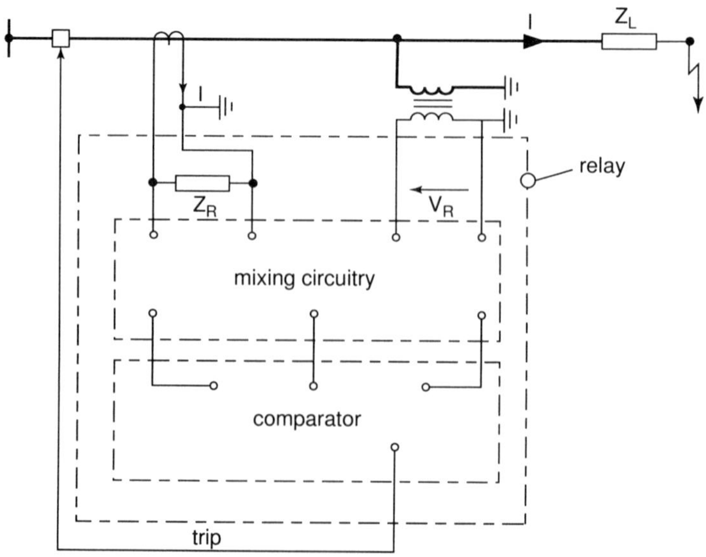

Figure 14.1 Basic principles of distance protection

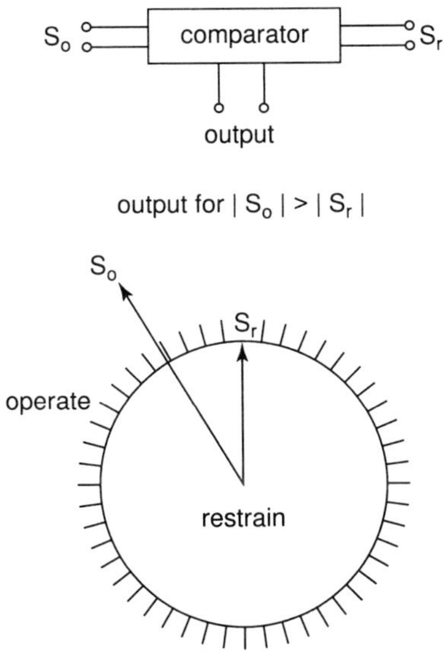

Figure 14.2 Amplitude comparator

14.5 Directional characteristics

From Figure 14.1, a directional characteristic is obtained by making

$$S_1 = V_R \qquad S_2 = IZ_R$$

where V_R is the restraint voltage. Dividing by I, we obtain

$$S_1 = \frac{V_R}{I} = Z_L \qquad S_2 = Z_R$$

Figure 14.4 shows the resulting characteristic. Z_R is taken as the reference quantity whose angle with respect to the R axis is typically 75°; this is known as the characteristic angle and is usually arranged to mimic the angle of the line to be protected. Z_R is also frequently termed the 'replica impedance'. The boundary of operation refers to the locus of Z_L with respect to Z_R. Values of Z_L to the right of the shaded area result in operation. Directional characteristics are unsuitable for use as distance-protection characteristics in their own right, but may form part of a composite characteristic, or assist in proving the directional feature of a polarised 'mho' characteristic (see Section 14.34).

14.6 Plain impedance characteristic

With reference again to Figure 14.1, the comparator inputs are derived as follows:

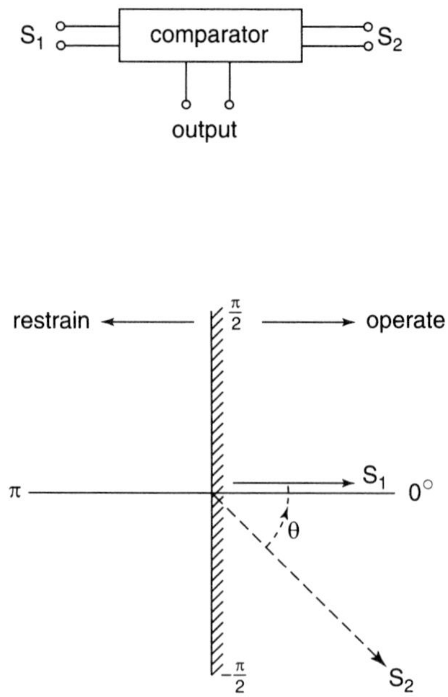

Figure 14.3 Phase-angle comparator

Let θ=angle by which S_1 leads S_2
Output for $-\pi/2<\theta<\pi/2$

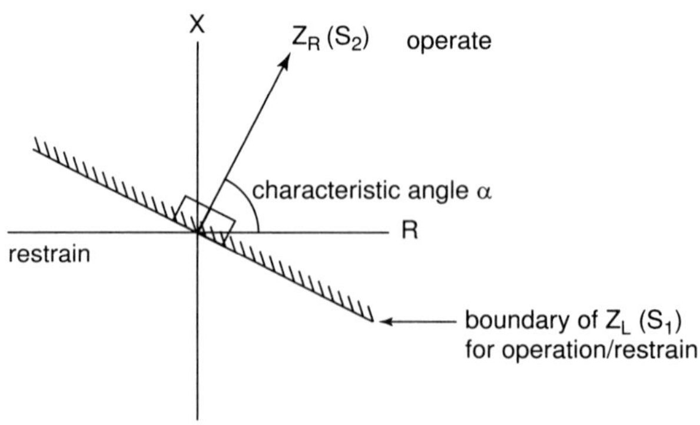

Figure 14.4 Directional characteristic

$$S_1=IZ_R-V_R \qquad S_2=IZ_R+V_R$$

Dividing by I we obtain

$$S_1=Z_R-Z_L \qquad S_2=Z_R+Z_L$$

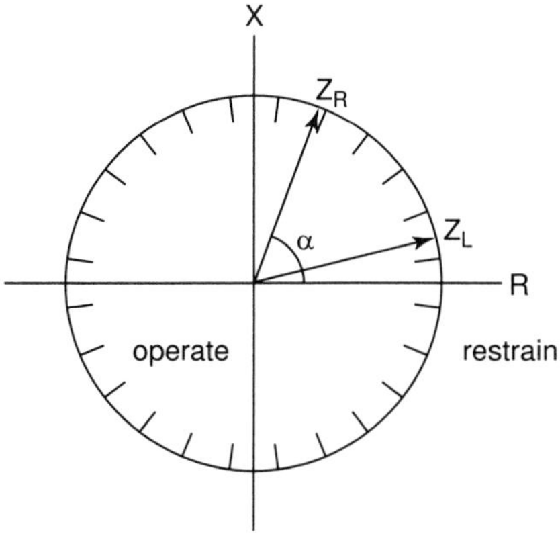

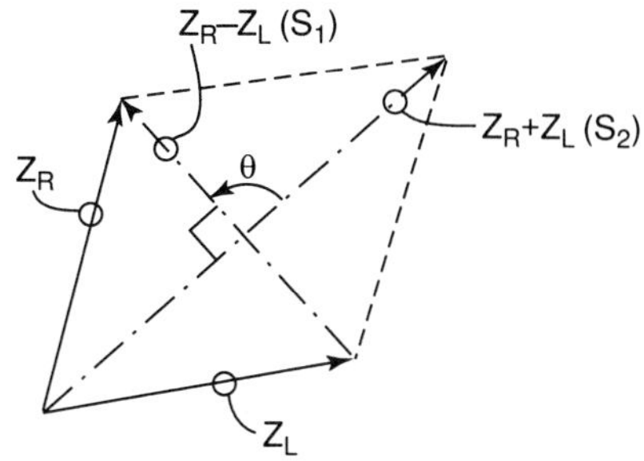

Figure 14.5 Plain-impedance characteristic

a Characteristic boundary
b Proof

The resulting characteristic boundary, as shown in Figure 14.5*a*, is a circle of radius Z_R. For the magnitude of Z_L less than Z_R, operation of the comparator results. Figure 14.5*b* illustrates the proof. Phasors $Z_R - Z_L$ and $Z_R + Z_L$ are the diagonals of a rhombus (since $Z_R = Z_L$ at the boundary), which always cross at right angles. For Z_L greater in magnitude than Z_R, θ exceeds 90° and stability results. Z_R is the relay-reach setting.

14.7 Polarised-'mho' characteristic

The plain impedance characteristic has the disadvantage that it does not contain a directional feature, i.e., it will operate for faults both in front of and behind the relaying point (origin of the X–R diagram). The polarised-'mho' characteristic, as shown in Figure 14.6, remedies this situation. The majority of the characteristic is in quadrant 1(+X, +R), with none in quadrant 3 (reverse fault). Feeder fault impedances do not, in practice, exist in quadrant 2 (+X,-R) or quadrant 4 (-X, +R). Again with reference to Figure 14.1, the comparator inputs consist of

$$S_1 = V_P$$

where V_P is known as the polarising voltage (see Section 14.10) and

$$S_2 = IZ_R - V_R$$

For the special case of $V_P = V_R$,

$$S_1 = V_R \qquad S_2 = IZ_R - V_R$$

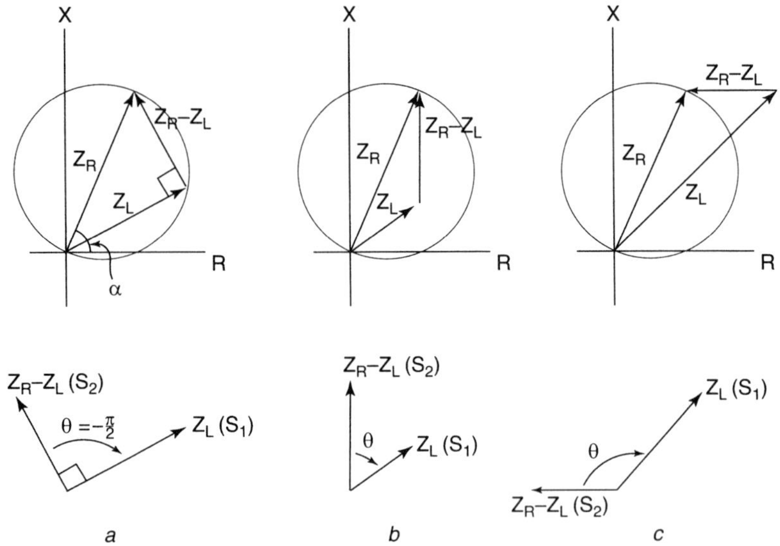

Figure 14.6 Polarised-mho characteristic

a Boundary
b Operation
c Stability

Multiply by $1/I$. Then

$$S_1 = Z_L \qquad S_2 = Z_R - Z_L$$

With reference to Figure 14.6, Z_R is the diameter of a circle the circumference of which is the boundary of operation. Fault impedance Z_L, which lies on the boundary, is at right angles to phasor $Z_R - Z_L$, since this is the angle subtended in a semicircle. Fault impedances which fall within the circle result in θ being within ±90°, resulting in comparator operation.

Fault impedances which fall outside the circle result in θ being greater than ±90°, so causing comparator restraint.

14.8 Offset-mho characteristic

The polarised-mho characteristic has the disadvantage of not operating for a fault at the relay position (i.e. the origin of the *X-R* diagram). This is demonstrated by making the V_R input, into S_1 and S_2 in Section 14.7, equal to zero. S_1 then becomes zero and the comparator cannot function. The offset-mho characteristic overcomes this problem. Again, taking the basic inputs to the relay, as shown in Figure 14.1, the comparator inputs are derived as follows:

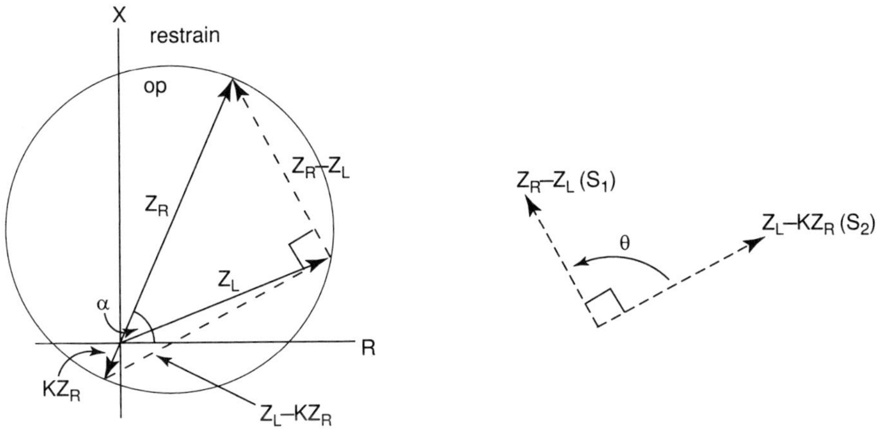

Figure 14.7 Offset-mho characteristic

$$S_1 = IZ_R - V_R \qquad S_2 = V_R - KIZ_R$$

where K is a number, less than 1, which determines the degree of offset (K is frequently -0.1)

Multiply by $1/I$. Then

$$S_1 = Z_R - Z_L \qquad S_2 = Z_L - KZ_R$$

The resulting characteristic is shown in Figure 14.7. The boundary of operation is a circle with a diameter equal to $Z_R + KZ_R$.

14.9 Practical relay

A practical composite distance-protection-relay arrangement is shown in Figure 14.8. This shows the characteristics for any one of the six permutations of comparators (R–E, Y–E, B–E, R–Y, Y–B, B–R), and it consists of three zones:

Zone 1: This zone is required to protect as much of the feeder A–B as possible. However, CT and VT errors require the relay Z_1 to be set to 80% of the impedance of A–B (see Section 14.18), to ensure that the relay at A will not operate for faults external to feeder A–B. The operating time for a zone-1 protection would typically be under 40 ms.

Zone 2: Zone 2 serves two purposes: to cover the length of A–B not covered by zone 1, and to provide backup protection to the busbars at substation B. Its setting Z_2 is usually of the order of 150% of the impedance of feeder A–B. The operating time of zone 2 must discriminate with, and therefore be longer than, a zone 1 fault clearance; it therefore has a typical time delay of the order of 0.3–0.5 s.

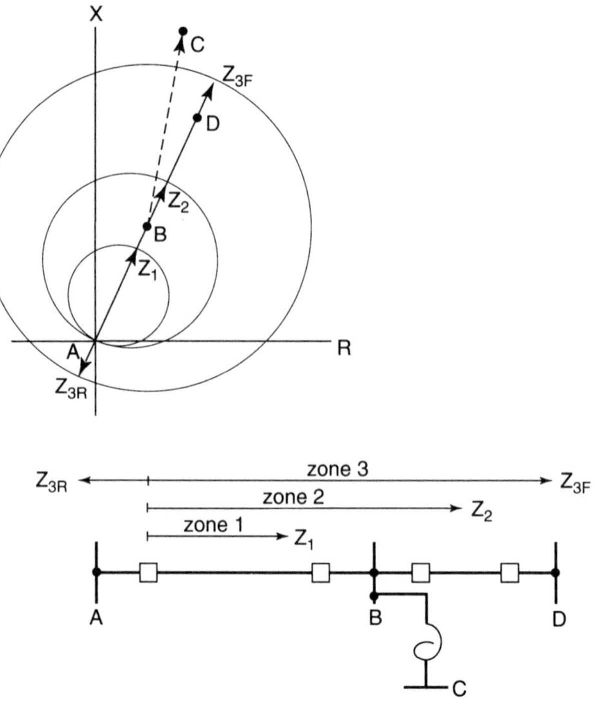

Figure 14.8 Practical relay arrangement

Zone 3: This provides backup protection to the power system in general, and therefore has a long forward reach, Z_{3F}. It also provides backup protection for reverse faults, which necessitates an offset-mho characteristic. The operating time delay is typically of the order of 0.8–1.5 s to allow other faster protections nearer to the fault to operate and disconnect the faulty plant.

Figure 12.5 illustrates the essential elements of a modern distance relay suitable for a transmission system. Each zone has six comparators, one each for the fault permutations listed above. Some older relays have the same comparators for zones 1 and 2. This is accomplished by the zone 3 comparator initiating the zone 2 timer which, if the zone 2 time elapses, (i.e. the fault is not within zone 1) changes the comparator-reach setting from zone 1 to zone 2. Operation of the zone 2 comparator now causes an immediate trip output signal.

14.10 Polarisation

Section 14.7 stated that the S_1 input to the polarised comparator was V_P, which was assumed to be equal to V_R, the restraint voltage. This is an approximation. In practice V_P tends to comprise the following:

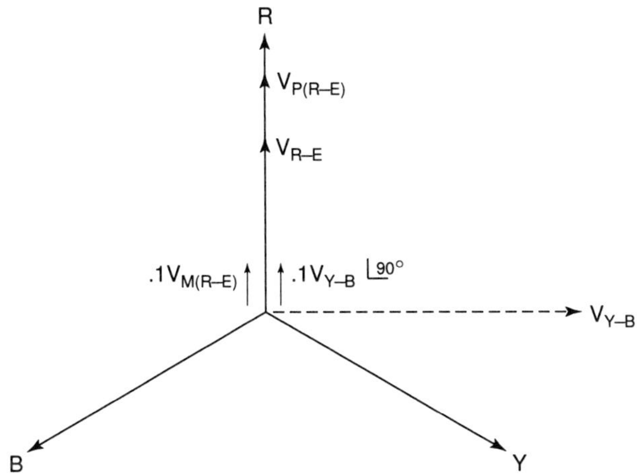

Figure 14.9 Polarising voltage for R–E fault comparator

V_P	$= V_R$	$+ V_{Pol}$	$+ V_M$
polarising voltage	= restraint	+ cross-polarised	+ memory

V_{Pol} is known as the crosspolarised or sound-phase-polarised voltage. It consists of a small percentage of a healthy (nonfaulted) phase voltage which is phase shifted to be in phase with V_R. This ensures that, when a phase-to-earth or phase-to-phase fault takes place close up to the relay point, causing V_R to fall to zero, V_P does not in turn become zero. This is important because, if V_P and hence S_1 falls to zero, the comparator cannot function. V_{Pol} also has an influence on the characteristic shape, as discussed in Section 14.11.

If the fault is a three-phase fault close-up to the relay point, then, since all three phase voltages drop to zero, V_{Pol} becomes zero. To avoid V_P, in turn, becoming zero, some relays utilise a quantity V_M, a small percentage of the prefault voltage which is phase-shifted to be in phase with V_R, and time delayed to maintain a voltage for a few cycles after fault inception, to provide a voltage to enable the comparator to operate. This feature is frequently termed a 'memory feature'.

A typical example of how the polarising quantity is made up for a red-earth-fault comparator is as follows:

$$V_{P(R-E)} = V_{R-E} + 0.1\ V_{Y-B}\,\underline{/90^\circ} + 0.1\ V_{M(R-E)}$$

The phasor diagram is shown in Figure 14.9. Note that some relays omit the memory feature and accept that the relay will fail to operate for a close-up three-phase fault. In such instances, other types of relay will have to cover this shortfall.

14.11 Source impedance

Source impedance has a two-fold effect on distance-relay performance:

(i) It changes the characteristic of the polarised-mho relay, as will shortly be described.
(ii) It affects the accuracy of the relay, as described in Section 14.13.

The characteristic described in Section 14.7, where V_P is made equal to V_R, should be properly termed the 'mho' comparator, because it contains no crosspolarising voltage. The effect of the crosspolarising voltage is to change the comparator characteristic significantly at all fault angles except the characteristic angle. Consider again the inputs to the polarised-mho comparator, but this time utilising V_P as described in Section 14.10 (but leaving out the memory voltage to simplify the analysis). With reference to Figure 14.10, let

$$S_1 = V_R + V_{Pol} \qquad S_2 = IZ_R - V_R$$

Now let

$$V_{Pol} = KV_{HF}$$

where V_{HF} is the healthy-phase voltage and K is a multiplier which determines the magnitude of V_{Pol} (typically$=0.1\ V_R$), and causes KV_{HF} to be in phase with V_R. Thus

$$S_1 = V_R + KV_{HF} \qquad S_2 = IZ_R - V_R$$

Now S_1 can be further developed such that

$$KV_{HF} = K_1 V$$

where V=source voltage in Figure 14.10 and K_1=a multiplier with a typical magnitude, again, of 0.1. Thus

$$S_1 = V_R + K_1 V \qquad S_2 = IZ_R - V_R$$

Now, from Figure 14.10

$$V_R = \frac{Z_L}{Z_L + Z_S} V \text{ and } I = \frac{V}{Z_L + Z_s}$$

so

$$S_1 = \frac{Z_L}{(Z_L + Z_S)} V + \frac{(Z_L + Z_S)}{(Z_L + Z_S)} K_1 V \qquad S_2 = \frac{V}{(Z_L + Z_S)} Z_R - \frac{Z_L}{(Z_L + Z_S)} V$$

Multiply by $\dfrac{(Z_L + Z_S)}{V}$

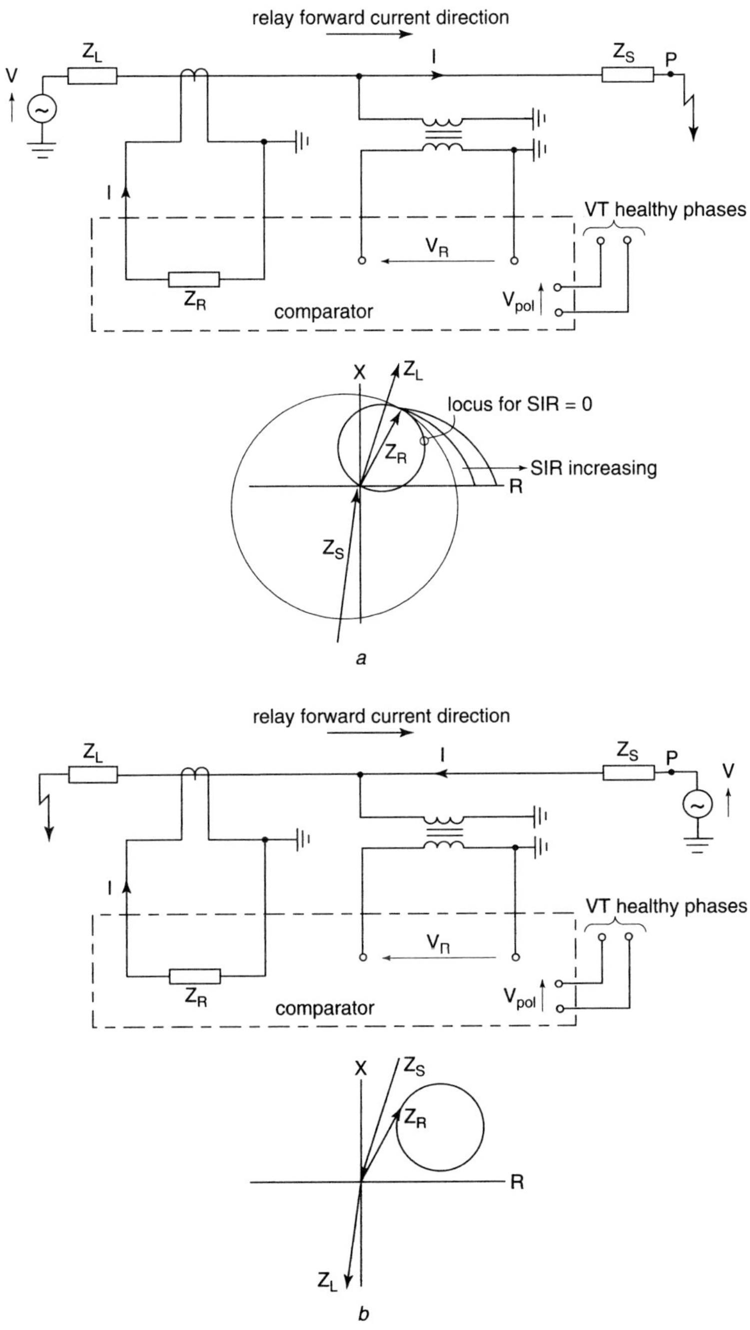

Figure 14.10 Polarised-mho characteristics for forward and reverse faults

a Fault in forward direction
b Fault in reverse direction

Then $S_1 = Z_L + (Z_L + Z_S) K_1$ and $S_2 = Z_R - Z_L$

or $S_1 = Z_L(1 + K_1) + K_1 Z_S$ and $S_2 = Z_R - Z_L$

Note that the above expression differs from that in Section 14.7 in that it now includes the source impedance Z_S. It is this factor which modifies the characteristic. The comparator defines the boundary of operation of a variable fault impedance Z_L, with reference to a fixed setting impedance Z_R, and now, additionally, a fixed source impedance Z_S, which has two possible values depending on whether the source Z_S is behind the relay (fault in the forward direction) or in front of the relay (fault in the reverse direction).

From the characteristic shown in Figure 14.10*a*, it can be seen that the polarised-mho characteristic has distorted to become an offset-mho characteristic encompassing the origin; thus it may seem that it will operate for faults behind the relaying point. This is not so. Figure 14.10*a* is the characteristic for fault impedances in front of the relaying point, the shape of which is caused by Z_S being behind the relaying point. Any fault impedance in the third quadrant ($-X$, $-R$) relates to an impedance in front of the relay with values of $-X$, $-R$. In practice a $-R$ cannot exist in front of the relay (i.e. we cannot purchase a negative resistance). In fact, in practice, only quadrant 1, with fault-impedance values of $+X$, $+R$, can exist; thus quadrants 2, 3 and 4 represent hypothetical fault impedances. Note that load currents can produce impedances which appear to have a $-R$ component but, generally, such impedances will have a magnitude far in excess of the boundary of operation of the characteristic.

Figure 14.10*b* illustrates the characteristic for faults behind the relaying point, the shape of which is caused by Z_S being in front of the relaying point. Note that the fault impedances we are considering here are those which are behind the relaying point, as seen by a relay looking in the forward direction. Thus impedances which would fall within the characteristic, and cause operation, would be capacitive. However, feeder fault impedances tend to be inductive and not capacitive, and therefore for most practical applications relay operations for reverse faults will not occur. Note that some literature, when analysing this characteristic (i.e. source impedance in front of relay point) assumes the forward direction to be behind the relaying point; this then causes Z_R in the comparator inputs to be negative, i.e.

$$S_1 = Z_L(1 + K_1) + K_1 Z_S \qquad S_2 = -Z_R - Z_L$$

This approach causes the characteristic shown in quadrant 1 of Figure 14.10*b* to fall into quadrant 3. In this instance, the $-X$ axis now implies a capacitive impedance, so the same conclusions can be drawn in terms of the relay not operating for faults behind the relaying point.

Note that, as $Z_S \rightarrow 0$, the two characteristics shown in Figure 14.10*a* and *b* move closer to that shown in Figure 14.6, and when $Z_S = 0$ the two merge into one with the boundary passing through the origin.

A question which may reasonably be asked, in connection with commissioning and maintenance, is, which of the two characteristics would operate, as restraint voltage is reduced, if the comparator is secondary injected with voltage and current. The comparator has, of course, no knowledge of Z_S and Z_L, but the input quantities V_R, I, V_{Pol} can be interpreted within the comparator in terms of Z_S and Z_L. Except where the two characteristics may intercept, values of Z_S and Z_L in both the forward or reverse directions are defined by unique values of V_R, I, V_{Pol}. Thus the characteristic adopted is dependent on the quantities injected.

No attempt has been made to describe the mathematical derivation of the characteristics, since the complexity is beyond the scope of this text, and beyond the needs of commissioning and maintenance engineers. For those wishing to acquire a deeper understanding, reference should be made to the paper by L. M. Wedepohl (see Bibliography).

The only characteristic of practical interest is therefore quadrant $1(+X, +R)$ and the effect of increasing source impedance on this quadrant is further to distort the characteristic, as illustrated in Figure 14.10*a*. It is usual to describe the effect of source impedance in terms of a quantity called 'system-impedance ratio' (SIR). This is defined as:

$$\frac{\text{source impedance}}{\text{line impedance at relay reach setting}} = \frac{Z_S}{Z_R}$$

As can be seen, the effect of increasing SIR (i.e. increasing source impedance) is to increase the distortion of the characteristic in the direction of the resistive axis. At first sight, this might seem disadvantageous, but this is not so; this type of characteristic has the advantage of assisting comparator operation for faults with significant arc, or fault, resistance. It is particularly advantageous for short lines where the ratio of arc resistance to line impedance is high.

Distortion of the characteristic is due to the fact that V_R and V_{Pol} are only arranged to be in-phase at the characteristic angle (at which point there is no distortion). As fault impedance becomes more resistive, a phase angle arises between V_R and V_{Pol}. In addition, as SIR increases, V_R becomes smaller in magnitude and thus V_{Pol} has a greater influence on the characteristic. These factors result in the characteristic extending into the resistive axis. Note that, for a three-phase fault, both V_R and V_{Pol} are equally affected and there is therefore virtually no distortion.

14.12 Minimum line length

When distance relays are called on to operate with very small voltages and currents, accuracy is impaired. This arises because all relays have a small inherent restraint bias, equivalent to friction in an electromechanical relay, which must be overcome to achieve operation. At small voltages and currents,

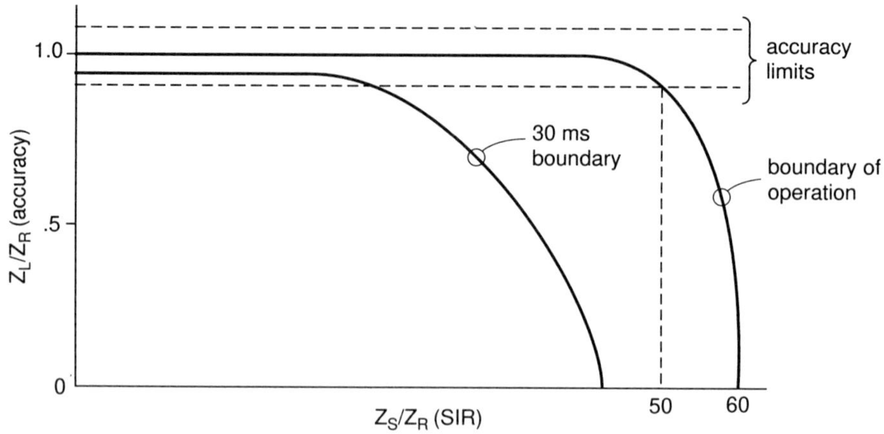

Figure 14.11 Effect of source impedance on operating characteristic

this restraint bias has a greater influence on the operating characteristic, which is greatest for the instance of a three-phase fault, when the polarising quantity of the polarised-mho comparator is also depressed.

Now if the source impedance Z_S increases, both the relay voltage and current decrease. It is instructive to demonstrate the impact of source impedance by plotting relay accuracy Z_L/Z_R (where Z_L is the line impedance to cause operation, and Z_R is the relay-reach-setting impedance) against system impedance ratio Z_S/Z_R, as illustrated in Figure 14.11. As can be seen at high values of SIR (i.e. low values of restraint voltage and fault current) the line impedance, to cause operation, must reduce below the setting impedance. This characteristic enables the minimum length of line to which a distance protection can be applied to be determined. If the acceptable limits of accuracy are say ±5%, it can be seen that these are exceeded with an SIR of about 50. Thus at limit of accuracy

$$\frac{Z_S}{Z_R} = 50 \quad \text{or} \quad Z_R = \frac{Z_S}{50}$$

If $Z_S = 2\%$ impedance (equivalent to 5000 MVA source infeed at 400 kV)

then $$Z_R = \frac{2\%}{50} = 0.04\%$$

If Z_R is set to, say, 80% line impedance, then line percent impedance

$$= \frac{0.04\%}{0.80} = 0.05\%$$

This is equivalent to about 3 km of 400 kV line.

The minimum voltage sensitivity of the relay for an SIR of 50 is

$$\frac{Z_R}{Z_S+Z_R}V=\frac{Z_R}{51Z_R}63.5=1.2\text{ V}$$

Figure 14.11 also illustrates the characteristic for a constant operating time of 30 ms. Again it can be seen that, as SIR increases, the magnitude of Z_L must decrease to maintain constant operating time.

In practice, those circuits in the UK power system to which distance protection is applied, tend, in the main, to exhibit SIRs of below three, with many applications below one. Higher SIRs can be exhibited for the instance of a single transformer feeding a short feeder, where SIRs as high as 15 may be experienced (e.g. a 275/132 kV transformer feeding 5 km of 132 kV overhead line). Given that modern distance protections are capable of maintaining accuracy up to SIRs of at least 50, it can be applied successfully to all but the very shortest of feeders.

When secondary injecting distance protections, it is unusual to inject voltage and current of such a low magnitude as to push the apparent SIR to a value which exceeds the relay accuracy limits. However, if the tests do indicate that the relay is out of limits it is worth bearing the effect of high SIR in mind.

14.13 Impedance measurement

The preceding sections have discussed the fault impedance Z_L in terms of a simple V/I ratio. However, it was explained in Chapter 3 that the line impedance comprises the three-sequence components of impedance Z_1, Z_2, Z_0 and thus comparator operation needs to be interpreted in the context of these quantities.

The difficulty with a simple V/I impedance measurement is illustrated in Figure 14.12 for the instance of a phase-to-earth fault. If we measure V/I we obtain

$$Z=\frac{V_R}{I_R}=\frac{V_1+V_2+V_0}{I_1+I_2+I_0}$$

i.e.

$$Z=\frac{I_1Z_1+I_2Z_2+I_0Z_0+(V_{FI}+V_{F2}+V_{FO})}{I_1+I_2+I_0}$$

Now in the fault $I_{F1}=I_{F2}=I_{F0}$, and also $V_{F1}+V_{F2}+V_{F0}=0$; but in the line, although I_1 will always be the same as I_2 (since positive-sequence impedance always equals negative-sequence impedance), it may not be the same as I_0

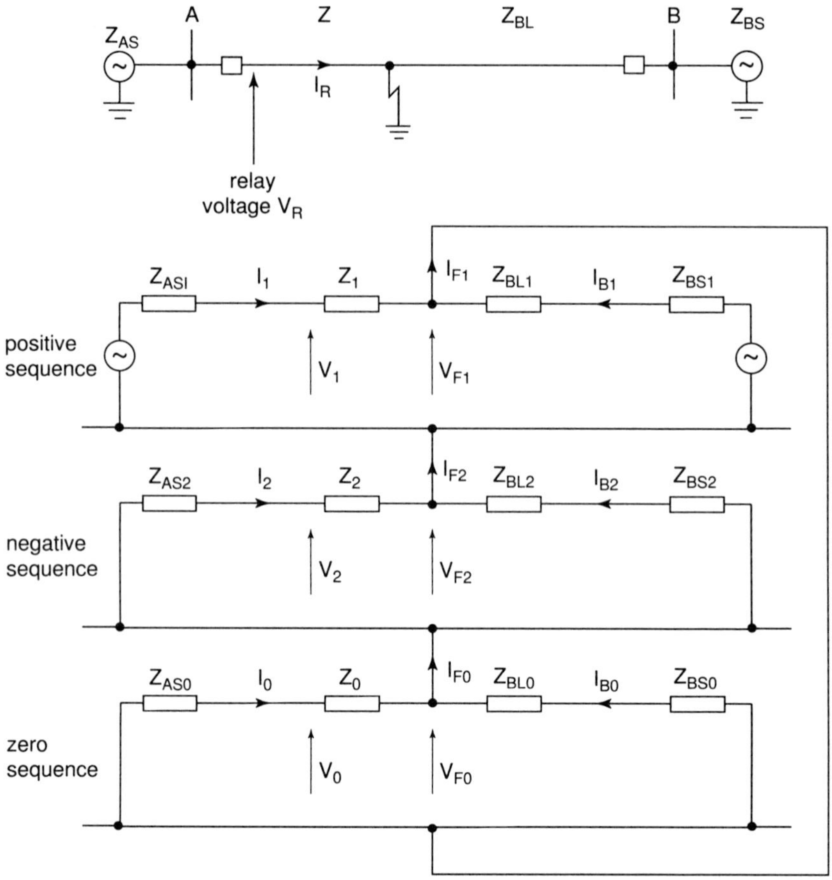

Figure 14.12 Circuit for analysing a single-phase-to-earth fault

(since this depends on the value of the zero-sequence impedances on either side of the fault, and these are dependent on the number of earthed neutrals). Now

$$Z_1 = Z_2,\ I_1 = I_2,\ \text{and let } I_0 = NI_1$$

Then

$$Z = \frac{I_1(2Z_1 + NZ_0)}{I_1(2+N)}$$

i.e.

$$Z = \frac{2Z_1 + NZ_0}{2+N}$$

Now, to make a consistent assessment of fault position, the impedance measured must be proportional to the distance between the relaying point and the fault. The above ratio is therefore clearly unsatisfactory, since the measured impedance is dependent on the relative values of the positive- and zero-sequence impedances.

To simplify matters, with the objective of measuring an impedance which is proportional to distance, distance relays are designed to measure the positive impedance to the fault. To achieve this for earth faults, a technique termed 'residual compensation', and for phase-to-phase faults a technique termed 'phase-fault compensation', is employed.

14.14 Residual compensation

With reference to Figure 14.12 and considering a red-phase-to-earth fault, let the impedance to be measured equal Z_1, the positive-sequence impedance between the relay point and the fault, where

$$Z_1 = \frac{V_R}{I_R + I_C} \tag{14.1}$$

where I_c is a compensating current (to be determined) and V_R and I_R are red-phase quantities, i.e.

$$Z_1 = \frac{V_1 + V_2 + V_0}{I_R + I_C}$$

so

$$Z_1 = \frac{I_1 Z_1 + I_2 Z_2 + I_0 Z_0 + (V_{FI} + V_{F2} + V_{FO})}{I_R + I_C}$$

Now

$$V_{F1} + V_{F2} + V_{F0} = 0$$

and

$$Z_1 = Z_2$$

Thus

$$Z_1 = \frac{2I_1 Z_1 + I_0 Z_0}{I_R + I_C}$$

But

$$I_R - I_0 = I_1 + I_2 = 2I_1 \qquad (\text{since } I_1 = I_2)$$

Therefore

$$Z_1 = \frac{(I_R - I_0) Z_1 + I_0 Z_0}{I_R + I_C}$$

or

$$I_C = \frac{I_0 (Z_0 - Z_1)}{Z_1}$$

$$= I_0 \left(\frac{Z_0}{Z_1} - 1 \right)$$

Now eqn. 2.5 gives

$$I_0 = \frac{I_R + I_Y + I_B}{3}$$

Thus

$$I_C = \frac{I_R + I_Y + I_B}{3} \left(\frac{Z_0}{Z_1} - 1 \right)$$

Thus, inserting I_C into eqn. 14.1,

$$Z_1 = \frac{V_R}{I_R + \frac{I_R + I_Y + I_B}{3} \left(\frac{Z_0}{Z_1} - 1 \right)} \qquad (14.2)$$

or

$$Z_1 = \frac{V_R}{I_R + I_0 \left(\frac{Z_0}{Z_1} - 1 \right)} \qquad (14.3)$$

The quantity $1/3\ (Z_0/Z_1 - 1)$ is termed the 'residual compensation', and is usually a setting provided on the relay. Note that this setting facilitates the impedance-measuring process, which mut not be confused with the relay-reach setting (i.e. Z_R, as shown in Figure 14.1).

The basic CT circuitry is illustrated in Figure 14.13. The transformer arrangement provides the residual compensation. For example, if the ratio of $Z_0/Z_1 = 3$, the residual compensation is 0.67 (a typical value).

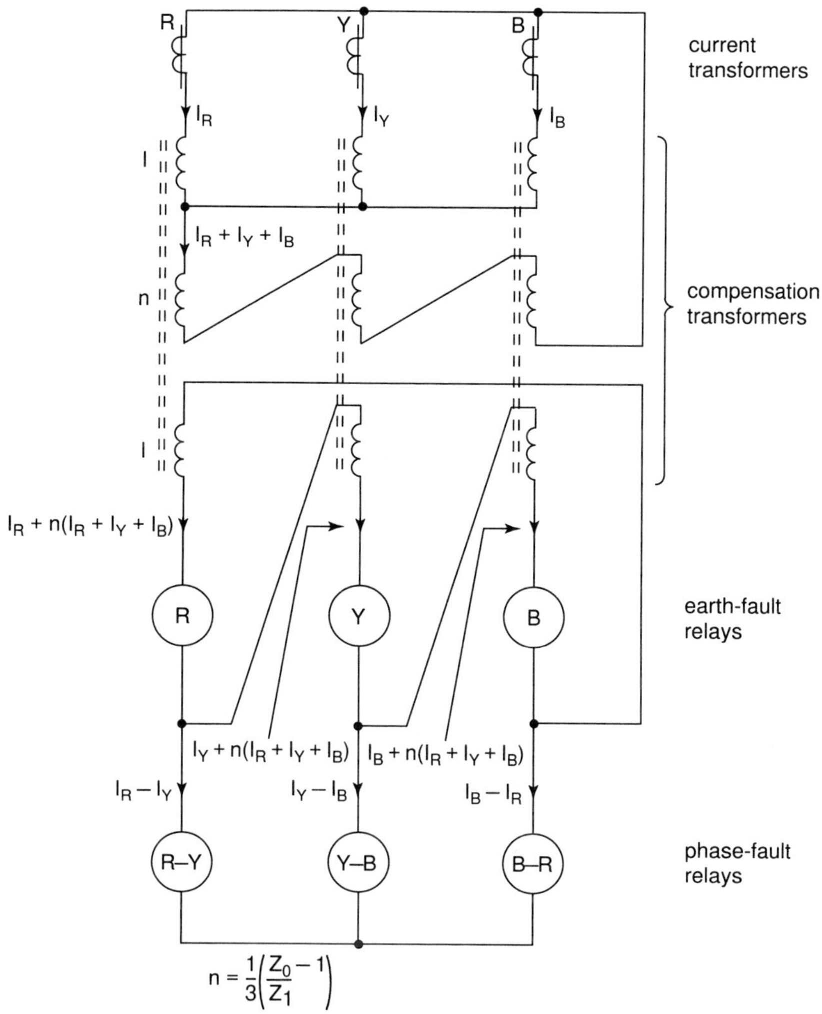

Figure 14.13 Basic CT circuitry

14.15 Phase-fault compensation

The quantities injected into, and the impedance measured by, the Y–B-phase fault comparator are arranged to be

$$Z = \frac{V_Y - V_B}{I_Y - I_B} \tag{14.4}$$

where V_Y and V_B are the voltages at the relay point, i.e.

$$Z = \frac{(V_{Y1} + V_{Y2}) - (V_{B1} + V_{B2})}{(I_{Y1} + I_{Y2}) - (I_{B1} + I_{B2})}$$

$$= \frac{(a^2 V_{R1} + aV_{R2}) - (aV_{R1} + a^2 V_{R2})}{(a^2 I_{R1} + aI_{R2}) - (aI_{R1} + a^2 I_{R2})}$$

$$= \frac{(a^2 I_{R1} Z_1 + aI_{R2} Z_2) - (aI_{R1} Z_1 + a^2 I_{R2} Z_2)}{(a^2 I_{R1} + aI_{R2}) - (aI_{R1} + a^2 I_{R2})}$$

Now $I_{R1} = -I_{R2}$ for a phase-to-phase fault, and $Z_1 = Z_2$. Thus

$$Z = \frac{I_{R1} Z_1 \{(a^2 - a) - (a - a^2)\}}{I_{R1}\{(a^2 - a) - (a - a^2)\}} = \frac{2I_{R1} Z_1 (a^2 - a)}{2I_{RI}(a^2 - a)}$$

so

$$Z = Z_1$$

The CT circuitry to achieve this is again shown in Figure 14.13.

It is instructive to note which comparators operate for a three-phase fault. In this situation only positive sequence quantities flow, which, if inserted into eqns. 14.3 and 14.4, produce in each instance Z_1. Thus all six comparators operate.

14.16 Line-impedance variation

The value of the positive-, negative- and zero-sequence impedance of a double-circuit overhead line feeder is not fixed, but dependent on the magnitude and direction of current in the parallel circuit. The effect of current in the parallel circuit is to cause mutual coupling (inductance) from one circuit to the other. From Figure 14.14, the resulting impedance variation for 100 km of a typical 400 kV overhead line is given in Table 14.1. As can be seen, the variation of zero-sequence impedance is much greater than that of positive-sequence (note that negative-sequence impedance is always equal to positive-sequence impedance). This arises because the mutual fluxes from positive-sequence currents tend to cancel, whereas the mutual fluxes from zero-sequence currents are additive, since they are in-phase quantities. Note that the terms Z_{1DC}, Z_{0DC} merely refer to the sequence impedances of a double-circuit line with the currents of both circuits flowing in the same direction, whereas Z_{1CD}, Z_{0CD} refer to currents flowing in opposing directions.

14.17 Zone 1-setting policy

The question arises of the value of positive-sequence impedance to which to set the relay-reach setting, and the value of zero- and positive-sequence

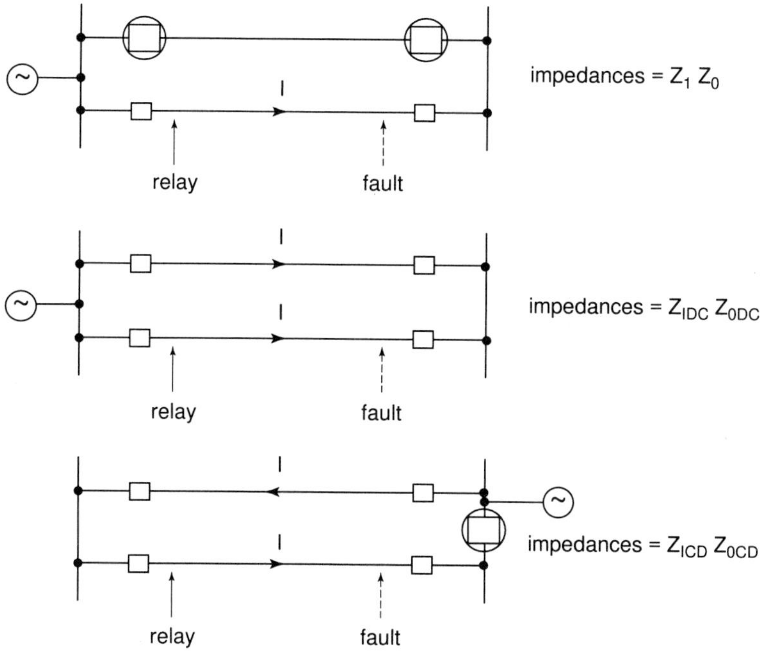

Figure 14.14 Line-impedance variation

Table 14.1 Line-impedance variation

Positive-sequence impedance (%)	Zero-sequence impedance (%)
$Z_1 = 2.1$	$Z_0 = 5.3$
$Z_{1DC} = 2.0$	$Z_{0DC} = 8.1$
$Z_{1CD} = 2.2$	$Z_{0CD} = 2.5$

All impedances are on a 100 MVA base.

impedance to which to set the residual compensation setting. The criterion to be adopted is that of achieving maximum cover of the protected feeder, while not suffering overreach resulting in tripping for faults beyond the protected feeder. The latter can arise if, for example, we set to Z_{1CD} and a Z_{1DC} type of fault occurs. In addition CT and VT errors must be taken into account.

Investigation has shown that the following best satisfy this criterion:

Zone 1 E/F = 65% Z_1 line length

Zone 1 P/F=80% Z_{1DC} line length

Residual compensation $=\frac{1}{3}\left(\frac{Z_0}{Z_1}-1\right)$; this is a common setting for all zones.

The 65% Z_1 figure stated for earth faults is very much a worst theoretical figure, and thus some setting engineers adopt a higher compromise setting which may be as high as 80% Z_1.

14.18 Zone 2-setting policy

Zone 2 is usually required to detect, and provide backup protection for, a busbar fault on the busbars at the far end of the feeder (i.e. the whole of the feeder is covered). From Section 14.18, if a setting of 65% Z_1 line length is required to ensure nonoperation for faults just beyond the feeder, then to ensure operation for a busbar fault (which is just beyond the feeder), a setting of 1/65% is required, i.e. approximately 150% line length. For commonality of settings, then,

Zone 2 E/F=150% Z_1 line length
Zone 2 P/F=150% Z_1 line length

Refinements may be required to the above basic setting. For example, the zone 2 setting should not exceed the zone 2 setting of the shortest feeder out of the facing substation, as otherwise discrimination will be lost.

The zone 2 time delay must discriminate with fault clearance times for faults on other circuits. Typical zone 2 time delays are in the range 300–500 ms.

14.19 Zone 3-setting policy

The function of zone 3 is usually to provide backup protection for faults which are remote from the relay, e.g. faults at the end of feeders out of the next substation. This is shown diagrammatically in Figure 14.8. A frequent constraint on the zone 3 setting is that it should not operate for faults on the LV side of a transformer (e.g. point C in Figure 14.8). In addition, load currents can sometimes infringe the circular zone 3 characteristic, and thus may necessitate a lemon- or peanut-shaped characteristic, as shown in Figure 14.15.

The zone 3 time delay usually has to discriminate with zone 2 or IDMT-relay clearances. This usually necessitates time delays in the range 0.8–1.5 s.

When a circuit is energised onto a close-up three-phase fault, the zone 1 polarised-mho comparators will not operate, since all three-phase voltages are at zero and there is no memory voltage to feed into the comparator. To

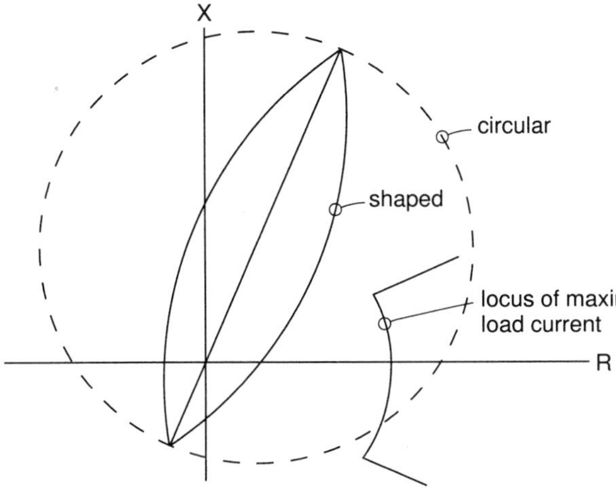

Figure 14.15 Shaped zone-3 characteristic

ensure an instantaneous fault clearance for such a fault, the offset-mho zone 3-comparator time delay is usually bypassed for the first 200 ms or so after energisation. This is known as the 'switch on to fault' feature, as illustrated in Figure 12.5.

14.20 Primary to secondary ohms

Distance protection is supplied from voltage and current transformers. The impedance seen by the distance relay, termed the 'secondary ohms', therefore differs from the primary system ohms. Consider primary impedance

$$Z_P = \frac{V_P}{I_P}$$

where the P suffix indicates a primary quantity.

If the CT ratio $= N_C$ (e.g. 2000/1)
and the VT ratio $= N_V$ (e.g. 400 kV/110 V)

then current into the relay $= \dfrac{I_P}{N_c}$

and the voltage supplied to the relay $= \dfrac{V_P}{N_V}$

Then the impedance seen by the relay $Z_S = \dfrac{V_P}{N_V} \times \dfrac{N_C}{I_P} = \dfrac{N_C}{N_V} Z_P$

where the S suffix indicates a secondary quantity.

For the ratios stated above, $Z_S=0.55\ Z_P$

14.21 Setting example

A distance protection is required to protect 80 km of 400 kV overhead line, which consists of:

$$Z_1 = 28\ \Omega$$
$$Z_{1DC} = 26.5\ \Omega$$
$$Z_0 = 80\ \Omega$$
$$\text{CT ratio} = 2000/1$$
$$\text{VT ratio} = 400\ \text{kV}/110\ \text{V}$$

The settings are determined as follows:

$$\text{The CT/VT ratio} = \frac{110}{400\ 000} \times 2000 = 0.55$$

$$\text{Zone 1 E/F} = \text{CT/VT ratio} \times 0.65 \times Z_1$$
$$= 0.55 \times 0.65 \times 28 = 10\ \Omega$$

$$\text{Zone 1 P/F} = \text{CT/VT ratio} \times 0.8 \times Z_{1DC}$$
$$= 0.55 \times 0.8 \times 26.5 = 11.7\ \Omega$$

$$\text{Zone 2 E/F} = \text{CT/VT ratio} \times 1.5 \times Z_1$$
$$= 0.55 \times 1.5 \times 28 = 23.1\ \Omega$$

Zone 2 P/F = Zone 2 E/F = 23.1 Ω

Zone 3 E/F = CT/VT ratio × (primary ohms as determined in Section 14.20)

Zone 3 P/F = as Zone 3 E/F

$$\text{Residual compensation} = \frac{1}{3}\left(\frac{Z_0}{Z_1} - 1\right)$$

$$= \frac{1}{3}\left(\frac{80}{28} - 1\right) = 0.62$$

14.22 Acceleration schemes

Figure 12.5 illustrates the basic distance-protection scheme, to which circuitry necessary for an accelerated trip has been added. For most applications the

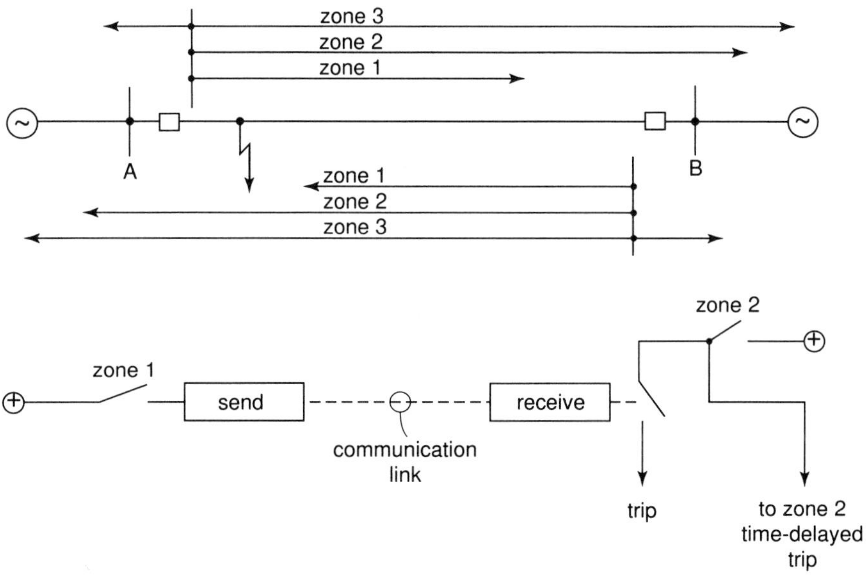

Figure 14.16 Accelerated distance protection

basic scheme is unacceptable, since it results in faults between the limit of the zone 1 reach and the end of the feeder being cleared by zone 2, which is a time-delayed trip. This limitation can be overcome by use of an acceleration scheme. Figure 14.16 illustrates the acceleration circuitry required for both ends of the feeder. The fault is detected by zone 1 at end A and a signal is sent to end B. The fault is also detected by a zone 2 comparator at end B which closes a contact in series with the received signal to trip end-B circuit breaker. The advantage of an acceleration scheme is that it is relatively fast. With a typical zone 1 operating time of 25 ms (for transmission-system relays) and a typical acceleration send-to-receiver time of 15 ms, the total accelerated-protection-operating time would be 40 ms. A disadvantage is that if the line stands energised from one end only, and a fault occurs beyond zone 1, the acceleration feature is lost. It is also difficult to apply to three-ended feeders.

14.23 Blocked distance protection

Blocked-distance-protection circuitry is illustrated in Figure 14.17. The circuitry shown is in addition to the basic scheme circuitry shown in Figure 12.5. A necessary requirement of blocked distance protection is a reverse-zone comparator. Some schemes utilise the zone 3 comparator as the reverse-zone; other schemes utilise a separate comparator.

Consider the external fault at position X in Figure 14.17. It is detected by a zone 2 comparator at end B, which starts the delayed-operate timer TD. It

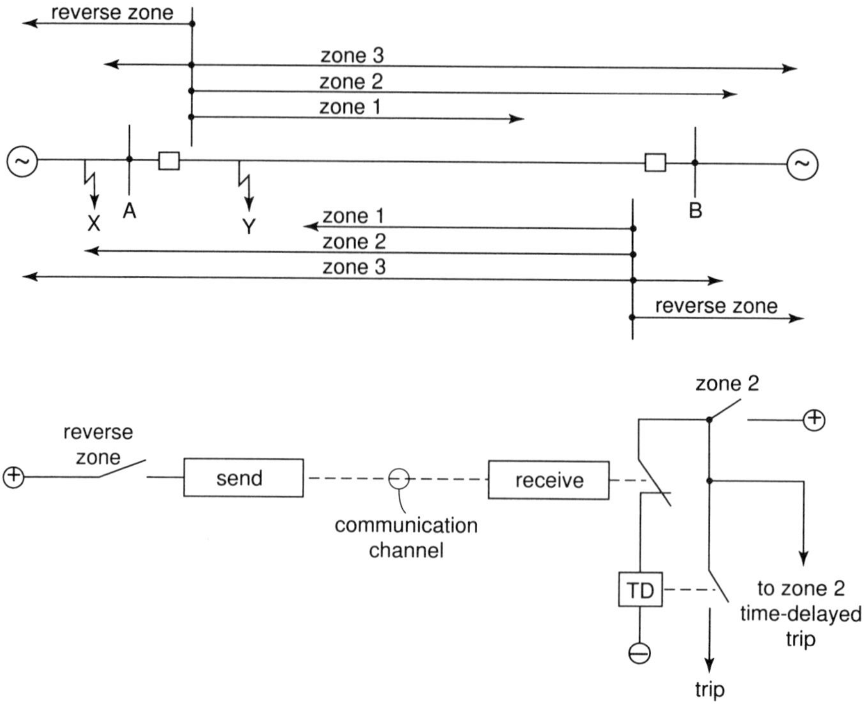

Figure 14.17 Blocked distance protection

is also detected by the reverse zone at end A which sends a signal to the receive circuitry at end B, which in turn opens a contact to de-energise TD and correctly prevent a trip.

Consider now the internal fault at position Y, which is beyond the zone 1 reach of the relay at B. It is detected by zone 2 at end B, which starts TD. This time there is no received signal from end A, and thus TD times out and causes a trip. The time delay to which TD is set must exceed the longest possible time for receiving a signal from end A (to ensure that a false trip does not occur). Typical times for TD are between 30 and 60 ms. With a typical zone 2-comparator pickup time of 25 ms, this would give a blocked-protection-operating time of between 55 and 85 ms.

The blocked scheme effectively makes distance protection into a unit protection. It can also be applied to very short feeders, since a (near) instantaneous trip can be achieved using a zone 2 comparator, which in turn will result in a lower SIR which is beneficial in maintaining relay accuracy (see Section 14.13). The disadvantage of blocked distance protection is that, when fast operating times are an imperative to maintaining system stability, protection operating times up to 85 ms may be too slow. However, for many applications this is not the case. Note that if the block signal communication link becomes faulty for any reason, it is necessary to inject an artificial block signal into the receive circuitry, to prevent the scheme becoming unstable for external faults.

14.24 Three-ended circuits

Three-ended circuits are often very difficult to protect, and distance protection is one of the few protections which can usually be successfully applied. The objective is to ensure an instantaneous trip for an in-zone fault, and stability for an out-of-zone fault. Although it is possible to protect some three-ended circuits with an accelerated scheme, this is not always the case. For example, it is necessary to ensure with an accelerated scheme that every part of the feeder is covered by a zone 1 reach from at least one of the three ends. In some circuit configurations, this is not possible (e.g. with a T point which connects two short lines to a very long line). Blocked schemes, which are dependent on the longer reach of zone 2 comparators, are better suited to protecting three-ended circuits. Some of the major considerations are as follows:

(*a*) With reference to the example in Figure 14.18*a*, the zone 1 reach setting must be such that it does not operate for a fault just beyond the end of any line connected to the T point. As can be seen, in simple terms, the comparator at end A measures an apparent impedance to fault of 8 Ω. The zone 1 setting at end A should then be set to 80% (P/F) or 65% (E/F) of the apparent impedance.
(*b*) The zone 2 comparator must be able to detect a fault which is just beyond the end of the protected feeder under the worst throttling conditions. Throttling is the creation of an apparent higher impedance, due to current infeed into the T point, from the other infeeding leg of the circuit. For the simple example shown in Figure 14.18*b*,

$$\text{voltage at relay} = 4I + 18I = 22I$$

$$\text{impedance measured at relay} = \frac{\text{voltage at relay}}{\text{current through relay}}$$

$$= \frac{22I}{I} = 22\ \Omega$$

Therefore zone 2 setting $= 1.5 \times 22 = 33\ \Omega$
(*c*) With reference to the fault and the circuitry shown in Figure 14.18*c*, it can be seen that current is flowing out of the circuit at end C. It will therefore incorrectly send a blocking signal to ends A and B to prevent tripping. In such an instance, it is necessary that the fault be detected by zone 1 from at least one end. For example, if zone 1 at end B operated and tripped the end-B circuit breaker, current would change direction at end C, the blocking signals

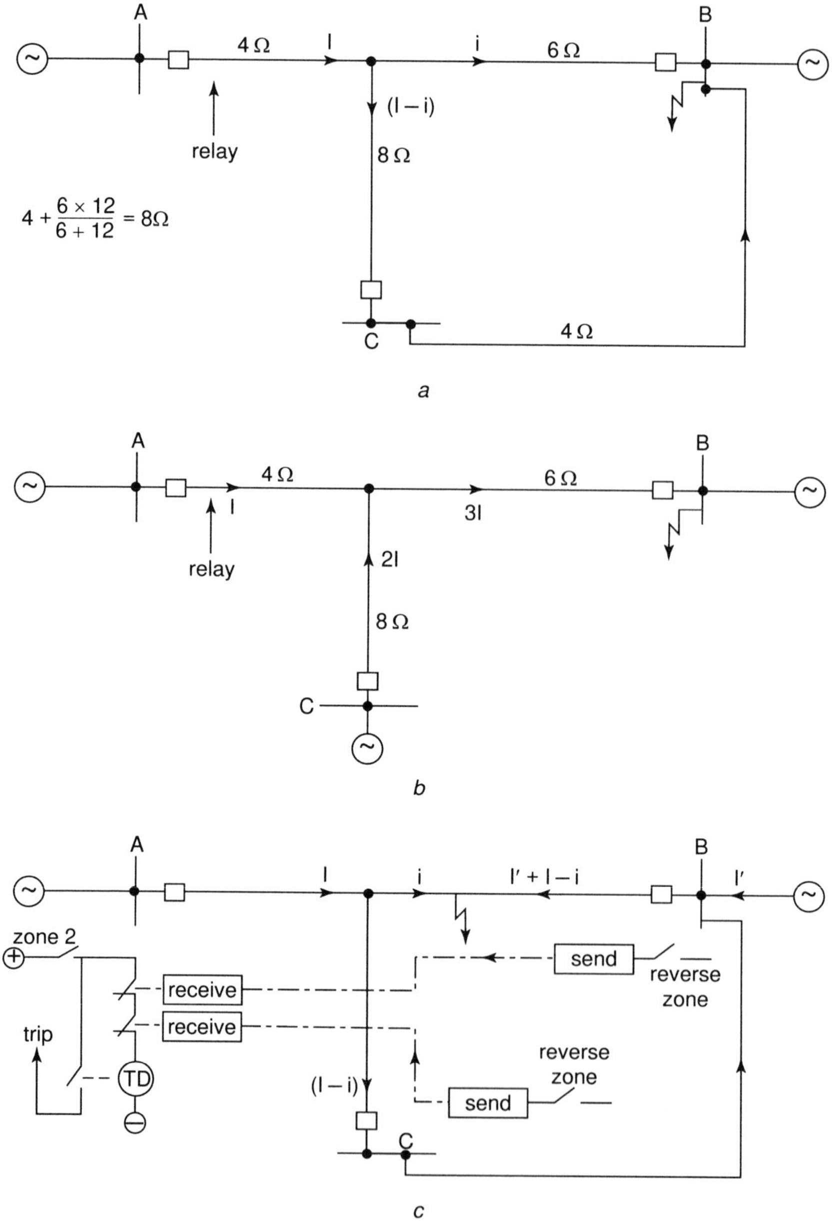

Figure 14.18 Blocking-scheme criteria for a three-ended circuit

a Relay must not detect a fault beyond the end of the line
b Zone 2 must detect a fault beyond end of line under worst conditions
c Zone 1 from at least one end must detect this type of fault

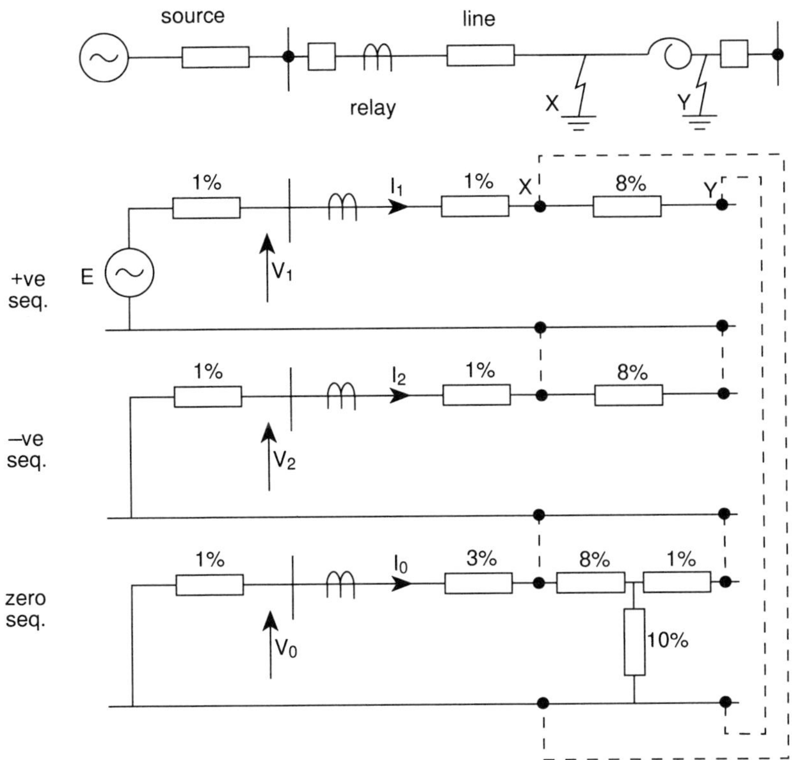

Figure 14.19 Application to a transformer feeder

would then be removed, and ends A and C would trip via the blocking circuitry.

14.25 Transformer feeders

When a distance protection is applied to a transformer feeder as illustrated in Figure 14.19, advantage may be taken of the high impedance of the transformer to enable the zone 1-comparator reach to cover all of the feeder plus part of the transformer. Care needs to be taken when selecting the earth-fault setting, since the ratio of Z_0/Z_1 for the transformer is significantly different from that of the line, and this provides a dilemma for the selection of Z_0/Z_1 for residual compensation. It is usual, however, to use the Z_0/Z_1 ratio for the line, but this causes a false measurement of positive-sequence impedance for certain faults; for example a fault at the end of the line (position X) results in a correctly measured impedance of 1%, but a fault on the LV side of the transformer (position Y) results in a measured impedance of 6.1%, i.e. 6.1/9 or 68% of the actual positive-phase-sequence impedance.

Table 14.2 Comparison of line and cable residual compensation

	Overhead line	Cable
Z_1 (Ω) =	$0.07 + j0.7$	$0.08 + j0.36$
Z_0 (Ω) =	$0.31 + j2.0$	$0.84 + j0.3$
residual compensation =	$0.62\underline{/10°}$	$0.3\underline{/-55°}$

In each instance, relationship 14.3 is used to determine the impedance, i.e.

$$Z = \frac{V_1 + V_2 + V_0}{(I_1 + I_2 + I_0) + I_0\left(\dfrac{Z_0}{Z_1} - 1\right)}$$

Generally, if the relay reach setting does not exceed 65% of the feeder positive-phase-sequence impedance +50% transformer positive-phase-sequence impedance, overreach is not experienced; however each case should be evaluated on its own merit.

14.26 Composite feeders

Composite feeders refer to mixed overhead and cable circuits. Such circuits can also be problematic for measuring the correct impedance for earth faults, because the Z_0/Z_1 ratio for cables is significantly different from that for overhead lines, particularly at the higher voltages. Therefore if the Z_0/Z_1 ratio for overhead lines is used for residual compensation, the true positive-phase-sequence impedance to the fault is not measured. Generally, if the cable comprises less than 10% of the feeder length its effect on impedance measurement can be ignored. However, with longer lengths of cable, the impact must be taken into account. Table 14.2 compares values of residual compensation, for 15 km of cable and overhead line of similar cross-section area. As can be seen, the overhead line impedances have much higher reactive components than those of the cable. This is due to the close proximity of the cable conductors and sheath and the impact on induced voltages. There can be a wide variation in cable impedances.

From the above it can be seen that there is a significant difference between the two values of residual compensation. However, distance relays are usually not calibrated to accommodate residual-compensation settings which have a phase angle other than 0°. Therefore if distance protection is applied to a cable circuit, the residual-compensation setting may have to be set to zero, or just 0.3. This in turn will require some compromise reach setting to avoid unacceptable relay overreach or underreach.

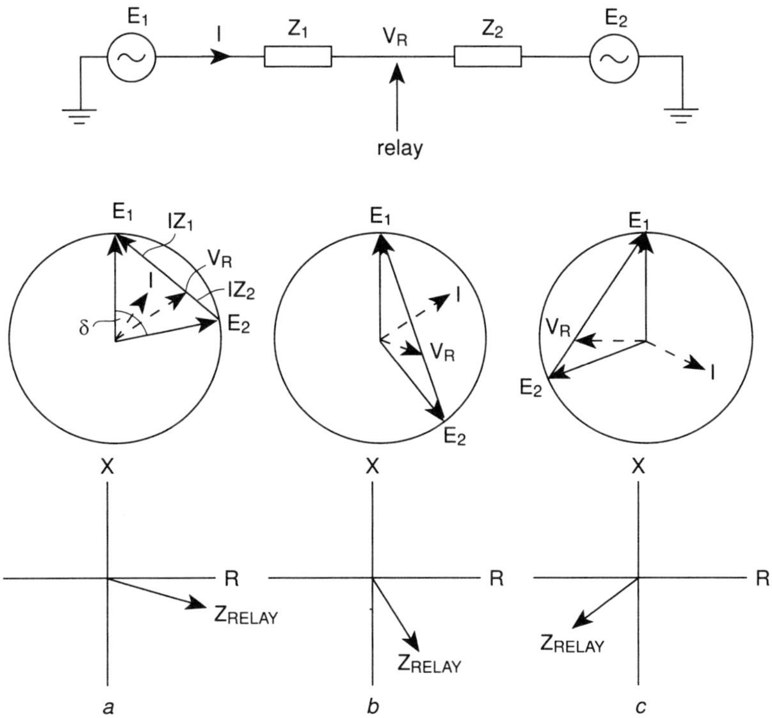

Figure 14.20 Impedance measured during pole slipping

14.27 Pole slipping

Pole slipping occurs when synchronism is lost between generators (or generators and synchronous motors) on the system. This results in the voltage of one generator, or a number of generators, slipping with respect to the rest of the system, causing wide variations of voltage and current. The effect can be damaging to generators as a result of pulsating mechanical forces, and the rotating flux between stator and rotor induces currents into the rotor body resulting in overheating.

Pole slipping can cause operation of protections on the power system but, since both voltage and current are affected, the effect is perhaps more significant to distance-protection performance.

Figure 14.20 illustrates the basic concept. If we consider voltage E_2 rotating with respect to voltage E_1, the current is given by

$$I = \frac{E_1 - E_2}{Z_T}$$

where $Z_T = Z_1 + Z_2$ and is assumed to be reactive. The impedance measured by

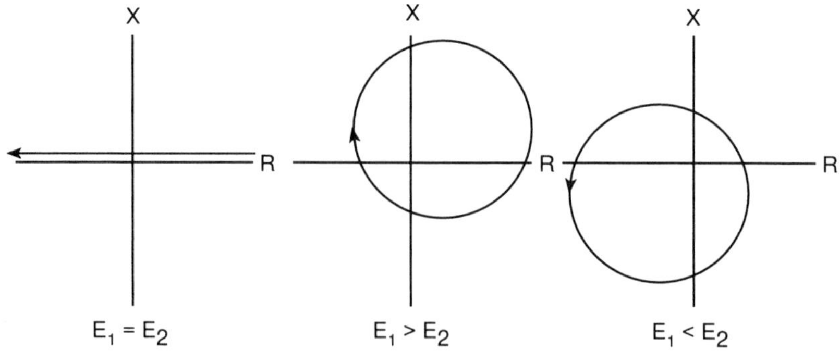

Figure 14.21 Pole-slipping impedance locus

the distance relay is then

$$Z_{RELAY} = \frac{V_{RELAY}}{I}$$

where $V_{RELAY} = E_1 - IZ_1$.

Note that pole slipping is a three-phase phenomenon; therefore, distortion of the polarised-mho characteristic as described in Section 14.11 does not occur. The impedance measured by the distance-protection comparators follows a rotating circular characteristic as a complete cycle of slip occurs. Figure 14.21 illustrates the impedance locus experienced depending on whether E_1 is over or under excited with respect to E_2. The direction of rotation depends upon whether E_1 moves faster or slower than E_2. If the system centre (to be described shortly) is close to a source of generation, the generator terminal voltage can be significantly affected and the AVR endeavours to adjust rotor excitation accordingly. This results in the characteristics shown in Figure 14.21 becoming spiral in shape.

Once during each cycle of slip, the RMS voltage at a certain point in the system becomes zero. This occurs when E_1 and E_2 are 180° out of phase, as illustrated in Figure 14.22. This point, known as the system centre, appears from a distance-protection viewpoint as the location of a three-phase fault, and therefore the minimum impedance during the pole-slip cycle presented to the distance protection. Operation of distance protection is more likely to occur in zone 3, with its longer reach than zone 1 or 2. However, the pole-slip impedance characteristic needs to remain within the zone 3 setting for the duration of the zone 3 time delay. It is generally advantageous for distance protection to trip on pole slipping, to sever the system and limit possible damage to generation, and to stabilise voltage supplies.

Some generators have pole-slip protection fitted to disconnect generation, but this is very much the exception.

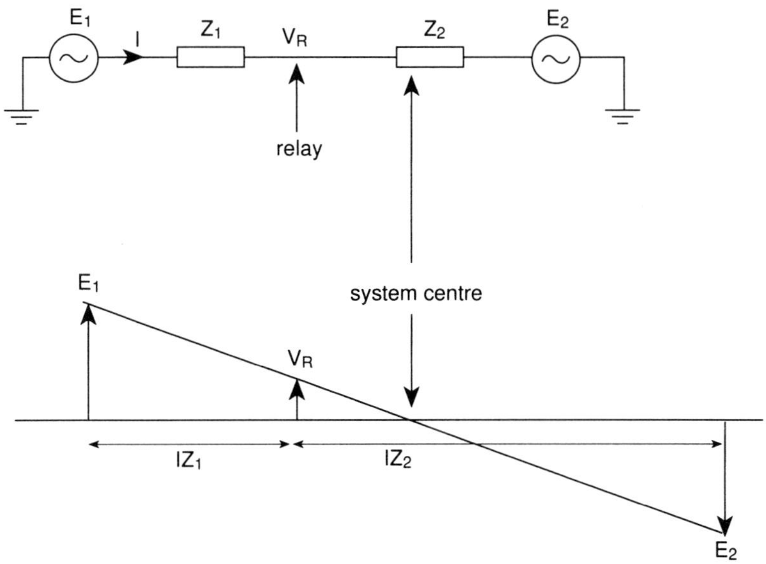

Figure 14.22 *System centre*

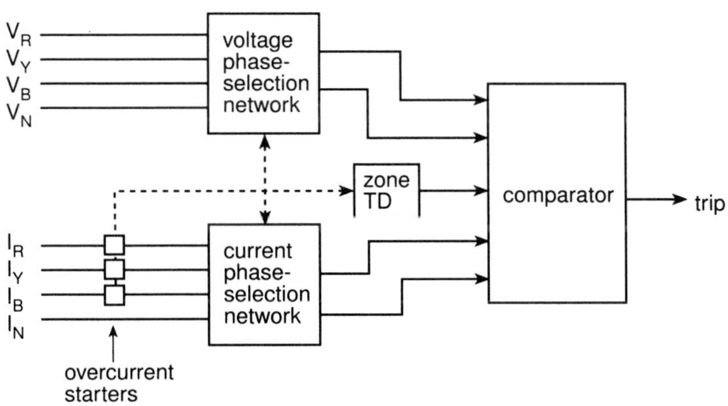

Figure 14.23 *Switched distance relay*

14.28 Switched distance relays

The distance-protection schemes discussed so far have been of the type which utilises six comparators per zone (i.e. one per fault type per zone). At lower voltages a more cost effective version is often installed, consisting of a single comparator to which the voltages and currents are switched, in accordance with the type of fault. This is illustrated in Figure 14.23. The disadvantage of this form of relay is that extra time delays are incurred in the overcurrent starters and phase-switching circuitry. Switched distance relays therefore tend to be applied to distribution networks (33 kV and some 132 kV) where fault-clearance times are less critical. The following sections will concentrate on the six-comparators-per-zone schemes.

COMMISSIONING

14.29 Commissioning tests

The commissioning tests which require examination are:

- secondary injection
- on-load tests

All other tests are identified in Section 12.6.

14.30 Secondary injection

It is usual to secondary inject distance protection to prove only the settings with which the equipment will enter service. In general, and with reference to Figure 12.5, the following should be carried out:

(*a*) measurement of the zone 1 setting reach at the boundary of operation, at the characteristic angle. This consists of reducing the impedance presented to the relay incrementally until it operates, as indicated by closure of contacts 1 and 2 of the zone 1 repeat relay ZIR. Note that monitoring of the output relay is inappropriate for this test since slowly decreasing the test impedance would result in zones 2 and 3 operating before zone 1. This test should be carried out for all six comparators (i.e. R–E, Y–E, B–E, R–Y, Y–B, B–R);
(*b*) measurement of zone 2, zone 3 forward and zone 3 reverse setting reach as described for zone 1 above;
(*c*) measurement of the zone 1 operating time for all six comparators. This would consist of presenting the relay with an impedance of about 80% of the reach setting, at characteristic angle. The fault should be applied at the same time as a timer is started. The timer should be stopped from the trip output relay. Prior to applying the fault, all three voltages should be at full nominal voltage, with zero current injected;
(*d*) measurement of zone 2, zone 3 forward and zone 3 reverse operating times, as zone 1 above;
(*e*) measurement of the operating time for a close-up fault. This would consist of carrying out a timing test with the voltage of the comparator to be tested switched to zero. The objective is to prove the cross polarising (healthy phase) voltage;
(*f*) measurement of the operating time for a close-up three-phase fault. This

would consist of starting the timer coincident with removal of all three phase voltages. The objective is to prove the memory feature;
(*g*) measurement of the operating time for a switch onto fault (SOTF). This consists of switching on current and zero voltage, coincident with starting the timer;
(*h*) measurement of VT-supervision sensitivity. Many distance protections have an integral feature which gives an alarm if VT supplies to the relay are lost (e.g. a blown VT fuse). This is described in Section 10.42. Many distance protections have an option to block tripping (trip snatching) if a VT-supervision alarm is output. This is usually not employed in the UK, where it is thought preferable to allow tripping to occur, thus minimising the risk of system instability through a failure to clear a system fault;
(*i*) injection of the various comparators to prove blocking or acceleration circuitry, tripping to trip relays and alarms.

Secondary injection is usually carried out by use of a distance-protection test set, simulated transmission line, or an automatic test device, as described below.

14.31 Distance protection test set

This is one of the oldest types of injection equipment for distance protections. Essential features are illustrated in Figure 14.24. Selector switches on the test set route the injected quantities to the selected comparator. Measurement of the zone 1 reach is determined by injecting a current, say 1 A, into the protection and then reducing the voltage variac until the comparator operates. The measured voltage at the point of operation is then compared with the calculated voltage. For example, for red-phase-to-earth fault, eqn. 14.2 gives

$$Z_1 = \frac{V_R}{I_R + \frac{I_R + I_Y + I_B}{3}\left(\frac{Z_0}{Z_1} - 1\right)}$$

Now if we inject current only into the red-phase comparator, $I_Y = I_B = 0$. Thus

$$Z_1 = \frac{V_R}{I_R\left\{1 + \frac{1}{3}\left(\frac{Z_0}{Z_1} - 1\right)\right\}} = \frac{V_R}{I_R(1+n)} \tag{14.5}$$

where n=residual compensation. Thus $V_R = I_R(1+n)Z_1$.

Thus with reference to the example in Section 14.22, where it was determined that the reach setting for zone 1 E/F was 10 Ω, and the residual-compensation setting was 0.62, if we now inject, say, 1 A, then

$$V_R = 1 \times (1 + 0.62) \times 10 = 16.2 \text{ V}$$

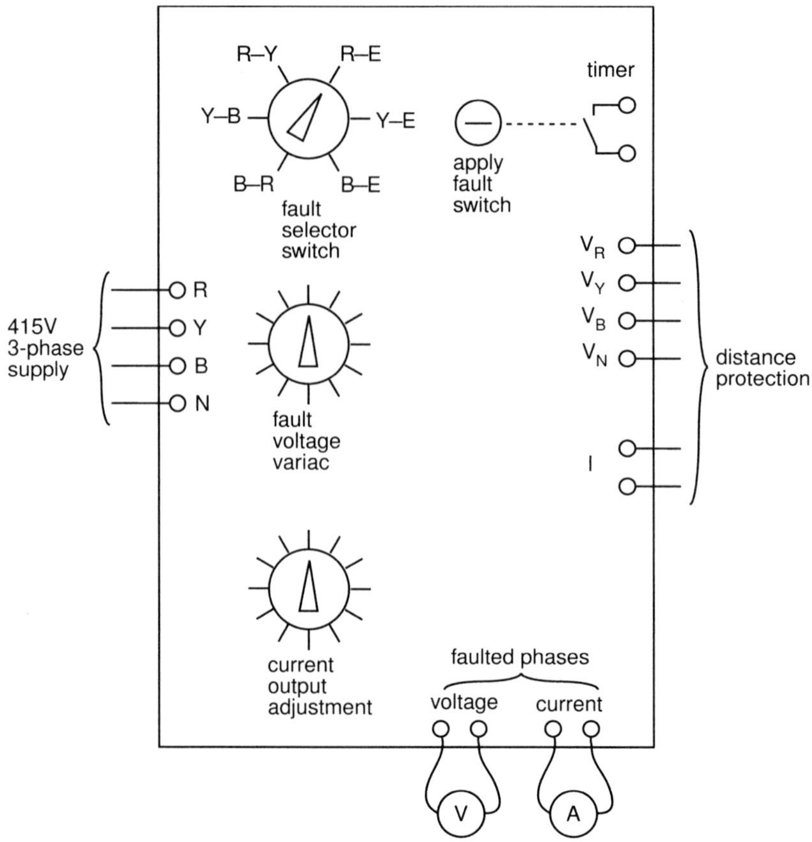

Figure 14.24 Distance-protection secondary-injection set

If the measured voltage for zone 1 E/F is within ±10% (normal limits of accuracy) of 16.2 V, then the comparator has passed the test.

Similarly, from eqn. 14.4 the phase-to-phase-fault relationship for a Y–B fault is given by

$$Z_1 = \frac{V_Y - V_B}{I_Y - I_B}$$

Now $I_Y = -I_B$ for a phase-to-phase fault; thus

$$Z_1 = \frac{V_Y - V_B}{2I_Y} \tag{14.6}$$

$$V_Y - V_B = 2I_Y Z_1$$

Again with reference to the example in Section 14.21, zone 1 P/F = 11.7 Ω.

Thus if we inject 1 A into the Y–B comparator, then

$$V_{Y-B} = 2 \times 1 \times 11.7 = 23.4\ \text{V}$$

Again if the measured voltage is within ±10% of 23.4 V the comparator has passed the test.

14.32 Simulated transmission line

This consists of a simulated three-phase transmission line comprising source and fault impedances. The fault impedance is reduced progressively until the comparator operates. The measured impedance at which operation occurs is given by eqns. 14.5 and 14.6 as follows:

For single-phase-to-earth faults:

$$Z\ \text{measured} = \frac{V_R}{I_R(1+n)}$$

$$= \frac{Z_{Transmission\ Line}}{(1+n)}$$

For phase-to-phase faults:

$$Z\ \text{measured} = \frac{V_Y - V_B}{2I_Y}$$

$$= \frac{Z_{Transmission\ Line}}{2}$$

14.33 Automatic testing

Both the distance-protection test set and the simulated transmission line are manual methods of injection, with which it may take well over a day to test a distance protection fully. Automatic test sets can reduce this time considerably. Automatic testing programmes generally tend to be designed to facilitate routine maintenance tests, and thus more exhaustive commissioning tests may require some 'manual driving' of the automatic test equipment, resulting in testing times of two to three hours. A point to note with automatic testing is that, if the residual-compensation setting n is not entered into the automatic test set, the true measured positive sequence impedance Z_1 is given by

$$Z_1 = \frac{Z}{(1+n)}$$

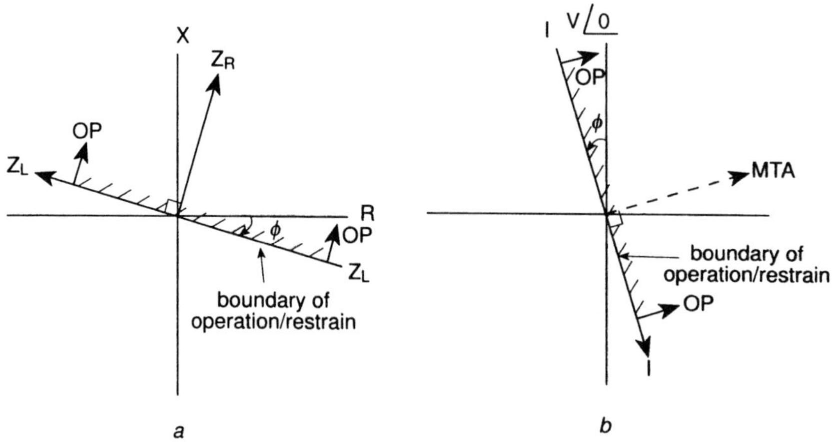

Figure 14.25 Polarised-mho comparator converted into a directional comparator

a Directional characteristic
b Voltage and current equivalent of Figure 14.25*a*

where Z is the measured impedance stated by the automatic test set.

It is also possible with an automatic test set quickly to carry out a full polar plot of each characteristic. This may well result in the distorted polar characteristic described in Figure 14.10, which implies a loss of directionality but which, as we have seen in Section 14.11, is not the case.

14.34 On-load tests

The purpose of a distance-protection on-load test is to prove, with the use of circuit-load current, that the relay is looking in the correct direction, i.e. it will operate for faults on the feeder to be protected, and not for faults behind the relay point. However, given that load current usually presents a high impedance to distance protection, it is usually not possible to determine directly the directionality of polarised and offset-mho comparators. In recognition of this, distance protections are fitted with a facility for removing the restraint voltage of the polarised-mho comparators. This results in the comparators being converted into a directional characteristic. For example, with reference to the comparator inputs described in Section 14.7, they now become

$$S_1 = V_P \qquad S_2 = IZ_R$$

but, for V_P approximately $= V_R$, $S_1 = V_R$.

This results in the directional characteristic illustrated in Figure 14.25*a*. The advantage of this type of comparator is that the operating characteristic is now independent of the magnitude of the impedance presented by the load current, and dependent only on the direction of the impedance. It is, however, more convenient to work in terms of voltage and current than in terms of impedance. Figure 14.25*b* illustrates the equivalent of Figure 14.25*a* in terms of an operation/restrain boundary of current with respect to a voltage reference. The relationship between the two Figures is given by

$$I\underline{/\Phi} = \frac{V\underline{/o}}{Z_l\underline{/-\Phi}}$$

The term MTA stands for maximum torque angle, which is a throwback to the days of electromagnetic relays. It refers to the most sensitive angle in the operating characteristic.

Figure 14.25*b* can be translated into Figure 14.26, which shows the directional-comparator relationship for all three phases. Consider now the situation where we have converted the polarised-mho characteristic into its directional characteristic. We are now ready to prove the directionality of the relay. If, for example, the current flow into the feeder is equivalent to 300 MW import and 100 MVAR import, the angle of the current phasor with respect to red-phase voltage V_R is shown by red current I_R. As can be seen, this is in the restraint direction for the red phase, and thus the comparator should restrain. It is necessary to prove the comparator in both the operate and restrain directions. To achieve operation for the current phasor shown, one of the following may be carried out:

- reverse the direction of the current into the relay
- rotate the voltages feeding the relay

If the CT wiring into the relay consists of two wires per phase, it is possible to reverse the connections to obtain a reversal of current into the relay, i.e. the current phasor becomes $-I_R$. However, some relays have a four-wire input, (i.e. three phases and a common neutral), and thus CT-wiring reversal is not possible. This necessitates the alternative of rotation of the voltage connections. In Figure 14.26 it can be seen that I_R is in the operate characteristic of the yellow-phase comparator. Thus, if the red-phase comparator is fed with yellow-phase, as opposed to red-phase, voltage, operation should occur. Rotation of the voltages is shown in Figure 14.27. The connections can usually be made by removing the VT links and strapping across the link holder as shown.

14.35 Practicalities

Figure 14.25 shows a somewhat idealised directional characteristic which takes no account of the inherent stability bias of a practical comparator. The effect of bias is illustrated in Figure 14.28 for a typical modern distance-protection relay. This shows a maximum sensitivity of about 100 A at a

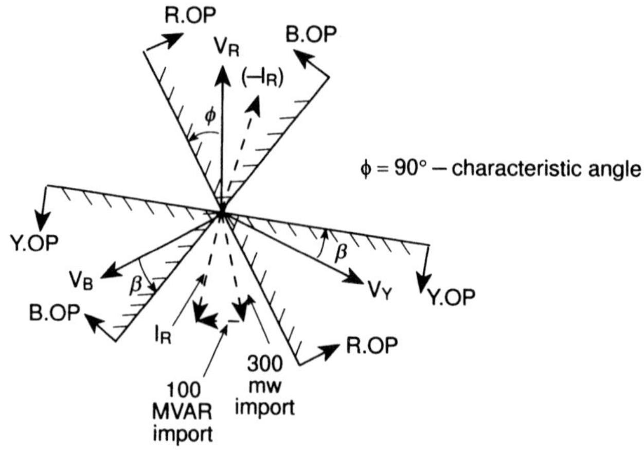

Figure 14.26 Directional comparator: three-phase relationships

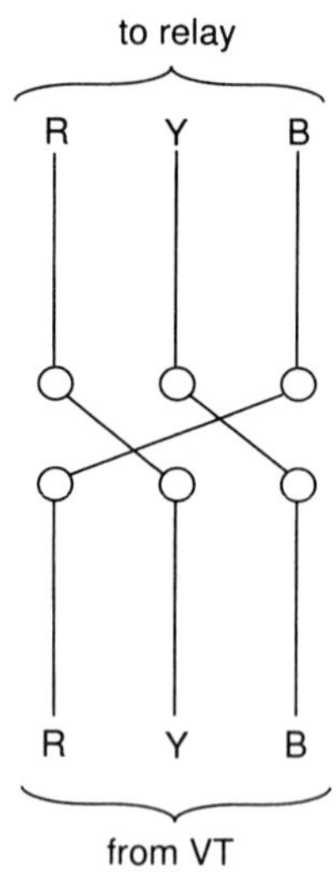

Figure 14.27 Voltage vector rotation

nominal voltage of 400 kV at the characteristic angle. Sensitivity is dependent on the relay-reach setting and, with long transmission lines, sensitivities of the order of 30 to 40 A at 400 kV can be obtained. This is about the order of magnitude of line-charging current when the line is energised from one end only. Figure 14.29 illustrates the phasor position of line-charging current which, when reversed, is within about 20° of the maximum torque angle. This,

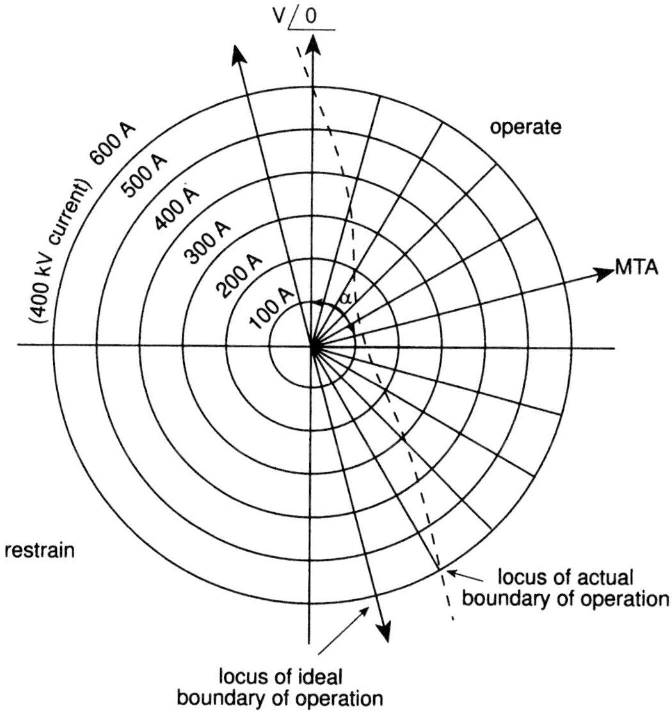

Figure 14.28 Directional-comparator minimum-current operating characteristic

α = relay-characteristic angle = MTA

together, with the fact that line-charging current has a fixed magnitude, i.e. it does not fluctuate as does load current, makes it advantageous to utilise it for proving directionality.

If relay sensitivity is too low to enable line-charging current to be used, and the magnitude of load current is also too low to operate the directional comparator, a number of options are still open:

(i) increase the relay-reach setting to increase sensitivity;
(ii) rotate the relay characteristic angle (for those relays with an angle setting), to make the load current coincide with the maximum torque angle (i.e. maximum sensitivity);
(iii) reduce the CT ratio, to increase the current into the relay.

Ideally we would not want to utilise any of the above, since it means tampering with final settings, but it may be a necessary course of action.

Some distance relays have a facility to enable the zone 3 offset-mho comparator to be changed to a directional comparator. However, due to the offset feature, the resulting characteristic is more complex than that for the converted polarised-mho comparator. As a result, comparator operation is

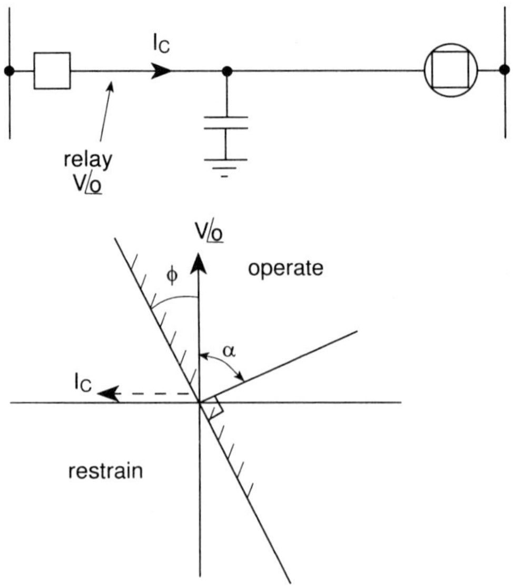

Figure 14.29 Line-charging-current phasor relationship

α=MTA
ϕ=90° − α

not so conclusive in proving directionality; it is therefore recommended that offset comparators be ignored for this purpose.

MAINTENANCE

14.36 Maintenance

Maintenance tests are generally a repeat of the commissioning tests, except that there is no need to verify directionality again. For many years maintenance was carried out every two years; however, with the advent of self-monitoring relays the modern tendency is towards maintenance intervals of four years or longer.

Chapter 15

Unit protection of feeders

15.1 Introduction

Many feeder unit protections work on the principle of measuring current at both ends of the feeder, and comparing those currents via a communication medium, to conclude whether the current flow is into an internal or external fault. Two of the most common types of unit protection are

- pilot-wire systems
- phase-comparison power-line-carrier systems

Theoretical and commissioning considerations for each will be examined. Current-comparison systems utilising fibre-optic communication links are slowly being introduced, but as yet they are few in number.

PILOT-WIRE PROTECTION: THEORY

15.2 Background

Pilot-wire protection systems have their origins in the earliest days of the transmission and distribution system, and utilise a pilot cable as the communication medium. At distribution voltages of 33 kV and below, private pilots are mainly employed, particularly when the pilot can utilise the same trench as the primary cable to be protected. At voltages of 132 kV and above, where the bulk of the system is overhead, it has been common practice, for many years, to utilise pilots rented from the public telephone company. However with telephone companies moving away from metallic pilots, to fibre and microwave systems, conventional pilot-wire protection is now being replaced by other systems.

15.3 Types of pilot-wire protection

There are two fundamental types of pilot-wire protection:

- current balance
- voltage balance

Figure 15.1*a* illustrates the basic principles of a current-balance system.

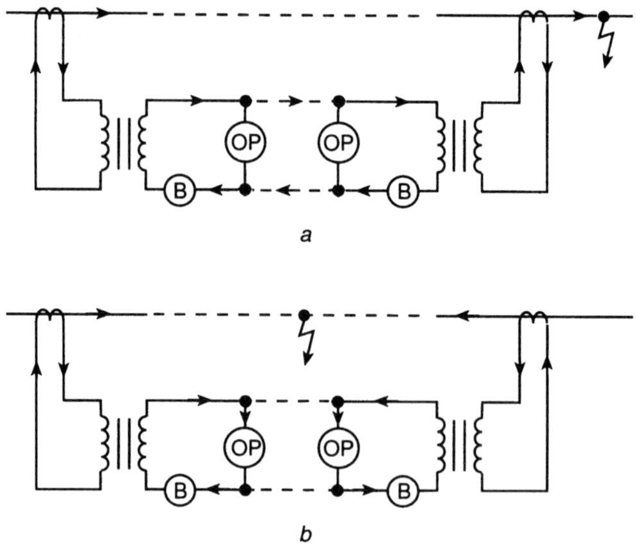

Figure 15.1 *Current-balance system*

a External fault
b Internal fault

OP = operate coil
B = bias coil

During external-fault conditions, current flows around the CT circuitry, but during internal fault conditions the current flows into the low-impedance operate winding. However, with the use of a pilot wire as the communication medium perfect current balance cannot be guaranteed and spill currents may flow into the relay during external faults, particularly when the fault current contains a DC offset component causing saturation, or partial saturation, of one CT. This can be overcome by adding a restraining, or bias, winding to the relay, the effect of which is to utilise the circulating current to cause relay restraint. It can be seen, by examining Figure 15.1, that an open circuit of the pilot loop can cause false operation of the protection under external-fault conditions, while a pilot short-circuit can cause failure of the protection to operate under internal-fault conditions

A simple illustration of a voltage-balance system is shown in Figure 15.2. In this instance the outputs from the summation transformers are connected in opposition around the pilot loop during through-load or external-fault conditions. Although voltage-balance systems tend to be less susceptible than current-balance systems to pilot-circuit imbalance, they still necessitate a bias coil. Both the operate and bias coils have low impedances.

When CTs are connected in opposition, as with a voltage-balance system, dangerously high voltages can arise across the secondary circuitry. To avoid this situation, it is usual to convert the CT current source into an equivalent

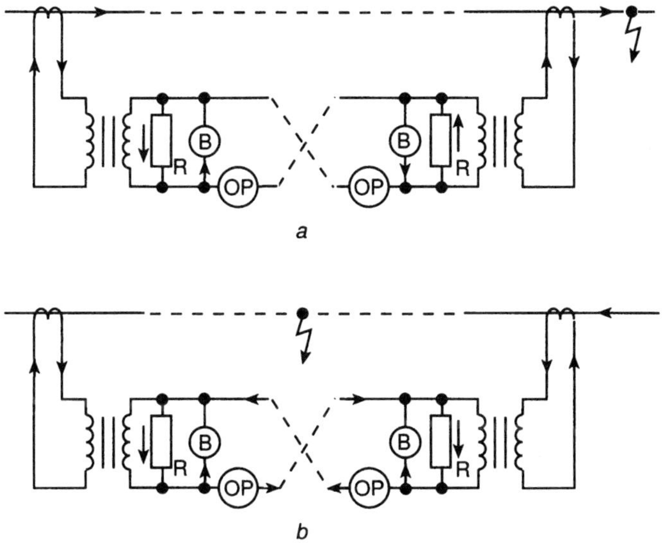

Figure 15.2 *Voltage-balance system*

a External fault
b Internal fault

OP = operate coil
B = bias coil
R = voltage-reference resistor

voltage source. This is accomplished by connecting a relatively low resistance across the CT secondary (typically 50 to 100 Ω).

It can be seen from Figure 15.2 that a short-circuited pilot can cause false operation of the protection under an external fault condition, while an open-circuit pilot can cause a failure of the relay to operate under internal-fault conditions.

Table 15.1 identifies some of the more widely applied pilot-wire protection systems in the UK.

15.4 Bias

Figure 15.3*a* illustrates the operating characteristic of an unbiased relay. As the magnitude of current for an external fault is increased, the spill current into the operate coil due to the natural unbalance and limitations of the protection system increases, until the relay operates.

Figure 15.3*b* shows how the effect of a bias winding increases the relay setting, requiring an increase in the magnitude of external fault current to cause operation, hence increasing relay stability. Some systems also contain additional bias windings which are responsive to harmonics associated with power-transformer magnetising inrush current.

Table 15.1 Pilot wire protection types in use in the UK

	Private pilot system	Telephone-company-type pilot system
Current balance	Reyrolle Solkor 'R'	Reyrolle HRPW GEC Translay 'S'
Voltage balance	AEI Translay 'H' Reyrolle Solkor 'A' Reyrolle Solkor 'B' AEI 'DSE'	GEC 'DSC8'

15.5 Summation transformer

Figures 15.1 and 15.2 illustrate the basic principles on a single-phase basis, whereas practical circuitry involves three phases. In most designs, the circuitry is optimised by use of a summation transformer, as shown in Figure 15.4. The summation transformer resolves three basic problems:

(*a*) It reduces three-phase circuitry to single-phase, thus economising on the number of pilot cores required.
(*b*) It typically has a turns ratio of 1:2 between primary and secondary; thus the secondary impedance when reflected into the primary is reduced by a factor of 4, thereby reducing the burden imposed upon the CTs.
(*c*) It provides an insulation barrier between CT and pilot (see Section 8.45).

The summation-transformer output is dependent on the type of system fault, as shown in Table 15.2. The output for earth faults is considerably greater than for phase-to-phase faults; this is advantageous on resistance-earthed systems where the earth-fault current is limited. A typical value for n is 3.

The summation-transformer tappings ratio needs to be considered carefully by the manufacturer at the design stage, since complex fault conditions can result in zero output. A 1.5/1/13.5 tapping arrangement is used in some designs and has proved reasonably satisfactory in practice. The knee point of the summation transformer must also be considered, and must be high enough to accommodate maximum fault currents with accompanying DC offset, without saturation.

15.6 Pilot characteristics

The pilot acts as the communication medium between the two ends of the circuit. Physical characteristics of pilots, and their susceptibility to induced

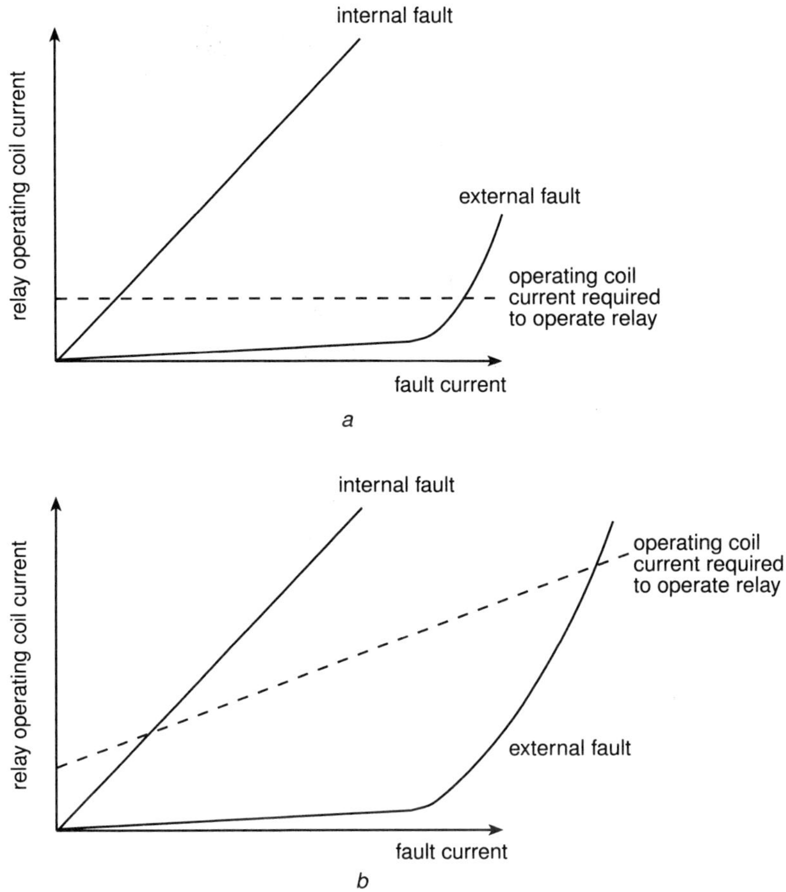

Figure 15.3 *Effect of bias coil*

a Unbiased-relay characteristic
b Biased-relay characteristic

voltages, are discussed in Chapter 8. Figure 15.5*a* shows the equivalent circuit of a pilot which can be represented as a multiple-tee network comprising resistance and shunt capacitance; the inductance is usually negligible. This can be approximated by the simpler equivalent circuit of Figure 15.5*b*.

Pilot characteristics are best described with reference to pilot input impedance in terms of the open-circuit and short-circuit characteristics, as shown in Figure 15.6. The two cable types shown are a multicore cable (7/.067 metric or 7/.029 imperial) and a multipair cable (1/0.8 mm), the latter being a British Telecom type of cable. As can be seen, with increasing pilot length, both the open- and short-circuit impedance tend towards the same value. Under such conditions, the relay at one end is unable to evaluate

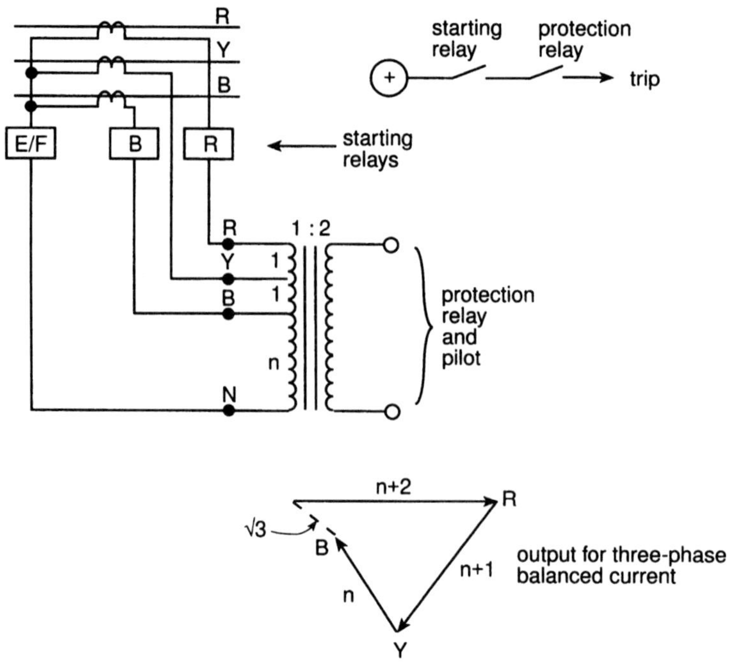

Figure 15.4 Summation transformer

Table 15.2 Summation-transformer outputs

	Comparative output for equal fault currents	Comparative output for identical fault currents when $n=3$
R–N	$n+2$	5
Y–N	$n+1$	4
B–N	n	3
R–Y	1	1
Y–B	1	1
B–R	2	2
3-phase	$\sqrt{3}$	1.73

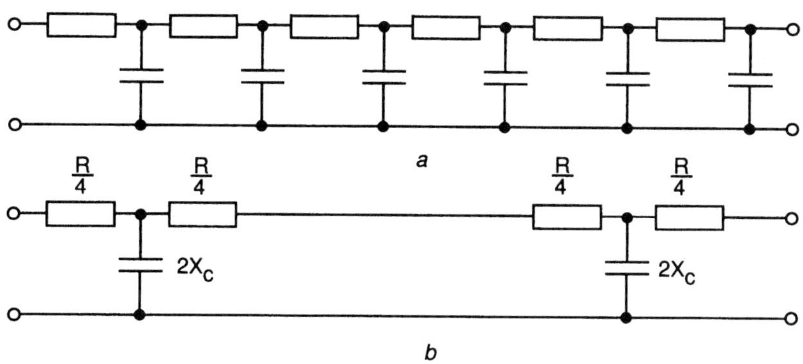

Figure 15.5 *Pilot equivalent circuit*

a Pilot equivalent circuit
b Simplified pilot equivalent circuit

the input at the other end. Pilot length is therefore a limiting factor in the application of pilot-wire protection.

To obtain consistency of performance from pilot-wire protection it is usual to specify a maximum pilot-loop DC resistance and, in those instances where the loop resistance is less than the maximum, padding resistors are provided in the relay equipment to pad up to the maximum. A typical maximum pilot-loop resistance for protections using British Telecom type pilots would be between 2000 and 3500 Ω.

Some protections are also designed to minimise the effect of the pilot shunt capacitance, particularly voltage balance systems which are more susceptible to the effects of pilot capacitance. Two such methods are:

(*a*) pilot tuning by use of an inductance connected across each end of the pilot as shown in Figure 15.7*a*. The distributed nature of the pilot resistance and capacitance means that this method cannot be 100% effective;
(*b*) use of a pilot replica impedance at each end of the pilot which provides an input into the operate winding to counteract that of the pilot.

The above methods are more likely to be utilised on protections which utilise telephone-type pilots.

15.7 Voltage limiters

It was pointed out in Section 8.45 that British Telecom specifies that the maximum voltage applied to its pilots should not exceed 130 V peak. This is usually accomplished by use of a metrosil (nonlinear resistor) connected across the summation transformer, as shown in Figure 15.7. The metrosil influences the performance of the relay at higher fault currents. For

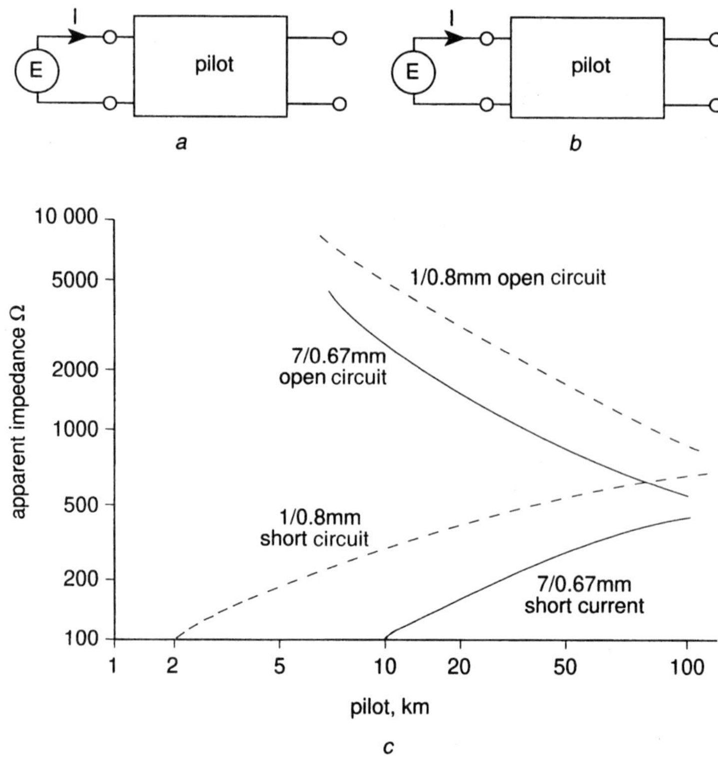

Figure 15.6 Pilot open- and short-circuit characteristic

a Open circuit
b Short circuit
Apparent impedance = E/I
c Apparent impedance under open and short-circuit conditions

consistency of design and performance, the metrosil is often utilised on all telephone-type pilots, irrespective of the company from which the pilot is rented.

15.8 Protection characteristic

One of the most meaningful methods of displaying the protection operated/restraint characteristic is illustrated in Figure 15.8*a*. The ratio of the two summation-transformer inputs I_A and I_B is plotted on a complexor diagram with real and imaginary axes. The locus shows the boundary of operation of the relay for the instant of, say, I_B being held constant at co-ordinates 1, 0 and the magnitude and phase angle of I_A varied. As can be seen, the relay compares both current magnitude and phase angle. However, as the magnitudes of both I_A and I_B increase, the limiting effect of the metrosil takes

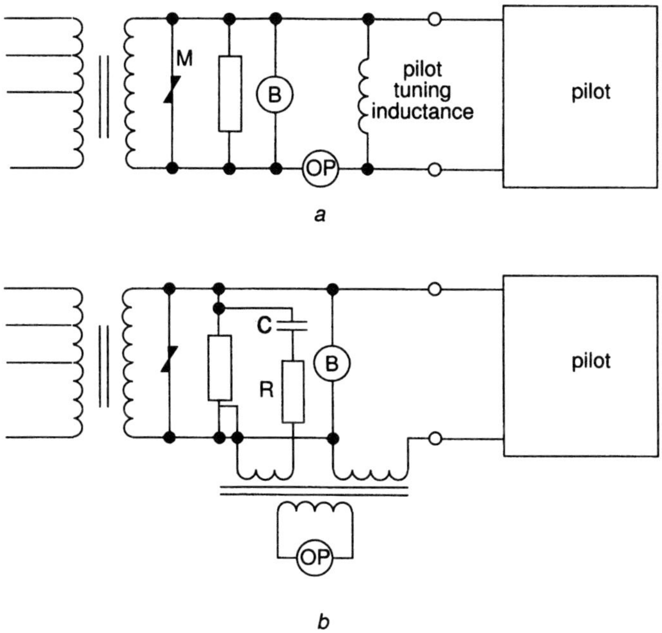

Figure 15.7 Pilot-tuning methods

a Pilot tuning by parallel inductance
M = metrosil
b Pilot tuning by replica impedance
CR = replica impedance

effect. This limits the amplitude of the voltage applied to the relay, so that the relay becomes, in effect, a phase-angle comparator only. This is illustrated in Figure 15.8*b*.

The above characteristic is somewhat idealised, since it assumes that relays at both ends of the circuit operate simultaneously. In practice, slight differences in relay performance, and the fact that the pilot will not be completely homogeneous throughout it length, result in a characteristic for each end of the circuit, as shown in Figure 15.8*c* (e.g. I_B is held constant, I_A is varied, and the characteristics of the relays at each end of the circuit are plotted).

15.9 Pilot supervision

Pilot supervision was originally required, on pilots rented from British Telecom, to provide a pilot-condition-monitoring system in case of inadvertent short- or open-circuiting of the pilot by British Telecom. Figure 15.9 illustrates the basic principle. A direct current of about 5 mA is circulated

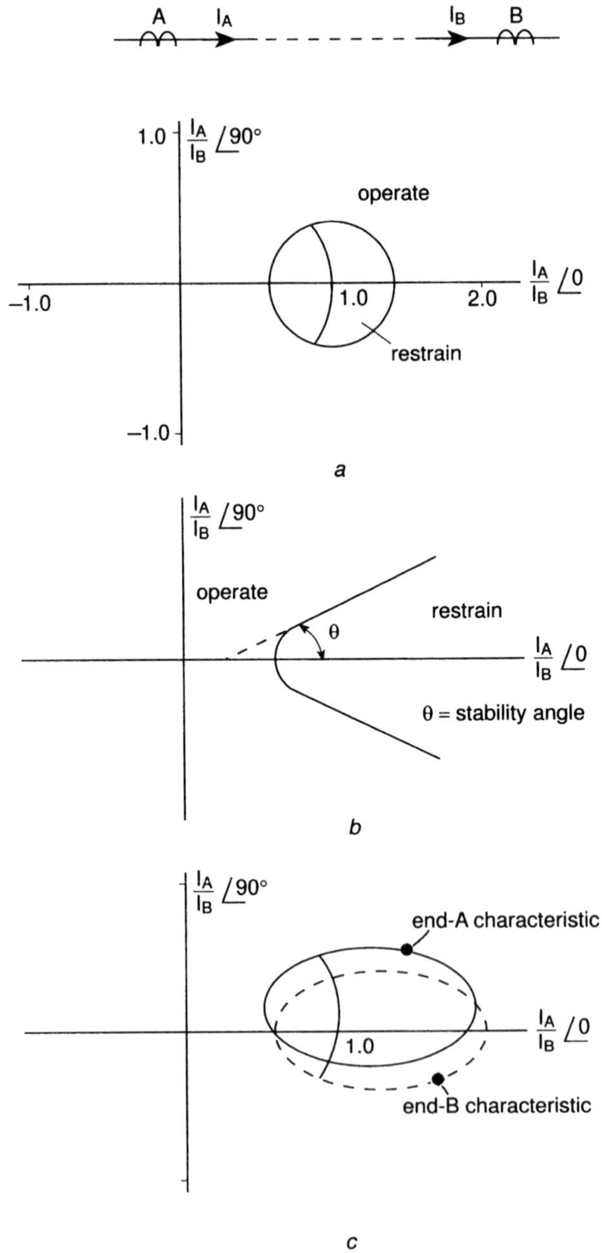

Figure 15.8 Protection operate/restraint characteristic

a Basic polar characteristic
b Effect of limiting voltage magnitude
  θ = stability angle
c Practical characteristics (without limiting)

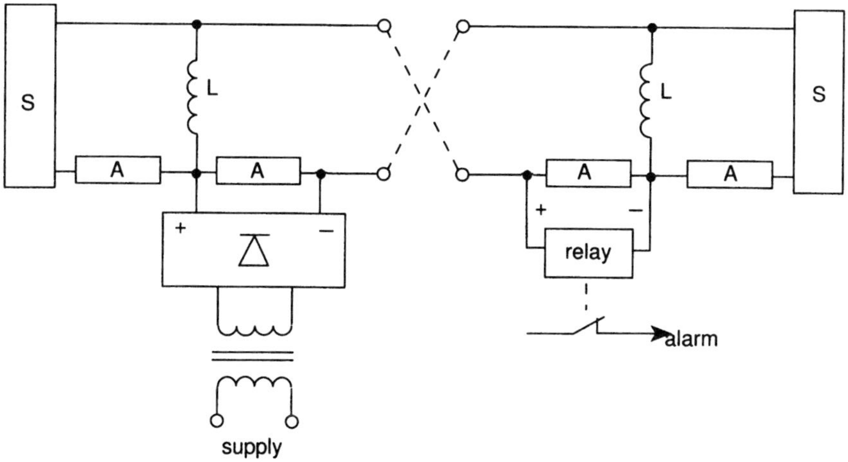

Figure 15.9 *Typical pilot-supervision arrangement*

S = summation transformers and relays
L = tuning inductance
A = 50 Hz acceptor circuit, DC rejector

around the pilot loop, consisting of tuning inductances at both ends of the circuit and a polarity-conscious supervision relay at the end remote from the DC source. Normal functioning of the protection system is enabled by the use of 50 Hz acceptor circuits (series inductance and capacitance) which prevent 50 Hz voltages being imposed on the supervision circuitry, and also prevent the DC supervision voltage being imposed on the protection relay. It is important that the supervision relay is polarity conscious to monitor the situation of the pilots inadvertently being crossed.

Schemes which do not employ pilot tuning may alternatively have a normally closed-low set relay contact short-circuiting the pilot terminals. During fault conditions the low-set contact is removed to enable current comparison to take place.

Pilot supervision is now also increasingly common on private pilots.

15.10 Starting relays

Figure 15.4 illustrates the location and application of starting relays. Starters ensure that, if the pilot becomes faulty, maloperation of the relay resulting in an unwanted trip will not take place under normal load conditions. This provides time for remedial action.

Starting relays were traditionally only applied to rented pilot systems, but more recently have also been applied to private pilot systems.

Table 15.3 Typical differential-relay sensitivity

Fault type	Fault setting (%)	Relay rating (A)	Main CT ratio	Primary fault setting (A)
R–E	20	1	1200/1	240
Y–E	25	1	1200/1	300
B–E	35	1	1200/1	420
R–Y	100	1	1200/1	1200
Y–B	100	1	1200/1	1200
B–R	50	1	1200/1	600
R–Y–B	58	1	1200/1	700

15.11 Settings

Settings fall into two distinct areas, the differential relay and the starting relay.

(*a*) *Differential relay:* The differential-relay settings are usually factory set. The fault setting is dependent on the fault type, since the summation transformer produces different outputs for each type of fault. The settings are usually very sensitive, and expressed as a percentage of relay rating. Table 15.3 shows a typical example relating to the 275 kV system.

To ensure satisfactory operation, the minimum internal fault current to which the relay is exposed should be twice the above sensitivities (i.e. factor of safety of two).The sensitivity of the differential relay should also exceed the out of balance caused by feeder capacitance current. Figure 15.10 illustrates the most onerous situation, of an external blue-phase-to-earth fault.

(*b*) *Starters:* Starter settings need to be higher than those of the differential relay since they are there to prevent circuit tripping, in the event of pilot failure resulting in operation of the differential relay.

The overcurrent starter would normally be set to about 15% higher than the full-load current of the circuit. This is to ensure the starter does not remain 'picked up' after clearance of a fault on another circuit (i.e. after a through fault). It should also be set to less than half the minimum phase-to-phase fault current (factor of safety of 2).

The earth-fault starter should be set higher than the standing residual current (i.e. 3 × zero-sequence), which may be as high as 15% of full-load current. It should also be set to less than half the minimum phase-to-earth fault current.

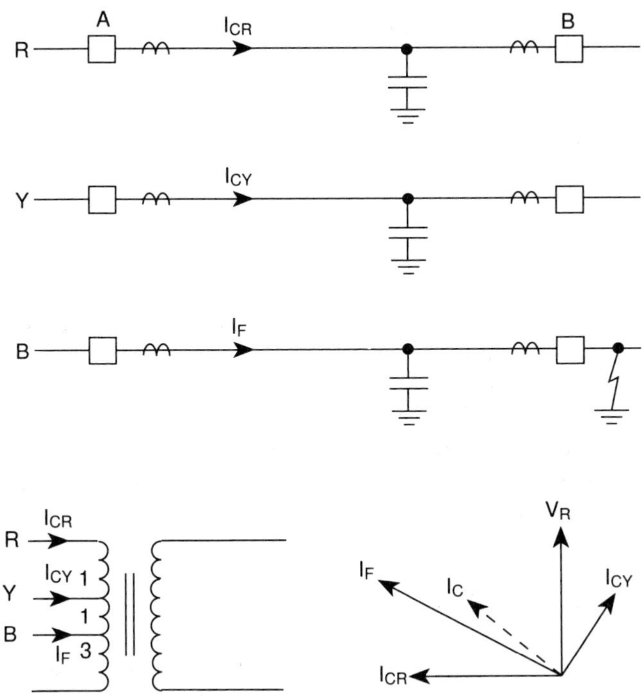

Figure 15.10 *Effect of capacitance current*

Output due to capacitance current = $5I_{CR} + 4I_{CY} = 4.6I_C$
Output due to fault current = $3I_F$
End-A output = $3I_F + 4.6I_C$
End-B output = $3I_F$

PILOT-WIRE PROTECTION: COMMISSIONING

15.12 Commissioning tests

Commissioning tests typically consist of the following:

(*a*) summation-transformer ratio, polarity and knee point;
(*b*) pilot characteristics:

- loop resistance
- open-circuit characteristic
- short-circuit characteristic

(*c*) pilot tuning (as per manufacturer's requirements);
(*d*) secondary injection:

- prove bias feature (at each end)
- minimum sensitivity with pilots normal (from each end)

- end-to-end stability-angle test
- timing test
- starter tests

(*e*) tripping tests;
(*f*) on-load tests.

15.13 Summation transformer tests

The ratio of a summation transformer can be obtained by injecting current into each phase in turn of the primary and measuring the corresponding current flowing into a short-circuit on the secondary. The polarity can be found from a flick test.

The knee point can be found by obtaining a magnetisation curve (see Section 10.15) with the secondary on open circuit. Modern schemes may have knee points as high as 700 V.

15.14 Pilot characteristics

The pilot-loop resistance should be obtained by disconnecting the pilot from the rest of the equipment, placing a short-circuit at one end and measuring the loop resistance from the other end. Figure 15.6 shows the method of determining the open- and short-circuit characteristics. These tests are important in facilitating investigations (following maloperation) to check whether the pilot has been interfered with and changed.

15.15 Secondary injection

The bias feature and minimum sensitivity should be verified in accordance with the manufacturer's recommendations.

The end-to-end stability-angle test may be carried out by injecting rated current (usually 1 or 5 A) simultaneously into the summation transformers at each end of the circuit, on the R–N tap, with the equipment and pilot normal. At one end, the injection should be via a phase shifter. With reference to Figure 15.8, the phase-shifter angle should be varied slowly to obtain the stability angles (i.e. the angle whereby the relays restrain). Typical angles would be

end A = $90 \pm 15°$
end B = $90 \pm 15°$
ends A and B both stable = $65°$

This information is useful for comparison with maintenance and fault-investigation results.

The starters should be injected with current to obtain the pickup and drop-off values.

A further useful test to aid fault investigations is to inject 1 A (say) into each phase of the summation transformer primary, in turn, and record the voltage to the earth link of every terminal which can be accessed.

15.16 Tripping

Commissioning of tripping merely involves operating the protection to prove it operates the trip relay.

15.17 On-load tests

The on-load tests are required to prove

- relay stability with pilots normal
- relay operation with pilots reversed

The manufacturer's handbook recommendations should be followed. A typical on-load test schedule is as follows:

(*a*) Short-circuit and disconnect all CTs at one feeder end, and short-circuit and disconnect all but red-phase CTs at the other end of the feeder. This simulates a red-phase-to-earth fault fed from one end.

Check that the relays at both ends tend to operate and measure the current in the operate coils.

(*b*) Connect and remove the short-circuit from the red-phase CT at the other end of the feeder and check both relays for stability. Measure the current in the operate coils; it should be negligible.

(*c*) Reverse the pilots and check that the relays now tend to operate. Again measure the current in the operate coils. Reconnect the pilots as normal.

(*d*) Repeat above for the yellow and blue phases and finally with all three phases connected.

PILOT-WIRE PROTECTION: MAINTENANCE

15.18 Maintenance

Maintenance will usually consist of a repeat of:

- loop-resistance measurement
- secondary injection tests
- tripping tests

POWER-LINE-CARRIER PROTECTION: THEORY

15.19 Background

The maximum length of feeder to which pilot-wire protection can be applied, as a result of design limitations, is about 35 km. This approximates to the maximum feeder length for distribution networks, but many transmission feeders are considerably longer.

Power-line-carrier phase-comparison protection is a unit protection which has been successfully applied to 275 and 400 kV transmission lines up to 200 km in length. This type of protection is comparatively expensive because of the coupling equipment required, but it is fast, sensitive and reliable, with the added advantage that the communication medium (the overhead line) is not subject to interference by personnel, as with pilot-wire protections.

15.20 Basic principle

Figure 15.11 illustrates the basic principle. 50 Hz power-system current is used to modulate a high-frequency carrier signal, comprising 10 ms blocks of carrier and 10 ms blocks of no carrier, which is injected via coupling equipment onto the transmission line, and transmitted to the remote end of the feeder. The locally and remotely derived blocks of carrier signal are combined and passed into a phase discriminator. If the gap in the carrier signal (which is a function of the phase-angle difference of the current at both ends) exceeds a predetermined setting (usually equivalent to 30°), the phase discriminator outputs a signal to trip the circuit breaker.

15.21 Essential features of scheme

The essential features of a power-line-carrier phase-comparison protection scheme, as shown in Figure 15.12, are described in Sections 15.22 to 15.40.

15.22 Sequence network

Whereas pilot-wire protection utilises a summation transformer to derive a single output from a three-phase current input, phase-comparison protection utilises a sequence-component network. A feature of power-line carrier protection is that over the frequency range utilised (70–700 kHz), UK Telecommunications Regulations prohibit continuous-carrier transmission, with a restriction of 2 s of carrier per 2 min. The sequence-network approach is a superior design in accommodating this restriction.

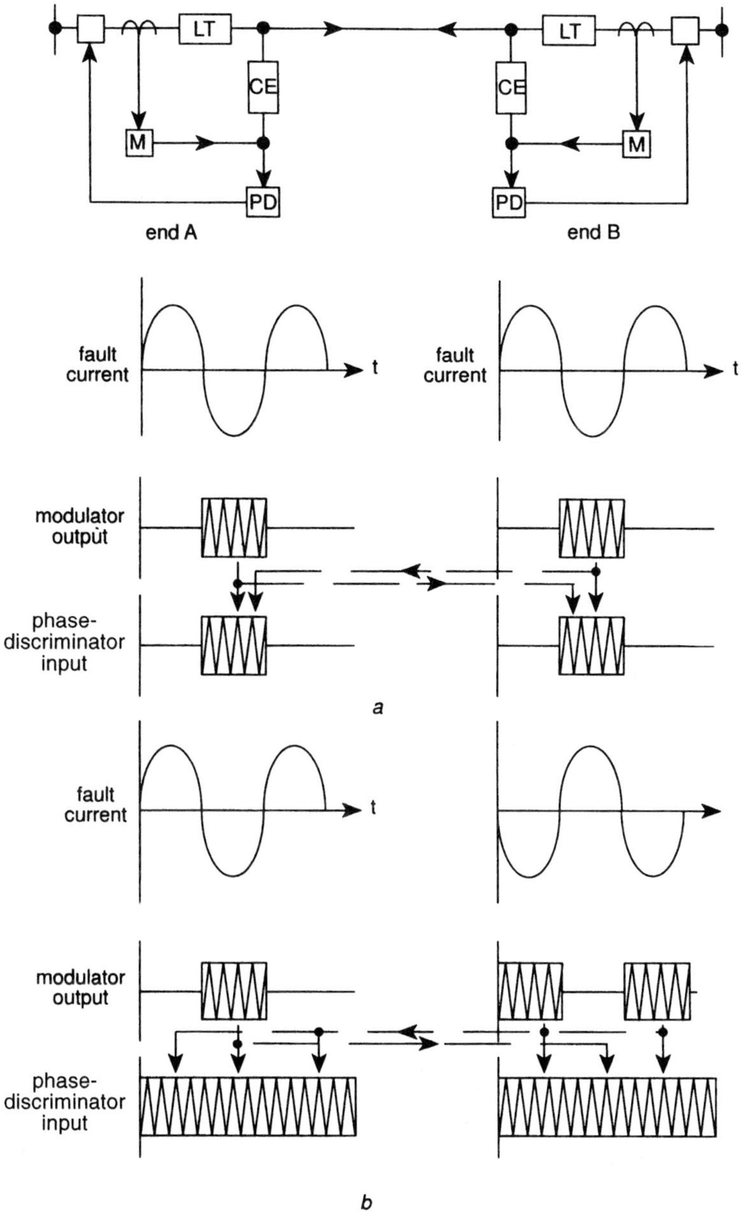

Figure 15.11 *Basic principles of phase-comparison protection*

a Internal fault
b External fault

LT = line trap
CE = coupling equipment
M = modulator
PD = phase discriminator

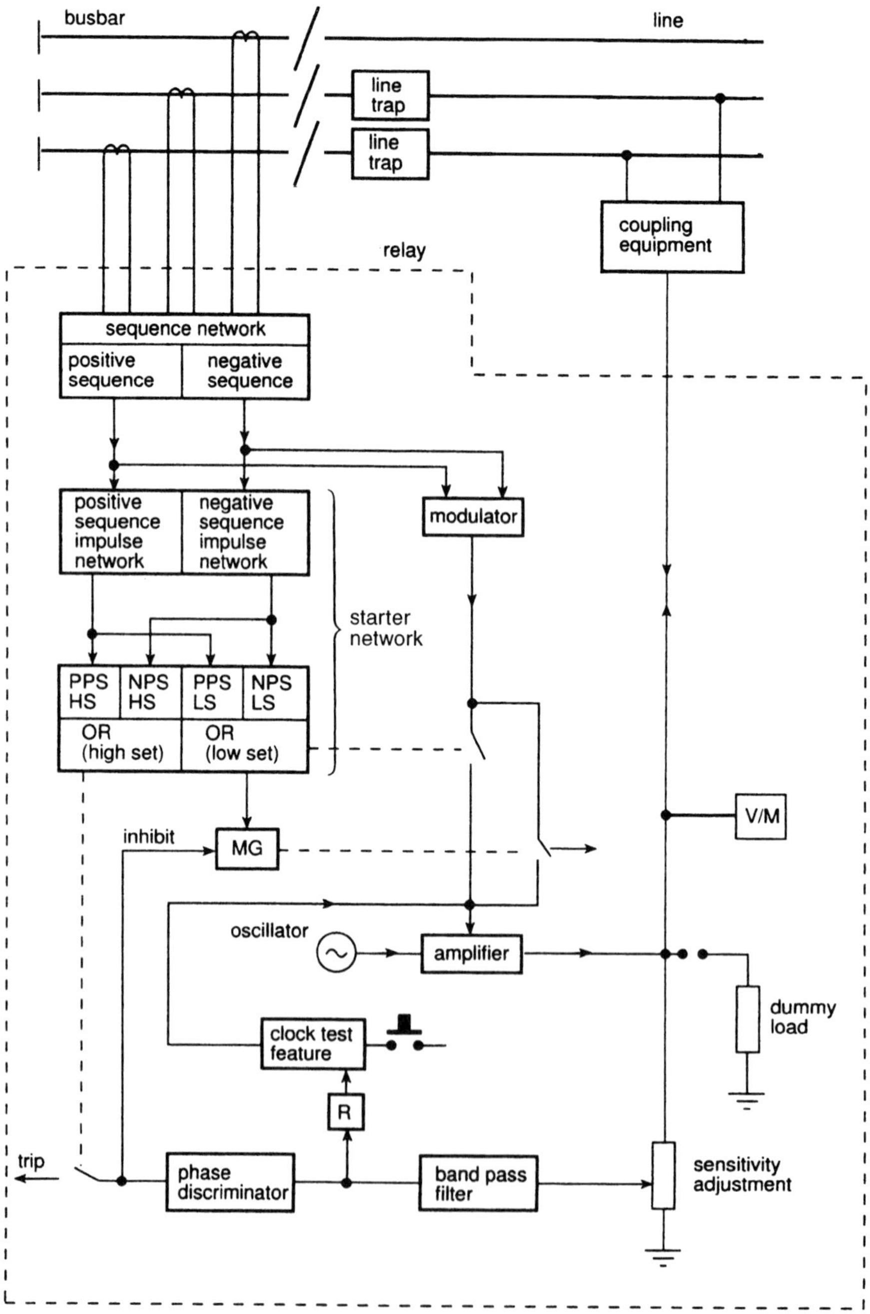

Figure 15.12 Block diagram of phase-comparison scheme

The sequence network derives positive- and negative-phase-sequence quantities from the three-phase currents. This usually involves passing the currents through a delta-connected interposing CT to filter out the zero-sequence component. In some older designs the main CTs themselves are delta connected.

15.23 High- and low-set starters

Starters are required to ensure that the carrier signal is only transmitted under fault conditions and not continuously. From Figure 15.12, the sequence network feeds positive- and negative-phase-sequence quantities (PPS and NPS) to the respective impulse networks. These, in turn, provide an output, proportional to the step change in input current, to the high- and low-set starters. If the level (magnitude) of the output from the impulse network exceeds the level setting of a starter, the starter produces a pulsed output of fixed duration (frequently termed the 'dwell time').

The starter settings are termed:

- PPS impulse high-set (PPS HS)
- PPS impulse low-set (PPS LS)
- NPS impulse high-set (NPS HS)
- NPS impulse low-set (NPS LS)

The high-set current setting for both PPS and NPS starters has a higher setting than the low-set. Typically, high-set = 1.5 × low-set. The low-set starter initiates transmission of carrier and the high-set starter interlocks the tripping. Comparison of carrier quantities must be assured (i.e. low-set starters must be picked up at both ends of the circuit) before tripping is allowed; hence the need for both types of starter.

Early designs of these equipments had an NPS continuous starter in place of the NPS impulse starter, but the NPS impulse starter has superior performance.

15.24 PPS impulse starter

The PPS impulse high-set-starter settings must be sensitive enough to facilitate protection operation and tripping of the circuit breaker for a fault at the far end of a feeder, as shown in Figure 15.13*a*. The most onerous condition associated with such a fault, which the starter must accommodate, is shown in Figure 15.13*b*, and relates to a change from maximum load current to minimum fault current. The starter setting must be of a smaller value than this transition. Once operated, the starter outputs a pulse, typically 500 ms for high-set starter and 600 ms for low-set starter (since carrier transmission must not cease before tripping is inhibited). The ratio between high-set and low-set starter settings must cater for the effect of feeder capacitance current. Figure

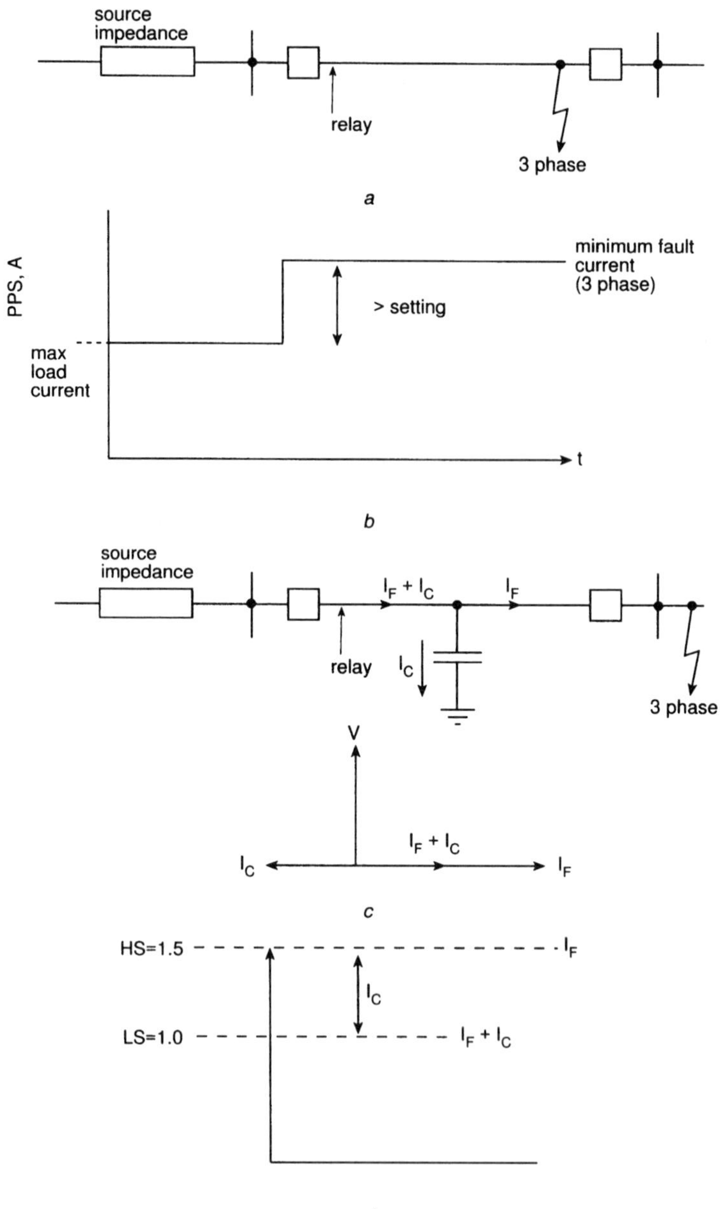

Figure 15.13 Criterion for PPS high-set impulse-starter setting

a Fault at far end of feeder
b Change from maximum load current to minimum fault current
c Fault limited by high source impedance
d Implication for the starters

15.13*c* illustrates an external three-phase fault whose magnitude is limited by a high source impedance, resulting in the phasor diagram shown. Figure 15.13*d* shows the resulting implication for the starters. The maximum value of I_C is equal to 0.5, which is 0.33 of the HS setting. If we assume a 5% error on the HS- and LS-starter pickup, then the maximum permitted magnitude of I_C as a ratio of the HS setting becomes

$$\frac{(0.95\times 1.5) - 1.05}{1.5} = 0.25$$

Thus I_C must be less than 0.25 HS.

15.25 NPS continuous starter

The NPS continuous starter as found on older protections requires feeder NPS current to exceed the starter-setting threshold. A continuous starter output, resulting in a continuous carrier, is maintained until the NPS current falls below the setting again. The NPS component of load current due to the natural unbalance of the system can, however, be as high as 10% (of positive-phase sequence load current). This high standing level of NPS frequently causes conflict with the required sensitivity of the setting, and thus an NPS impulse starter is preferred.

15.26 NPS impulse starter

The NPS impulse high-set starter setting is based on similar criteria to the PPS impulse starter, as illustrated in Figure 15.14. The NPS setting should also be in excess of four times the feeder NPS capacitance current, for similar reasons to that shown in Figure 15.13. The NPS setting is usually more sensitive than the PPS, since the levels of current considered for the setting are less.

15.27 Marginal guard

The low- and high-set starters reset after the dwell times expire. If the starters are picked up due to an external fault, it is important that the low- and high-set starters do not reset at one end of the feeder before the other; otherwise false tripping will occur. If precautions are not taken, this could occur as a result of inaccuracies in the dwell timers. The marginal-guard (MG) feature overcomes this problem by sustaining a continuous carrier signal for some 350 ms after the low-set starter resets. To prevent carrier being transmitted in the event of an internal fault, it is usual to inhibit (mute) marginal guard on operation of the protection.

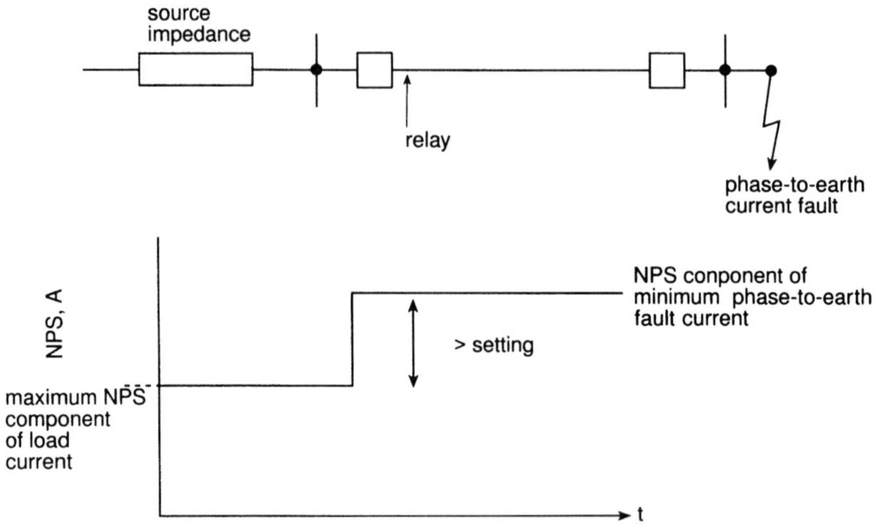

Figure 15.14 Criterion for NPS high-set impulse-starter setting

15.28 Reflex test

All carrier protections have an inbuilt facility termed the reflex test, which proves the integrity of the communication link between transmitter and the remote-end receiver. This test may be initiated either manually or via an inbuilt clock, which usually initiates the test every half hour. When it is initiated, a burst of continuous carrier of about 100 ms duration is transmitted to line. If this is correctly received at the remote end, a signal is transmitted in return, which must be received within about half a second of the initial signal being sent, otherwise an alarm is generated implying that the communication link is faulty. The reflex-test receiver sensitivity (relay R in Figure 15.12) is arranged to have a higher setting than the phase discriminator, to ensure that an alarm is initiated in advance of the received voltage level falling to a magnitude less than the phase-discriminator setting.

15.29 Modulator

The modulator receives positive- and negative-sequence currents from the sequence network and combines them to form a single output. A commonly used modulating quantity is proportional to:

$$5I_2 - I_1$$

or, more generally, $$MI_2 - NI_1 \qquad (15.1)$$

where I_1 = positive-sequence current, I_2 = negative-sequence current and M and N are constants.

Table 15.4 Modulating quantities for different fault types

Fault type	Fault current	Sequence currents	Modulator output ($5I_2 - I_1$)
3-phase	I_F	$I_1 = I_F$	I_F
Phase-to-earth	I_F	$I_1 = 0.33 I_F$ $I_2 = I_1$	$1.32 I_F$
Phase-to-phase	I_F	$I_1 = 0.58 I_F$ $I_2 = -I_1$	$3.48 I_F$

The modulating quantity chosen must be such that an output is provided for all permutations of faults. Table 15.4 gives some comparative values for the instance of the fault-current magnitude being identical for each type of fault.

Feeder capacitance current has a pronounced effect on the design of the modulator. Consider the situation following clearance of an external fault, when the starters have not reset and with little or no load current. At this instant, feeder capacitance current has a maximum impact on modulator performance. The worst situation occurs when capacitance current is fed from both ends of the feeder, as shown in Figure 15.15. This results in the largest possible angle θ between the currents at the two ends of the feeder, which may exceed the angle setting of the phase discriminator, resulting in an incorrect trip. By arranging the modulator characteristic to produce gaps in the carrier signal only after the capacitance-current magnitude has been exceeded (i.e. 0.5 total feeder capacitance current), stability for this condition can be achieved.

Figure 15.16 shows how the stability feature is achieved. In practice, the modulation threshold is usually made equal to the total feeder capacitance current (factor of safety of 2). At magnitudes of current less than the threshold, no gaps in the carrier are produced. As current magnitude increases, the gaps increase until, with very high currents, 180° modulation is achieved.

Finally, consideration needs to be given to the ability to detect high-resistance earth faults in the presence of positive-phase-sequence load current, as shown in Figure 15.17.

Now for earth faults $I_1 = I_2 = I_F/3$. Recalling eqn. 15.1,

$$I_{MOD} = MI_2 - NI_1$$

At end A

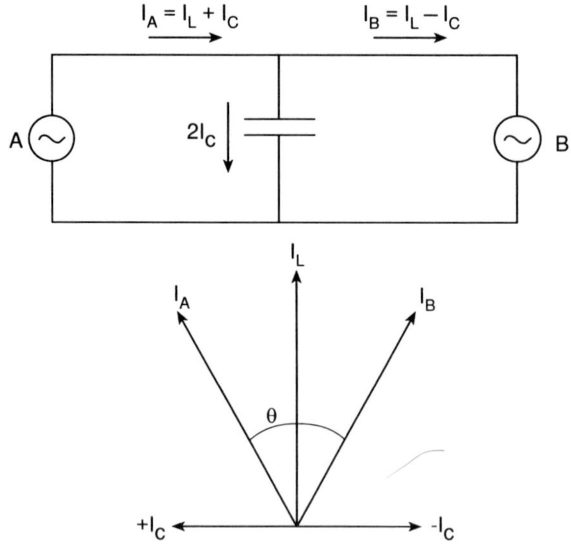

Figure 15.15 Effect of capacitance on modulator characteristic

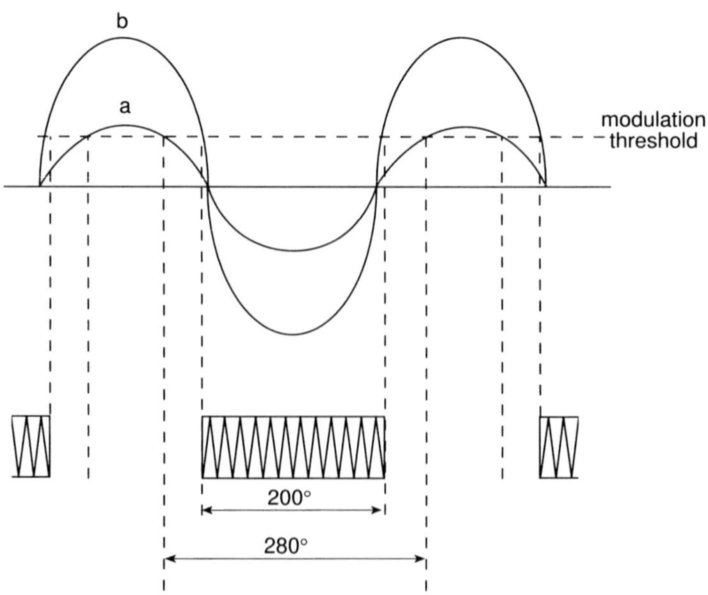

Figure 15.16 Modulation threshold

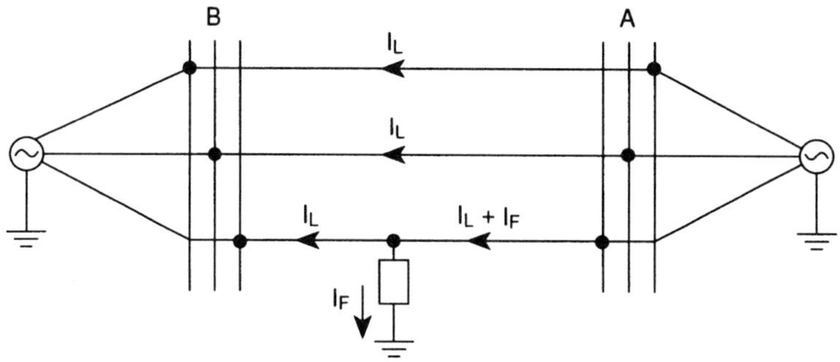

Figure 15.17 Effect of load current on modulation

$$I_{MOD} = \frac{MI_F}{3} - \frac{NI_F}{3} - NI_L$$

$$= \frac{I_F}{3}(M - N) - NI_L$$

To achieve carrier modulation, $I_{MOD} > 0$, i.e.

$$\frac{I_F}{3}(M - N) - NI_L > 0$$

For $M = 5$, $N = 1$

$$I_F > \tfrac{3}{4} I_L$$

Thus the minimum earth-fault current should exceed 0.75 of the maximum load current. Note that modulation should always be present when the low-set starter picks up.

15.30 Oscillator and amplifier

The modulator produces a 50 Hz input into the oscillator/amplifier unit, which produces carrier in accordance with the criterion described in Figure 15.16. The power output of the amplifier is usually in the range of 10–20 W. This ensures a reasonable level of signal, without causing an undue radiated interference signal to other equipment on the system. This unit is frequently termed the transmitter.

15.31 Bandpass filter and phase discriminator

The received signal from the remote end of the feeder and the locally transmitted carrier signal pass through a narrow passband filter to the phase discriminator. The phase discriminator (the receiver) measures the relative phase angle of the current between the two ends of the feeder by looking at the total carrier signal. If the measured gap angle exceeds a predetermined setting, a trip signal is output.

The trip-angle setting needs to be as small as possible, with allowance being made for the natural phase shifts of the feeder, which are briefly as follows:

(i) Transmission time of carrier signal over, say, 150 km of transmission line = 0.5 ms = 9° phase shift
(ii) Phase shift between feeder ends due to capacitance current under external fault conditions = 10°
(iii) Allow 11° for equipment errors = 11°
(iv) Total = 30°

The stability angle is thus ±30° (it may be 40° in some equipment) and the trip angle is 300°.

15.32 Carrier-coupling equipment

Coupling equipment provides the route for the carrier signal, from the protection relay to the transmission line. A typical installation is shown in Figure 15.18. It is usual to utilise two phases of the transmission line as the carrier transmission medium, as this results in lower attenuation of the signal. This is usually the top and bottom phase.

The coupling equipment achieves four basic requirements:

(*a*) It provides the connection between the phase-comparison protection and transmission line via the shunt and series arms.
(*b*) It blocks the signal flow into the busbars, via the line trap, and thus directs it towards the far end of the feeder.
(*c*) It provides the capability for more than one carrier equipment (e.g. protection plus intertripping) to be transmitted down the same transmission line, via the wideband unit.
(*d*) It matches the impedance of the protection equipment and coaxial cable to the transmission line, via the matching transformer, to achieve maximum power transfer.

15.33 The series arm

The series arm consists of capacitances of C_T and C_B, and inductance L_4. It provides a resonant low-impedance circuit at carrier frequency and a high impedance to 50 Hz voltages. It enables the carrier signal to be transmitted to the transmission line. C_T and C_B comprise the capacitances of the capacitor voltage transformer (see Section 10.37). L_4 is usually adjustable to achieve resonance at the carrier frequency. Note that the non-carrier-connected phase would only be equipped with the capacitor voltage transformer.

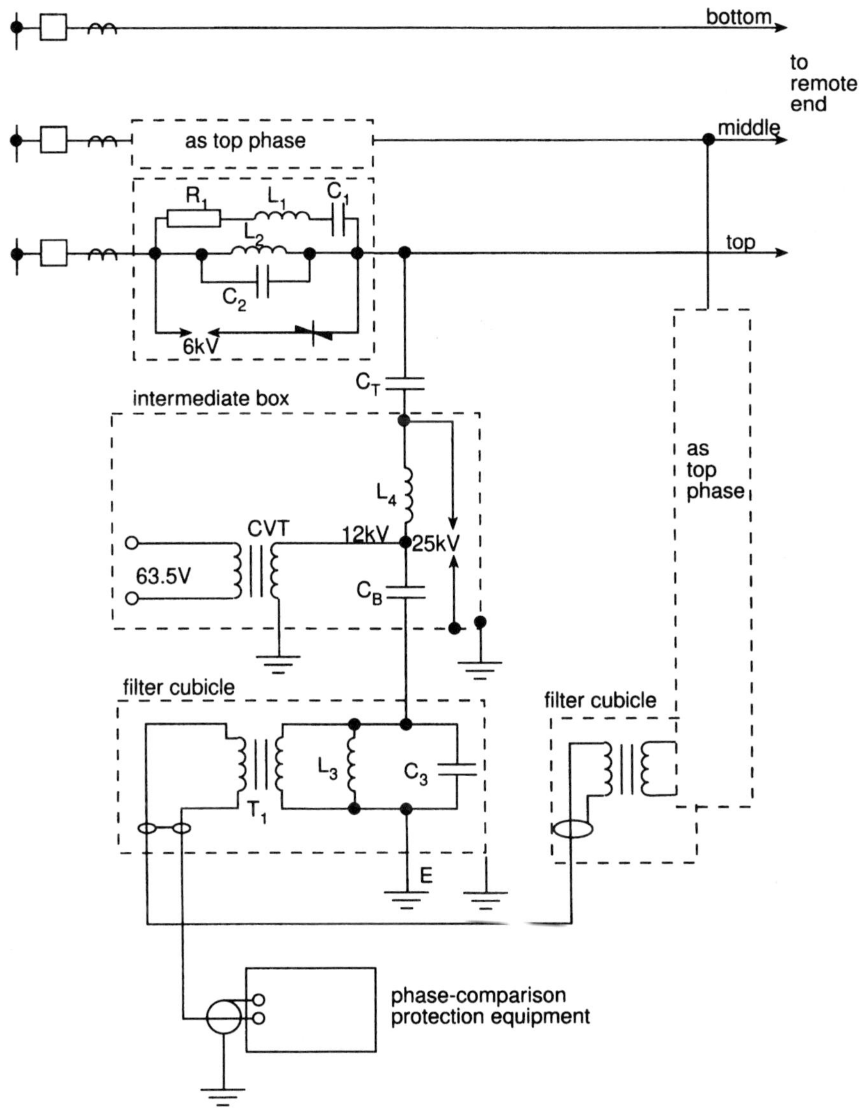

Figure 15.18 Typical carrier-coupling-equipment arrangement

15.34 The shunt arm

The shunt arm consists of capacitance C_3 and inductance L_3. It is resonant, and exhibits a low impedance at 50 Hz and a high impedance at carrier frequency. It prevents 50 Hz voltages from being imposed on the carrier protection.

15.35 The line trap

The line trap comprises capacitance C_2 and inductance L_2. It is resonant, and exhibits a high impedance at carrier frequency and a low impedance at 50 Hz. At 50 Hz L_2 has an impedance less than 0.1 Ω and carries the load and fault currents. The line trap blocks the carrier signal from flowing into the busbars. C_2 is varied to achieve the resonant frequency. The line traps may be suspension or post mounted. It is to be noted that the performance of coupling equipment may be analysed in detail from classical filter theory.

15.36 Matching transformer

The matching transformer T_1 serves two purposes:

(*a*) It matches the surge impedance of the line, typically 600–800 Ω, to the surge impedance of the co-axial cable, 75 Ω.
(*b*) It separates the earth system of the carrier protection from that of the coupling equipment, to avoid any overvoltages or surges being transferred to the carrier protection.

15.37 High frequency earth

To ensure that any surges and overvoltages are conducted directly to earth, it is usual to connect from the shunt unit to earth via a high frequency earth. This is a rod driven vertically into the ground, connected to the base of the shunt unit.

15.38 The wideband unit

The wideband unit consists of resistor R_1, inductance L_1 and capacitance C_1. It serves two functions:

(*a*) In practice, it is difficult to tune the line trap perfectly to carrier frequency. This leaves the possibility of the busbar capacitance acting in conjunction with the line-trap inductance to form a low-impedance path, thus reducing the signal available to be transmitted to line. The wideband unit acts as a controlled-loss circuit, which limits the loss of carrier signal to 50% of the power available. A typical value for R_1 is 300 Ω.
(*b*) Figure 15.19 illustrates the effect of the wideband unit on the resulting apparent impedance of the combined line trap and wideband unit, as frequency is varied. As can be seen, the profile has become wider with a flatter top. This enables two equipments of differing carrier frequencies f_1 and f_2 to be transmitted simultaneously. The bandwidth is given from classical filter

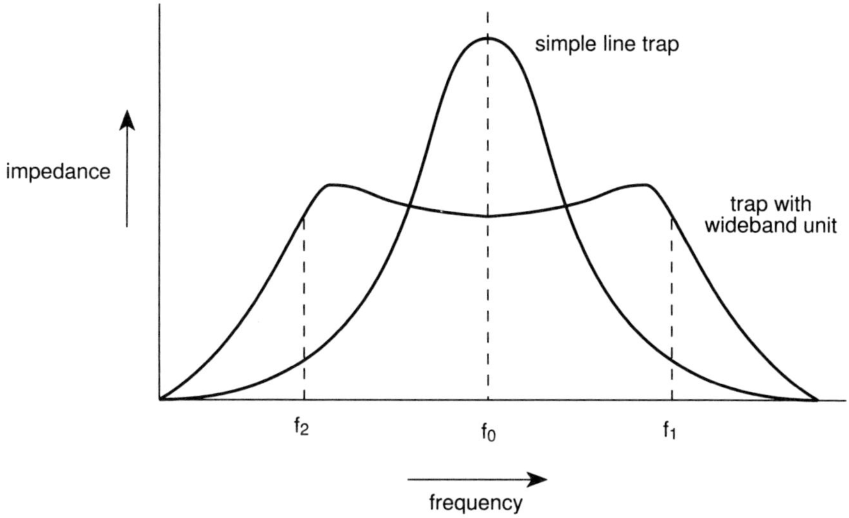

Figure 15.19 Effect of wideband unit

Table 15.5 Effect of load impedance on reflected voltage

Load condition Z_L	Reflected voltage V^-	Total voltage at the load $V^+ + V^-$
$Z_L = \infty$ (open circuit)	V	$2V$
$Z_L = 0$ (short circuit)	$-V$	0
$Z_L = Z_0$ (matched line)	0	V

design as:

bandwidth $= f_1 - f_2$
midband frequency $f_0 = \sqrt{(f_1 f_2)}$ (see Table 15.6)

C1 and L1 are a low-impedance resonant path at carrier frequency.

15.39 Capacitor voltage transformer

Figure 15.18 shows the position of the capacitor-voltage-transformer tapping. It is preferable for inductance L_4 to be positioned above the tapping, as this theoretically results in a lower carrier voltage being applied to the CVT and hence a lower carrier power loss. More modern designs of equipment,

Table 15.6 Coupling bands

Band	Frequency range $(f_1 - f_2)$ (kHz)	Midband frequency $\sqrt{(f_1 f_2)}$ (kHz)
1	70–81	75
2	80–95	87
3	90–110	99.5
4	100–125	112
5	110–140	125
6	120–158	138
7	130–175	151
8	150–214	179.5
9	180–280	225
10	250–500	354
11	350–700	495

however, have repositioned L_4 between the shunt arm and capacitor C_B, and located in the filter cubicle. Performance with this theoretically less efficient arrangement seems equally satisfactory.

15.40 Matching

Transmission lines consist of a distributed impedance over the length of the line, resulting in a finite time for voltage to propagate down the line. Transmission lines therefore need to be interpreted in terms of wave theory. Classical wave theory identifies a forward transmitted voltage wave V^+ from the voltage source and a reflected voltage wave V^- from the load, the resulting voltage at any point on the line being the sum of the two. Corresponding transmitted and reflected current waves also exist and are given by the expressions:

$$\frac{V^+}{Z_0} \text{ and } \frac{V^-}{Z_0}$$

respectively. Z_0 is known as the surge, characteristic or intrinsic impedance of the line and, for the simple case of a loss free line,

$$Z_0 = \frac{\sqrt{L}}{\sqrt{C}}$$

where L and C are the distributed inductance and capacitance of the line.

Transmission-line theory gives the value of the reflected wave at the load as being

$$V^{-} = \left(\frac{Z_L - Z_0}{Z_L + Z_0}\right) V^{+} \qquad (15.2)$$

where Z_L is the impedance of the load.

Based upon eqn. 15.2, Table 15.5 identifies values of reflected voltage waves for certain load conditions for a transmitted voltage of magnitude *V*. In practice, of course, the voltages are sinusoidal, and the effect of a sinusoidal forward and reflected wave is to create a standing-wave pattern along the transmission line, as shown in Figure 15.20. The implications of a termination which produces a standing wave pattern are:

(*a*) the existence of a reflected wave which implies that not all of the incident power is absorbed by the load. This results in an inefficient power-transfer medium;

(*b*) the source is subject to an input impedance, whose value is dependent on the line length. Some equipments are not designed to operate into unreasonably high or low impedances and damage may result.

Power-line carrier equipments are designed so that the terminating impedance Z_L is made equal to the (reflected) line surge impedance Z_0 (i.e. 75 Ω) such that the system is perfectly matched. This maximises the transfer of power and provides a fixed input impedance of 75 Ω irrespective of line length.

15.41 Matching of transmitter to receiver

To ensure perfect matching the receiver must present a 75 Ω load to the incoming carrier signal. Equally, the transmitter needs to generate into a 75 Ω input impedance. Figure 15.21 illustrates a typical transmitter/receiver junction which satisfies this requirement; the coaxial cable itself has a 75 Ω surge impedance.

15.42 Coupling bands

Table 15.6 shows the coupling bands employed. To achieve the coupling band, the various component values within the line-trap, shunt and series arm, passband filter etc. must be appropriately chosen or adjusted.

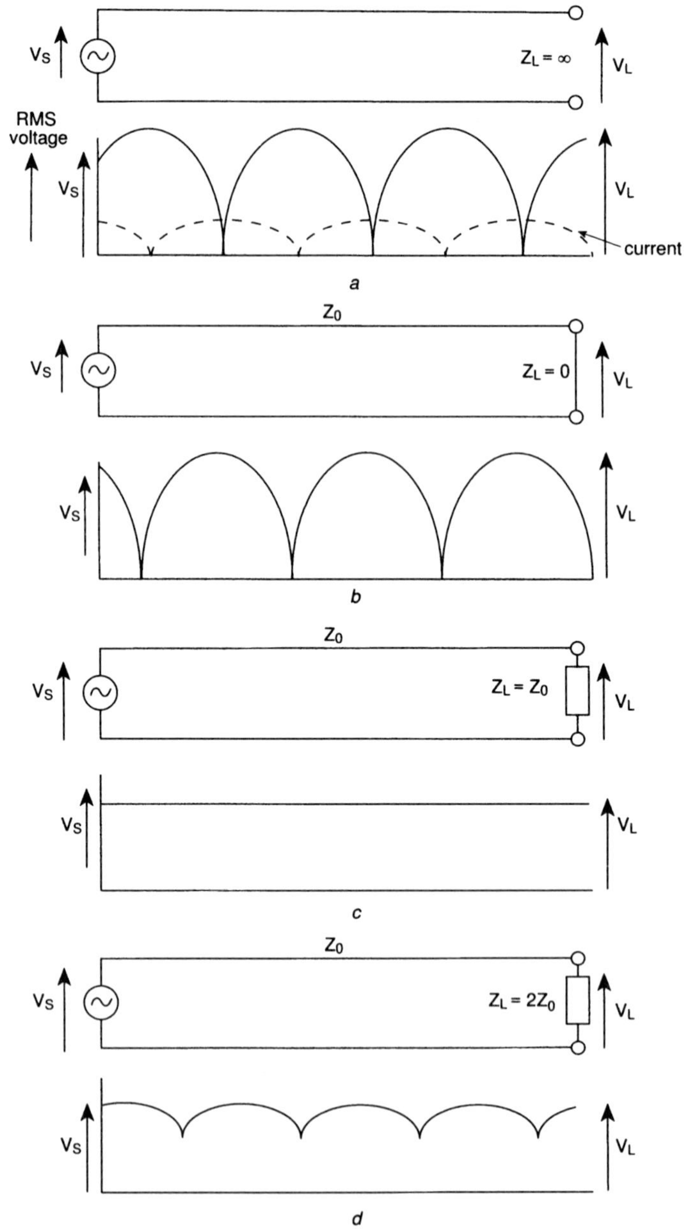

Figure 15.20 Standing-wave patterns

a Open circuit
b Short circuit
c Matched line
d Loaded line

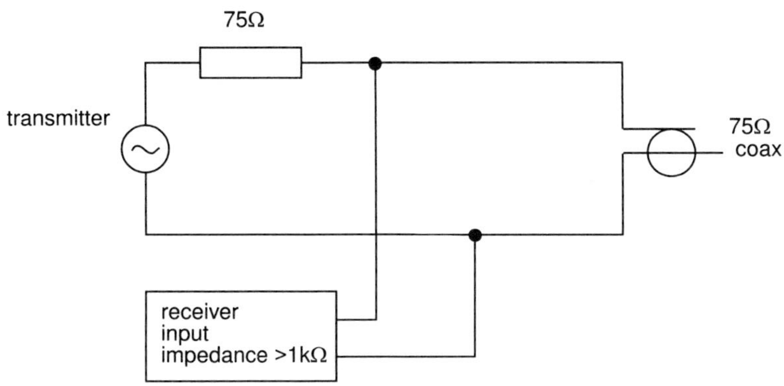

Figure 15.21 Transmitter/receiver junction

15.43 Application

Power-line-carrier phase-comparison protection can be applied successfully to transmission lines of up to about 200 km in length. Above this length, the attenuation of the signal becomes too great, and regulations prohibit increasing the transmitted power to compensate for losses. In addition, as already described, the angle difference caused by capacitance current becomes adversely more significant. To minimise attenuation, the lower-frequency allocations should ideally be applied to the longer feeders.

Power-line carrier cannot usually be applied to cable circuits because the losses are too great. It is also not usually applied to mixed line and cable circuits, as transient reflections from the line/cable junction can fill the carrier gap. For the same reason, it also cannot usually be applied to T circuits.

15.44 Combining filters

With reference to Figure 15.22, a combining filter is employed as the junction between two power-line-carrier equipments, with the objective of preventing one equipment being a source of signal loss to the other (i.e. to enable each equipment to operate independently of the other). The two equipments are usually a power-line-carrier phase-comparison protection equipment and a power-line-carrier protection signalling or intertripping equipment. Some equipments are designed to function without the need for a combining filter.

Figure 15.22 also illustrates the earthing arrangements. It is usual to earth coaxial-cable armouring at all locations, thereby providing electric- and magnetic-field screening. The coaxial-cable outer screen is usually earthed at only one point. This prevents stray currents flowing in the coaxial conductors which could fill any carrier gaps.

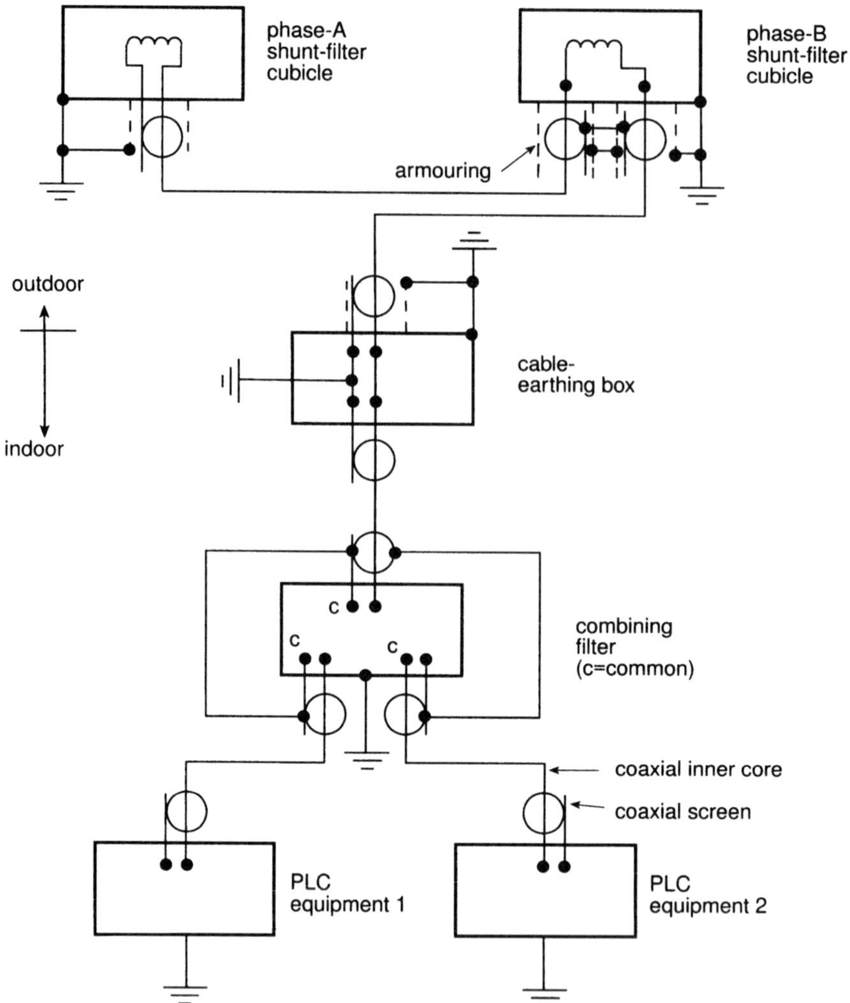

Figure 15.22 Combining filter and system earthing

POWER-LINE-CARRIER PROTECTION: COMMISSIONING

15.45 Commissioning tests

Commissioning tests are required on the phase-comparison protection and the carrier-coupling equipment as follows:

(*a*) *Phase-comparison protection secondary injection*

- starter-current settings
- dwell time of starters

- marginal guard duration
- receiver-voltage sensitivity
- receiver trip angle
- overall operating time
- equipment-specific tests

(*b*) *Primary injection*
See Chapter 10.

(*c*) *Carrier-coupling equipment*

- tuning of line trap, wideband unit, series arm and shunt arm
- checking of matching transformer polarity
- insertion loss
- return loss

(*d*) *End-to-end tests*

- prior to going on-load

(*e*) *On-load tests*

- to prove stability

These tests will be examined in more detail in the following sections.

15.46 Example illustrating settings

To place the commissioning tests in context, the settings for a typical transmission line will be calculated.

The feeder data are:

- Maximum load current (PPS) = 4000 A
- NPS component of load current (NPS) = 400 A
- Minimum three-phase fault current = 5000 A (for fault at far end of feeder)
- Minimum single-phase-to-earth fault current = 3600 A (for fault at far end of feeder)
- Total feeder capacitance current = 40 A (per phase at nominal voltage)
- CT ratio = 2000/1

(*a*) *PPS high-set impulse-starter setting*
From Figure 15.13,
PPS HS < 0.8 (minimum three-phase fault current – maximum load current)
< 0.8 (5000–4000)
< 800 A
< 0.4 A secondary

A typical available equipment setting = 0.3 A.

Notes: 1 The 0.8 used above is a factor of safety.
2 For a 1 A relay, setting would be 30%.
3 For a high-set/low-set ratio of 1.5 the low-set setting = 0.2 A.

Thus PPS HS impulse = 0.3 A (= 600 A primary)
PPS LS impulse = 0.2 A (= 400 A primary)

(*b*) *NPS high-set impulse-starter setting*
From Figure 15.14,

NPS HS <0.8 (NPS component of minimum earth-fault $-$ NPS component of load current)
<0.8 (3600/3–400)
<640
<0.32 A secondary

Typical nearest setting is 0.12 A.

Notes: 1 The 0.8 used above is a factor of safety.
2 For a 1 A relay, setting would be 12%.
3 For a high-set/low-set ratio of 1.5 the low-set setting = 0.08 A.

Thus NPS HS impulse = 0.12 A (= 240 A primary)
NPS LS impulse = 0.08 A (= 160 A primary)

(*c*) *Modulator-threshold setting*
From Section 15.29,

(i) Modulation threshold >0.5 (capacitance current)
$>0.5 \times 40$
>20 A
>0.01 A secondary

Typical setting would be 0.012 A
(i.e. 12% on a 1 A relay).

(ii) Ratio of $\dfrac{\text{minimum earth-fault current}}{\text{maximum load current}}$

$$=\frac{3600}{4000}$$

$$=0.9$$

This exceeds 0.75 (see Section 15.29); therefore modulation will take place.

15.47 Termination in dummy load

Before commencing testing it is necessary to disconnect the relay from the coaxial cable and terminate it in a 75 Ω noninductive resistor, to simulate a perfectly matched line. Most equipments have inbuilt facilities for this purpose.

15.48 Starter-current settings

It is usually difficult to obtain an adequate balanced three-phase-current source on site; therefore it is usual to utilise single-phase and phase-to-phase current injection. Table 15.7 summarises the required measurements, using the setting data determined in Section 15.46, and the subsequent text illustrates how the injected current is determined.

R–E injections

From Chapter 3,

$$I = I_1 + I_2 + I_0 \text{ (where } I = \text{injected current)}$$

$$\text{and} \quad I_1 = I_2 = I_0$$

$$\text{so} \quad I_1 = I_2 = I/3$$

Therefore, for operation:

$$\frac{I}{3} = \text{setting}$$

or $I = 3 \times \text{setting}$

Thus the current I to be injected for A, B, C and D in Table 15.7 is 0.9 A, 0.6 A, 0.36 A and 0.24 A, respectively.

Y–B injections

From Chapter 3,

$$I = I_Y = -I_B \text{ (where } I = \text{injected current)}$$

$$\text{and} \quad I_{R1} = -I_{R2}$$

$$\text{Now} \quad I_Y = I_{Y1} + I_{Y2}$$

$$= a^2 I_{R1} + a I_{R2}$$

$$= (a^2 - a) I_{R1}$$

$$\text{therefore} \quad I_Y = \sqrt{3}\, I_{R1} = 1.74\, I_{R1}$$

$$\text{so} \quad I_1 = I_2 = I/1.74$$

$$\text{Therefore} \quad \frac{I}{1.74} = \text{setting}$$

$$\text{or} \quad I = 1.74 \times \text{setting}$$

Thus the current I to be injected for E, F, G and H in Table 15.7 is 0.522, 0.348, 0.208 and 0.139 A, respectively.

Although only R–E and Y–B have been cited above, it would be usual to inject R–E, Y–E, B–E, R–Y, Y–B, B–R.

Table 15.7 Required injections

Settings to be checked	Phases injected	
	R–E	Y–B
PPS impulse HS op current (0.3 A)	A	E
PPS impulse LS op current (0.2 A)	B	F
NPS impulse HS op current (0.12 A)	C	G
NPS impulse LS op current (0.08 A)	D	H

(where A, B, C, D, E, F, G, H represent the injected current as described below)

The technique to be employed is to switch on the injected current at increasingly higher levels until the starters operate. Pickup of the low-set starter can be verified by observing a burst of carrier on an oscilloscope connected across the dummy load. Pickup of the high-set starter is verified by the protection tripping (as a result of single end transmission, 180°-modulated carrier). The NPS starters have a more sensitive setting and need to be inhibited when testing the PPS starters. Facilities are usually provided on the equipment for this purpose.

15.49 Starter dwell time

The starter dwell times can be obtained by initiating a timer from the same switch that switches on the current, and stopping the timer from spare contacts on the high-set and low-set relays, respectively. The injected current should be at least three times the setting current.

15.50 Marginal-guard duration

Marginal-guard duration can be obtained by monitoring carrier transmission on one channel of an oscilloscope, and monitoring low-set operation/reset by connecting from a spare low-set-relay contact, through a battery, to the other channel of the oscilloscope. Switching on current will result in a low-set impulse operation and reset after the dwell time. The duration of carrier (which should be unmodulated) after low-set resets is the marginal-guard time.

15.51 Receiver sensitivity

With reference to Figure 15.12, this normally consists of three tests:

- check on receiver sensitivity R (alarm sensitivity)
- check on phase-discriminator sensitivity (tripping sensitivity)
- check on bandpass filter

Using an external oscillator (with 75 Ω output impedance), or the in-built facilities of the relay, apply 1.5 V (usual receiver sensitivity) across the dummy load. Vary the sensitivity adjustment until the relay R just operates. Leave the sensitivity adjustment in this position. This causes the reflex-test-alarm facility to be sensitive to transmitted carrier signals down to a magnitude of 1.5 V.

The high-set starter should now be initiated manually and the oscillator voltage reduced until the phase discriminator produces a trip output. The voltage at which this occurs is usually 1.0 V (i.e. trip-to-alarm ratio is usually 1.0/1.5); this setting should be preset. Note that these settings are with a continuous carrier applied.

The filter sensitivity is usually checked by applying the full carrier voltage from an oscillator across the dummy load and sweeping through the frequency range 50–700 kHz, and noting that the test receiver R is only picked up in the passband.

15.52 Receiver trip angle

The test of receiver trip angle would usually consist of monitoring the transmitted carrier signal across the dummy load with an oscilloscope, manually operating the high-set and low-set relay and then slowly increasing the modulation angle until the relay operates. Modulation-angle variation can be achieved by:

- use of inbuilt facilities
- slowly increasing injected current above modulation threshold

The modulation gap at which tripping occurs (MOD) measured in milliseconds should be recorded, and used to determine the trip angle from:

$$\text{trip angle} = 360° \times \frac{\text{MOD ms}}{20\text{ ms}}$$

For a trip angle of 30°, MOD would be 1.67 ms

It is usual to carry out this test at two levels: at full carrier-signal strength and at a carrier-signal strength equal to the receiver-sensitivity-setting voltage (i.e .1.5 V).

15.53 Overall operating time

The overall operating time is simply the operating time as measured on a timer from the instant of current injection (at say 3 × high-set setting) to protection operation. A typical time would be 30 ms for a modern protection.

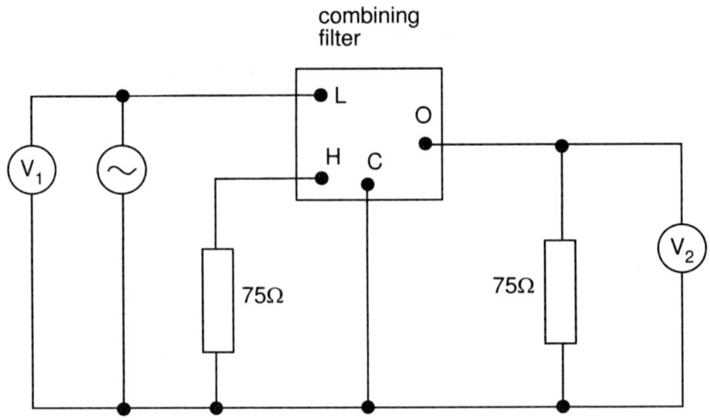

Figure 15.23 *Testing a combining filter*

L = lowpass input
H = highpass input
O = output
C = common

15.54 Equipment-specific tests

These relate to facilities specific to a type of equipment. For example, some equipments disable and lock out the protection if the transmission time exceeds 2 min.

15.55 Combining filter

Figure 15.23 shows the test method for the combining filter. A frequency run should be carried out. At the passband frequency of the input under test (i.e. input L in Figure 15.23) the voltage on V_2 should be a maximum with a magnitude of at least 80% of the applied voltage V_1. While sweeping through the passband frequency of the input not under test, the voltage on V_2 should not exceed about 30% of the input voltage. The test should be repeated for the other input with the test connections of the signal generator and the 75 Ω load (to terminals L and H) transposed.

15.56 Coupling-equipment tuning

A few general points are worthy of note:

(i) Care must be taken that the unit under test does not have other components of the coupling equipment inadvertently in parallel with it. For this reason, the line trap should normally be earthed on the feeder side.

(ii) Test connections should be as short as possible to minimise stray capacitance.
(iii) When the line trap is tuned, the down droppers from the overhead line should be disconnected to minimise the effect of induced voltages in the line affecting the test quantities.
(iv) If single-channel working is employed and a wideband unit is not fitted, the coupling equipment should be tuned to the single channel in use; if a wideband unit is fitted, the coupling equipment should be tuned to the midband frequency.
(v) The following test method utilises an oscillator-amplifier unit. The oscillator-amplifier system should have a high open-circuit voltage (typically 50–100 V) to mask possible interference voltages. Alternative methods are based on an impedance-measuring device in conjunction with a selective level meter.
(vi) At higher frequencies (above 500 kHz) stray capacitances may corrupt the results slightly.

Specific tests are as follows:

(*a*) *Line trap*
The test arrangement is shown in Figure 15.24*a*. The 30 000 Ω resistor provides a high source impedance so that the amplifier output voltage is relatively unaffected by line-trap-impedance variation. The resistor should be noninductive and of about 0.5 W rating. The wideband unit should be disconnected. It is usually necessary to carry out this test from a mobile bucket positioned next to the line trap. Inevitably, it will rain, so be prepared.

The technique is to carry out a frequency run. At resonant frequency, the maximum voltage will be obtained on the voltmeter. It may be necessary to vary the shunt capacitance to obtain the resonant frequency. This can be a very fiddly job. It should be noted that, at resonance,

$$\frac{1}{2\pi fC}=2\pi fL$$

or

$$f=\frac{1}{2\pi\sqrt{LC}}$$

Thus, to raise the resonant frequency, the value of the capacitance should be decreased.

(*b*) *Shunt unit*
The shunt unit should be disconnected from the matching transformer and series arm, and tested for parallel resonance as for the line trap (see Figure 15.14*b*).

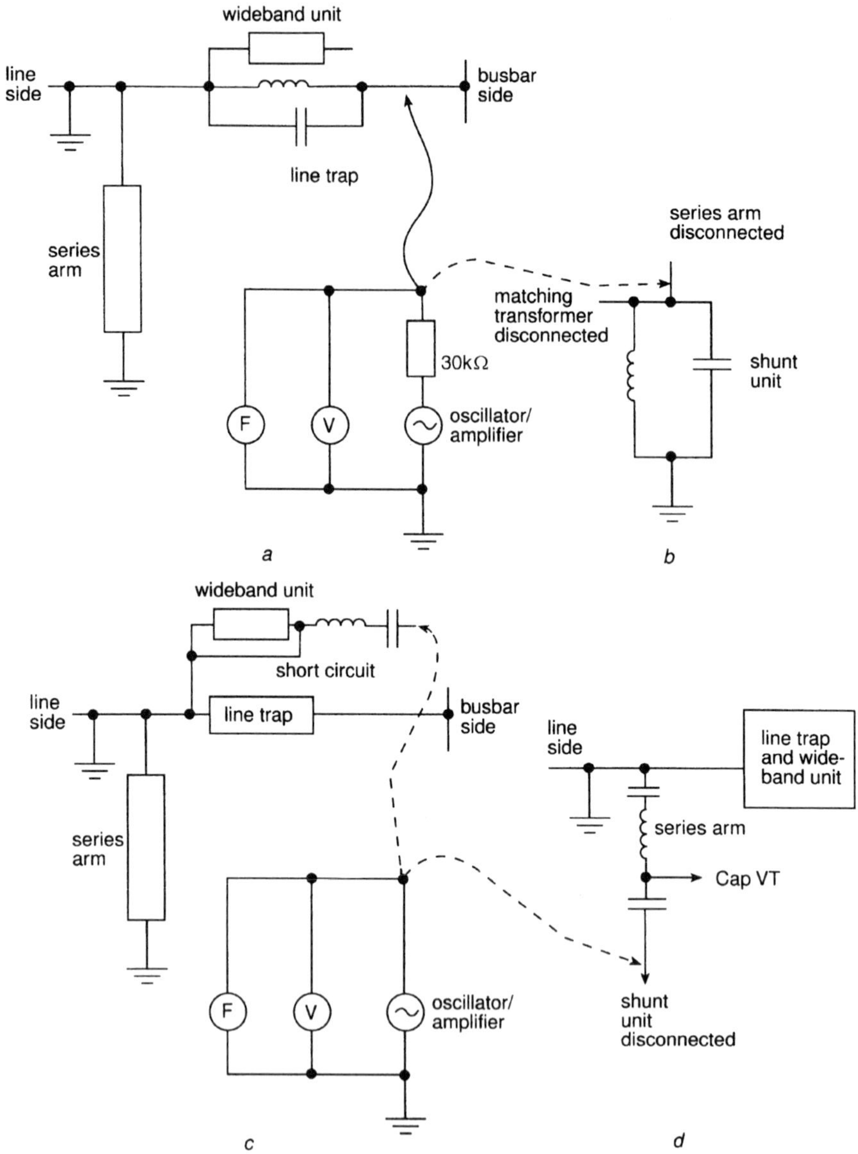

Figure 15.24 Coupling-equipment tuning connections

a Tuning line trap
F = frequency meter
V = voltmeter
b Tuning shunt unit
c Tuning wideband unit
d Tuning series arm

Table 15.8 Some actual tuning results

Frequency (kHz)	Line trap (V)	Shunt unit (V)	Wideband unit (V)	Series arm (V)
95	0.5	0.9	9.0	8.0
100	0.6	1.1	8.5	5.0
106	1.0	3.5	8.0	3.0
108	1.5	4.5	6.0	2.7
110	4.0	8.0	4.5	4.0
115	6.0	6.0	4.8	6.8
120	3.0	4.0	6.0	9.0
125	0.9	1.3	9.0	10.0

Target resonance frequency = 110 kHz

(c) Wideband unit
The wideband unit should be disconnected from the line trap, the resistor short-circuited and the circuit tested as a series resonance circuit. The voltmeter reading is a minimum at resonance. The inductance is usually adjustable to vary the resonant frequency (see Figure 15.24*c*).

(d) Series arm
The series arm should also be tested as a series resonant circuit. Adjustment of resonance frequency can normally be obtained by a slug in the series tuning unit (see Figure 15.24*d*). Table 15.8 gives some typical results.

15.57 Matching transformer

To check the matching-transformer polarity, a signal, at the midband frequency of the coupling equipment, is injected into the coaxial cable, as shown in Figure 15.25. For correct polarity, the signal voltage will be depressed to a low value when the short-circuit is applied. Polarity is important in ensuring that the two line conductors are at positive and negative polarity, respectively.

15.58 Overall Tests

The preceding sections have described commissioning tests on component units of the coupling equipment. Tests therefore need to be carried out on the whole, connected system. These consist of an insertion-loss test to prove the local end and a return-loss test to prove the whole communication link.

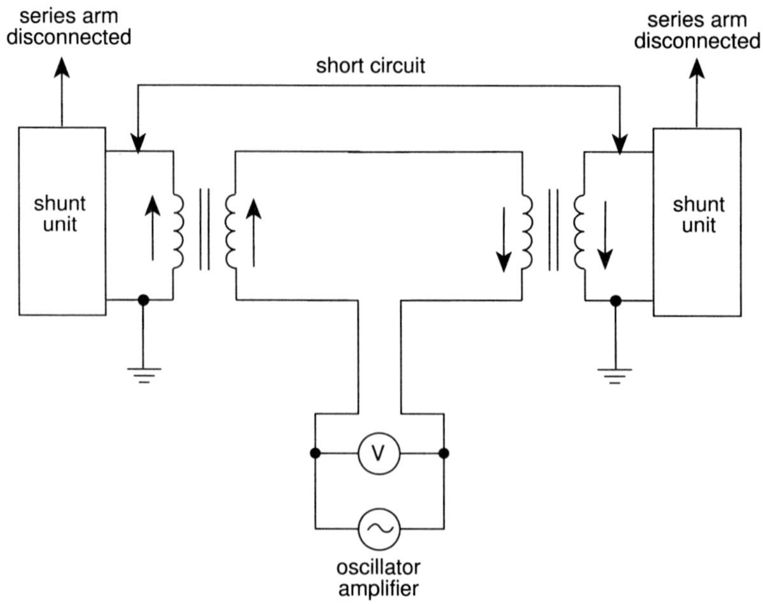

Figure 15.25 Matching transformer polarity

The tests are interpreted in terms of a unit of measurement termed the decibel. This is defined as:

$$\text{decibel (dB)} = 10 \log_{10} \frac{V_1}{V_2}$$

where V_1 and V_2 represent the voltage input and output of a two-port network. The advantage of using this unit is that the attenuation of a succession of stages in an electrical system can, if expressed in logarithmic form, be added (as opposed to being multiplied).

It is more usual to work in terms of power than in terms of voltage. Power is proportional to the square of the voltage. Thus

$$\text{dB} = 10 \log_{10} \frac{V_1^2}{V_2^2}$$

$$\text{or dB} = 20 \log_{10} \frac{V_1}{V_2}$$

The decibel is commonly used in telecommunication practice, but rarely in power engineering; it is not a term towards which power engineers naturally gravitate.

It is also usual to relate the power level to some datum. This is generally 1 mW into 600 Ω, which corresponds to an applied voltage of 0.775 V; this is

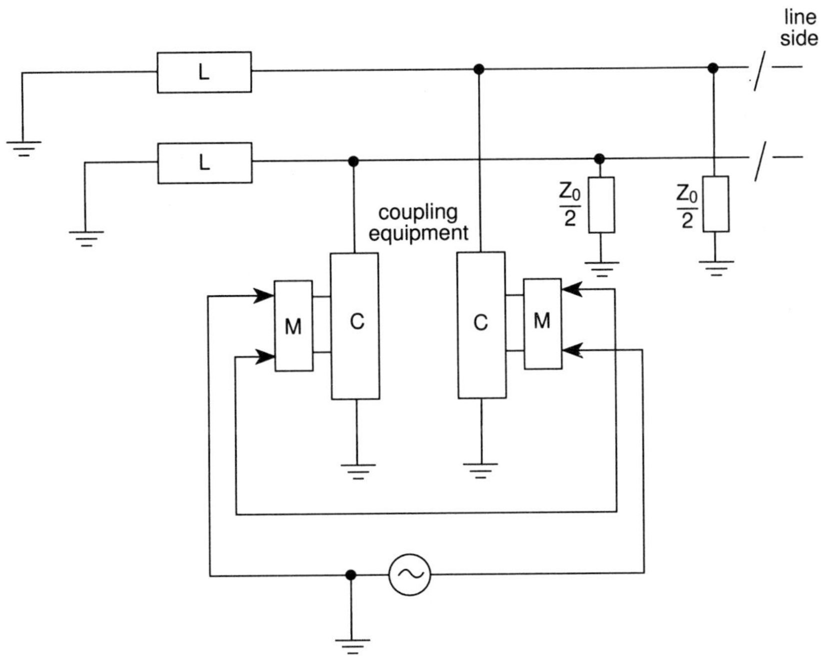

Figure 15.26 Insertion-loss test

L = line trap + wideband unit
C = series + shunt arm
M = matching transformer

equivalent in power terms to 0.274 V into 75 Ω (known as 0 dBm). Thus a generated power of 10 W into 600 Ω is expressed as

$$10 \log_{10} \frac{10}{0.001} = 40 \text{ dB}$$

15.59 Insertion loss

Insertion loss is the simple ratio of the voltage V_2 out of a network divided by the voltage V_1 into the network, expressed in the form:

$$\text{insertion loss} = 20 \log_{10} \frac{V_2}{V_1}$$

In effect, it is a measure of the loss caused by the network.

Figure 15.26 shows the connections required for an insertion-loss test on a single end of coupling equipment; the oscillator should be connected to the coaxial cable, in place of the phase-comparison relay. The oscillator should have a 75 Ω output impedance to provide a matched system. Before the test is carried out, the following should be checked:

- all connections must be very clean
- the injection system must only be earthed at one point
- Z_0 is equal to the line surge impedance, e.g. 600 Ω
- if a combining filter is in service, the test should be carried out from each input of the combining filter in turn, with the other input terminated in 75 Ω.

The oscillator output should be set to about 20 V with the frequency selected to the operating frequency of the carrier protection. The insertion loss is calculated from

$$20 \log \frac{V_2}{V_1 \times T} \text{ dB}$$

where V_2 is the arithmetic sum of the voltages across each test impedance $Z_0/2$, V_1 is the oscillator voltage, and the T is the matching-transformer ratio.

The insertion loss should generally not exceed 5 dB for systems without a combining filter, or 7 dB for systems with a combining filter. More sophisticated testing techniques than the above may be employed, utilising selective level meters.

15.60 Return loss

Figure 15.27 illustrates the test connections required to carry out a return-loss test. The objective is to measure the degree of mismatch between the ideal characteristic impedance and the actual measured value. It is defined by the relationship

$$\text{return loss } (RL) = 20 \log_{10} \frac{(R+Z)}{(R-Z)} \text{ dB}$$

where:

R= the ideal characteristic impedance of 75 Ω
Z= the actual impedance across which voltage V_L is dropped

Examples of return-loss values are:

for $Z=R$, RL= infinity (the ideal)
for Z= open circuit (i.e. infinity), $RL=0$
for Z= short-circuit (i.e. zero), $RL=0$

Some carrier equipments are capable of working into mismatched lines, but others may be damaged unless the return loss is sufficiently high.

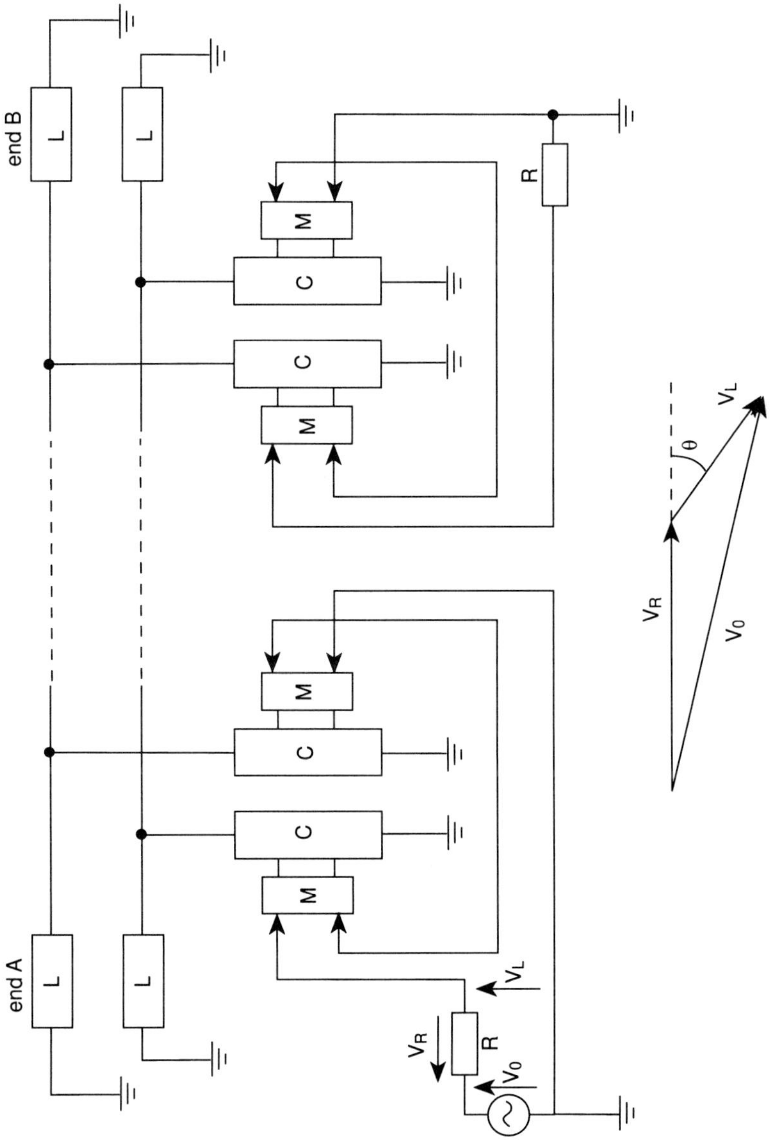

Figure 15.27 Return-loss test

L = line trap + wideband filter
C = series + shunt arm

M = matching transformer
R = 75 Ω noninductive resistor

From Figure 15.27

$$\theta = 180^\circ - \cos^{-1}\frac{V_R^2 - V_L^2 - V_0^2}{2V_R V_L}$$

Now $$Z = \frac{RV_L}{V_R}$$

Therefore $$r = Z\cos\theta$$

and $X = Z \sin \theta$ (where r and X are the resistive and reactive components of Z)

Therefore $$RL = 20 \log_{10}\frac{\sqrt{\{(75+r^2)+X^2\}}}{\sqrt{\{(75-r^2)+X^2\}}}$$

The test should be carried out from both ends of the circuit in turn. The technique consists of carrying out simultaneous readings of V_R, V_L and V_0 and calculating *RL* from the above. It is usual to carry out the test at a number of frequencies in addition to the carrier frequencies in use. The oscillator output voltage should be about 10 V. For satisfactory matching, the *RL* at the carrier frequency in use should be 10 dB or better.

An alternative method to the above is to utilise an impedance-measuring bridge in conjunction with a selective level meter.

This test can be problematic, and unusual results are occasionally obtained. This may be due to sharp deviations in the line route, changes in the type of construction, or relatively large-amplitude high-frequency noise 'standing' on the line.

15.61 End-to-end test

The end-to-end test consists of earthing the line traps on the busbar side, at both ends of the circuit, transmitting unmodulated carrier signal from the protection equipment, at each end in turn, and measuring the send and receive voltages. For a transmitted power of, say, 10 W, the corresponding transmitted voltage would, for perfect matching, be:

$$V = \sqrt{(10\ \text{W} \times 75\ \Omega)} = 27.4\ \text{V}$$

Typical and acceptable transmitted voltages experienced in practice are between 19 and 32 V. Typical, and acceptable, received voltages in practice are between 5 and 12 V. The highest transmitted voltage does not necessarily correspond to the highest received voltage. A received voltage of 3 V or less would cause concern.

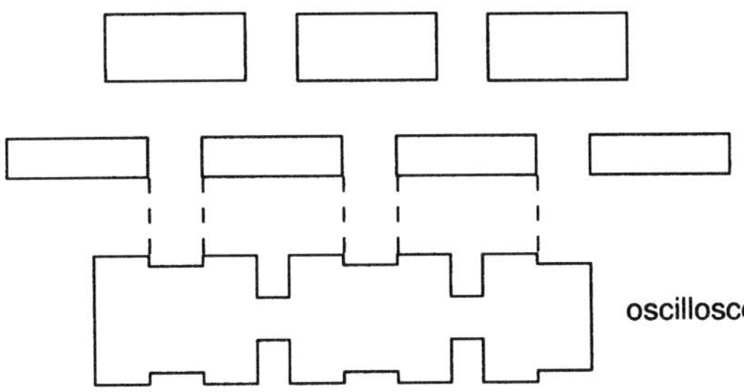

Figure 15.28 On-load test oscilloscope trace

15.62 On-load tests

With the circuit placed on load, a repeat of the end-to-end test should be carried out, to verify that the amplitude of the received voltage is acceptable during in-service conditions.

The objective of the on-load test is to prove that the CTs and equipment are correctly installed such that the protection will stabilise for an external fault. In brief the test consists of:

(*a*) connecting an oscilloscope, with a memory feature, across the coaxial-cable connection to the phase-comparison relay;
(*b*) picking up the low-set starter at both ends of the circuit to generate modulated carrier, and observing on the oscilloscope a continuous carrier signal with no gaps, thereby proving stability for through current. Facilities are usually provided in the phase-comparison relays to assist with this process.

It is necessary to have levels of load current which exceed the modulator threshold setting significantly; otherwise continuous, or near continuous, carrier will be generated. To maximise the modulation, the yellow- and blue-phase CTs at each end should be short-circuited and disconnected. Recalling that modulation is equal to:

$$5I_2 - I_1$$

with three-phase load current, modulation $= I_1 = I$ (where $I =$ load current)

but with only red-phase current entering the relay the equivalent of a red-earth fault is simulated; thus:

$$I_2 = I_1 = I/3$$

Therefore modulation $= 5I/3 - I/3 = 4I/3$ which is 33% more current, for modulation, than with three-phase load current. It is helpful to have a plot of modulation angle against single-phase-to-earth injection current, to give an appreciation of how much modulation will be achieved with the load current available.

Figure 15.28 gives an illustration of an actual on-load test oscillogram. The shape may at first sight seem puzzling. The local-end transmitted carrier appears at full magnitude, whereas the remote-end transmitted carrier is attenuated by the communication channel. A number of zones exist on the oscilloscope trace:

- local-end carrier only
- remote-end carrier only
- local end + remote end

Note that the local- and remote-end oscillators will be functioning at marginally different frequencies. As a result, the zone consisting of the summation of local- and remote-end carrier may in some instances appear as an increase in carrier signal (phase angles of oscillators coincide) and at other instances a decrease (phase angles of oscillators are antiphase).

POWER-LINE-CARRIER PROTECTION: MAINTENANCE

15.63 Maintenance

Maintenance tests would usually consist of a repeat of the commissioning secondary-injection tests, and a repeat of the end-to-end carrier-signal level measurements, both with earth switches closed behind the line traps and on-load. These tests were, for many years, carried out two yearly, but experience with the equipment has dictated that four to six years is acceptable.

It is not usual to carry out routine maintenance tests on the coupling equipment. If the received carrier levels are satisfactory, the coupling equipment is healthy. A visual inspection of the line trap should be carried out every four to six years.

15.64 Fault investigations

Although operating experience with this type of equipment is very good, maloperations do occasionally occur. Points worthy of checking are:

(*a*) whether the equipment is correctly earthed;
(*b*) whether a disconnector, in close proximity to the equipment, was making or breaking charging current at the time of the incident. Interference voltages from disconnector arcing can cause filling in of gaps in the carrier signal;
(*c*) the weather conditions at the time of the failure. The signal attenuation in the transmission line increases significantly as a result of conductor icing.

The starting point for any investigation is to obtain the carrier send- and receive-level voltages. It is also helpful if the low- and high-set outputs were monitored on a fault recorder at the time of the incident. As a final resort, full recommissioning tests should be repeated. In addition, tests on protective spark gaps should be carried out and the effect of disconnector arcing during the opening and closing sequence observed on the equipment.

VOICE-FREQUENCY PHASE-COMPARISON PROTECTION

15.65 Voice-frequency phase-comparison protection

With the phased withdrawal of metallic pilot wire cables by British Telecom, phase-comparison protection operating at voice frequencies (VF) was brought into service in the 1980s. This protection is capable of functioning over British Telecom communication channels or other similar systems which are either metallic (now rapidly being replaced) or nonmetallic.

The principle of operation is almost identical to that of power-line-carrier protection. However, whereas the transmission time of the carrier signal, over power transmission lines, is virtually that of the speed of light, the transmission time through the public telephone system can be much longer. As a result, the locally derived carrier signal is held in a memory until such time as the remotely transmitted carrier signal arrives (typically 5-10 ms). Both signals are then combined for evaluation in the phase discriminator. To determine the length of time the carrier signal is held in memory, the equipment carries out a continuous process of measuring the carrier transmission between the two circuit ends. As a result this equipment tends to be more complex than its power-line-carrier equivalent.

15.66 Fibre-communication channels

Suffice it to say that unit protections capable of operating over a complete fibre route have already been installed, working on a biased differential principle; such systems point the way to the future.

Chapter 16
Unit protection of substation plant

16.1 Introduction

This chapter examines the theory, application, commissioning and maintenance of unit-protections for items of substation plant such as generators, transformers, reactors, capacitors, busbars and mesh corners. This concerns two basic types of unit protection:

(*a*) high-impedance circulating-current protection (also known as high-impedance differential protection);
(*b*) biased differential protection.

HIGH-IMPEDANCE CIRCULATING-CURRENT PROTECTION: THEORY

16.2 Stability voltage

Figure 16.1 illustrates the basic principles of a circulating-current scheme. The current distribution is shown for the out of zone fault F_1. This causes current to circulate around the CTs with no current through the relay. For the internal fault F_2, the current I_F through CT B changes direction. This results in a current through the relay of $2I_F^1$, causing the relay to operate correctly.

In practice, fault currents contain transient DC components (see Section 10.10), which can result in the saturation of CT cores and mismatch of the secondary currents. This causes, for fault F_1, a spill current through the relay

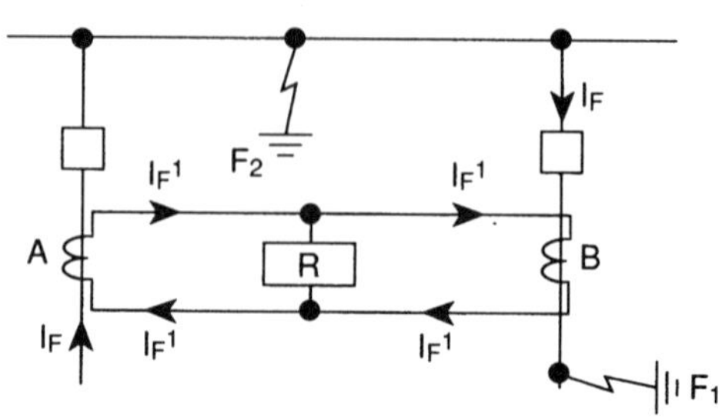

Figure 16.1 Circulating-current protection

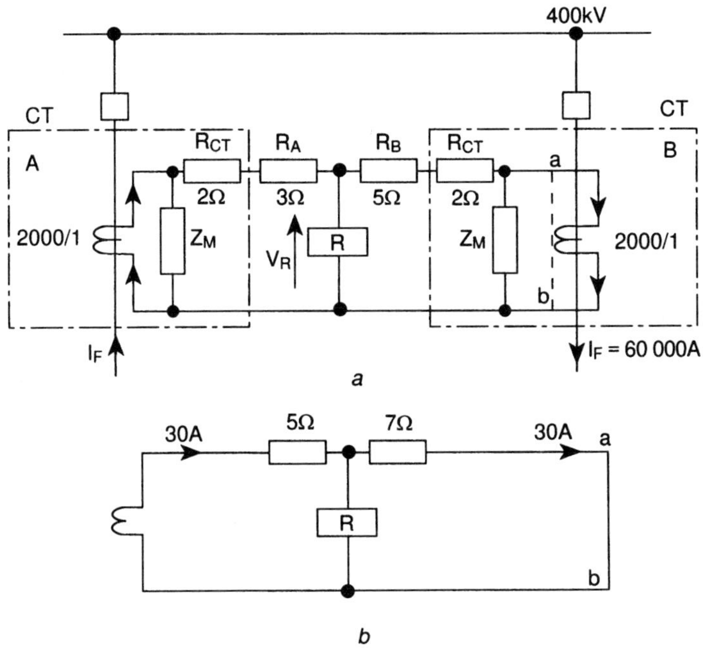

Figure 16.2 Derivation of stability voltage

which may result in incorrect operation. Saturation effects can be minimised by increasing the CT core size significantly, which in turn increases the cost. The relay *R* in Figure 16.1 is assumed to be a low-impedance relay; however, by utilising a high-impedance relay, it is possible to obtain stability while retaining the lower-cost CTs which are prone to saturation.

Figure 16.2*a* illustrates the same arrangement as Figure 16.1 but with the CT equivalent circuit (see Section 10.2) and the interconnecting wiring resistances R_A and R_B shown. Consider relay *R* now as a high-impedance relay, and consider also the situation where the CT of circuit *B* is driven into saturation by the fault current I_F (which we shall assume to be the worst case of system maximum fault current), while the CT on circuit *A* retains its accuracy. The CT on circuit *B* thus appears as a short-circuit between points *a* and *b*. This is shown more simply in Figure 16.2*b* (where the magnetising impedance of CT *A* is ignored because of its relatively high value). Now the voltage experienced across relay *R* is 30 A × 7 Ω = 210 V (assuming that *R* is of such a high impedance that it takes negligible current). Thus, if the relay setting is arranged to be in excess of 210 V, it will stabilise for this worst-case condition.

A corresponding calculation can be carried out for the instance of CT *A* saturating and CT *B* retaining full accuracy. In this instance, the voltage developed across the relay would be

$$30\ \text{A} \times 5\ \Omega = 150\ \text{V}$$

This is a less onerous condition, and thus the setting of the relay should be based on CT *B* saturating, i.e. 210 V.

16.3 Fault-current setting

The fault-current setting (i.e. the current which just causes operation of the relay) is determined from the current taken by the parallel-connected elements, at the relay stability-voltage setting. In this instance the parallel-connected elements are the relay *R* and the CT magnetising impedances. The CT and connecting wiring resistances are so small in magnitude compared with the relay resistance and CT-magnetising impedance that they are ignored. Based on the stability voltage of 210 V, determined above, let us assume that the relay setting is 250 V, i.e. the next available setting above the stability voltage, and let us further assume that the relay current at this setting is 40 mA. In addition, with reference to the magnetisation curve in Figure 10.9, the magnetisation current at 250 V is approximately 15 mA. Thus the total secondary current required to cause operation is

$$\underset{\text{CTs}}{(2 \times 15\ \text{mA})} + \underset{\text{relay}}{40\ \text{mA}} = 70\ \text{mA}$$

This translates into a primary fault current of

$$\frac{2000}{1} \times 70\ \text{mA} = 140\ \text{A}$$

16.4 The high-impedance relay

The high-impedance relay may be one of two possible types:

(*a*) A voltage-calibrated relay, with typical tappings somewhere in the range 25–300 V. To determine the fault-current setting, it is necessary to know the relay operating current for each tapping.
(*b*) A current-calibrated relay, with typical tappings somewhere in the range 20–400 mA. To convert this relay into a high-impedance relay, it is necessary to connect a resistance in series. This resistance is known as the 'stabilising resistance'. For the example above, the value of relay-stabilising resistance would need to be

$$\frac{250\ \text{V}}{40\ \text{mA}} = 6250\ \Omega$$

These relays are required to be immune to DC transients in the fault current.

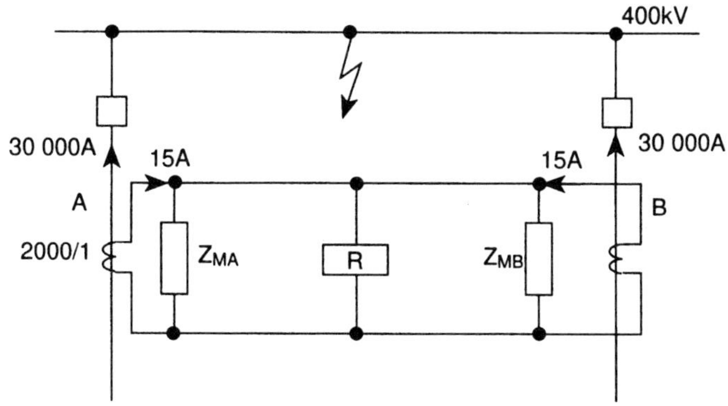

Figure 16.3 In-zone fault

16.5 Knee point voltage

Consider the situation in Figure 16.3 where an in-zone fault has taken place and fault currents as shown flow through feeders *A* and *B* (the relatively low-value CT and wiring resistance have been ignored). The corresponding secondary currents for ideal transformation, assuming no DC transient component, would be

$$\frac{30\,000}{2000} = 15 \text{ A}$$

Each CT would then supply 15 A into a parallel combination of magnetising impedance (typically 10 000 Ω impedance in the linear portion of the *B*/*H* curve) and relay resistance (say, 6250 Ω as in Section 16.4). This would result in a voltage across the relay and associated wiring exceeding 80 kV. In practice, the CT cores saturate and the voltage is limited to a much lower value (see Section 10.3). It is, however, very important that the CT does not saturate fully, otherwise, a voltage would not be developed across the relay to cause operation. For this reason, it is usual to arrange the knee point voltage of the CT to be at least twice the stability-voltage setting of the relay.

16.6 The general case

Figure 16.4 illustrates the general case for a high-impedance scheme. The salient components are:

(*a*) a number of infeeding circuits (in this instance, three A, B and C), with CT magnetising impedances Z_{MA}, Z_{MB} and Z_{MC} and CT resistances R_{CTA}, R_{CTB} and R_{CTC}. The CTs need to be of the same ratio, and for modern schemes they are usually class X, type B (or similar), as described in Section 10.7;

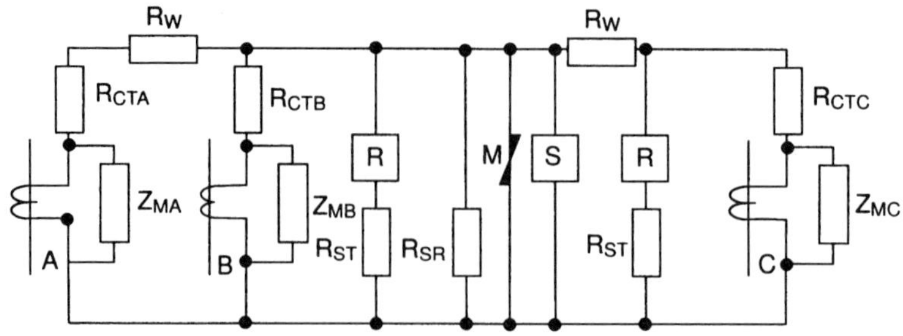

Figure 16.4 General high-impedance circuitry

(*b*) interconnecting wiring R_W;
(*c*) relays *R* with stabilising resistors R_{ST} (for current-operated relays). Many high-impedance schemes have more than one relay, as will be discussed shortly;
(*d*) metrosil *M*. This is installed to ensure that the relays, resistors, CTs and wiring are not damaged by the high voltages that can be generated during an in-zone fault, by limiting the peak voltage to 3 kV;
(*e*) setting resistor R_{SR}. This is used to increase the minimum operating current. It is not always required. Its application will be examined shortly;
(*f*) supervision relay *S*. This is used to detect an open circuit in the CT connections, which would manifest itself as a rise in voltage across the wiring. The relay is usually a high-impedance relay with a sensitive setting of between 2 and 10% of minimum operating current of the scheme. Some of the simpler high-impedance schemes dispense with the supervision relay, but their installation is usual if disconnector auxiliary contacts form part of the CT circuitry. Some schemes use the supervision relay to short-circuit the CT wiring, thus disabling the scheme and rendering it safe from continuous high voltage, in addition to giving an alarm.

16.7 Three-phase circuitry

So far we have examined the scheme from a single-phase standpoint, whereas in practice it is connected as a three-phase scheme. This is simply illustrated in Figure 16.5, where the R, Y and B boxes represent the relays, setting resistors, metrosils and supervision relays shown in Figure 16.4.

16.8 Settings summary

Two settings require consideration:

- through-fault stability voltage setting (i.e. the setting on the relay)
- in-zone fault current setting (i.e. the minimum primary fault current to cause operation of the busbar protection).

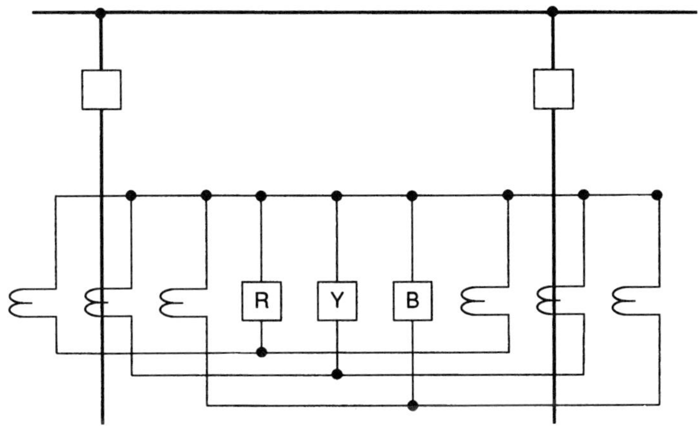

Figure 16.5 *Three-phase circuitry*

(*a*) *Stability voltage:* This is the setting of the relay and is determined from:

$$\frac{I_F}{\text{CT ratio}}(R_{CT}+R_W)$$

where I_F=system maximum through fault current (specific instances will be discussed shortly)

$(R_{CT}+R_W)$=the highest combination of CT (R_{CT}) and wiring (R_W) measured from the relay to the saturated CT

The relay setting should be the next highest setting above the calculated stability voltage. The CT knee point must be at least twice the magnitude of the stability voltage.

If necessary, the stability voltage can be reduced by increasing the wiring-conductor size or paralleling of multicore-cable cores.

(*b*) *Fault current:* The primary-fault-current setting is given by:

$$\text{CT ratio }(mI_R+nI_e+I_{SR}+I_V+I_M) \qquad (16.1)$$

where

m=number of connected relays
I_R=relay current
n=number of connected CTs
I_e=CT magnetising current
I_{SR}=setting-resistor current
I_V=supervision-relay current
I_M=metrosil current

All currents are those which flow when the relay setting voltage is applied.

The above settings criterion is somewhat empirical, and frequently requires an iterative process to obtain a compromise between the two settings requirements. However, factors of safety are in-built by making pessimistic assumptions such as:

(i) the relay currents and CT magnetising impedance currents are in-phase when in practice they are likely to be quadrature;
(ii) the metrosil current is sinusoidal;
(iii) utilising the worst-case system maximum fault currents in the stability-voltage calculation;
(iv) one CT totally saturates, while another retains full transformation capability.

Experience has proved that this is a very successful protection and, since it utilises relatively simple relays, it has significant inbuilt reliability.

HIGH-IMPEDANCE PROTECTION: APPLICATION

16.9 Busbar protection

Busbar protection is probably the most common and yet the most complex application of high-impedance protection, particularly when applied to distributed relay-room (blockhouse) substations, as commonly found on 400 kV systems.

Figure 16.6 illustrates a typical but simplified application to a blockhouse type of substation. Only one phase is shown to minimise the complexity of the Figure: for illustrative purposes the F1 circuit shows the position of the two other CTs. All wiring with the letter C is common wiring, and is connected together (usually on a per-zone basis) and earthed. It is usual to arrange the per-zone buswire as a ring for security purposes; again, to minimise complexity, these connections are omitted. Note that the major difference between Figures 16.6 and 10.16 is that the blockhouse type of substation has an individual discrimination and check relay per circuit, and not individual relays per busbar zone.

16.10 Busbar-protection discriminating zone

By way of example, consider first, with reference to Figure 16.6, the discrimination zone. It can be seen that the number of relays and CTs connected to a particular discrimination zone varies with the number of circuits connected to the zone, which in turn causes a variation in the fault-current setting of the zone. Typical criteria for determining the fault-current setting are as follows:

(*a*) For uniformity of application, the per-zone busbar-protection fault-current setting with all possible circuits connected to a busbar section should

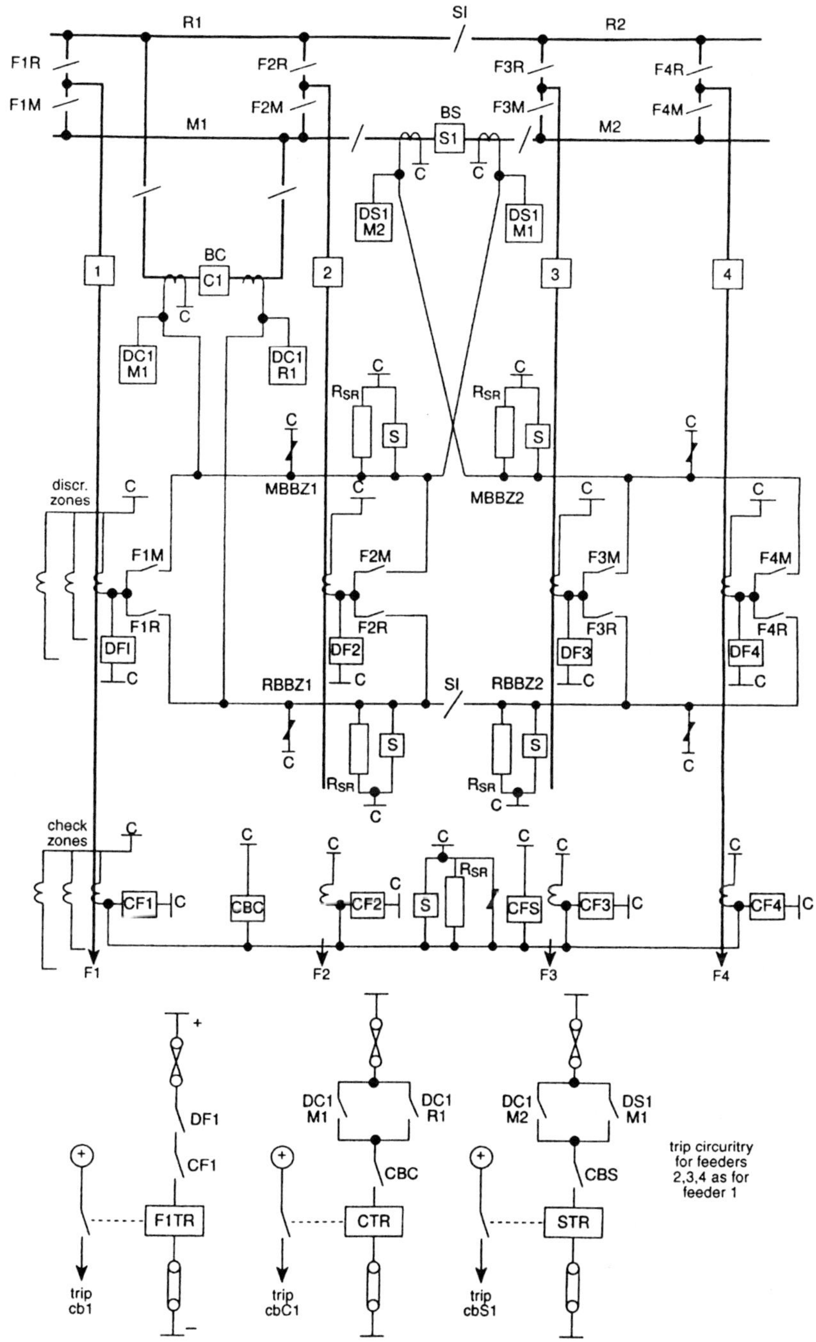

Figure 16.6 Busbar-protection scheme: blockhouse substation

not exceed 100% of the busbar rated current. If, when the reserve busbar sections are paralleled, the combined setting exceeds 100% of busbar rated current, the setting should be reduced below 100%, by increasing the setting resistance.

(*b*) The per-zone busbar-protection fault-current setting with one circuit connected should be made equal to 50% of the busbar rated current, but this can be reduced to satisfy the requirements of (*a*) above.

(*c*) The highest busbar-protection fault-current setting must be within 10–50% of the minimum busbar fault current. This settings range caters for periodic changes to the busbar configuration.

This would result in the following typical settings:

(*a*) *Stability voltage*

(i) This should be based on a maximum through fault current equal to switchgear maximum fault-current rating. This, for the 400 kV system in Figure 16.6, is 63 000 A. (NB. For simplicity 60 000 A is assumed.)

(ii) The highest combination of CT and wiring resistance is usually that of circuits connected to reserve busbars R1 and R2 with disconnector SI closed. For example, with reference to feeders F1 and F4 in Figure 16.6, assume that the CT connected to F4 saturates, then the voltage developed across relay DF1 would be due to current in the resistances associated with RBBZ1+RBBZ2+F4 CT resistance. Suppose that this total loop resistance is 9 Ω and the CT ratio is 2000/1, then:

$$\text{stability voltage} = \frac{I_F}{\text{CT ratio}}\,(R_{CT}+R_L)$$

$$= \frac{60\,000}{2000}\,(9) = 270\ \text{V}$$

Assuming a voltage-operated relay, the setting should thus be the next highest tapping, typically 300 V. This setting should be applied to all relays. The CT knee-point voltage should therefore exceed 600 V (i.e. twice the setting voltage).

(*b*) *Fault-current setting*

Assuming a busbar rating of 4000 A:

(i) Consider first the worst-case situation [see Section 16.10 item (*a*)] of the reserve zones paralleled. For this condition, the settings resistors R_{SR} for both reserve zones will be in parallel. Thus for all circuits selected to the reserve zone and utilising eqn. 16.1 we require that

$$\text{CT ratio}\,(mI_R + nI_e + 2I_{SR} + 2I_V + 2I_M) < 4000\ \text{A}$$

Thus at the setting voltage of 300 V and utilising the magnetisation curve in Figure 10.9,

$$2000\,(5\times 0.03 + 5\times 0.017 + 2I_{SR} + 0.02 + 0.02) < 4000\ \text{A}$$

(where 0.01 is a typical value for supervision and metrosil currents and 0.03 A is a typical relay current at 300 V setting)

$$\text{i.e. } 2000\ (0.275+2I_{SR})<4000\text{ A}$$

$$\text{i.e. } I_{SR}<0.86\ A$$

This would require R_{SR} to be 300/0.86=349 Ω. Thus utilising 350 Ω resistors, the fault setting would be just less than 4000 A.

(ii) Consider now the case of one feeder being connected to a section of busbar [see Section 16.10 item (*b*)]. Again, utilising eqn. 16.1, we require:

$$\text{CT ratio } (mI_R+nI_e+I_{SR}+I_V+I_m)>2000\text{ A}$$

$$\text{i.e. } 2000\ (0.03+0.017+0.86+0.01+0.01)>2000\text{ A}$$
$$\text{i.e. } 1854>2000\text{ A}$$

In this instance, the calculated minimum operating current of 1854 A is less than the ideal minimum current of 2000 A, but we shall just have to accept this lower figure as being less than ideal.

(iii) If the system minimum fault current available [see Section 16.10 item (*c*)] is, say, 10 000 A, the ratio of the highest fault-current setting of 4000 A to minimum fault current is

$$\frac{4000}{10\,000}\times 100\%=40\%$$

This is between 10 and 50%, and therefore acceptable.

16.11 Busbar-protection check zone

The check zone CTs, as shown in Figure 16.6, are permanently connected and thus the zone has a single fault-current setting.

(*a*) *Stability voltage*

The determination of the check-zone stability voltage follows the same format as that for the discrimination-zone stability voltage and usually involves the same combination of healthy and saturated CTs. This usually results in a setting identical to the discriminating zone, i.e. 300 V in this instance.

(*b*) *Fault-current setting*

This follows the same format as the discrimination-zone setting.

16.12 Autotransformer settings

Figure 16.7 illustrates the application of high-impedance protection to an autotransformer. Metrosils, setting resistors and supervision relays have been omitted, for the sake of clarity. Note that the ratios of all three sets of CTs are

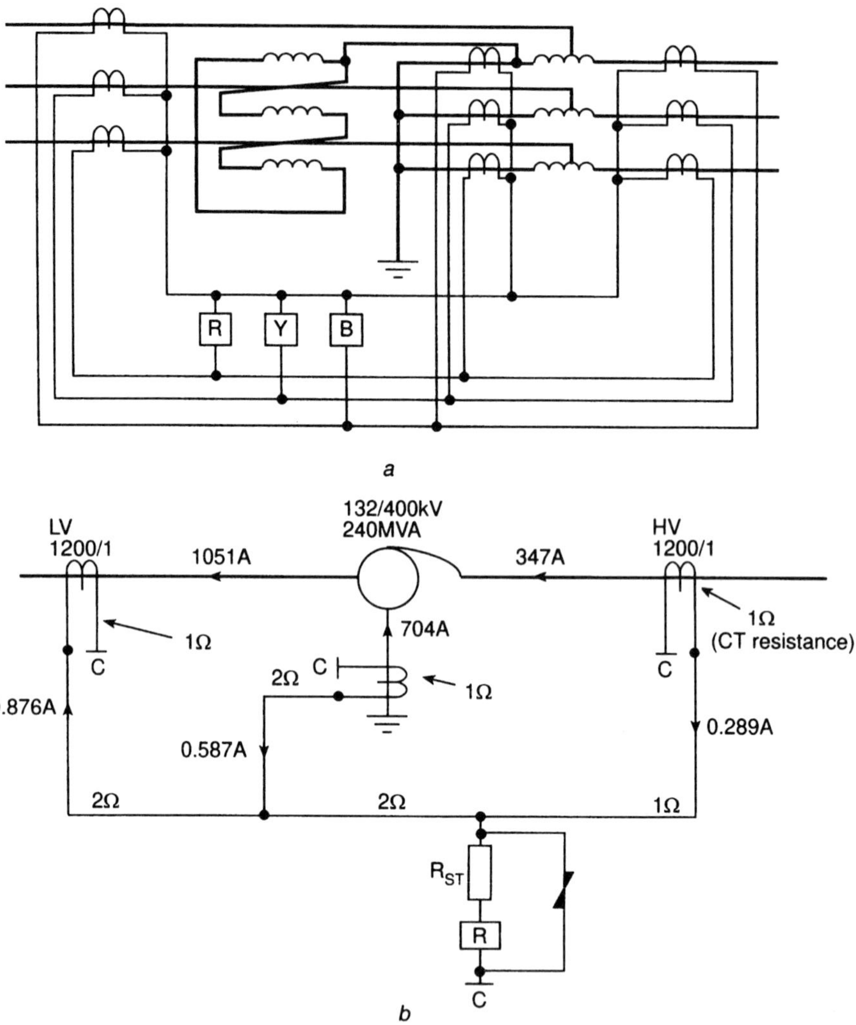

Figure 16.7 Autotransformer protection

a Full circuitry
b Simplified circuitry

identical. This is necessary to create a balanced circulating current for the instance of through load or fault current. Figure 16.7*b* shows how the currents balance when the transformer is subject to rated load current (only one phase is shown for simplicity).

Figure 16.7*a* shows the delta tertiary connected to the autotransformer neutral. This is necessary to provide protection for the instance of a tertiary earth fault. This connection is unnecessary if the tertiary is provided with its own earthing system and protection.

Note that, with the simpler schemes, supervision relays are frequently omitted, especially when there are no disconnectors in the protected zone.

(*a*) *Stability voltage*
The stability-voltage calculation is usually based on 16 times the rated transformer LV current (see Section 16.14) or the transformer short-circuit current if this is greater. Note that the earth-fault short-circuit current may be greater than that due to a three-phase fault, due to the influence of the tertiary zero-sequence impedance.

From Figure 16.7*b* it can be seen by inspection that the highest voltage developed across the relay is due to saturation of the LV CT:

$$\text{i.e. } V = 16\{(0.876\times 3)+(0.289\times 2)\} = 51.3\ \text{V}$$

Let us assume the next highest relay-setting tapping is 100 V. The CT knee point should therefore exceed 200 V.

(*b*) *Fault-current setting*
For a solidly earthed neutral, the minimum fault-current setting is usually required to be between 10 and 60% of the rated current of the high-voltage winding, i.e. from Figure 16.7 between 35 and 208 A. Thus, assuming the use of a current-operated relay and utilising eqn. 16.1, with the stability-setting voltage of 100 V, as determined above, then:

fault-current setting $= 1200\ (I_R + 3I_e + I_{SR} + I_M)$
Assuming that $I_e = 12$ mA at the setting voltage,
and $I_M = 5$ mA at the setting voltage,
then setting $= 1200\ (I_R + 0.036 + I_{SR} + 0.005)$.

By inspection, if we set I_R, the relay-setting current, to 0.1 A, the setting resistor (causing current I_{SR}) can be dispensed with.

So, setting $= 1200\ (0.1 + 0.041) = 170$ A
i.e. within the range 35–208 A.

(*c*) *Stabilising resistor*
Assuming that the relay resistance is negligible (which is not always the case), the value of the stabilising resistor is given by

$$\frac{\text{setting voltage}}{\text{setting current}} = \frac{100}{0.1} = 1000\ \Omega$$

16.13 Generator settings

The high-impedance protection arrangement for a synchronous machine is shown in Figure 16.8. Metrosils, setting resistors and supervision relays have been omitted for the sake of clarity.

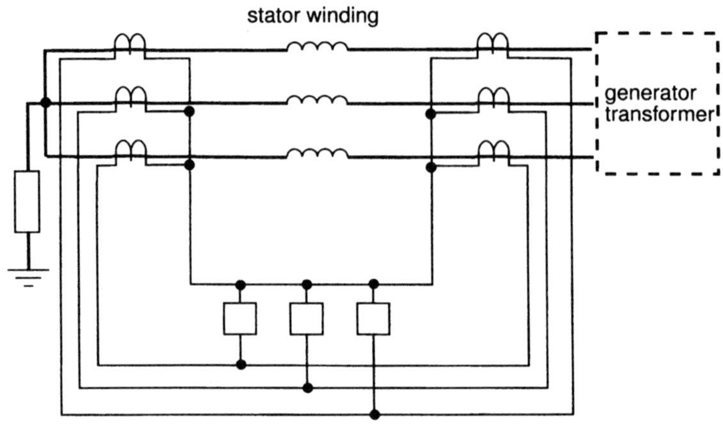

Figure 16.8 Generator protection

The stability voltage is usually determined from the maximum through fault current, for a fault on the HV side of the generator transformer. Within this context, the phase-to-earth fault current may differ from that of a three-phase fault. To obtain a worst-case situation, the generator subtransient reactance should be used in the calculation and any earthing resistor assumed to be short-circuited.

The fault-current setting should be less than 10% of the rated current of the generator winding being protected.

16.14 Two-winding-transformer settings

Figure 16.9 illustrates the application of high-impedance protection to a two-winding transformer, in this instance a 33/11 kV distribution transformer. This type of protection scheme is known as 'restricted-earth-fault' protection (the HV scheme is sometimes termed 'balanced-earth-fault' protection). The two schemes provide earth-fault protection for the HV and LV windings, respectively. The termed 'restricted' relates to the fact that the protection capability is restricted since it is usually not sensitive enough to protect the full winding. On both the HV and LV sides of the transformer, the CTs are paralleled. This results in both the positive- and negative-sequence currents summing to zero. The zero-sequence currents are additive and thus each relay is only responsive to zero-sequence, i.e. earth-fault currents. Figure 16.9 illustrates the current distribution for an out-of-zone earth fault on the LV side.

The derivation of the stability voltage is usually based on a through fault current equal to 16 times the rated current of the protected winding. This arises when the worst-case (highest) transformer-neutral current, for the fault shown in Figure 16.10, is considered (note that neutral current=3×zero-sequence current). As can be seen, the magnitude of transformer neutral

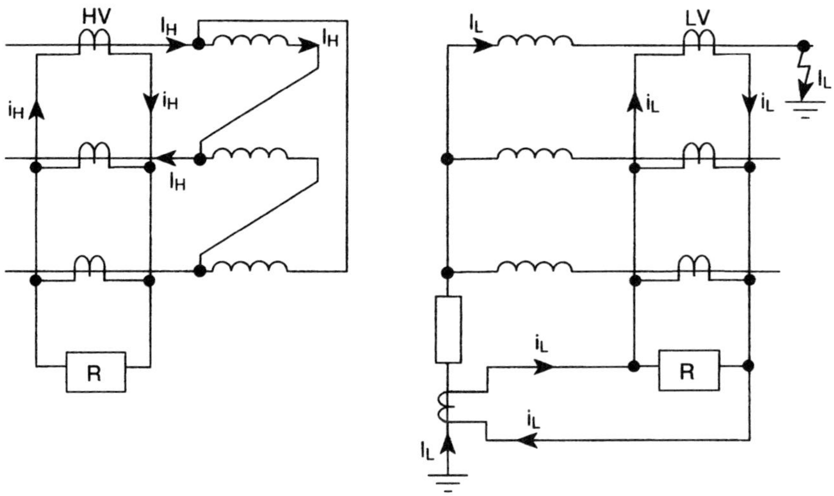

Figure 16.9 33/11 kV-transformer protection

current (i.e. zero-sequence current in the LV CT) is dependent on the transformer-to-system-impedance ratio of the various system components of impedance. Detailed investigation shows that the worst case in practice is equal to the 16 times factor specified above. Note that, for the purpose of the calculation, it is assumed that any earthing resistor is short-circuited.

The fault-current setting of the relay is usually required to be:

- 10–60% of the rated current of the protected winding for a solidly earthed power system;
- 10–25% of the minimum current available for an earth fault at the transformer terminals for a nonsolidly earthed power system.

Figures 16.16 and 16.19 illustrate high-impedance restricted-earth-fault protections, fed from the same CTs which feed an overall biased-differential scheme. In these instances, the burdens to be considered when determining the stability voltage need to include the reflected resistances of the interposing transformers and associated wiring. Note that, when determining stability voltage, phase-to-phase faults also need to be considered, since they also cause voltage drop around the wiring and saturated CT, and these voltages are imposed on the relay.

16.15 HV connections and mesh-corner settings

Figure 10.13 shows the general arrangement for high-impedance protection applied to a mesh corner. The setting criteria are usually identical to those for busbar protection.

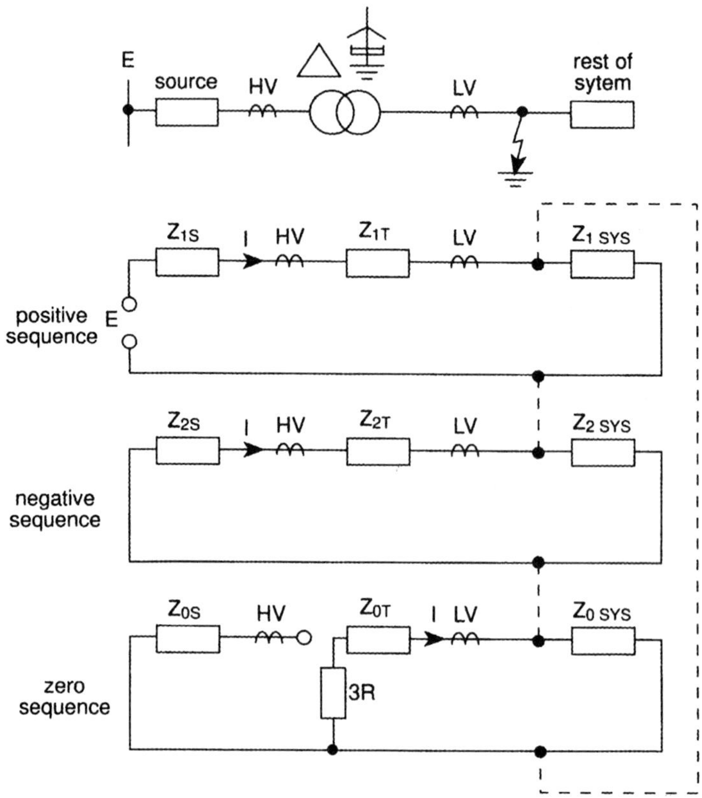

Figure 16.10 Sequence-components analysis of 33/11 kV transformer

HIGH-IMPEDANCE PROTECTION: COMMISSIONING

16.16 Busbar protection

Figure 16.6 illustrates the type of busbar protection scheme found at distributed-relay-room (blockhouse) substations, as installed on 400 and 275 kV systems. As can be seen, it requires a discrimination and check-zone operation to achieve tripping, and it is therefore a two-out-of-two system. In recognition that this type of system does not give the required level of reliability, a further common relay per zone was added (i.e. one each for the four discrimination zones and the check zone) in the early 1980s. This is not shown to avoid overcomplicating the Figure. The common relays, when operated, are arranged to trip each circuit breaker connected to the faulted busbar; the trip circuitry used is similar to that shown in Figure 10.17 and it is usual to share the buswiring with a circuit-breaker-fail scheme.

The required commissioning tests usually consist of:

(*a*) a measurement of burden (i.e. resistance) of the CTs and wiring resistances (see Section 10.16);
(*b*) CT magnetisation curve (see Section 10.15);
(*c*) secondary injection of each relay;
(*d*) DC operation to prove the tripping circuitry logic;
(*e*) primary injection, the principles of which are described in Section 10.23. Note that since the relay (and any associated stability resistor) presents a high impedance, it is usual, to facilitate most tests, to place an ammeter across the relay (and resistor). The operation shot with the relay operating in its normal high-impedance mode is usually accomplished via the secondary injection;
(*f*) secondary injection of the whole scheme, as described below, to verify:

- relay-setting (stability) voltage
- fault-current setting
- disconnector auxiliary switches

(*g*) on-load tests (see below).

Secondary injection of the whole scheme
It is much easier to secondary inject the whole scheme if the whole substation is built prior to initial energisation. In such an instance, the following (see Figure 16.6) should typically be carried out:

(i) Select all circuits that can be selected to busbar M1.
(ii) Inject a test voltage onto buswires MBBZ1 (i.e. main 1 discrimination-zone bus wires) and slowly increase the voltage until its value is 10% lower than the relay-setting voltage. Check that no relays have operated.
Note that:

- the supervision relay will operate, but it is usual with this type of scheme to arrange an alarm only, and not buswire short-circuiting;
- the injection should not be applied for too long. Consideration needs to be given to the rating of any setting and stabilising resistors.

(iii) Increase the test voltage to 10% above the setting voltage, and check that all relays connected to that zone only have operated.
(iv) Measure the injected current and ensure that it corresponds to the calculated value based on eqn. 16.1. Note that a slight discrepancy will occur between calculated and measured values due to the relay, setting resistor, supervision relay and metrosil currents being resistive, whereas the CT magnetising current is reactive.
(v) Repeat the above test with the same circuits selected to busbar R1.
(vi) Similarly, select the remaining circuits to facilitate the injection of MBBZ2 and RBBZ2, respectively; subsequently, select all circuits to a solid reserve bar (disconnector SI closed) and inject onto RBBZ1; and finally inject the check zone.

If, alternatively, the substation is built and energised circuit by circuit, the appropriate sections of the scheme ideally need to be proved by testing as above each time a circuit is added. If the substation is energised, there is always a risk, while testing, of inadvertently tripping an in-service circuit (e.g. due to incorrect wiring), and it is therefore preferable to de-energise the busbar whose zone is to be tested—although this is obviously not possible for the check zone.

If it is necessary to test a zone which must remain in service, it is essential to apply the test voltage through a high impedance (e.g. typically 2 kΩ resistor). This ensures that, should an actual busbar fault occur during the test, the test source does not unduly influence the zone operating current.

Note that, if it is necessary to break into the buswire rings to insert an additional circuit, the inherent factors of safety in the design of the busbar-protection system are such that no special precautions are usually necessary to prevent incorrect operation should an out-of-zone fault occur during this period.

Figure 10.16 illustrates the typical busbar protection schemes found in common relay-room sites, with the accompanying tripping circuitry shown in Figure 10.17. As can be seen, these schemes utilise a single relay per busbar zone, although, as stated above, transmission schemes are usually designed with duplicate relays per zone for security purposes. These schemes also employ supervision relays and metrosils, in the same way as Figure 16.6, but they are omitted from Figure 10.16 to minimise the complexity of the drawing. For the type of scheme shown in Figure 10.16 the supervision relay, when operated, is usually arranged to short-circuit its associated zone-tripping contacts, as well as give an alarm, such that the scheme is retained in service, with reduced security, as a one-out-of-one system. The commissioning process follows a similar format to that for Figure 16.6. These schemes tend to be found at 275 kV and 132 kV and at some 33 kV substations. Most 33 kV substations, however, are not of the double-busbar type and therefore do not contain disconnector auxiliary switches.

On-load tests

These consist of placing a voltmeter across each high-impedance relay once the circuit is energised and load current is flowing. For a correctly installed scheme, the measured voltage should be negligible.

16.17 Disconnector auxiliary switches

Disconnector auxiliary switches should be of a type specially approved for switching current transformers.

Busbar-protection schemes of the single-relay-per-zone type, as shown in Figure 10.16, have an extra set of normally closed auxiliary contacts per discrimination-zone feeder, as illustrated in Figure 16.11 (these are not shown in Figure 10.16 to avoid overcomplicating the diagram). These are required to short-circuit the CT to prevent high voltages arising for the combination of circumstances of both busbar-selector isolators being open, maintenance earths placed either side of the CT stack, and stray fault currents flowing

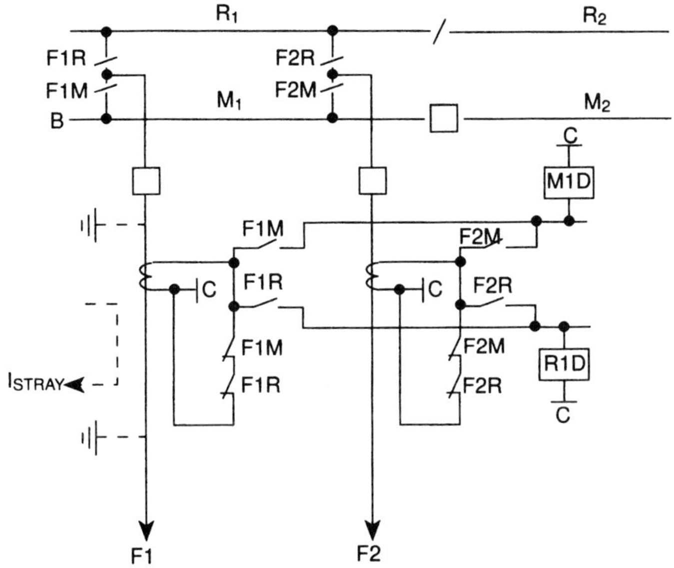

Figure 16.11 Busbar-protection auxiliary switches

through the earth loop. These contacts are not required in Figure 16.6 because the per-circuit relay will provide a low enough impedance to limit the voltage.

The timing of the auxiliary contacts is very important, and during commissioning the following should be checked:

(*a*) The normally open busbar-selector-disconnector auxiliary contacts F1M, F1R etc. should make before the main contacts reach the pre-arcing distance between fixed and moving contact when the disconnector is closed, and should break after the main contacts part when the disconnector is opened. This is to avoid instability of the discrimination zone, resulting from a current imbalance between the high-voltage busbars and the corresponding busbar-protection buswires during movement of the disconnectors.

(*b*) The normally closed series-connected contacts F1M, F1R should be arranged to make after the main contacts have opened, and to break before the main contacts have closed. This is to prevent the pertinent zone of busbar protection from being short-circuited during movement of the disconnector.

16.18 Other schemes

Commissioning of schemes other than busbar protection follow the same principles as those for busbar protection. However, many of these schemes do not involve auxiliary switches, which simplifies the commissioning requirements.

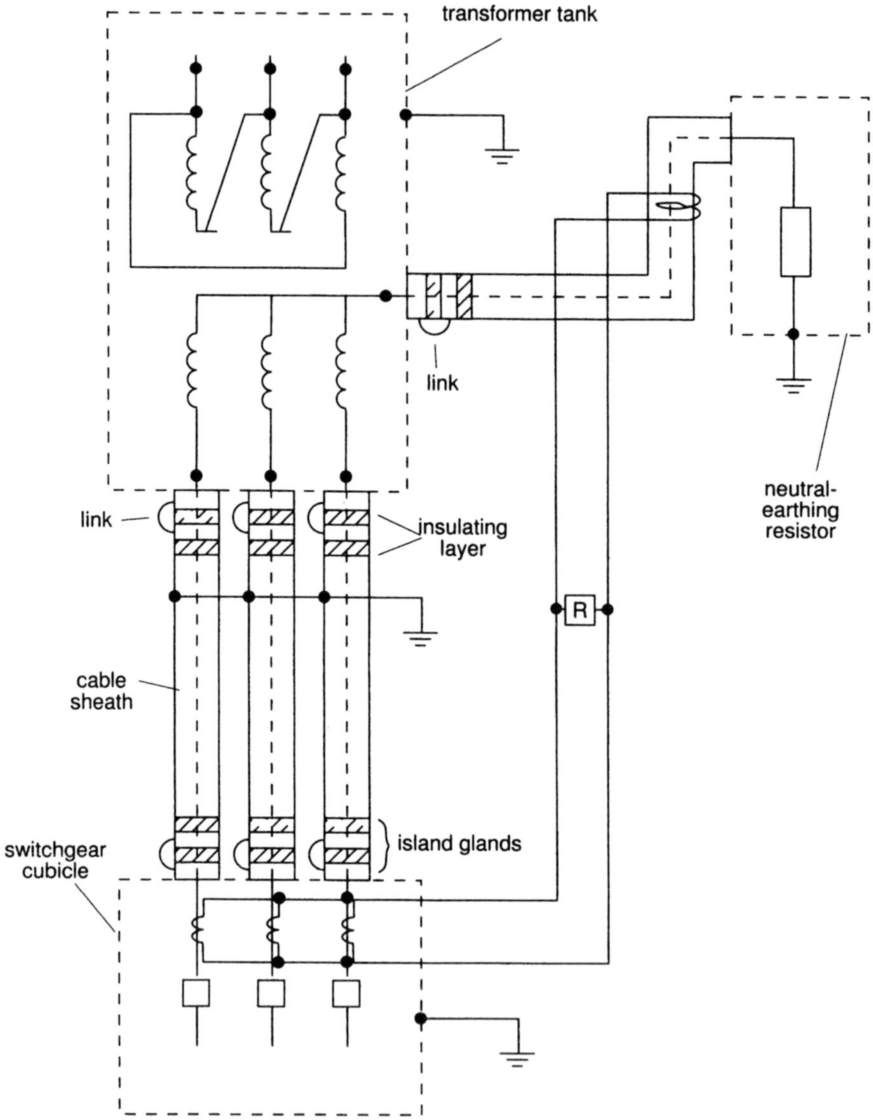

Figure 16.12 The island gland

16.19 The island gland

Figure 16.12 illustrates the application of a high-impedance protection scheme, in the form of restricted-earth-fault protection, to the 11 kV side of a 33/11 kV transformer (as Figure 16.9), which is connected to 11 kV metalclad switchgear via single-core cables. This is a common distribution-voltage application. To avoid cable-sheath circulating currents arising from induced voltages, and to ensure that fault current returns to the transformer neutral via the neutral-cable inner conductor (i.e. to prevent flow of current in the neutral-cable sheath which would result in a cancellation of flux in the neutral-cable CT), it is necessary to connect the cables to the cable boxes via double-insulating cable glands, usually known as an island gland. Two insulated layers are utilised to enable testing of the insulation. Testing would normally be carried out by removing the link temporarily and carrying out an insulation test with a 1000 V 'megger' between the inner metallic layer and earth.

HIGH-IMPEDANCE PROTECTION: MAINTENANCE

16.20 Maintenance

Maintenance of high-impedance protection would normally require a repeat of the secondary-injection tests and DC-operation tests carried out during commissioning.

During maintenance, care needs to be taken in earthing the CTs to avoid the effects of stray currents, as shown in Figure 16.11. Busbar-protection schemes for single-busbar substations are more at risk from this problem, since they usually consist of a single discrimination zone (i.e. a one-out-of-one system), with no auxiliary switches in the CT circuitry (as Figure 16.5) and, unlike Figures 10.16 and 16.16, the CTs remain connected to the buswires at all times. This may necessitate disconnection from the scheme buswires and short-circuiting of the CTs of the circuit which is switched out of service.

16.21 Protection operating time

During in-zone faults, high-impedance-protection CTs endeavour to feed current through the high-impedance relays. The CTs are therefore forced into partial saturation and the voltage is limited by the metrosils. However, operating experience has shown that fast operating times of the order of 20–30 ms are the norm, even though the voltage applied to the relay is nonsinusoidal and saturated. Experience, however, has also shown that the design of relay employed needs to be immune to DC transient components in the fault current. This latter point should be investigated if there is a failure to operate or slow operation is experienced.

BIASED-DIFFERENTIAL PROTECTION: THEORY

16.22 Basic principles

Biased-differential protection is an older design of protection than high impedance circulating current. It is widely applied to two-winding transform-

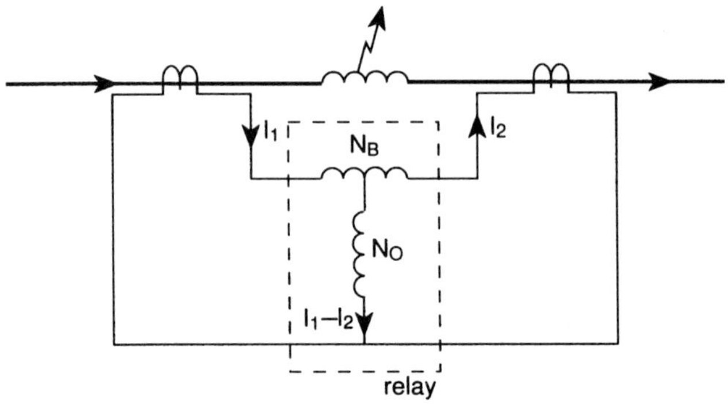

Figure 16.13 Biased-differential relay

ers except those connected to the lower distribution voltages. Until the advent of the high-impedance design, it was also applied to busbar protection, although the last few years have witnessed a new generation of biased-differential busbar-protection schemes. The basic features and principles of operation are illustrated in Figure 16.13. The biased-differential relay consists of two windings, an operate winding with turns N_o and a restraint, or bias, winding with turns N_B. The relay compares the operating force due to the out-of-balance current $I_1 - I_2$, arising from an in-zone fault with the restraint force due to the circulating current through the bias winding, arising from an external fault or load current. For operation,

$$(I_1 - I_2) N_0 > \frac{I_1 N_B}{2} + \frac{I_2 N_B}{2}$$

$$\text{i.e. } (I_1 - I_2) N_0 > N_B \frac{(I_1 + I_2)}{2}$$

$$\text{or } (I_1 - I_2) > \frac{N_B}{N_0} \frac{(I_1 + I_2)}{2}$$

The term $I_1 - I_2$ is known as the operating or differential current I_0,

the term $\frac{(I_1 + I_2)}{2}$ is known as the bias current I_B

and the term $\frac{N_B}{N_0}$ is the bias quantity B.

Thus, for operation,

$$I_0 > BI_B$$

For completeness, the above expression requires the addition of a term I_F to represent the current to overcome friction, or natural bias, of the relay. Thus

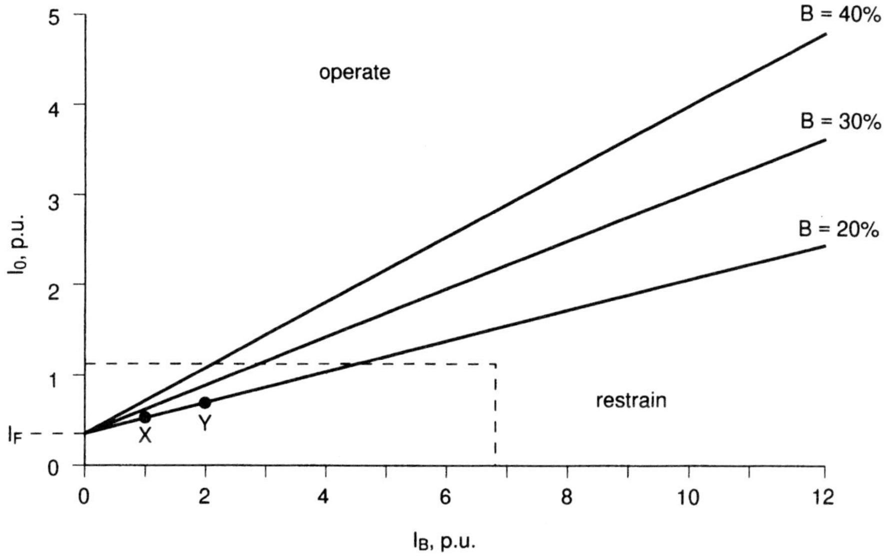

Figure 16.14 Typical bias characteristics

$$I_0 > BI_B + I_F \tag{16.1}$$

Figure 16.14 illustrates operate/restraint characteristics for various values of bias *B*. Modern relays embody the above principles within electronic circuitry.

Note that the above applies to relays whose characteristic is termed 'mean bias'. An alternative approach termed 'load bias' analyses the relay characteristic in terms of the differential current I_0 (which also flows through half of the bias winding) and a bias current I_B which flows through both halves of the bias winding.

Manufacturers' literature tends to analyse relay performance in terms of one of the above. The remainder of this chapter will assume the use of a 'mean bias' characteristic.

The biased-differential relay is a low-impedance relay and compares with the high-impedance relay as follows:

(*a*) *Advantages of biased scheme*:

(i) can operate with CTs with a lower knee point and less accuracy, since the relay is of a low-impedance type. The bias feature compensates for any CT inaccuracy or saturated CTs;

(ii) relay-setting calculations are less complex than with a high-impedance scheme.

(*b*) *Disadvantages of biased scheme*:

(i) not as sensitive as the high-impedance scheme (as will be shown shortly);

(ii) the scheme circuitry is more complex than that of a high-impedance scheme.

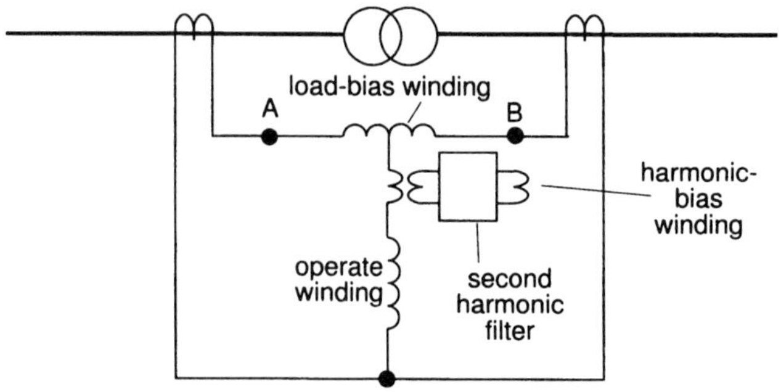

Figure 16.15 Application to a two-winding transformer

16.23 Application to a two-winding transformer

Figure 16.15 illustrates the application of a biased-differential scheme to a two-winding transformer (to simplify the illustration only one phase is shown). The ratio of the current transformers must be chosen carefully so that the secondary current I_S circulates around the bias winding, with no spill current into the operate winding. Two-winding transformers, however, produce spill currents as part of their natural operation, as follows:

(*a*) *Magnetising inrush current*: Section 3.36 described the phenomenon of magnetising inrush current which occurs when a transformer is energised. This current flows only in the winding connected to the source of supply, and thus causes a spill current into the biased-differential-operate winding. Magnetising inrush has a large second-harmonic component of current and therefore, as shown in Figure 16.15, it is usual to arrange an additional bias winding, which is responsive to second-harmonic currents, to offset the effect of magnetising inrush current in the operate winding.
(*b*) *Tap changing*: Most large power transformers are equipped with tap changers. This results in the ratio of the current transformers feeding the biased-differential relay matching the ratio of the power transformer at one tap position only (usually arranged to be the nominal tap). At other tap positions, the current-transformer secondary currents will not balance. This causes spill currents in the relay-operate windings. The relay bias must thus be set to compensate for the worst case of spill current. This occurs when the tap changer is selected to maximum (or minimum) tap, coincident with the highest out-of-zone fault current flowing through the transformer.

As an example, consider a 132/33 kV transformer with a maximum tapping of 15%. If the transformer has a percentage impedance of, say, 30% on a 100 MVA base, the HV fault current for an out-of-zone short-circuit close to the LV terminals is approximately 1450 A and, assuming an HV CT ratio of 200/1, the nominal secondary current is 7.25 A. The corresponding LV secondary current referred to the HV side, taking into account the maximum tap position of 15%, is 6.16 A. Therefore, for a 1 A relay, the bias current is

$$I_B = \frac{7.25+6.16}{2} = 6.7 \text{ p.u.}$$

and differential current I_0=7.25 − 6.16=1.1 p.u.
From Figure 16.14, a percentage bias of 20% would ensure stability for this condition.

16.24 Practical application

Figure 16.16 illustrates a practical application of a biased-differential protection to a 132/33 kV transformer. The key points of this application are:

(*a*) the biased-differential scheme functions in conjunction with HV and LV restricted-earth-fault schemes. This is for reasons of sensitivity, as will be described shortly;
(*b*) the secondary wiring is arranged to mimic the primary connections; thus a through current in the primary results in a corresponding balance of current in the secondary, with no spill current in the operate winding (apart from any due to the tap changer being off nominal tap);
(*c*) the star-delta-star interposing-transformer arrangement may seem somewhat puzzling. Since earth-fault zero-sequence current cannot pass from the 33 kV to the 132 kV side of the transformer (i.e. it cannot flow through a delta-star winding), it must be filtered out of the interposing transformer secondary winding. The star-delta-star transformer achieves this. Earth-fault currents flowing into the P1 winding of this transformer consist of positive, negative and zero-sequence currents, all of which are in phase (i.e. usual earth-fault criteria). The zero-sequence current is unable to flow in the unearthed star winding 1S1, 1S2 and thus, to obtain current (flux) balance across the transformer, it circulates around the delta winding 2S1, 2S2;
(*d*) the zig-zag, or interconnected-star, earthing transformer is the usual system for earthing the 33 kV network. It presents a high impedance to balanced three-phase (positive-phase-sequence) voltages, but a low impedance to in-phase (zero-sequence) voltages. This is because the windings produce opposing fluxes in each core, under zero-sequence conditions;
(*e*) Figure 16.16 shows how current balance is achieved for an out-of-zone earth fault and Figure 16.17 illustrates the corresponding phase-sequence network. If the fault had, alternatively, been in-zone (between the 33 kV winding and the 800/1 CT), there would be current flow through the 800/1 neutral CT only, so causing operation of the LV restricted-earth-fault relay. Furthermore, there would be virtually no current flow through the star-delta-star interposing transformer. This therefore causes the currents flowing out of the star-delta interposing transformer to flow into the biased-differential-relay operate winding. It is worth recognising that the biased-differential relay is

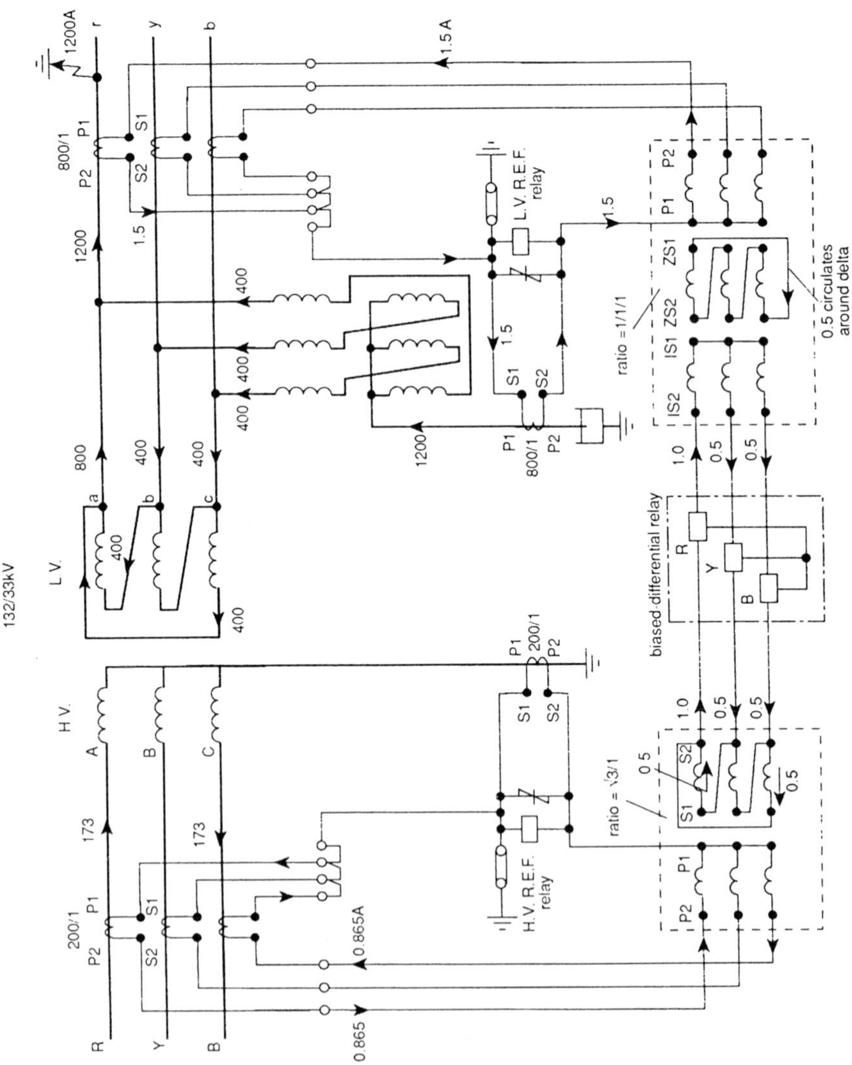

Figure 16.16 Practical application to a 132/33 kV transformer

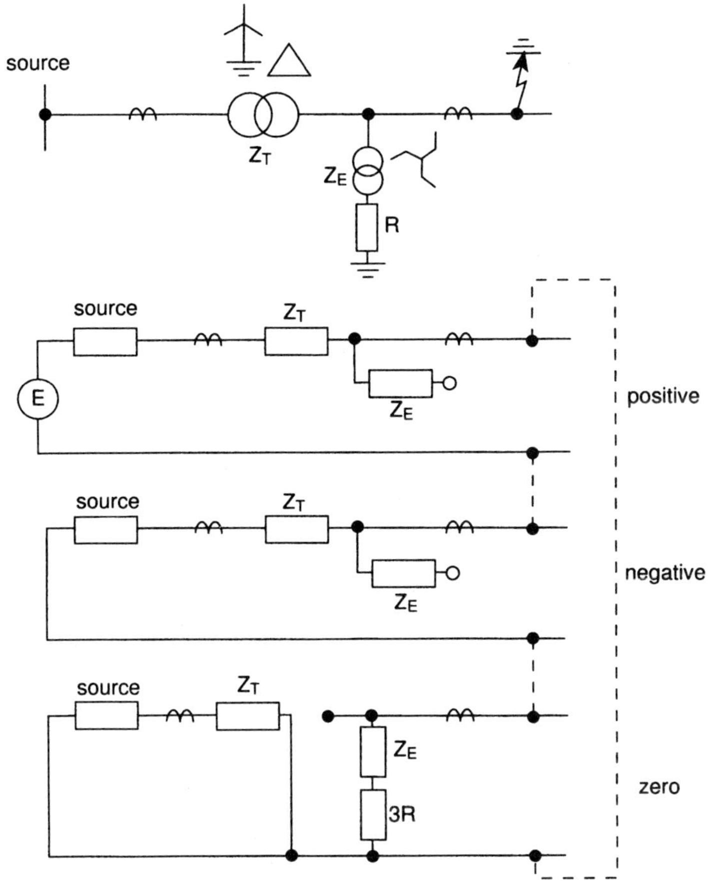

Figure 16.17 Phase-sequence network for fault in Figure 16.16

connected so that it will only respond to positive- and negative-sequence currents (because there is no zero-sequence return path from the relay star point).

16.25 Relay sensitivity

When a transformer is resistance earthed, the current resulting from a transformer-winding-to-earth fault may be relatively low and the percentage of winding protected may be unacceptably low. From Figure 16.18*a*,

$I_S = \dfrac{xV}{R}$ where V = nominal secondary voltage

x = fault position as a percentage of full winding

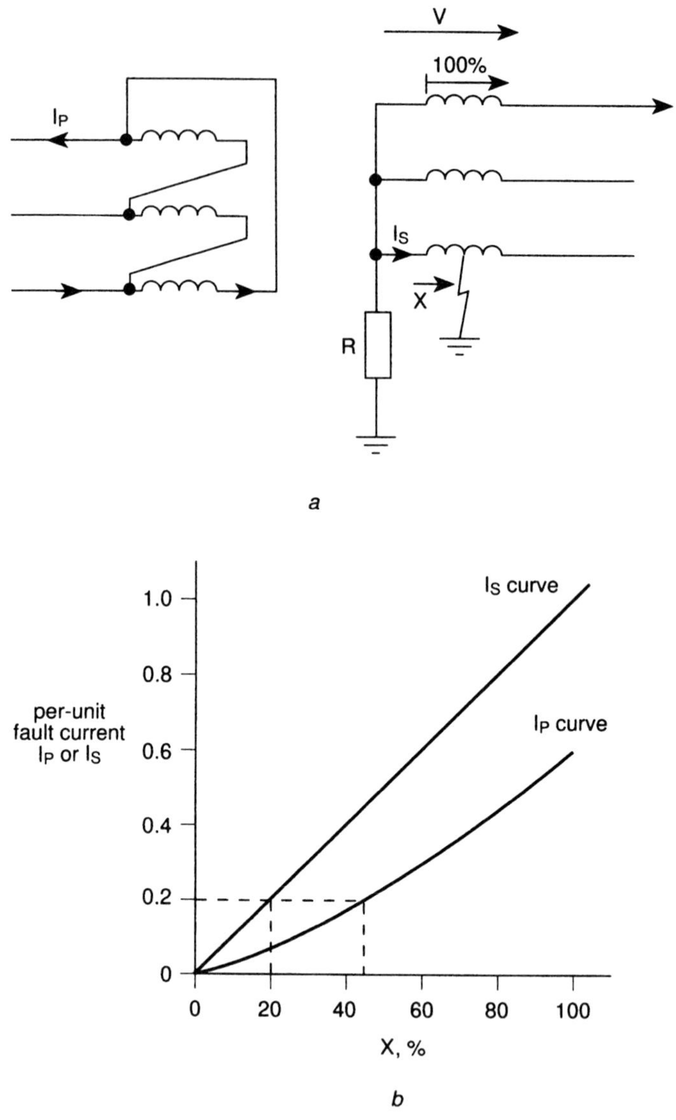

Figure 16.18 Winding-to-earth fault

a Circuit
b Fault current

or $\quad I_s = xI_{FLSEC}$ where I_{FLSEC} = secondary full-load current (this assumes that R limits the fault current to full load).

$$\text{Now } I_P = \frac{xN_s}{N_P} I_S$$

where N_S, N_P=number of turns on each winding

or
$$I_P = x^2 \frac{N_S}{N_P} I_{FLSEC}$$

But
$$I_{FL\,PRI} = \sqrt{3} I_{FLSEC} \frac{N_S}{N_P}$$

where I_{FLPRI} = primary full-load current expressed as a line quantity

so
$$I_P = x^2 I_{FLPR_I} \frac{1}{\sqrt{3}}$$

Taking I_{FLPRI} and I_{FLSEC} as one per-unit current and plotting I_P and I_S against x, Figure 16.18*b* results. Now, for this type of fault, I_S would flow into the restricted-earth-fault relay connected to the star winding, and I_P into the transformer biased-differential relay. It can be seen that, for the same relay sensitivity, the restricted-earth-fault relay protects a greater percentage of the winding than does the biased-differential relay, e.g. at a sensitivity of 0.2 p.u. the restricted-earth-fault relay protects down to 20% of winding while the biased-differential relay protects only down to 45% of winding. Somewhat similar conclusions also pertain to the more complex arrangement shown in Figure 16.16 and to resistance-earthed systems in general. For systems which are not resistance earthed, the case in favour of restricted-earth-fault protection tends to be less strong as the available fault current is now not restricted by a resistance-earthed neutral.

BIASED-DIFFERENTIAL PROTECTION: COMMISSIONING

16.26 Commissioning

Commissioning of a biased-differential-protection scheme would typically consist of the following:

- verification of the correctness of the schematic diagrams
- flick and ratio tests of interposing transformers (see Chapter 10)
- measurement of burdens (wiring and CT resistance) if a high-impedance restricted-earth-fault relay is included in the scheme (see Chapter 10)
- secondary injection of the relay
- primary injection
- DC-operation tests
- on-load tests

Salient features of these tests will now be examined.

16.27 Schematic diagrams

Before commencing any testing, it is important to verify that the schematic diagram is correct. This can be done by applying an external fault to the diagram, as shown in Figure 16.16, and calculating the magnitude and direction of all other currents, to prove that there is a current balance across the protection circuitry. This should ideally be done for faults on both the HV and LV sides of the transformer. A high level of confidence in the circuitry can be obtained by considering a single-phase-to-earth fault (since all three sequence components of current are present), followed by a three-phase fault (since current can be observed flowing in most wires). An example of the latter is illustrated, for a generator transformer, in Figure 16.19: an HV current of 1200 A is assumed, for ease of computation. To determine current flow through the power transformer, it is necessary to define the current relationship between the windings, and, from Figure 16.20, this can be determined as follows:

$\sqrt{3}\, V_{PL} I_{PL} = \sqrt{3}\, V_{SL} I_{SL}$ (relationship during balanced three-phase conditions)

where V_{PL}, I_{PL}, V_{SL}, I_{SL} are all line quantities. But

$$I_{SL} = \sqrt{3}\, I_{SP}$$

Thus

$$\sqrt{3}\, V_{PL} I_{PL} = \sqrt{3}\, V_{SL} \sqrt{3}\, I_S$$

or

$$I_{PL} = \sqrt{3}\, I_{SP} \frac{V_{SL}}{V_{PL}} \tag{16.2}$$

e.g., for a 132/33 kV transformer,

$$I_{PL} = \frac{\sqrt{3}}{4} I_{SP}$$

Note that in Figure 16.16 the interposing-transformer ratio $\sqrt{3}/1$ relates to the current ratio between the transformer windings and not to line quantities.

16.28 Secondary injection

This would normally consist of:

(*a*) *Determination of minimum operating current* by injecting 50 Hz current between terminals A and C, and B and C, of Figure 16.21 in turn, of the biased-differential relay, until the relay operates.
(*b*) *Determination of load bias*: from Figure 16.21, this requires a fixed 50 Hz current (say 1 A) to be injected through the bias winding, as indicated on

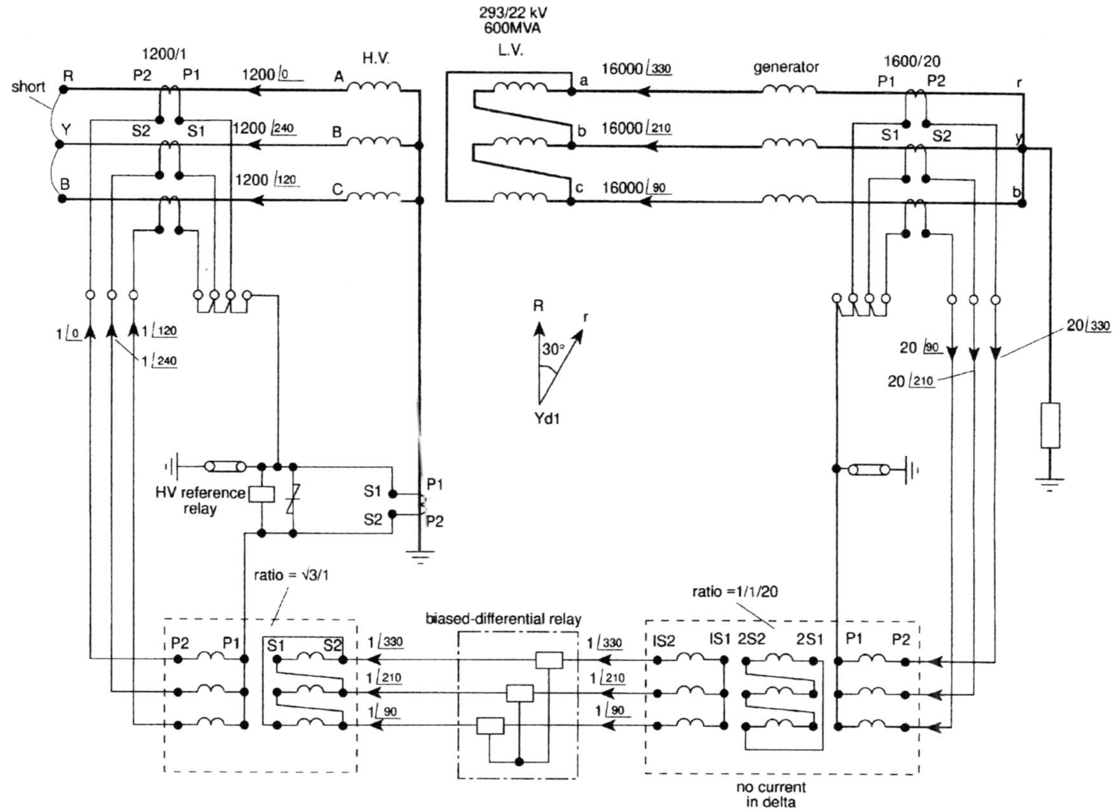

Figure 16.19 Current flow for a three-phase fault

Note that 293/22 kV is nominal tap

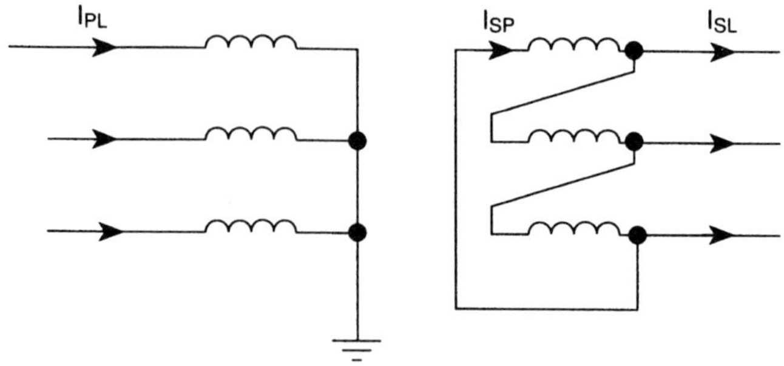

Figure 16.20 Transformer-current relationships

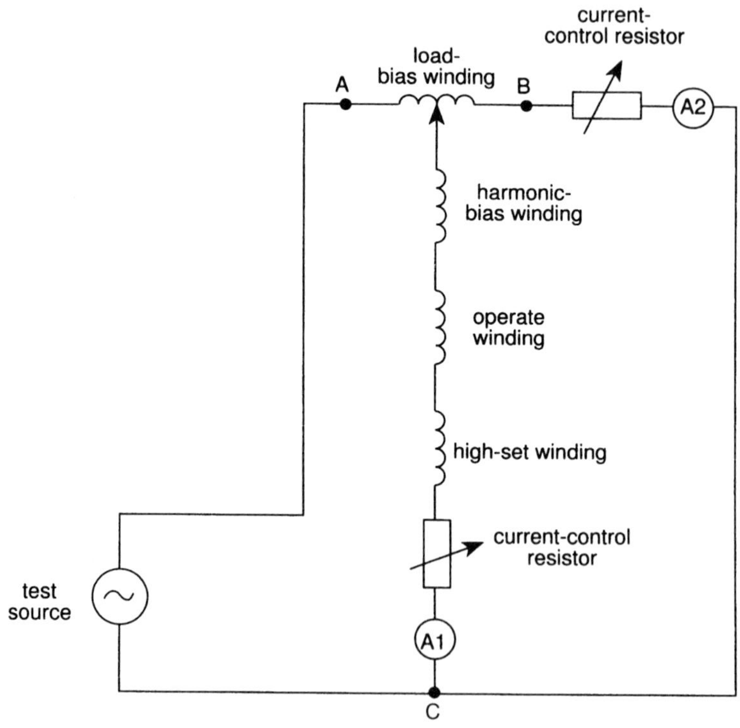

Figure 16.21 Secondary-injection circuitry

ammeter A2, while the current through ammeter A1 is incrementally increased until the relay operates. To determine the bias accurately (i.e. the slope of the graph), it is necessary to carry out the test at two values. With reference to Figure 16.14, points *X* and *Y* show two typical points; bias is thus calculated from

$$\text{bias} = \frac{I_{O(Y)} - I_{O(X)}}{I_{B(Y)} - I_{B(X)}}$$

where $I_{O(Y)}$ and $I_{O(X)}$ are the readings on ammeter A1 at points X and Y, respectively, and $I_{B(Y)}$ and $I_{B(X)}$ are the summed readings of ammeters (A2 + 0.5 ammeter A1) at points X and Y, respectively (i.e. mean-bias criterion).
(*c*) *Harmonic bias:* reference needs to be made to manufacturers' literature. For some designs of relay, it may be necessary to inject 100 Hz currents; on other designs an evaluation is possible utilising 50 Hz currents.
(*d*) *High-set:* a high-set element is usually included to facilitate relay operation during high internal fault currents which contain large restraining harmonic components, thereby preventing the biased-differential relay from operating. The high-set trips the circuit independently of the biased relay. The high-set is tested by incrementally increasing current until operation of the high-set element occurs.

16.29 Primary injections

Primary injections will accord with the principles set out in Section 10.22. All tests should be carried out at the transformer nominal tap.

16.30 On-load tests

On-load tests usually involve the following:

(*a*) energisation of the transformer to prove stability of the relay to magnetising inrush current. This should be carried out at least three times;
(*b*) a measure of the currents in the load bias windings and the operate winding at transformer nominal tap, i.e. at points A, B and C, respectively, in Figure 16.15. For a correctly installed scheme, there should be no more than a trace of current (a few milliamps) in the operate winding. The tap changer should then be raised and then lowered to top and bottom tap in turn (if system conditions permit) and the increase in current in the operate winding noted;
(*c*) the voltage across the restricted-earth-fault relays should be measured; for a correctly balanced scheme the value should be virtually zero.

BIASED-DIFFERENTIAL PROTECTION: MAINTENANCE

16.31 Maintenance

Maintenance of a biased-differential-protection scheme would generally consist of a repeat of the secondary-injection and DC-operation tests.

Chapter 17

Generator protection

17.1 Introduction

This chapter will examine the design, application, commissioning and maintenance of generator-protection systems. To ensure comprehensive coverage of the subject, generators up to 660 MW rating will be considered. Most of the protection systems to be discussed have already been examined in previous chapters, and therefore the text on commissioning will be limited to generator-specific protections. Emphasis will be placed on electrical protection of the generator, although brief mention will also be made of mechanical protection of the boiler and turbine.

With reference to Figure 17.1, which illustrates typical connection and protection arrangements for a 660 MW generator, the following protection systems will be examined:

- earth-fault protection and method of earthing
- overall biased-differential protection
- negative-phase-sequence-protection
- loss-of-excitation protection
- 400 kV and generator-transformer HV protection
- tripping and the low-forward-power interlock

17.2 Generator earthing

Generator earthing tends to fall into one of two categories: low-impedance or high-impedance. Figure 17.2*a* illustrates the type of low-impedance earthing developed in the UK by the CEGB in the 1950s. This type of earthing was designed to limit the fault current to about 300 A, which was seen as a compromise current between minimising the damage to the stator core at the point of fault, and limiting transient overvoltages. This type of earthing was employed on generators up to a typical maximum rating of 150 MW. The resistor is a liquid earthing resistor, as described in Chapter 5. Two earth-fault relays are employed. The instantaneous relay usually has a setting of about 10% (30 A) of the maximum fault current, and is capable of detecting faults in the top 90% of the winding. A more sensitive setting is not recommended, since there is a risk of unwanted operation for faults on the HV side of the generator transformer, which can result in current flow through the

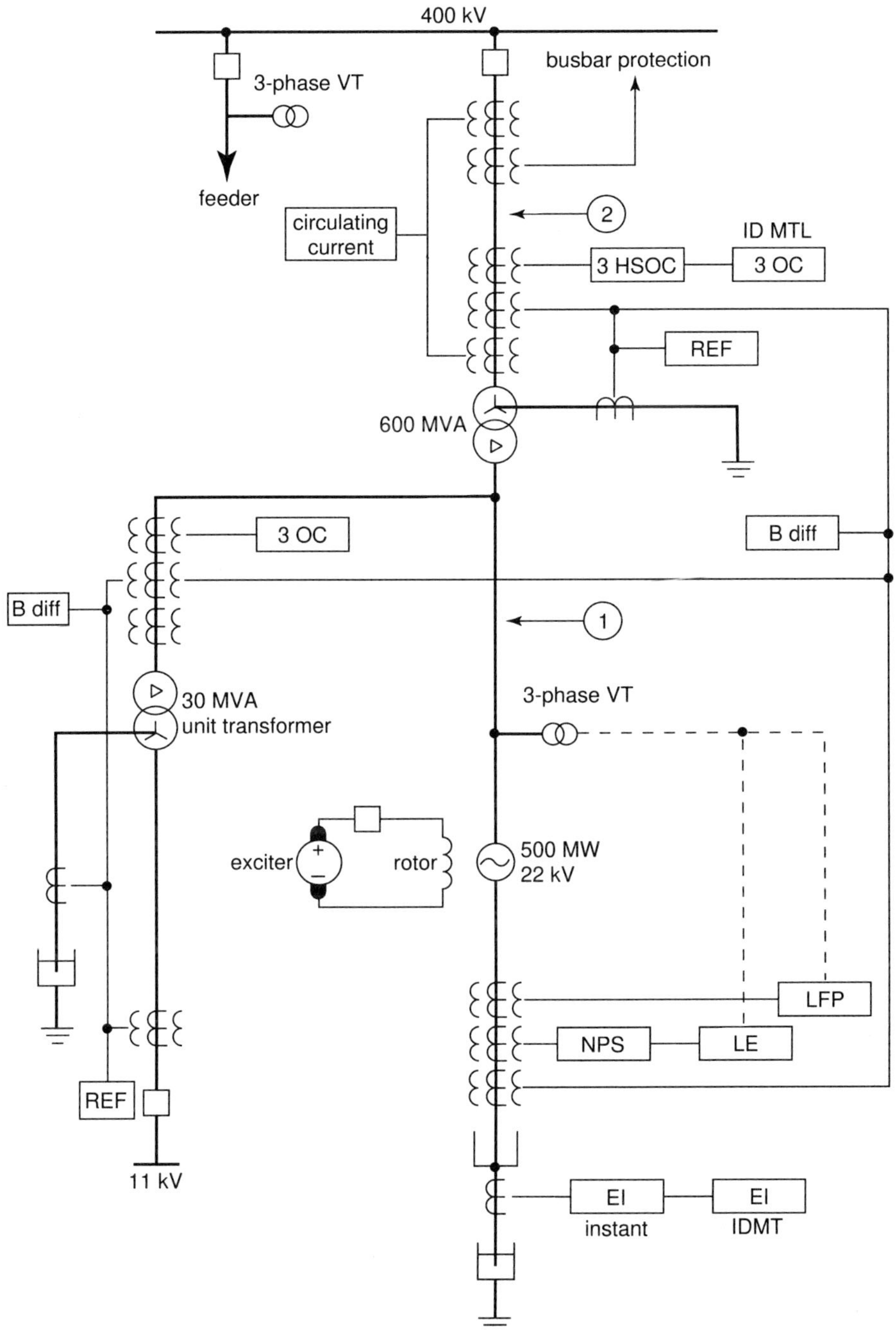

Figure 17.1 Typical generator-protection arrangement

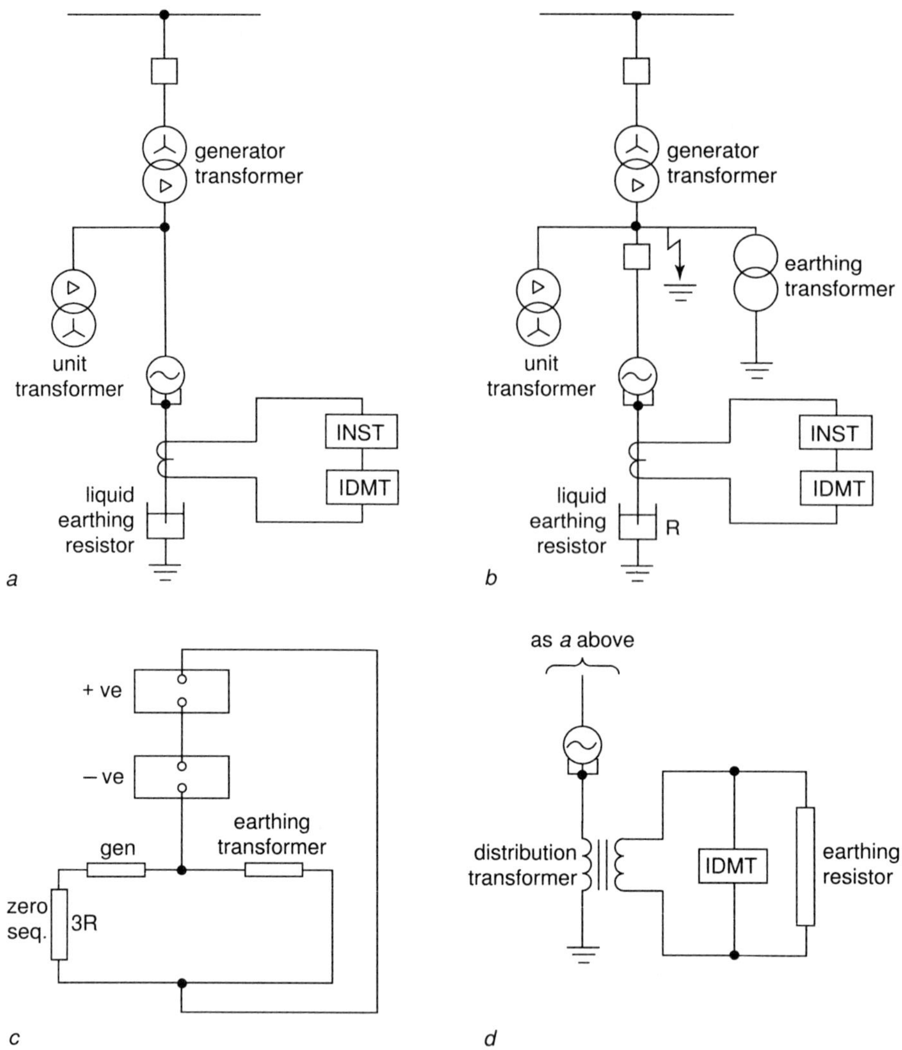

Figure 17.2 Generator-earthing arrangements

a Low-impedance earthing
b Low-impedance earthing: generator-switch arrangement
c Sequence-impedance diagram for earth fault in installation in Figure 17.2*b*
d High-impedance earthing

generator neutral via the transformer-interwinding capacitance.

An IDMT relay is usually additionally employed with a setting of 5% (15 A) of maximum fault current, to protect up to 95% of the winding.

Figure 17.2*b* illustrates low-impedance earthing for the instance of a generator circuit-breaker arrangement. Such an arrangement was installed on numerous large generators in the 1970s. With this arrangement, it is necessary to provide an earthing transformer to provide a system earth when the generator circuit breaker is open. It is usual to design the earthing arrangements so that 50% of fault current (i.e. 150 A) flows up each of the two earths. Figure 17.2*c* illustrates the resulting phase-sequence diagram.

Figure 17.2*d* illustrates an example of high-impedance earthing. This was employed on 500/660 MW machines in the 1970s and 1980s. The objective of this system is to reduce the current at the point of fault to a maximum of 10 A, to virtually eliminate any burning of the stator core. To obtain a high enough value of earthing resistance, the generator neutral is earthed through a distribution transformer which is loaded with a suitable value of resistance. The low value of earth-fault current precludes the use of an instantaneous-earth-fault relay. Protection is therefore afforded by use of a sensitive IDMT relay with a typical setting of 5% (i.e. 0.5 A). This relay needs to be immune to third-harmonic currents which can flow through the system capacitance to earth. High-impedance-earthing arrangements have also been utilised on systems with generator circuit breakers.

Note that all of the above relays are operated only by the flow of zero-phase-sequence currents. Since zero-sequence currents cannot pass through any of the transformers connected to the generator, there is no requirement for relay grading with relays connected to the HV side of the transformer.

17.3 Biased-differential protection

The biased-differential-protection scheme (see Figure 17.1) provides phase-to-phase fault protection for the generator and its connections. The principle of operation is described more fully in Chapter 16, and a complete scheme is illustrated in Figure 16.19. It is important to ensure that the relay-harmonic-restraint feature is provided, since transformer-HV inrush current can occur following clearance of an external fault. Note that some arrangements utilise the biased-differential scheme solely for the generator transformer, with a separate high-impedance circulating-current scheme for the generator: this assists in determining whether a fault is on the generator or on the generator transformer.

A biased-differential scheme is similarly employed for the unit transformer. Protection of the generator-transformer HV connections and windings, and the unit-transformer LV connections and windings, is via restricted-earth-fault protection (see Chapter 16).

17.4 Negative-phase-sequence protection

Sustained unbalance of the stator current can result in damage to a generator. This is caused by negative-phase-sequence (NPS) currents which rotate in the

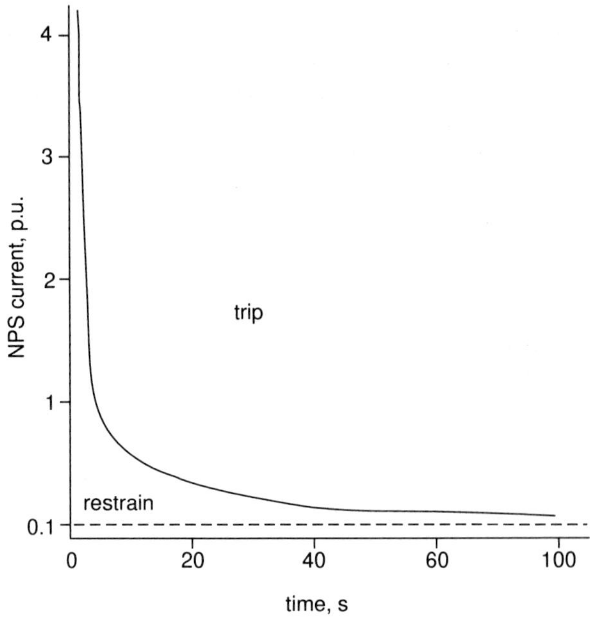

Figure 17.3 Typical NPS-relay characteristic

opposite direction to the rotor, so inducing eddy currents into the rotor body. These eddy currents, which are at twice the system frequency, give rise to overheating. The ability of a generator to withstand this heating will depend on the magnitude and duration of the negative-phase-sequence current. The negative-phase-sequence capability of a generator is therefore expressed by:

(*a*) the maximum continuous NPS current I_{2C}
(*b*) the maximum short-duration value of I_2^2t (i.e. heating effect) which can be withstood, when t is less than 100 s.

For a 500 MW generator, typical values would be:

$$I_{2C}=0.1$$
$$I_2^2t=3.0$$

where I_2 is measured in per-unit values (i.e. one per-unit of generator full-load current) Figure 17.3 illustrates in graphical form a typical generator NPS capability.

To protect the generator from the effects of NPS, an NPS relay, which has a characteristic set as closely as possible to that of the machine (e.g. as in Figure 17.3), is employed.

The relay is fed from three-phase CTs as shown in Figure 17.1 and derives, usually via phase-shifting networks, the negative-phase-sequence component of current. These relays would typically have a trip sensitivity (i.e. trip in infinite time) of about 10% of the machine rating, and an alarm sensitivity of 7% of the machine rating which is usually output after a 5 s delay.

The power system itself subjects generators to low levels of NPS current, primarily caused by the lack of full phase transposition of overhead lines. This is typically of the order of 0.5–3% of a 500 MW machine current rating. Abnormal conditions such as an open circuit or fault on the transmission system, or generator-transformer per-phase tap changers not being in step, will usually result in levels of NPS current exceeding 15%, so causing a tripping time of about 60 s or less. An uncleared fault can lead to a worst-case scenario of generation close to the system fault being tripped, causing a spread of NPS current to generators further away from the fault, and so resulting in cascade tripping.

Secondary injection of the NPS relay can be carried out by injecting single-phase-to-earth (R–E, Y–E, B–E) currents, and phase-to-phase currents (R–Y, Y–B, B–R) into the relay in turn, and comparing the measured operating time with that of the manufacturer's data. With reference to Chapter 3,

$I_2 = 0.33I$, where I is the injected phase-to-earth current

and $I_2 = 0.57I$, where I is the injected phase-to-phase current

Some relays contain an NPS-current meter. This is very useful for assessing standing levels of NPS.

17.5 Loss-of-excitation protection

Loss of rotor excitation will usually cause the generator to operate as an induction generator, at a speed of a few percent above synchronous speed (termed asynchronous running). This will result in the system supplying reactive power at a level which may well exceed the generator-stator continuous rating. In addition, the rotor will heat up rapidly because of the currents induced by the stator flux. To prevent damage to the generator, it is therefore imperative that it is disconnected from the system within seconds. This is achieved by the loss-of-excitation relay.

Modern loss-of-excitation relays utilise an offset-mho characteristic similar to that used by distance protection. Figure 17.4 illustrates a typical characteristic, where the $+R$ represents MW output from the generator, and lagging and leading-power-factor operation correspond to the $+X$ and $-X$ axes of the diagram, respectively. The characteristic is located largely on the $-X$ axis. This reflects the fact that the large levels of reactive-power flow which exist during asynchronous running, as an induction generator, appear to the relay as an impedance in the $-X$ direction.

The output of the relay is usually delayed by 2 s. This ensures nonoperation of the relay for pole-slipping conditions. Commissioning of the relay is similar to that required for distance protection, as described in Chapter 14.

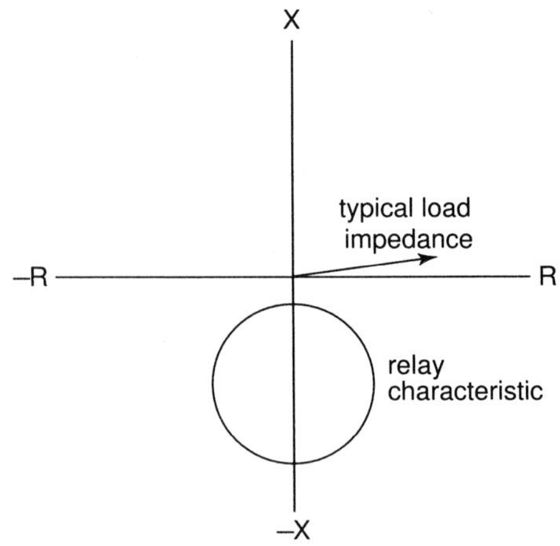

Figure 17.4 Loss-of-excitation characteristic

17.6 400 kV protection

For the purposes of this discussion, the 400 kV protection (with reference to Figure 17.1) consists of:

- 3-phase high-set overcurrent (3 HSOC)
- 3-phase IDMTL overcurrent (3 OC)
- circulating-current protection (HV connections)
- busbar protection

The purpose of the HSOC is to provide fast-acting protection for faults on the transformer HV terminals, or near the top of the winding, which are fed from the system. Its setting should be arranged to be insensitive to faults on the system, fed from the generator.

The IDMTL overcurrent is a backup relay for faults on the generator, generator transformer and associated connections. The author recalls a situation where the generators at a particular station were pole slipping with respect to the rest of the system and the only protection capable of detecting this pole slipping was the IDMTL overcurrent, which operated and saved the four generators from potential damage.

The high-impedance circulating-current protection protects the HV connections between the generator transformer and 400 kV circuit breaker. The principle of operation is described in Chapter 16.

The busbar protection provides protection for the 400 kV busbars; on smaller generators it is arranged to trip the HV circuit breaker only, to permit resynchronising to the other busbar without delay. On 500 MW generators and above, it is usually impossible to maintain the boiler plant in operation after a full-load trip, and therefore the busbar protection trips the full generator unit.

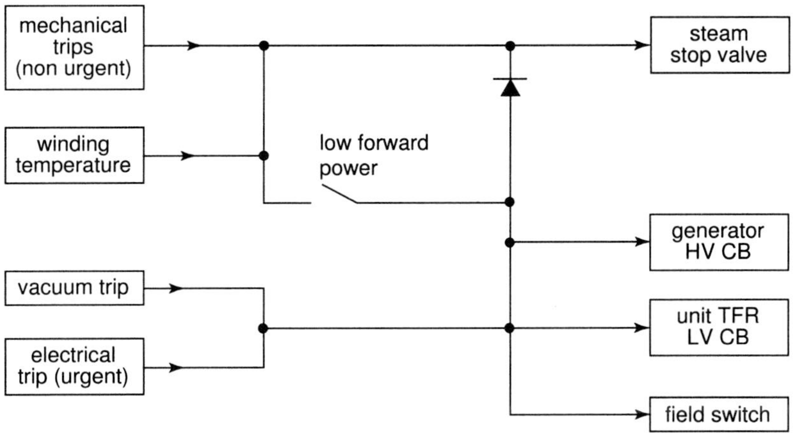

Figure 17.5 *Generator tripping*

17.7 Tripping

A fault on the generator, generator transformer and associated connections requires the following plant items to be tripped:

- HV (400 kV) circuit breaker
- unit-transformer LV circuit breaker
- excitation-system field switch
- turbine-steam stop valve, and boiler-firing system

If the steam stop valve fails to close, overspeeding of the set may occur, with the risk of severe damage. To avoid this situation, it has become usual practice on the larger generators to interlock the trip signal to the electrical side (i.e. HV circuit breaker, unit-transformer LV circuit breaker and field switch) via a contact from a power-measuring relay, such that tripping can only occur when the power output from the generator falls below a predetermined value, so implying that the steam stop valve must be closed. Figure 17.5 illustrates the typical tripping arrangements. Operation of electrical protection usually results in a direct trip, since any delay may result in significant damage to the generator; mechanical trips, usually considered as nonurgent, are usually routed via the power-measuring-relay interlock. Table 17.1 categorises the tripping arrangements.

17.8 Mechanical-trip initiations

The following identifies the purpose of some of the major mechanical trips:

(*a*) *Overspeed trip*
If a generator suddenly becomes disconnected from the system, it is suddenly

Table 17.1 Generator-tripping arrangements

Direct trips	Trips interlocked via power relay
Generator/transformer biased diff	Overspeed
Instantaneous earth fault	Loss of boiler water
IDMT earth fault	Loss of induced/forced fans
Generator transformer HV REF	Boiler excessive overfiring
Generator transformer HSOC	Low steam-inlet pressure
Generator transformer IDMT OC	High condenser-water level
BB Protection	Loss of lubricating oil
Generator transformer Buchholz	Loss of speed governor
Unit transformer biased diff	Loss of excitation
Unit transformer LV REF	
Unit transformer OC	
Unit transformer Buchholz	
HV-connection circulating current	
Negative-phase-sequence	
Low-vacuum trip	

deprived of its load and it suffers a 'full-load rejection'. With smaller generators, the governor should respond rapidly enough to prevent the rotor speed exceeding 8% above normal. With larger machines, overspeed-limiting gear may be fitted, which has a faster response time than the governor, to control the flow of steam. Should the rotor speed exceed 110% of normal, a centrifugally operated latch, the overspeed bolt, which is fitted to the shaft operates and trips the steam-stop valve.

(*b*) *Loss of boiler water*

If feed flow to the boiler is lost, the boiler-drum water level will fall rapidly and, beyond a certain point, overheating of the furnace tubes will occur, requiring the boiler firing to be extinguished.

(*c*) *Loss of induced/forced fans*

Loss of either fan can necessitate a reduction in generator output to about 60% of full rating.

(*d*) *Low steam-inlet pressure*

This is initiated by a sensing device which monitors steam pressure. The cause is usually a lack of boiler firing.

(*e*) *Low vacuum*

Loss of vacuum in the condenser results in rapid overheating of, and damage to, the turbine blades, owing to the increased wetness of the steam. A monitoring device for this purpose is found on all generators. This is usually a direct trip, not conditioned by the power-measuring relay.

(*f*) *Loss of lubricating oil*
On many generators the turbine servosystem-relay fluid is provided by the turbine lubricating-oil system. Loss of lubricating oil is therefore also accompanied by loss of relay fluid pressure. The lubricating oil also supplies the turbine bearings; thus any loss requires the machine to be tripped.

17.9 Low-forward-power relay

The power-measuring relay discussed in Section 17.7 is usually termed the 'low-forward-power relay' (LFP). A feature of this relay is that it is required to operate at low levels of power and at low power factors—these requirements are very exacting in terms of relay design.

The excess forward power required to accelerate a turbine to a dangerous speed is only about 0.7% of generator rated power. The relay is therefore set to about 0.4% of rated power. At these low values of power, the power factor tends to be very low, and the angle between voltage and current approaches 90°. Generally speaking, the relay must retain accuracy at phase angles between 89.5° lag and 89.5° lead. At these large angles small errors in angle measurement distort the power measured significantly. This is illustrated as follows:

$$\text{power measurement} = VI\cos\Phi$$

Suppose the actual power corresponds to an angle between V and I of 60°, but the relay measures an angle of 60.5°. Then the error in measured power is

$$\frac{\cos 60^\circ - \cos 60.5^\circ}{\cos 60^\circ}\times 100 = \frac{0.5 - 0.492}{0.5}\times 100 = 1.5\%$$

Suppose, now, that the actual power has an angle between V and I of 89° but the relay measures an angle of 89.5°. Then the percentage error in measured power is

$$\frac{\cos 89^\circ - \cos 89.5^\circ}{\cos 89^\circ}\times 100 = \frac{0.017 - 0.009}{0.017}\times 100 = 47\%$$

Hence, when measuring power at these large angles, a very accurate relay is required. This usually necessitates the CTs and VTs which feed the relay to be of accuracy class 0.1 (see Chapter 10).

Secondary injection of this relay requires precision instrumentation. The usual test is to inject voltage and current at differing angles (close to 90°) and measure the power to cause operation of the relay on a sensitive wattmeter. A plot can then be made of relay measured operating power against phase angle (between V and I). The measured power should typically be within 10% of the relay setting.

The output of the relay is usually via a time delay of typically 3 s. The time delay is required to ensure that any machine oscillations have died away before tripping is allowed to take place.

On-load tests on the relay can be difficult and are dependent on the degree of control that can be exercised over the generator. The objective is to verify that the relay operates at a specific level of decreasing generated power, i.e. proof of polarity and setting.

17.10 Primary injection

Primary injection of those current transformers directly associated with the generator must be accomplished by making use of the generator itself. With reference to Figure 17.1, the following summarises typical tests and conditions.

(*a*) Open the 400 kV and unit-transformer circuit breakers.
(*b*) Apply a three-phase bolted short-circuit to earth at position 1 in Figure 17.1.
(*c*) Set the field to minimum excitation. This will probably require the AVR to be taken out of service.
(*d*) Remove tripping. Run the generator up to synchronous speed. Slowly increase excitation to increase the generated current into the short-circuit.

Ideally, the generated current should be made equal to the full-load current, but this will depend on the capability of the short-circuit. Note that the full-load current for a 500 MW machine generating at 22 kV is of the order of 13 000 A.

Also note that, as the excitation is increased slowly, the machine will exhibit a reactance equal to synchronous reactance, i.e. it is not in a current-decrement situation (see Chapter 3).

The current should be measured in all relevant CTs and relays and the CT ratio and relay operation confirmed.

(*e*) Run down the machine and replace the 3-phase short-circuit with a single-phase-to-earth short-circuit. Run the machine up again as in (*d*), and pass a current of between 1 and 5 A into the short-circuit (depending on the type and rating of generator earthing). Measure the current in all relevant CTs and relays including the generator neutral CTs to prove the ratio of the neutral CT and relay operation.
(*f*) Repeat the above tests with the three-phase and single-phase-to-earth fault short-circuits, respectively, repositioned to position 2 in Figure 17.1. Current should again be measured in all relevant CTs and relays to verify the stability of all unit-protection schemes.

Note that, for this test, the generator transformer should be on nominal tap with fans and pumps running.

17.11 Phasing test and synchronising

Chapter 19 discusses in some detail the general principles of phasing tests. With reference to Figure 17.1, a typical programme of tests would be as follows:

(*a*) Commencing with the circuit breaker at the remote end of the existing (and proven) feeder open, run the generator up to full voltage and energise through the generator transformer and 400 kV busbar (which is isolated from all other sources of supply) to the feeder.
(*b*) Phase out between the generator three-phase VT and the feeder three-phase VT. This proves the phasing of the high-voltage connections and the generator-VT-wiring connections. Note that, when interpreting the phasing-test results, account will need to be taken of the generator-transformer phase shift.
(*c*) Open the generator 400 kV circuit breaker and energise the feeder from the 400 kV system. Phase out again between the generator and feeder three-phase VTs. Note that the voltage differences between the phases of the three VTs will rise and fall simultaneously to reflect the generator slip. The generator slip should be adjusted until the two systems are in, or as nearly as possible in, synchronism. Checks should also be made at this time of the voltage at the generator synchronising relays and synchroscope.
(*d*) If the above tests are satisfactory, utilise the synchronising relay to close the generator circuit breaker.
(*e*) Prove the metering by increasing generation to increase active power, and increasing excitation to increase exported reactive power.

17.12 Maintenance

In most instances, the maintenance of generator protection (and switchgear) can be categorised as opportunity maintenance, i.e. it is made to coincide with generator outages. In general terms, generator-protection maintenance would consist of:

- insulation-resistance tests
- secondary-injection tests
- proving of alarms
- tripping tests

Generator-protection maintenance is probably the most complex to undertake of any circuit, since it involves the tripping of at least four major devices (HV circuit breaker, unit circuit breaker, field circuit breaker, steam stop valve) at four different locations, within the power station. As a result, great care must be taken to ensure safe working practices are adhered to.

Chapter 18

Intertripping

18.1 Introduction

The term 'intertripping' may be defined as a method in which operation of a protection equipment at one end of a circuit causes a signal to be transmitted to trip a circuit breaker at the remote end of the circuit. Intertripping is an unconditional-tripping system and must not be confused with 'protection signalling', the latter being a conditional-tripping system concerned with 'distance protection' blocking or acceleration schemes (see Chapter 14).

As a result of being an unconditional trip, intertripping equipment is usually designed with greater security safeguards than protection signalling, to minimise the risk of incorrect tripping. This usually results in intertripping being a marginally slower process.

The methods utilised for intertripping are:

- voice-frequency signals, often via rented public telephone circuits (NB: in addition, digital systems are now being introduced)
- power-line carrier
- direct-current surge proof
- fault throwers
- unstabilisation of feeder protection

Intertripping used to comprise a single way (direction) of signal transfer, but most modern intertripping is two-way. The application of intertripping may be categorised as

- direct
- cascade
- triangulated

The following sections will examine the above techniques, their application to the power system, and commissioning and maintenance considerations.

EQUIPMENT DESIGN

18.2 Voice-frequency intertripping

Voice-frequency (VF) intertripping injects a signal into the communication medium with a frequency range of 300–3000 Hz. In the UK, it is probably the

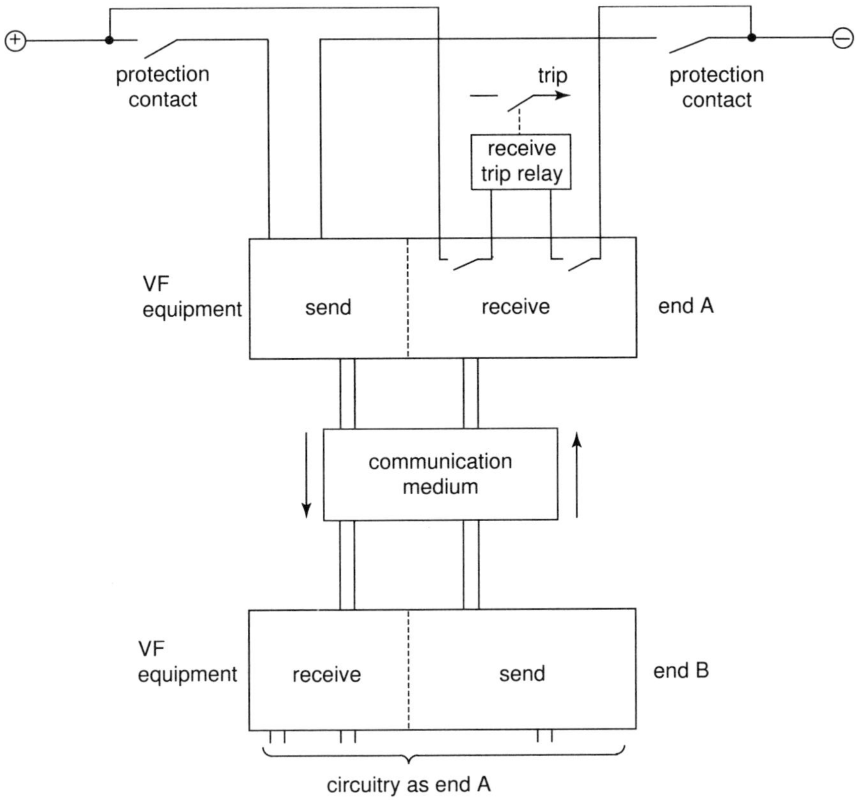

Figure 18.1 Basics of a VF intertripping system

most common type of intertripping system, with British Telecom providing the most common communication medium. At each end of the circuit, identical equipments will process the send and receive signals. Figure 18.1 illustrates a typical two-way-intertripping arrangement; salient points are:

(*a*) for reasons of security from induced voltages, it is usual to double-pole switch the send and receive circuitry (see Chapter 8).
(*b*) connection from the intertripping equipment into the communications medium is usually via twisted-pair cables.
(*c*) with modern equipments, the transmitted signals are likely to comprise one of the following formats:
(i) the transmission of two guard signals of frequencies f_1 and f_2 while in the quiescent state, changing to the transmission of a trip frequency f_3 when in the tripping state. This is termed a frequency-shift system;
(ii) the transmission of a digitally coded signal, the format of which changes when transferring from the guard to the tripping state.
(*d*) typical times from closure of the protection contact to closure of the remote-end-equipment receive contact are 25–35 ms.

(*e*) The purpose of the guard signal is continuous monitoring of the communication channel. An alarm is generated if the guard signal is not received.

The reliability of this type of equipment is generally very good, although experience has revealed that it may be prone to slow operation as a result of interference signals generated during system fault conditions.

18.3 Power-line-carrier intertripping

The circuitry used for power-line-carrier intertripping is similar to that shown in Figure 18.1, except that the connection to the intertripping equipment is a coaxial cable, and the communication medium is the power line itself (see Chapter 15). This is an expensive means of intertripping due to the cost of the carrier-coupling equipment, and is usually only applied to the 275 kV and 400 kV systems.

Power-line-carrier intertripping equipment installed in the 1960s and 1970s (some of which is in service today), was based on a frequency-shift system, consisting of the transmission of two guard and two tripping frequencies from each end, a total of eight different frequencies. A more recent design consists of a carrier-frequency-shift coded system.

Unlike power-line phase-comparison protection, the intertripping equipments described above have a continually transmitted guard signal. The operating frequencies are therefore constrained by communication regulations to specific frequency bands (to avoid interference with aircraft navigational beacons).

The power levels employed are typically 2 W while in the guard state and 10–20 W while in the intertrip state. A typical operating time from closure of the protection contact to closure of the remote-end-equipment receive contact is 20 ms—much faster than most other systems.

18.4 Surge-proof intertripping

Surge-proof intertripping was discussed in some detail in Chapter 8. This is a DC system, utilised mainly on distribution systems of 132 kV and below where speed of operation is less critical. This design has a typical fastest operating time of about 80 ms.

18.5 Fault throwers

This method of intertripping consists of applying a single-phase-to-earth fault to the power system via a fault-throwing switch; the latter is initiated by protection systems installed at the same end of the circuit, so causing

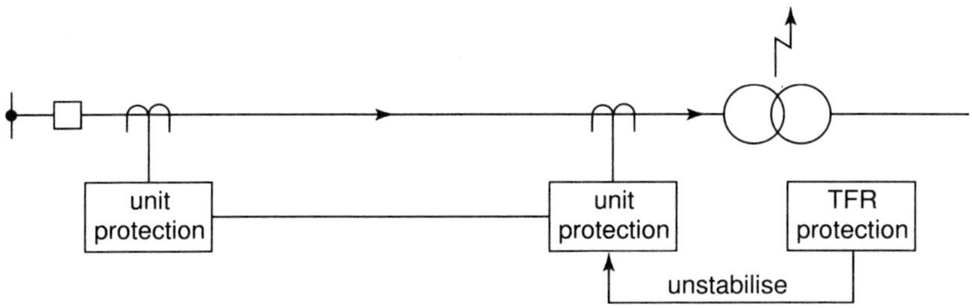

Figure 18.2 Principle of feeder-protection unstabilisation

operation of protection systems at the remote end, which in turn trip the remote-end circuit breaker. The fault-throwing switch is usually spring operated, initiated through the electrical release (via the protection) of a mechanical latch, and manually reset.

Typical operating times from initiation of the fault thrower to the application of the short-circuit are between 100 and 400 ms. Fault throwers are mainly installed on the 132 kV and 33 kV systems. They are advantageous at those locations where it is difficult, or costly, to install conventional intertripping systems.

Commissioning tests on fault throwers would normally consist of:

- measurement of contact resistance between the fixed and moving contacts, with the fault-thrower closed
- timing test from fault-thrower initiation to contacts closed
- operation of the fault thrower, once the circuit is in service

Operation of a fault thrower usually results in slight burning of the contacts, but no significant damage. The contacts should ideally be cleaned following operation onto a live circuit.

18.6 Unstabilisation

Unstabilisation consists of causing a feeder unit protection to operate for an out-of-zone fault. Figure 18.2 illustrates the principle of operation. For the transformer fault shown, the unit protection stabilises. Operation of the transformer protection causes a signal (usually a contact) to be input into the unit protection to cause 'unstabilisation' and operation. Typical methods of unstabilisation are:

- pilot-wire protection: short-circuit the pilot
- power-line phase-comparison protection: mute the locally generated carrier signal
- distance protections: initiation of an acceleration signal, or inhibition of a blocking signal

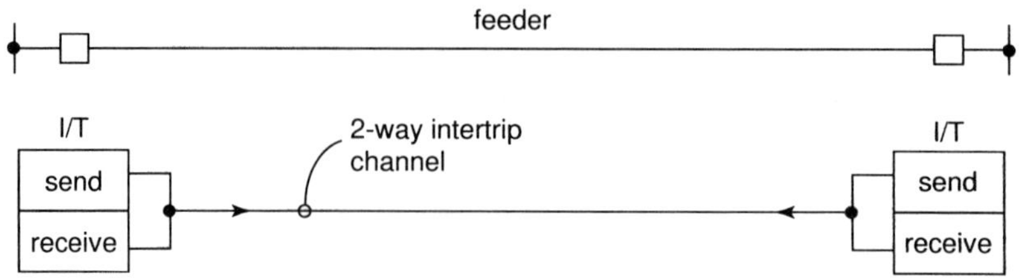

Figure 18.3 Intertripping of a supergrid feeder

It is usual to arrange the unstabilising contacts to prevent the local end of the unit protection from operating to avoid any confusion over the source of the fault, which would result from the simultaneous operation of transformer and feeder protection.

Note that for unstabilisation to be effective, any fault on the LV side of the transformer needs to result in sufficient current on the HV side to cause operation of the remote-end protection.

APPLICATION

18.7 Plain feeder

The application of intertripping to plain feeders is usually only justified on the 400 kV and 275 kV systems. This arises when the fault infeed at one end may be too low to operate the protection at that end, so requiring tripping of the associated circuit breaker by intertripping from the remote end; in addition, it facilitates the fast clearance of small zone faults. The intertripping is initiated from the feeder protection or busbar protection, and is usually two-way VF or power-line carrier. Figure 18.3 illustrates the requirement. It is also usual to employ unstabilisation of feeder protection from the busbar protection.

18.8 Supergrid transformer

By far the most common application of intertripping is associated with the protection of a transformer connected as a transformer feeder. Figure 18.4 shows a typical intertripping arrangement for a supergrid transformer connected as a transformer feeder. For security reasons, it is usual to employ two two-way intertrip systems, which should preferably be of different types working over different communication media, e.g. one system may be VF over a public or private communication system, and the other system may be a power-line carrier.

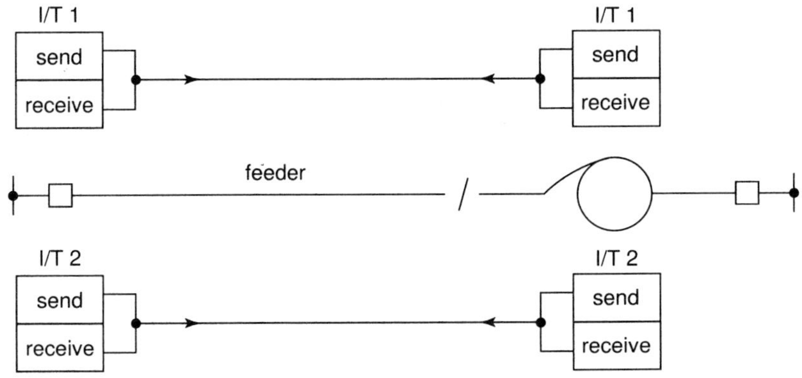

Figure 18.4 Intertripping of a supergrid transformer

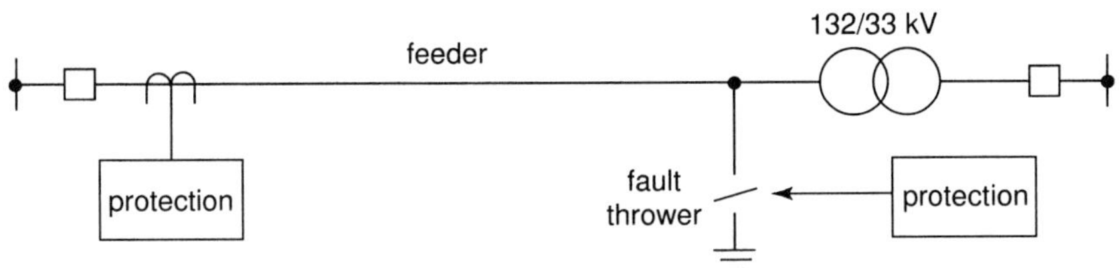

Figure 18.5 Intertripping of a 132/33 kV transformer via a fault thrower

If this is not feasible and two public communication channels are utilised then, for reasons of security, separate routes should be employed for each intertrip system. In addition, unstabilisation of the feeder protection from transformer protection would usually be employed.

18.9 132/33 kV transformer

Figure 18.5 illustrates a 132/33 kV transformer connected as a transformer feeder. In this instance, intertripping is frequently achieved via a fault thrower, although it could alternatively be achieved by use of a VF-intertripping system. Older VF-intertripping systems are usually one-way only, from transformer to remote-end circuit breaker.

18.10 Transformer HV disconnectors

At 400, 275 and 132 kV, it is usual to provide a disconnector, designed for off-load operation, on the HV side of the transformer, as shown in Figure 18.4.

This disconnector is caused to open automatically by operation of the transformer protection, after the circuit breakers have been tripped, to isolate the faulty transformer. At 400 kV it is usual to interlock the operation of this disconnector with the contacts of check relays, to ensure that the disconnector is not opened if the circuit is still energised (i.e. remote-end circuit breaker failing to trip). The check relays are usually associated with a measurement of voltage, transformer HV current and transformer neutral current.

At 275 and 132 kV, many existing installations have only a single VF or power-line-carrier intertripping facility. Under such situations, the interlocks are bypassed to enable the disconnector to be opened without any checks being made of system conditions. This means that, if the single intertripping facility fails to trip the remote end, the disconnector will open on an energised system. This is the intention, the reasoning being that it is preferable to risk damage to the disconnector, rather than risk further damage to a faulty transformer. This disconnector is frequently termed a 'fault interference' disconnector and personnel should be kept well removed from it while it is in service for safety reasons. This is usually accomplished by the creation of an exclusion zone around the disconnector. The author has observed a number of these disconnectors connected at 132 kV after they have opened in the fault-interfering mode, and the damage has been remarkably slight, being confined to pitting of the isolator blades; greater damage is sustained at 275 kV.

It is not usual to employ a fault-interfering disconnector at 400 kV. This is because the higher fault level would cause unacceptable damage, and also because it is usual to employ two intertripping facilities, so that the risk of intertrip failure to trip the remote-end circuit breaker is much reduced. Some transformer HV disconnectors are of the switch–disconnector type, suitable for switching load and magnetising current. These are not operated in the fault-interfering mode, since the interrupter is usually a sealed unit and thus the degree of damage, if so operated, would be difficult to assess. In general, operation of disconnectors in the fault-interfering mode is being phased out as rapidly as possible.

When commissioning or maintaining a fault-interference disconnector, it is not usual to subject it to opening on load.

18.11 33/11 kV transformers

33/11 kV transformers, when connected as transformer feeders, tend to utilise intertripping systems which are of the surge-proof design, or make use of fault throwers.

A protection system frequently utilised on such a transformer feeder is neutral-voltage-displacement (NVD) protection. From Figure 18.6, NVD consists of a capacitor connected to each phase, the star point of which is

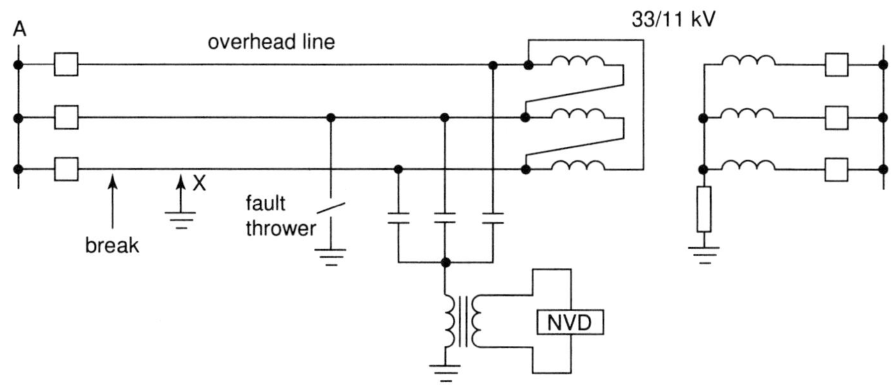

Figure 18.6 Neutral-voltage-displacement protection

connected to earth through the NVD relay. System earth faults resulting in zero-phase-sequence voltages will cause operation of the NVD relay. The relay is used to protect against the following fault conditions:

(*a*) a 33 kV earth fault at position *X* in Figure 18.6. This fault could result in the protection at substation A operating but, due to lack of fault infeed at substation B, no protection except NVD would detect the fault. The NVD protection is arranged to trip the 11 kV circuit breaker directly. Without NVD protection, the 33 kV system could remain energised from the 11 kV system, without a system earth.

(*b*) a break in an overhead line, as shown in Figure 18.6, resulting in the broken conductor falling to earth on the transformer side of the break. No protection except NVD may be sensitive enough to detect this condition. Once operated, the NVD relay would trip the 11 kV circuit breaker directly, and operate the fault thrower to cause operation of the protection at substation A. A limitation of this arrangement is that, if the open-circuit occurs on the phase with the fault thrower, the latter is no longer directly connected to the remote-end substation.

Off-load commissioning tests of an NVD scheme would usually consist of applying voltage from a high-voltage test source to each phase capacitor in turn, and measuring the current through the NVD relay to cause operation. The on-load test would consist of energising the circuit and measuring the current in the NVD relay, which should be negligible.

18.12 33 kV ring circuits

33 kV distribution systems are frequently operated as closed rings. Each circuit in the ring may be protected by a unit protection or more simply by directional overcurrent relays. Situations can arise whereby the fault infeed from one end of the circuit may be so low, due to the long route length back to the substation, that the protection does not operate. In such instances

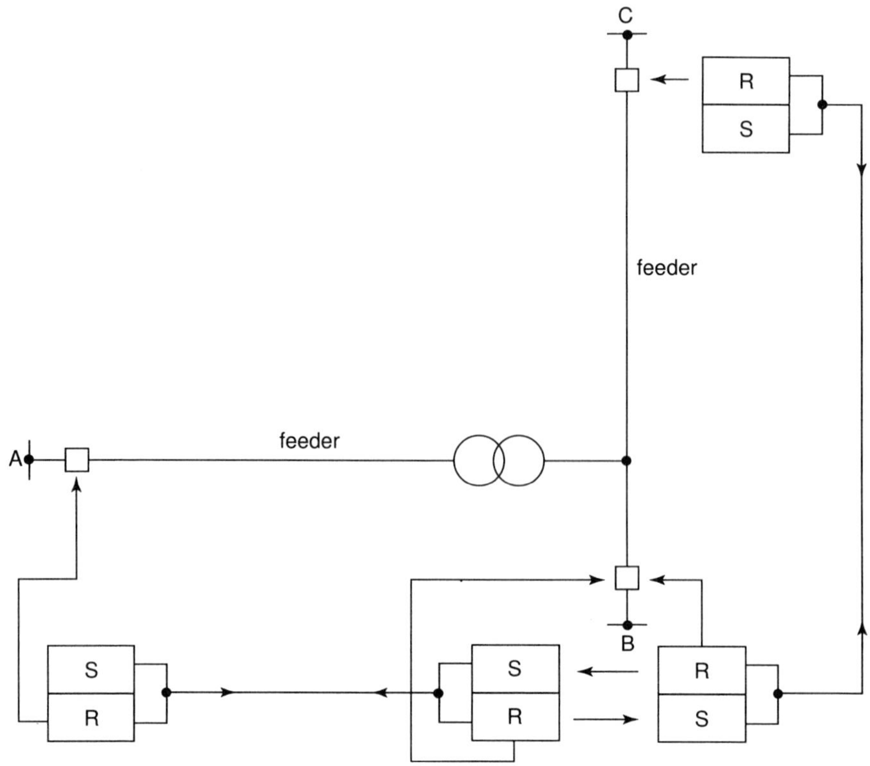

Figure 18.7 Cascade intertripping

intertripping, usually of the surge-proof type, is initiated by protection at the circuit end which is tripped to open the circuit end which fails to trip.

18.13 Cascade intertripping

Figure 18.7 illustrates an example of a cascade-intertripping arrangement. The receive circuitry of each intertripping system at the intermediate substation B initiates (i.e. cascades into) the send circuitry of the other intertripping system. This arrangement optimises the number of intertripping systems, but results in a doubling of the total intertripping time, since there is no direct intertrip from A to C.

18.14 Triangulated intertripping

Figure 18.8 illustrates an example of triangulated intertripping. Consider a protection operation at substation A. Intertrip signals are sent from substation A to substations B and C, and these are cascaded to substations C and B, respectively, and further cascaded back to substation A. The timer at

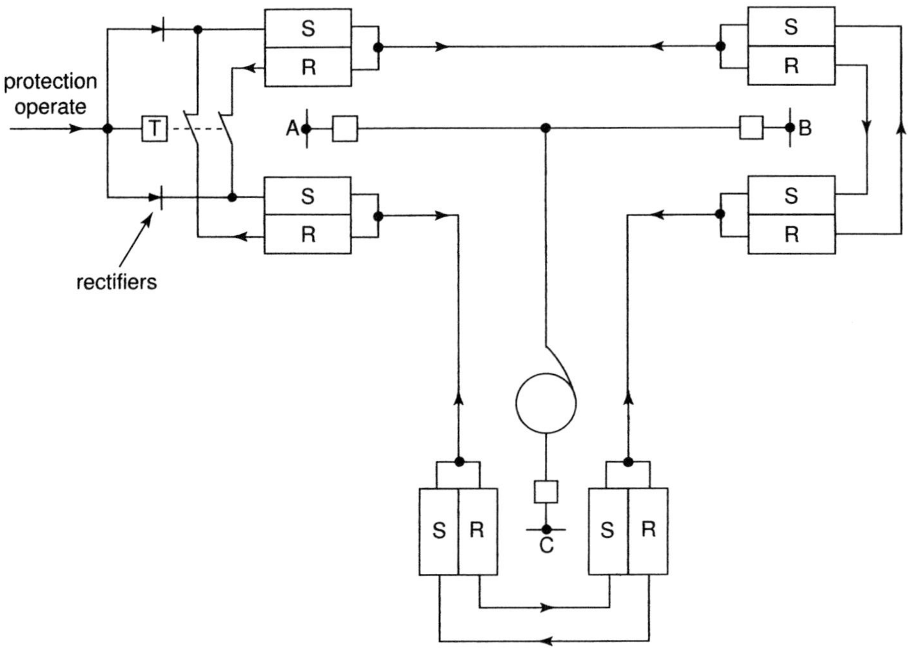

Figure 18.8 Triangulated intertripping

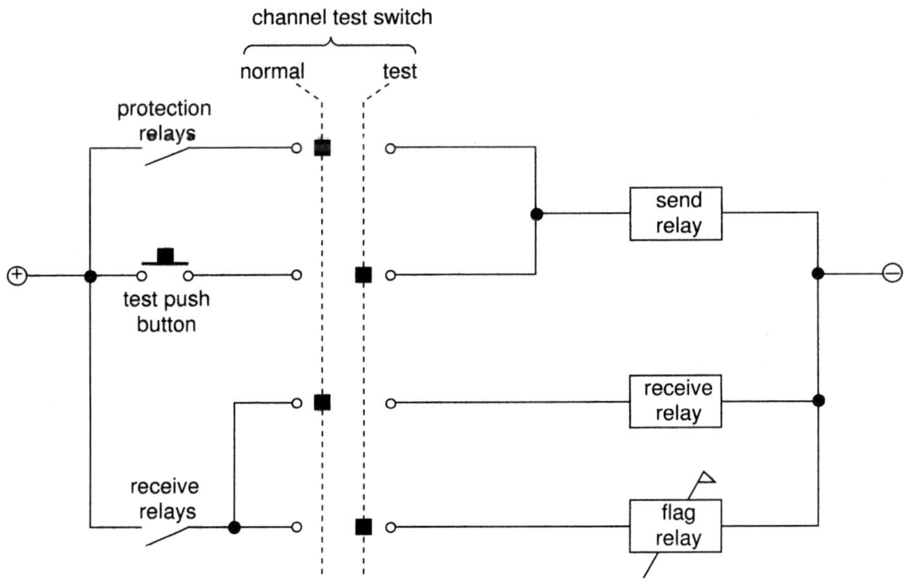

Figure 18.9 Simple illustration of intertrip-channel test facility

substation A has typically a 0.5 s time delay, and open-circuits the intertripping loop circuitry to prevent the system from locking up. The timer circuitry shown at end A would also exist at ends B and C, but is omitted from Figure 18.8 for reasons of clarity. This arrangement consists of a direct trip (typically 40 ms) and a cascaded trip (typically 80 ms) to each substation.

At supergrid voltages of 400 and 275 kV it is usual to duplicate the cascade arrangement shown in Figure 18.7, resulting in eight ends of intertripping equipment; however, it would be equally acceptable to utilise the circuitry in Figure 18.8, thus requiring only six ends of intertripping equipment. In both instances these systems could only be applied if the indirect intertrip times would not adversely affect system stability.

18.15 Persistent-intertrip feature

The persistent-intertrip feature is utilised in conjunction with delayed-autoreclose circuits, and described in Chapter 11; it is installed at 132 kV and above.

COMMISSIONING

18.16 Commissioning

Commissioning tests on VF or carrier intertripping would usually consist of the following:

(*a*) setting up the signal transmit levels, receiver sensitivity and alarm sensitivity;
(*b*) local DC logic-operation tests to prove the protection send- and receive-relay circuitry;
(*c*) off-load end-to-end tests to prove satisfactory receipt of signal;
(*d*) off-load reflex test to verify the operating time of the system. This is usually accomplished by arranging the remote-end intertrip-receive circuitry to be looped into the remote-end send circuitry to return the signal immediately; hence the term reflex. The operating time in one direction is taken as being half the loop time. Many installations have inbuilt facilities to assist this test; and
(*e*) on-load tests. Where appropriate these would usually be carried out in conjunction with autoswitching tests as described in Chapter 11.

MAINTENANCE

18.17 Maintenance

The maintenance of intertripping systems consists largely of a repeat of the commissioning tests.

Many intertripping systems possess an intertrip-channel test facility as shown in Figure 18.9. Use of this facility requires the selection of an intertrip test switch from the NORMAL to TEST position at both ends of the circuit. Operation of the test pushbutton causes an intertrip signal to be sent to the remote end, to energise an intertrip-receive-flag relay. During the test, the intertrip-receive-trip relay is disabled by the test switch. This facility can also be used when the circuit is in service, if it is acceptable to switch out the intertrip feature for about ten minutes—the time to carry out the test. Many utilities use it in this mode to facilitate routine (approximately six-monthly) tests. The test facility would also be used to prove the intertrip channel following temporary failure and subsequent repair.

Chapter 19

On-load tests

19.1 Introduction

On-load tests constitute the final activity in the commissioning process. They are carried out immediately after the newly installed equipment is energised from the power system, and they are the definitive tests in verifying that the equipment has been correctly installed, and is satisfactory for commercial operation.

The tests generally comprise some, or all, of the following:

(*a*) soak test
(*b*) power-system-phasing tests, i.e.
- voltage magnitude
- voltage direction of rotation
- voltage phase-angle

(*c*) synchronising and paralleling of supplies
(*d*) accuracy and polarity of metering equipment
(*e*) ratio and polarity of current transformers
(*f*) that the correct voltage- and current-transformer supplies are connected to the appropriate protection and control equipment
(*g*) on-load testing of protection equipments
(*h*) autoswitching-equipment tests
(*i*) automatic-voltage-control tests

A number of on-load tests have already been examined for specific equipments in previous chapters; more general tests will now be examined briefly.

19.2 Preparation prior to energisation

Before energising new equipment, the commissioning panel should prepare a switching programme which takes full account of all safety, operational and technical requirements. The switching programme should specify pre-energisation checks and precautions, as follows:

(*a*) All off-load-test documentation should be formally completed. Formal acknowledgment of this should be through completion of stage 1 of the Acceptance Certificate (see Chapter 1) or a similar type of document.

(*b*) A final check should be made to ensure that all CT test links are in the in-service position.
(*c*) Check that all alarms are reset.
(*d*) All safety documentation should be cancelled, all earths removed and all equipment left in the open or de-energised position. The latter point should be checked both visually and at all local and remote indication boards.
(*e*) When entering service, all protection systems should be normal and in service. Autoswitching and automatic-voltage-control equipment should be switched out of service. Consideration should be given to the transformer tap position (usually left on nominal tap).
(*f*) The identity and location of all personnel participating in the on-load tests should be clearly stated.

19.3 Energisation

At the time of energising a new circuit, neither the new circuit breaker nor its associated protection systems are fully proven. It is therefore usual to energise the new circuit from an already proven circuit. This is often accomplished by using a bus-coupler or bus-section circuit breaker which is equipped with a fast-operating nonunit protection (e.g. an instantaneous overcurrent and earth-fault relay), which should be set as sensitively as possible. The switching sequence should consist of the following:

(*a*) Select the bus-section (or coupler) circuit and the circuit to be commissioned to the same busbar—initially with both circuit breakers open—and with no other circuits selected to that busbar.
(*b*) Close the bus-section circuit breaker to energise up to the circuit breaker associated with the circuit to be commissioned.
(*c*) Close the circuit breaker on the circuit to be commissioned to energise the circuit.
(*d*) Energisation is now complete; soak tests and on-load tests can now commence.

When switching, to energise new equipment, the following should be adhered to:

(i) Ideally, all personnel should be at positions remote from the equipment to be energised, for safety reasons.
(ii) The minimum number of personnel should be located in the control/switching room. No-one carrying out switching desires an audience.
(iii) Personnel carrying out a switching instruction should read it twice; they must understand what is to be done, and what the implications will be; they should act slowly and deliberately. They should also consider in advance what will be heard or observed on energising the equipment, e.g. transformer hum, voltages and currents appearing on instruments.
(iv) Once energised, all high-voltage equipment (e.g. circuit breakers,

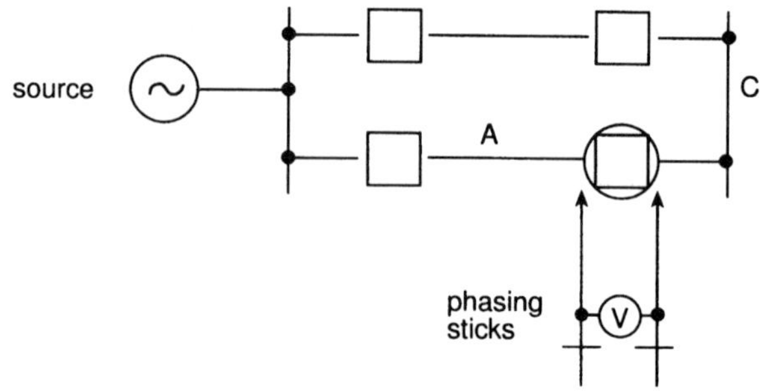

Figure 19.1 Phasing out using phasing sticks

disconnectors, tap changers) should be operated to prove satisfactory operation with the system voltage applied.

19.4 Soak test

Soak testing consists of leaving the new equipment energised from the system voltage for a period of time, before it is subject to load current. This is to observe that the equipment suffers no distress when subject to the system voltage.

A typical soak-test duration for a circuit breaker is about half an hour, and for a transformer a minimum of two to three hours, to observe that there is no accumulation of gas in the Buchholz. Note that there is a modern tendency to abbreviate the soak test to a brief visual inspection of the plant.

19.5 Energising a radial circuit

A typical example of a radial circuit is a 415 V distribution board. Once it has been energised, the only tests necessary are to check the magnitude of all voltages to the neutral and earth, and to each other, and their phase rotation.

19.6 Phasing via phasing sticks

Figure 19.1 illustrates the case of a new 11 kV distribution circuit A which is to be commissioned onto the busbars at C. Before the remote-end circuit breaker associated with A can be closed, a phasing-check test is required. This can be accomplished with the use of phasing sticks. These consist of two long insulated probes connected at the base by a conducting bond and fitted with an indicating voltmeter.

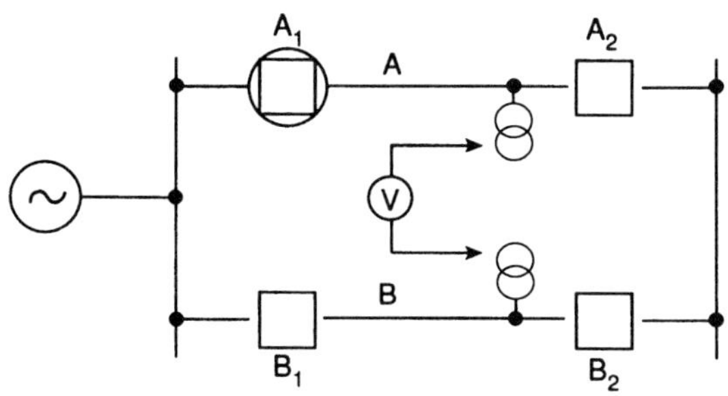

Figure 19.2 Phasing test via VTs

An operator wearing insulated gloves holds the two sticks, and inserts one up each of the busbar and feeder spouts, respectively, of metalclad circuit breakers, or across the feeder and busbar connections of open-terminal circuit breakers. Each phase is tested against each other phase. In-phase connections should result in zero volts on the voltmeter, while across-phase connections should indicate 11 kV. Phasing sticks are generally available for voltages up to 33 kV. Note that they should be proved before and after use using a proving device.

19.7 Phasing via VTs

With higher-voltage systems, it is necessary to carry out the phasing test with the use of voltage transformers. This consists of a two-stage process: first to prove the voltage transformer connected to the circuit whose phasing is to be proved, by phasing out against a supply whose phasing is already verified; and secondly to phase out to prove the phasing of the circuit under consideration. Figure 19.2 gives an example for the instance of circuit A requiring to be commissioned:

(*a*) Starting conditions are with circuit breakers Al and A2 both open, and an existing circuit B energised and in service.
(*b*) Circuit breaker A2 should be closed, thus energising feeder A from feeder B.
(*c*) Phasing tests should be carried out between the VTs on feeders A and B:
(i) Voltage magnitudes should be verified by measuring the voltage to earth (63.5 V) and between phases (110 V) at each VT box in turn.
(ii) Voltage phase rotation should be verified by applying a phase-rotation meter to each VT box in turn.
(iii) An insulated wire should then be run between the two VT boxes.
(iv) At one end, a voltmeter should be connected between the wire and VT terminals in turn (preferably the VT connected to the in-service circuit to minimise the risk of causing a short circuit on that VT), and at the other end the wire should be touched onto each VT terminal in turn. For correct phasing, the results matrix shown in Table 19.1 should then be obtained.

Table 19.1 Phasing-test results

		VT A			
		R	Y	B	N
VT B	R	0	110	110	63.5
	Y	110	0	110	63.5
	B	110	110	0	63.5
	N	63.5	63.5	63.5	0

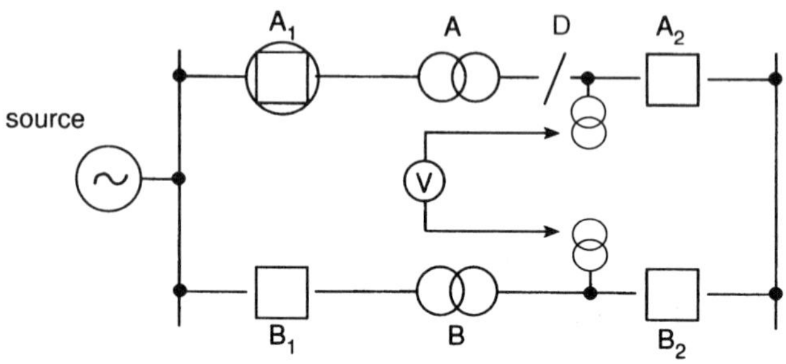

Figure 19.3 Phasing test of transformer supplies

The purpose of this test is to verify the voltage transformer connected to feeder A, by phasing against a known and proven source of supply (feeder B and the VT connected to B), prior to proving the phasing of the conductors connected to feeder A.
(*d*) Circuit breaker A2 should now be opened and circuit breaker Al closed. Phasing tests between the VTs on feeders A and B should then be carried out again, as described in (*c*) above. This test proves the phasing of feeder A's conductors. Circuit breaker A2 is now in a position to be closed.

19.8 Phasing of transformer supplies

Figure 19.3 shows the instance of a transformer A which is to be commissioned. Transformer B is existing and in service. The phasing tests are identical to those described in Section 19.7. In brief:

(*a*) The starting condition is with circuit breakers Al and A2 open, and disconnector D open.
(*b*) Close A2 and phase out between the A- and B-circuit VTs, i.e. from a known supply to prove the A-circuit VT wiring

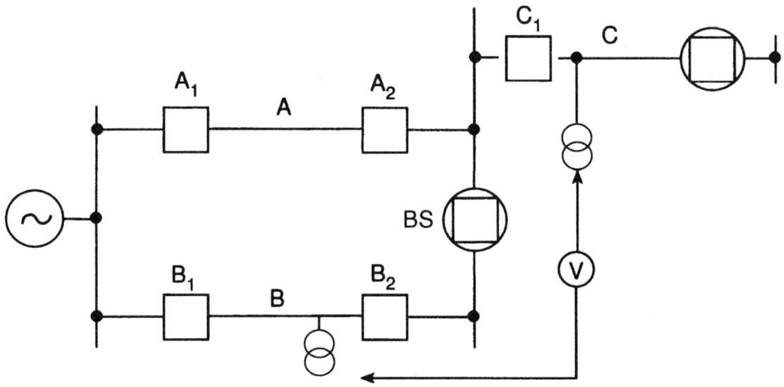

Figure 19.4 Phasing test to a VT on another circuit

(*c*) Open A2, close disconnector D and circuit breaker Al, and phase out across the A- and B-circuit VTs. The phasing of transformer A is now verified.

19.9 Absence of 3-phase VT

The above phasing tests are dependent on a three-phase VT being installed on each circuit. Some circuits are equipped with either a single-phase VT or no VT. In such an instance, the phasing must be carried out by utilising a three-phase VT on another circuit. For example, with reference to Figure 19.4, circuit A does not have a three-phase VT installed; therefore the three-phase VT on circuit C is utilised. Assuming that VTs on feeders B and C have been previously phased out, the procedure is as follows:

(*a*) Open the bus-section circuit breaker, to ensure that feeders A and B do not become connected.
(*b*) Close circuit breakers A1, A2 and C1.
(*c*) Phase out between VTs on B and C feeders.

The author has experienced instances of a new substation being constructed where the incoming supply feeder was the only circuit equipped with a three-phase VT. Phasing out was carried out between the VT and a 415 V supply derived from another part of the power system. Careful evaluation of the phasor position of the 415 V supply relative to the three-phase VT supply was required, utilising the principles set out in Chapter 5.

19.10 Voltage-reference point

The preceding sections on phasing tests have assumed that the three-phase VTs have their star points earthed, thus establishing a symmetrical and

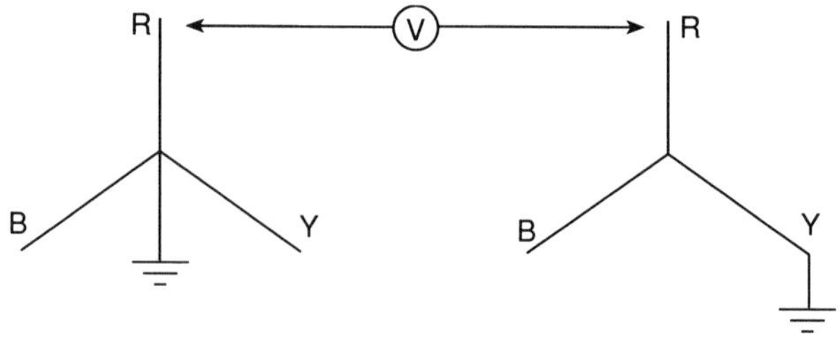

Figure 19.5 Phasing for different VT-earthing arrangements

common reference point for the phasing test. It must be noted, however, that the earthing of the star point must not be assumed. Some VTs, particularly those on the 33 kV system, are often earthed via the yellow phase.

In some instances it may therefore be necessary to carry out a phasing test between VTs with different earthing arrangements. When the results are interpreted, account must therefore be taken of the relative phase positions of the two VTs; see, for example, Figure 19.5. Note that before phasing tests are begun it is imperative that each phase voltage of the VT is measured to earth and neutral to determine the earthing arrangement.

19.11 Synchronising and paralleling of supplies

Once the phasing tests have been carried out, the operator is in a position to close the circuit breaker, to parallel supplies. At the lower voltages, the circuit breaker would be closed manually from the control board (or in some instances at the circuit breaker itself.) At higher voltages, a synchronising scheme is usually employed, as described in Chapter 11.

19.12 Metering

It is necessary to prove polarity-conscious metering, i.e. the direction of flow of active and reactive power, during the on-load tests. Once proven, the metering may be utilised to assist in proving polarity-conscious protection. Figure 19.6 illustrates the metering convention.

Active power and lagging reactive power flowing out of the busbars are termed export quantities, and active power and lagging reactive power flowing into the busbars (the latter being equivalent to leading reactive power flowing out of the busbars) are termed import quantities. Figure 19.7 illustrates methods for proving the metering. Figures 19.7*a,b,c* compare the unproven metering with proven metering. Figure 19.7*d* utilises a wattmeter, ammeter and voltmeter connected directly to the CT and VT circuitry, to enable the active and reactive quantities to be calculated.

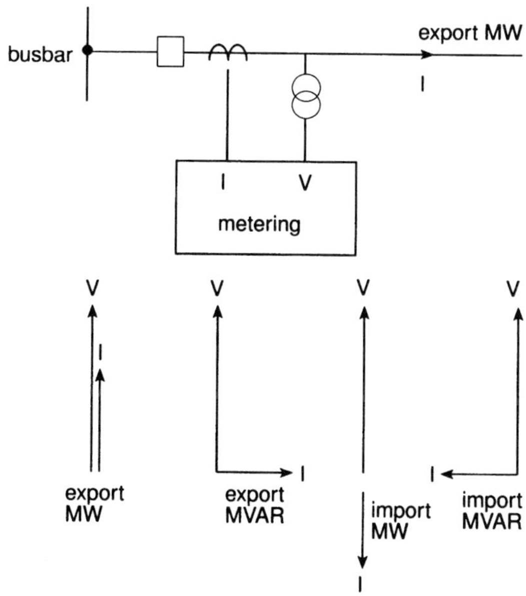

Figure 19.6 Metering convention

19.13 Ratio and polarity of protection current transformers

The current flowing in the secondary wiring of each CT can be measured with a miniature clip-on ammeter or, with some relays, by inserting a split plug directly into the relay. The measured current should be compared with that being indicated on the circuit metering (circuit ammeter), to prove the CT ratio. Current-transformer polarity is usually verified by carrying out on-load polarity tests on the relay, as described in Section 19.15.

19.14 Voltage-transformer supplies

Each item of protection and control equipment should be checked to ensure that it is connected to the correct VT phase. One method for achieving this is to run a cable, whose cores have been identified, from the VT terminal box to a terminal block in the relay room. Supplies from the terminal block to each relay can then be phased out. A phase-rotation meter should be connected to at least one item of equipment to confirm correct voltage-phase rotation.

19.15 Protection equipment

The on-load tests carried out on protection equipment are required to prove the following:

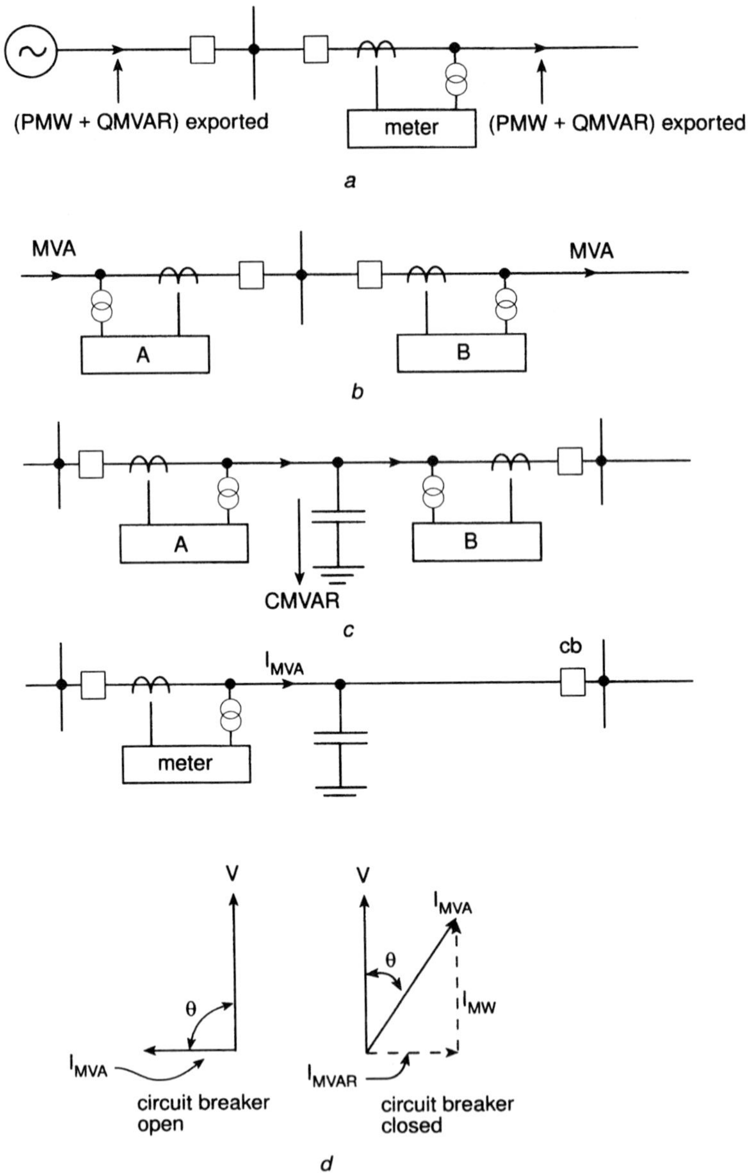

Figure 19.7 Methods of proving metering

a From a known source of generation
b Compare with local proved metering
Meter A (proved) = *P* MW imported + *Q* MVAR exported
Meter B (unproved) = *P* MW exported + *Q* MVAR imported
c Compare with remote proved metering
Meter A (proved) = *P* MW imported + *Q* MVAR exported
Meter B (unproved) = *P* MW exported + (*Q* + *C*) MVAR imported
d Using a phase-angle meter, ammeter and voltmeter

(*a*) that each relay is supplied from the correct ratio of CT. This is simply checked by measuring the current in each relay and comparing it with that indicated on the circuit metering. In effect, this test confirms the primary-injection test;
(*b*) that relays with directional features (e.g. distance protection, directional overcurrent) are looking in the correct direction. This is accomplished by checking whether the relay is tending to operate or restrain when current flow is in a known direction. This is also a check of correct CT polarity;
(*c*) that unit protections are exhibiting balance and stability with the flow of load current. This, again, is also a check of CT polarity;
(*d*) that, when the power system is used as the communication medium, such as with power-line-carrier systems, the magnitudes of the send and receive voltages are within acceptable limits.

With tests (*b*) and (*c*) above, it is usual with feeder protections to reverse the CT secondaries (or rotate the voltage phases) to prove that the relay tends to operate or restrain as appropriate with reversed CT polarity. The preceding chapters have discussed on-load tests for specific protection equipments.

19.16 Auxiliary supplies

Once the main power-system connections and protection and control supplies have been proved, any site auxiliary supplies fed from the commissioned equipment should be proved. This may require phasing out of low-voltage supplies, or observing the direction of rotation of transformer fans, pumps etc.

19.17 Load current

When carrying out on-load tests, it is advantageous to have as much load current as possible. In some instances, the power system may require manual intervention to create load current. Examples of this include generators running out of merit, adjusting transformer taps to cause circulating currents, or switching circuits to transfer load. Occasionally, it is just a matter of sitting it out until the peak daily load occurs. Long feeders which are lightly loaded can be problematic, since the feeder capacitance current causes a mismatch of current in unit-protection CTs.

Where sufficient load current cannot be obtained to enable the on-load tests on unit-protection systems to be carried out, consideration should be given to injecting a current directly through the high-voltage circuit from a primary injection set, or from a 415 V supply. This consists of injecting current at one end of the circuit and connecting to earth at the other. Usually a current of about 200 A is sufficient. This method of on-load testing can be carried out only when the primary-system impedance is relatively low, and where induced voltages into the circuit in question do not constitute a safety hazard.

19.18 Autoswitching and automatic-voltage-control tests

Finally, the autoswitching and automatic-voltage-control equipments should in turn be switched into service, and manually initiated tests carried out. These tests have been examined in more detail in Chapters 11 and 6, respectively.

19.19 Maintenance

It is not usual to repeat the on-load tests following maintenance. Some companies, however, check that currents and voltages are present at each protection relay following the return to service of a circuit after maintenance.

Bibliography and references

SAY, M. G.: 'The performance and design of alternating current machines' (Pitman, 1970)

COTTON, H., and BARBER, H.: 'The transmission and distribution of electrical energy' (The English Universities Press, 1970)

MORTON, A. H.: 'Advanced electrical engineering' (Pitman, 1971)

LAITHWAITE, E.R., and Freris, L. L.: 'Electric energy: its generation, transmission and use' (McGraw-Hill, 1980)

LYTHALL, R.T.: 'J & P switchgear book' (Butterworth, 1986)

RÜDENBERG, R.: 'Electrical shock waves in power systems' (Havard University Press, 1968)

ETA (Ed.): 'Power system protection, Vol.1, 2, 3' (Peter Peregrinus, 1990)

WEDEPOHL, L.M.: 'Polarised mho distance protection' Proc. IEE, **112**, (3), 1965

BUNGAY, E., and McAllister, D.: 'Electric cables handbook' (BSP Professional Books, 1990)

Various Authors : 'Protective relays application guide ' (GEC Alstom Measurements, 1990)

RYAN, H.M. (Ed.): 'High voltage engineering and testing' (Peter Peregrinus, 1994)

HILL, T.: 'Production operations management' (Prentice/Hall, 1983)

BS 3939 (1989): 'Graphical symbols for electrical power, telecommunications and electronics'

BS 5311 (1988): 'Specifiction for high-voltage alternating-current circuit breakers'

BS 171 (1970-1993): 'Power transformers'

BS 148 (1992): 'Specification for unused mineral insulating oils for transformers and switchgear'

BS 3938 (1992): 'Specification for current transformers'

BS 3941 (1992): 'Specification for voltage transformers'

BS 142 (1982): 'Specification for all-or-nothing electrical protection relays'

BS 5874 (1993): 'Method for determination of the electric strength of insulating oils'

BS 5737 (1979): 'Method for the measurement of relative permittivity, dielectric dissipation factor and d.c. resistivity of insulating liquids'

Electricity Association document ESI 09-6 (1988): 'Auxiliary multicore and multipair cables, Issue 6'

Electricity Association document ESI48-3 (1977): 'Instantaneous high-impedance differential protection, Issue 1'

Electricity Association document ESI 48-4 (1983): 'DC relays associated with a tripping function in protection systems, Issue 3'

NB: British Standards as well as foreign and International standards, can be obtained by post from BSI Customer Services, 389 Chiswick High Road, London W4 4AL

Electricity Association ESI documents are supplied by the Electricity Association, 30 Mill bank, London SW1P 4RD

Index